MW00814767

Rainer Hippler, Sigismund Pfau,
Martin Schmidt, Karl H. Schoenbach (Eds.)

Low Temperature Plasma Physics

Rainer Hippler, Sigismund Pfau,
Martin Schmidt, Karl H. Schoenbach (Eds.)

Low Temperature
Plasma Physics

Fundamental Aspects and Applications

Berlin · Weinheim · New York · Chichester · Brisbane · Singapore · Toronto

Editors:
Prof. Dr. Rainer Hippler, Institute of Physics, University of Greifswald, Germany
e-mail: hippler@physik.uni-greifswald.de
Prof. Dr. Sigismund Pfau, Institute of Physics, University of Greifswald, Germany
e-mail: pfau@physik.uni-greifswald.de
Dr. Martin Schmidt, Institute of Low Temperature Plasma Physics, Greifswald, Germany
e-mail: schmidtm@inp-greifswald.de
Prof. Dr. Karl H. Schoenbach, Physical Electronics Research Institute, Old Dominion University,
Norfolk, Virginia, USA
e-mail: schoenbach@ece.odou.edu

Cover:
Microwave-excited argon-hydrogen plasma (with kind permission of A.A. Meyer-Plath,
Institute of Low Temperature Plasma Physics, Greifswald, Germany)

1st edition

Library of Congress Card No: applied for

British Library Cataloguing-in-Publication Data: A catalogue record for this book is available from the British
Library.

Die Deutsche Bibliothek – CIP Cataloguing-in-Publication-Data
A catalogue record for this publication is available from Die Deutsche Bibliothek.

© WILEY-VCH Verlag Berlin GmbH, Berlin (Federal Republic of Germany), 2001

ISBN 3-527-28887-2

Printed on non-acid paper.

Printing: Betzdruck GmbH, D-63291 Darmstadt
Bookbinding: Wilhelm Osswald & Co., D-67433 Neustadt (Weinstraße)

Printed in the Federal Republic of Germany.

WILEY-VCH Verlag Berlin GmbH
Bühringstrasse 10
D-13086 Berlin

Preface

Recent developments in the fabrication of microelectronic and micromechanic devices have led to a renewed interest in plasma physics and plasma technology. In particular the advances in microelectronics are to a large extent due to plasma processing, the application of the low-temperature, non-thermal plasma for plasma etching and plasma assisted deposition procedures. Plasma is also the essential tool in the field of large area thin film deposition, the fabrication of hard coatings for tools, and the surface treatment of various biomaterials, to name only a few applications. Most of the new developments in plasma technology are driven by commercially important demands. Some examples are high-pressure discharge lamps for improved lighting, the emerging field of plasma displays in flat panel devices, vacuum and high-pressure switches for pulsed electrical power applications, and last, but not least, the continuing efforts towards the utilization of thermonuclear fusion as a viable future source of energy.

This book on low-temperature plasma physics and applications contains the contributions of distinguished plasma scientists and engineers, not only from academic institutions in Europe, the United States of America and Asia, but also from industry. An overview of the basic theory of plasmas, with emphasis on weakly ionized gases, is provided in the first three chapters. Plasma-wall interactions, an important topic with respect to plasma applications, are discussed in the two following chapters. Plasma diagnostics and the diagnostics of plasma treated surfaces are the topics of chapters six to ten. A wide range of plasma sources and plasma applications including plasma chemistry and environmental technologies, plasma displays, lighting, and surface processing is covered in chapters eleven to twenty. The concluding book chapter deals with markets for these emerging technologies. The effective application of plasmas in all these fields of research and development requires a sound knowledge of plasma physics and technology. Basic plasma theory and plasma applications are the topic of many elementary courses at the university level. A more detailed knowledge is provided through special courses, designed for graduate students and postgraduates. The purpose of this book is to provide supporting material for these special courses, but also to serve the scientists and engineers in research laboratories and industry as a reference book. It is our hope that it will be a useful companion for anyone working in this area of research and development, but also for those who want to know more about the fascinating applications of non-thermal plasmas.

We would like to thank all the contributors to this book. It was a pleasure working with them. We acknowledge gratefully the help of Mrs. H. Boldt in the technical preparation of the manuscript, and would like to express our special appreciation to Mrs. V. Dederichs of Wiley-VCH for her valuable advice and her patience during the process of editing this book.

R. Hippler, S. Pfau, M. Schmidt, K.H. Schoenbach

Greifswald and Norfolk, November 2000

Contents

1 Characteristics of low-temperature plasmas under non-thermal conditions — a short summary

Alfred Rutscher

Institut für Physik, Ernst-Moritz-Arndt-Universität Greifswald, Domstraße 10a
D-17487 Greifswald, Germany

1.1 Introduction

The concept of a plasma dates back to Langmuir (1928) and originates from the fundamental difference between regions of electrical gas discharges which are distant from boundaries (bulk of the discharge) compared with regions which adjoin the boundaries (sheaths).

1.1.1 Definition

Plasmas are quasi-neutral particle systems in the form of gaseous or fluid-like mixtures of **free** electrons and ions, frequently also containing neutral particles (atoms, molecules), with a large mean kinetic energy of the electrons and/or all of the plasma components (0.2 eV ... 2 MeV per particle) and a substantial influence of the charge carriers and their electromagnetic interaction on the system properties.

The interactions between the electric charges of the plasma components show two aspects:

▷ Coulomb interaction among the charge carriers. Owing to the long range of the Coulomb force in the case of large charge-carrier densities ($n_e \gg 1/\lambda_D^3$), each charge carrier interacts simultaneously with many others (collective interaction).

▷ Formation of macroscopic space charges (in the frame of quasi-neutrality) as a consequence of external influences and modification of charge-carrier movement in the electrical field of these space charges.

Related to the quasi-neutrality and the presence of free charge carriers, the most intrinsic attribute of the plasma state is its tendency to minimize external electric and magnetic fields inside the bulk, in contrast to its behaviour in the surrounding sheaths.

1.1.2 Types of plasmas

Plasmas are frequently subdivided into low- (LTP) and high-temperature plasmas (HTP). A further subdivision relates to thermal and non-thermal plasmas.

Table 1.1: Subdivision of plasmas.

Low-temperature Plasma (LTP)		High-temperature Plasma (HTP)
Thermal LTP	Non-thermal LTP	
$T_e \approx T_i \approx T \lesssim 2 \times 10^4$ K	$T_i \approx T \approx 300$ K	$T_i \approx T_e \gtrsim 10^7$ K
	$T_i \ll T_e \lesssim 10^5$ K	
e.g., arc plasma at normal pressure	e.g., low-pressure glow discharge	e.g., fusion plasmas

1.2   Starting point for modeling the plasma state

There are three different basic approaches towards a theoretical description of the many-particle plasma state: single-particle trajectories, kinetic and statistical theory, and hydrodynamic approximation.

1.2.1   Single-particle trajectories

This model is based on the motion of individual particles (e.g., under the influence of the Lorentz force). Problem: The electric and magnetic fields in the plasma must be regarded as given and cannot be obtained in a self-consistent manner from the cooperative movement of the particles. Using Monte Carlo simulations the study of single-particle trajectories could be extended to kinetic ensembles taking into account the effect of collisions. This technique is an alternative to the kinetic theory.

1.2.2   Kinetic and statistical theory

On the basis of kinetic criteria each particle ensemble of the plasma is analyzed taking into consideration the specific conditions and generalizing the kinetic theory of neutral gases to plasmas. The ultimate goal is to be able to calculate the space and time dependence of all the interesting distribution functions by solving the kinetic equations.

For non-thermal low-temperature plasmas the most important of these is Boltzmann's equation (1872) for the energy or velocity distribution of the electron component. This equation describes the balance of the particle density in phase space. The total time derivative of the distribution function is the sole outcome of particle collisions, contained in the so-called collision integral, which usually encloses a multitude of terms for different collisions of electrons (elastic, inelastic, collective...). After explicit replacement of the external forces by the general Lorentz force, which then appears as self-consistently given by electric and magnetic fields of space charges and moving charge carriers, the collision free approximation of the Boltzmann equation reduces to the Vlasov equation.

The kinetic theory is the strongest instrument of plasma theory, e.g., for handling extreme non-equilibrium conditions as well as deviations from the Maxwell distribution function in many realistic plasmas.

1.2.3 Hydrodynamic approximation

This model treats the plasma as a continuum and determines the interesting macroscopic characteristics (density, flow, pressure ...) from the balance equations of the number, energy and momentum of each particle species. The balance equations are obtained as integrals of the appropriate kinetic equation. A special form of this approximation is the so-called magneto-hydrodynamics (MHD) model, in which the plasma is considered as an electrically conducting liquid under the influence of magnetic fields.

1.3 The role of charge carriers

The existence of charge carriers as the dominating components of the plasma is connected with a series of characteristics which are also important in industrial applications. The most active component of a non-thermal low-temperature plasma (LTP) is the hot electron gas. The high mean kinetic energy of electrons results in the generation of electromagnetic radiation (lines and continua) and in the production of numerous ionized, excited and dissociated species of increased chemical activity. Applications are plasma light sources and plasma chemical reactors.

Furthermore, the existence of charge carriers manifests itself in the following:

- Occurrence of electrical conductivity,

- Screening of electric fields,

- Occurrence of a multitude of oscillations and waves, typical for the plasma (Langmuir oscillations, ion acoustic oscillations, cyclotron oscillations, drift waves, surface waves, Alfven waves ...), as well as corresponding instabilities (plasma turbulence),

- Interaction with magnetic fields. This is an important aspect of modern plasma physics. In the interaction with magnetic fields the whole spectrum and variety of the typical plasma properties become effective, and

- Formation of characteristic boundary sheaths due to the contact of plasmas with solid surfaces. This is of particular importance in the technology of plasma processing.

1.4 Facts and formulas

1.4.1 Electron energy distribution functions (EEDF)

Calculation of the distribution functions F for the velocity or energy of the electron component under existing conditions in each case is the central problem in non-thermal low-temperature plasmas. One approach is offered by solving the Boltzmann equation adapted to plasmas:

$$\frac{dF}{dt} = \frac{\partial F}{\partial t} + \vec{c}\,\frac{\partial F}{\partial \vec{r}} + \frac{\vec{F_L}}{m_e}\,\frac{\partial F}{\partial \vec{c}} = C(F), \tag{1.1}$$

Table 1.2: Values of a for given m.

m	1	2	3
a	1.128	0.970	0.798

where $\vec{r}$ is the position vector of the particle and $\vec{c}$ its velocity, and

$$\vec{F}_L = e_o (\vec{E} + \vec{c} \times \vec{B}) \tag{1.2}$$

is the Lorentz force. The treatment of this equation has shown considerable progress during the last third of the 20th century, especially with the handling of the complexity of the collision integral $C(F)$ and the time and space variable terms (see, e.g., chapter 2). However, the problem has not been mastered at all and will continue into the 21st century. Sometimes the approximation of distribution functions by simple formulas is desirable. On this occasion the following standard terms for the electron energy distribution have proved to be valuable:

$$\hat{f}_0 (U) = \frac{a}{U_e^{3/2}} \exp\left[-\frac{1}{m}\left(\frac{U}{U_e}\right)^m\right] \qquad \text{with } m > 0 \tag{1.3}$$

and where

$$a = (1/m)^{(3-2m)/2m}/\Gamma(3/2m) \tag{1.4}$$

(see table 1.2). A possible dependence on space and time of the kinetic energy $e_0 U$ may be incorporated in the distribution parameter $U_e = U_e(\vec{r}, t)$. With $m = 1$, equation 1.3 yields the Maxwell distribution, while $m = 2$ represents the Druyvesteyn distribution. Of course the application of standard energy distributions in plasma physics is restricted to special cases, e.g., if only electrons with kinetic energies near the mean energy are of importance.

1.4.2   Kinetic temperature of electrons

The kinetic temperature T_e of an electron gas is defined by means of the mean energy U_{em},

$$\frac{3}{2} kT_e = e_0 U_{em} \qquad \text{with} \qquad U_{em} = \int_0^\infty U^{3/2} \hat{f}_0 (U) \, dU . \tag{1.5}$$

This results in

$$T_e \text{ (Kelvin)} \,\hat{=}\, 7734 \, U_{em} \text{ (Volt)} .$$

Using equation 1.3, we obtain

$$U_{em} = \varepsilon \, U_e \qquad \text{with} \qquad \varepsilon = m^{1/m} \Gamma(5/2m)/\Gamma(3/2m) \tag{1.6}$$

(see table 1.3). For the Maxwell distribution we thus obtain

$$T_e \text{ (Kelvin)} \,\hat{=}\, 11600 \, U_e \text{ (Volt)} .$$

In most cases the approximation $1 \text{ V} \simeq 10^4 \text{ K}$ is sufficient.

Table 1.3: Values of ε for given m.

m	1	2	3
ε	1.5	1.046	0.856

1.4.3   Coefficients for particle and energy transport

The electron flow density depends on electron density n_e and drift velocity $\vec{v}_e$ and is given by

$$\vec{j}_e = n_e \vec{v}_e = -n_e \mu_e \vec{E} - \mathrm{grad}\,(n_e D_e). \tag{1.7}$$

The first term on the right side represents the electrical field drift, the second one combines the action of diffusion and thermodiffusion. Using the usual approximations the mobility μ_e and the diffusion coefficient D_e are given by

$$\mu_e = -\frac{1}{3}\left(\frac{2e_0}{m_e}\right)^{1/2}\int_0^\infty \lambda_e U \frac{\partial \hat{f}_0}{\partial U}\,dU, \tag{1.8}$$

$$D_e = \frac{1}{3}\left(\frac{2e_0}{m_e}\right)^{1/2}\int_0^\infty \lambda_e U \hat{f}_0\,dU. \tag{1.9}$$

The particle flow is connected with a flow of energy,

$$\vec{j}_e^* = n_e U_{em}\vec{v}_e^* = -n_e \mu_e^* U_{em}\vec{E} - \mathrm{grad}\,(n_e U_{em} D_e^*), \tag{1.10}$$

$$U_{em}\mu_e^* = -\frac{1}{3}\left(\frac{2e_0}{m_e}\right)^{1/2}\int_0^\infty \lambda_e U^2 \frac{\partial \hat{f}_0}{\partial U}\,dU, \tag{1.11}$$

$$U_{em}D_e^* = \frac{1}{3}\left(\frac{2e_0}{m_e}\right)^{1/2}\int_0^\infty \lambda_e U^2 \hat{f}_0\,dU. \tag{1.12}$$

Using a simple exponential approximation for the energy dependence of the free path λ_e for momentum transfer of electrons:

$$\frac{1}{\lambda_e N} = \sigma_m(U) = a_m U^n, \tag{1.13}$$

where N is the particle density of the neutral gas and $\sigma_m(U)$ the cross section for momentum transfer, the standard distribution Eq. 1.3 results in the following transport coefficients

$$\begin{aligned}
\mu_e &= \frac{\mu}{a_m N}\left(\frac{2e_0}{m_e}\right)^{1/2} U_e^{-(2n+1)/2} \\
D_e &= \alpha U_e \mu_e \qquad \text{Nernst} - \text{Townsend} - \text{Einstein relation} \\
D_e^* &= \alpha^* U_e \mu_e \\
\mu_e^* &= \gamma \mu_e
\end{aligned} \tag{1.14}$$

with

$$\mu = \frac{m^{\frac{2m-2n-1}{2m}}}{3} \frac{\Gamma\left(\frac{1+m-n}{m}\right)}{\Gamma\left(\frac{3}{2m}\right)} \quad ; \qquad \alpha = m^{\frac{1-m}{m}} \frac{\Gamma\left(\frac{2-n}{m}\right)}{\Gamma\left(\frac{1+m-n}{m}\right)} \tag{1.15}$$

$$\alpha^* = m^{\frac{1-m}{m}} \frac{\Gamma\left(\frac{3}{2m}\right)\Gamma\left(\frac{3-n}{m}\right)}{\Gamma\left(\frac{5}{2m}\right)\Gamma\left(\frac{1+m-n}{m}\right)} \quad ; \qquad \gamma = \frac{\Gamma\left(\frac{3}{2m}\right)\Gamma\left(\frac{2+m-n}{m}\right)}{\Gamma\left(\frac{5}{2m}\right)\Gamma\left(\frac{1+m-n}{m}\right)} . \tag{1.16}$$

1.4.4   Generalized Boltzmann equilibrium

In front of insulating walls or floating metallic surfaces a plasma shows, as a matter of principle, inhomogeneities which are similar to the Boltzmann equilibrium in a neutral gas under the
action of external forces. These inhomogeneities should not be confused with sheath regions
because there is no violation of the plasma conditions. In particular the quasi-neutrality remains in force. Of course important differences from the Boltzmann equilibrium with neutral
gases exist in the plasma:

- The origin of the forces are space charges in the plasma

- Deviations from the Maxwell distribution function must be taken into consideration

- The distribution function may vary in space.

The first condition for Boltzmann equilibrium is vanishing particle flow in the flow direction.
Then equation 1.7 yields

$$\vec{E} = -\frac{1}{\mu_e} \frac{\mathrm{grad}\,(n_e D_e)}{n_e} . \tag{1.17}$$

In the case of a Maxwell distribution which is constant in space we have:

$$\vec{E} = -U_e \frac{\mathrm{grad}\, n_e}{n_e} = -\mathrm{grad}\, V \tag{1.18}$$

and the concentration $n_e(\vec{r})$ shows the well-known exponential behaviour under the action of
the potential $V(\vec{r})$ according to the barometric formula:

$$\frac{n_e(\vec{r})}{n_e(0)} = \exp\left(\frac{V(\vec{r})}{U_e}\right) . \tag{1.19}$$

In this case the properties of the Maxwell distribution also provide for the energetic equilibrium of the electrons ($\vec{j}_e^* = 0$). Every generalization (e.g., $U_e = U_e(\vec{r})$) has to ensure
additionally that

$$\vec{j}_e^* = 0, \qquad \text{resulting in} \qquad \vec{E} = -\frac{1}{U_{em}\,\mu_e^*} \frac{\mathrm{grad}\,(n_e\,U_{em}\,D_e^*)}{n_e} . \tag{1.20}$$

Equations 1.17 and 1.20 determine the necessary conditions of the spatial variables $U_e(\vec{r})$,
$V(\vec{r})$, $n_e(\vec{r})$ for generalized equilibrium. Hence for the standard distributions (Eq. 1.3) and
with Eq. 1.13 it follows that:

$$\frac{\mathrm{grad}\,U_e}{U_e} = \frac{\delta\,\mathrm{grad}\,n_e}{n_e} \quad ; \qquad \delta = \frac{\alpha^* - \alpha\gamma}{(n-1/2)(\alpha^* - \alpha\gamma) - \alpha^*} \tag{1.21}$$

$$\vec{E} = -\alpha\left(1+\left(\frac{1}{2}-n\right)\delta\right)\frac{U_e\,\mathrm{grad}\,n_e}{n_e}\quad;\qquad \frac{U_e(\vec{r})}{U_e(0)}=\left(\frac{n_e(\vec{r})}{n_e(0)}\right)^\delta. \tag{1.22}$$

For the Maxwell distribution, which holds for $U_e = const.$, we have $\delta = 0$. For non-Maxwell distributions ($\delta \neq 0$) we obtain

$$\frac{n_e(\vec{r})}{n_e(0)} = \left(\frac{\delta}{\alpha(1+(1/2-n)\delta)}\frac{V(\vec{r})}{U_e(0)}+1\right)^{1/\delta}. \tag{1.23}$$

Consideration of the spatial variations of the electron distribution function is of utmost importance in the case of deviations from the Maxwell distribution. Such deviations are common in plasmas, mostly as a consequence of collisions with heavy particles. Consequently a detailed analysis of the energetic relations of the electron gas is necessary. Very recently the non-local complex nature of the power and momentum balance in space-dependent plasmas was studied for the first time, starting from the Boltzmann equation (Winkler 1996, see also chapter 2). Neglecting collisions, the predominance and stability of the Maxwell distribution (e.g., in low pressure discharges, Langmuir paradox) should be a consequence of an energetic quasi-equilibrium $\vec{j}_e^* \approx 0$.

Table 1.4 shows numerical values for the particle and energy transport, using standard distributions, while figures 1.1 and 1.2 contain some illustrations of the Boltzmann equilibrium and its generalization.

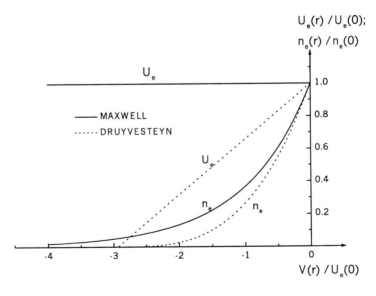

Figure 1.1: Equilibrium values of the electron concentration and temperature for given electrical potentials (full curves: Boltzmann equilibrium, i.e. Maxwell distribution with $U_e = const$; dashed curves: Generalization for a Druyvesteyn distribution with $U_e = U_e(r)$ and $n = 0$ in equation 1.13).

Table 1.4: Numerical values for particle and energy transport, calculated for three different standard distributions.

$m = 1$ (Maxwell distribution)						
	$n = -1$	$-1/2$	0	$1/6$	1	$3/2$
α	1	1	1	1	1	1
α^*	2	1.667	1.333	1.222	0.667	0.333
γ	2	1.667	1.333	1.222	0.667	0.333
δ	0	0	0	0	0	0
μ	0.752	0.5	0.376	0.354	0.376	0.667

$m = 2$ (Druyvesteyn distribution)						
	$n = -1$	$-1/2$	0	$1/6$	1	$3/2$
α	0.627	0.697	0.798	0.842	1.253	2.093
α^*	0.956	0.956	0.956	0.956	0.956	0.95
γ	1.797	1.667	1.524	1.476	1.198	1
δ	0.243	0.275	0.315	0.333	0.444	0.543
μ	0.647	0.5	0.406	0.383	0.324	0.333

$m = 4$						
	$n = -1$	$-1/2$	0	$1/6$	1	$3/2$
α	0.489	0.571	0.692	0.746	1.282	2.443
α^*	0.659	0.716	0.789	0.820	1.036	1.269
γ	1.713	1.667	1.616	1.579	1.498	1.427
δ	0.453	0.488	0.522	0.534	0.598	0.636
μ	0.593	0.5	0.429	0.409	0.334	0.307

1.4.5 Ambipolar diffusion

Within the plasma the movements of ions and electrons are interconnected via electric space charges. In the absence of external forces these space charges provide for equal electron and ion drifts. For instance in the direction towards isolating walls the steady state drift velocities of electrons and ions converge to the common velocity v_{am} of ambipolar diffusion ($n_e \approx n_i = n$)

$$v_{am} = -D_{am} \frac{\mathrm{grad}\, n}{n}, \tag{1.24}$$

$$\text{where} \quad D_{am} = \frac{\mu_e\, D_i + \mu_i\, D_e}{\mu_e + \mu_i}$$

is defined as the *Ambipolar Diffusion Coefficient*. Generally we have $\mu_e \gg \mu_i$, which results in

$$D_{am} \approx \alpha\, \mu_i\, U_e. \tag{1.25}$$

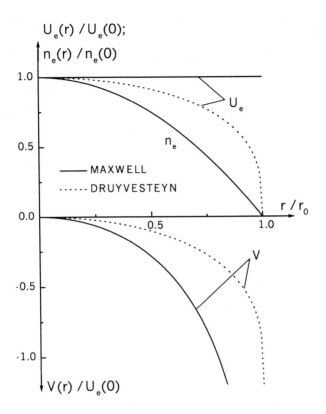

Figure 1.2: Equilibrium values of the electron temperature and the electrical potential for a preset concentration profile $n_e(r)/n_e(0) = 1 - (r/r_0)^2$; curves according to the conditions of Fig. 1.1.

The regime of ambipolar diffusion ($j_e = j_i$) shows some analogy to the Boltzmann equilibrium ($j_e = j_i = 0$). Therefore the formulas of section 1.4.4 are also valid approximately for ambipolar diffusion. Instead of formula 1.17 we obtain for the internal electric ambipolar field:

$$\vec{E}_{am} = -\frac{1}{\mu_e}\left(1 - \frac{\mu_i}{\mu_e}\right)\frac{\mathrm{grad}\,(n_e D_e)}{n_e}. \tag{1.26}$$

1.4.6 Condition of quasi–neutrality

In order to guarantee the status of free particles for electrons and ions in the plasma the field energy of space charges is limited to values much less than the kinetic energy of the charge carriers. This results in tolerable deviations $\Delta n_e = n_{eo} - n_e$ from exact neutrality $n_{eo} = n_{io}$:

$$|\Delta n_e|/n_{eo} \leq \lambda_D/L, \tag{1.27}$$

where λ_D is the Debye screening length (see section 1.4.7), and L a characteristic plasma length (e.g., the radius of a plasma column). Within the Debye length considerable deviations from neutrality may occur in plasmas. In this case the dynamics of such deviations is governed by the Langmuir plasma frequency.

1.4.7   Debye screening length

The electrical potential distribution of a charge carrier inside a plasma is different from the corresponding distribution in a vacuum. In a plasma each charge carrier polarizes its surroundings and thereby reduces the interaction length of the Coulomb potential V_c which is compensated in part by the space charge potential V_R (see Fig. 1.3). In the case of ions of charge Ze_0 and with $e_0 V_D \ll kT_i$ the screened potential is

$$V_D(r) = \frac{1}{4\pi\varepsilon_0}\frac{e_0}{r}\exp\left(-\frac{r}{\lambda_D}\right) \quad ; \quad \lambda_D^2 = \frac{\varepsilon_0\,kT_e}{e_0^2\,n_{e0}\,(1 + ZT_e/T_i)}. \tag{1.28}$$

Outside the Debye length λ_D the potential may be neglected ($V_D \approx 0$). This cut-off is typical for plasma conditions and of great importance for the interaction of charge carriers. Equation

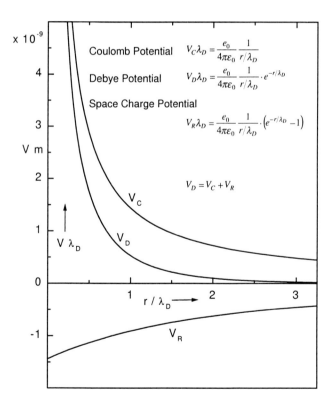

Figure 1.3: Variation of the electrical potential around an ion imbedded in a plasma.

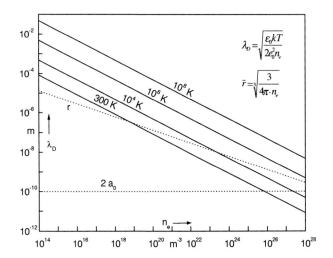

Figure 1.4: Debye screening length λ_D versus electron concentration ($\bar{r}$, mean distance of electrons; a_0, radius of the 1st Bohr orbit).

1.28 is based on the assumption that in a sphere of radius λ_D many charge carriers exist, i.e.,

$$\frac{4\pi}{3} n_{e0} \lambda_D^3 \gg 1 . \tag{1.29}$$

Then the plasma state is termed **ideal**. In this case the Coulomb interaction energy between two charge carriers at the mean distance is much smaller than the thermal energy. For **non-ideal** plasmas the electrostatic energies exceed the thermal energy. Fig. 1.4 displays calculated values of the Debye length.

1.4.8 Degree of ionization

Calculation of the degree of ionization in a closed form is only possible for plasmas in the exact equilibrium state. The Saha–Eggert equation then holds. For single ionization the following equation is valid:

$$\frac{x^2}{1 - x^2} = 2 \left(\frac{2\pi m_e}{h^3} \right)^{3/2} \frac{g_1 E_0^{5/2}}{g_0 p} \left(\frac{kT}{E_0} \right)^{5/2} \exp\left(-\frac{E_0}{kT} \right) , \tag{1.30}$$

where $x = n_e/(n_0 + n_e)$ is the degree of ionization, $p = (n_0 + 2n_e)kT$ the kinetic pressure, g_0 and g_1 are statistical weights, and E_0 is the ionization energy.

Fig. 1.5 shows according to equation 1.30 some curves of constant degree of ionization. Under the non-equilibrium conditions of LTP the calculation of n_e or x requires a detailed analysis of the corresponding balances. The energy balance of the electrons then results in a very simple and useful expression for the estimation of the degree of ionization. In the steady

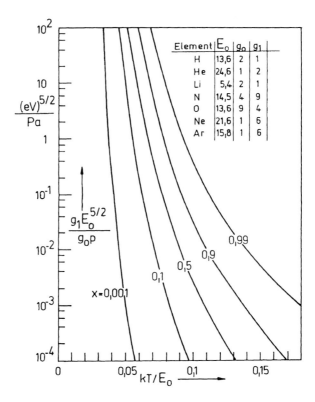

Figure 1.5: Curves for constant degree of ionization in conformity with the Saha–Eggert-equation.

state and for $n_e \ll n$ it reads as follows:

$$\frac{n_e}{n} \approx \frac{2}{3} \frac{1}{\delta_{loss}} \frac{\tau_e\, n}{e_0\, U_e} \frac{P/V}{n^2}, \tag{1.31}$$

where τ_e is the mean free time between electron collisions, P/V is the power density supplied to the plasma, and δ_{loss} is the mean fraction of energy that the electrons lose in a single collision. For elastic collisions, $\delta_{loss} = \delta_{el} = 2m_e/M$ is of the order of $\delta_{el} \approx 10^{-4} \cdots 10^{-5}$. The inelastic energy loss δ_{inel} is typically larger by one to two orders of magnitude. The quantity $\tau_e n/\delta_{loss}$ is a function of U_e, but often its variations are rather small. Compared to the range of P/Vn^2 it may be regarded as constant. This is a rather good approximation at higher gas pressures ($p \geq 10^3\, Pa$).

1.4.9 Electrical conductivity

Under the action of an electrical field $\vec{E}$ the free electrons and ions of the plasma gain drift velocities and generate an electric current density of

$$\vec{j} = -e_0\left(n_e\,\vec{v}_e - n_i\,\vec{v}_i\right) = e_0\left(n_e\,\mu_e + n_i\,\mu_i\right)\vec{E}. \tag{1.32}$$

Generally, since of $\mu_e \gg \mu_i$ and $n_e \approx n_i$, only the contribution of electrons determines the current density. Then the electrical conductivity σ of a plasma is given by

$$\sigma = e_0\,n_e\,\mu_e. \tag{1.33}$$

The concentration n_e and mobility μ_e of the electrons are given by, e.g., equations 1.8 and 1.31. The elementary kinetic approximation for the conductivity reduces to

$$\sigma = e_0^2 n_e \tau_e / m_e. \tag{1.34}$$

With regard to the degree of ionization two cases may be distinguished:

▷ **Weakly ionized plasmas.** The mean free time of flight τ_e is defined by electron-atom collisions and is independent of n_e. Consequently, $\sigma \propto n_e$.

▷ **Fully ionized plasmas.** τ_e is determined by Coulomb collisions with $\tau_e \propto 1/n_e$ and the conductivity is constant.

For a fully ionized thermal plasma the corresponding equation is named the Spitzer formula

$$\sigma = \frac{64\sqrt{2\pi}\,\varepsilon_0^2}{e_0^2\,\sqrt{m_e}}\,\frac{(kT_e)^{3/2}}{\ln\Lambda} \tag{1.35}$$

where $\ln\Lambda$ is the Coulomb logarithm; its numerical value for the majority of plasmas lies in the range from 15 to 20.

1.4.10 Plasma frequency

Representative of the multitude of dynamic processes in plasmas are the longitudinal electrical oscillations (Langmuir 1928). The occurrence of space charges generates in general a quasi-elastic coupling of the electrons to the ionic background and results in oscillations with a frequency ω_P given by

$$\omega_P^2 = \frac{e_0^2\,n_e}{\varepsilon_0\,m_e}. \tag{1.36}$$

The electron plasma frequency (Eq. 1.36) is critical for the propagation of electromagnetic waves in plasmas. In the range $\omega < \omega_P$ the damping of the waves is strong. For $\omega = \omega_P$ electromagnetic waves show strong reflection at the plasma interface. This is related to the refractive index $n^2 = 1 - (\omega_P/\omega)^2$ (Eccle relation).

2 Electron kinetics in weakly ionized plasmas

Rolf Winkler

Institut für Niedertemperatur-Plasmaphysik, Friedrich-Ludwig-Jahn-Straße 19
D-17489 Greifswald, Germany

2.1 Introduction

2.1.1 The active role of electrons in the plasma

Weakly ionized plasmas are complex systems. In the simplest case they consist largely of neutral gas atoms and, to a lesser extent, electrons and positive ions.

Electric fields acting on the charged particles, and especially on the electrons, are usually the main power source of the plasma. Owing to the small mass of the electrons, only a poor energetic contact of the electrons with the heavy gas particles of the plasma is established by elastic collisions. As a consequence, the electrons reach a mean kinetic energy which is much higher than that of the neutral and charged heavy plasma components, and the plasma becomes a non-thermal medium.

Thus, a significant proportion of the electrons is energetically capable of overcoming the threshold energies above which inelastic electron collision processes with the atoms or molecules take place. However, in each of these inelastic collision processes the corresponding electron loses at least the threshold energy for this process, and the electron is transferred from a region of higher kinetic energies to a low energy region. Thus, the occurrence of inelastic collisions causes an efficient depopulation of the electron energy space in the range of higher energies. As a consequence the electron population in the region of inelastic collisions decreases markedly with increasing energy.

This interplay between the field action and the elastic and inelastic collision processes causes the electron component generally to reach a state which is far from thermodynamic equilibrium. Thus, the nonequilibrium state of the electron component established in non-thermal plasmas or its temporal and spatial evolution can only be described on an appropriate microphysical basis.

A microphysical study of the electron behaviour becomes possible by following either of two quite different approaches. One method consists of formulating an adequate electron kinetic equation and its approximate solution. The other approach uses the techniques of particle simulation. Both approaches have their particular advantages and disadvantages. For example, only a limited number of electrons can be treated in any simulation and the statistics limits the accuracy reached particularly with respect to the energy distribution at higher energies. Such a limitation does not usually occur when solving the kinetic equation of the electrons. In any of these microphysical approaches the complex interplay of the action of an electric field and of elastic and various inelastic electron collision processes with atoms or molecules of the gas has to be taken into account in detail.

The primary purpose of the kinetic treatment of the electron component in non-thermal plasmas, briefly presented in the following, is the determination of its velocity distribution function and of its various macroscopic properties. The latter can be obtained by appropriate averages over the velocity distribution function.

The macroscopic properties of the electrons are very important for the global behaviour of the plasma. The electrons are the only plasma component that is capable of causing inelastic collisions with atoms and molecules, thus leading to excitation, dissociation, or ionization. These are usually the basic processes through which the primary activation of the working gas takes place. Subsequently other collision processes and chemical reactions between activated heavy particles of the plasma are often initiated.

The velocity distribution function depends to a large extent on the particular plasma conditions considered. In this connection the lumped frequency of various inelastic electron collision processes and the structure of the electric field must be mentioned. The spectrum of important electron collision processes is broad and includes elastic collisions as well as inelastic collisions causing excitation, dissociation, ionization and attachment. With respect to the inelastic collision processes remarkable differences between electron collisions with atoms and molecules generally have to be taken into account. Inelastic collisions with atoms are characterized by energetic thresholds of several electron volts. In addition to such electronic excitation processes, the rotational and vibrational excitation of molecules in collisions with electrons occurs. Both these inelastic processes have considerably lower energy thresholds of a few hundredths and a few tenths of an electron volt, respectively. Owing to the low energy loss in each collision event, the impact of the rotational excitation (and de-excitation) on the kinetics of the electrons is usually of less importance in non-thermal plasmas typically maintained by an electric field. For this reason, these inelastic collisions are often neglected.

Furthermore, the procedure ultimately used to determine the velocity distribution sensitively depends on the type of the plasma and is, for example, quite different when studying the electron kinetics in steady-state, time-dependent or space-dependent plasmas.

If the electric current and thus the density of both electrons and excited atoms and molecules grows in the plasma, electron collisions with excited atoms and molecules and the Coulomb interaction between the electrons become increasingly important and have to be included additionally in the kinetic study of the electron behaviour.

2.1.2   Action of electric fields and collision processes

The electrons undergo two basic impacts, (a) the action of an electric field and (b) binary elastic and inelastic collisions with heavy particles [1, 2, 3].

The electric field accelerates the electrons parallel to its direction. If the field in the plasma is parallel to a fixed space direction, as is often the case, the electric field causes a change in the component of the electron velocity parallel to this fixed direction. As a consequence of the field action, this velocity component plays an exceptional role in causing anisotropy of the velocity distribution function. Moreover, the sole action of the electric field causes an increase both in the individual and the mean electron energy.

However, in addition to the action of the field several types of binary collision processes between the electrons and the atoms or molecules occur in the plasma. Each collision event causes a change in the velocity direction of the colliding electron. As a result, each electron collision process leads to a pronounced, more or less isotropic scattering of the electrons and of their vectorial velocities into all space directions. Thus, the electron collision processes tend to reduce the anisotropy of the velocity distribution primarily produced by the action of the electric field. In addition to the scattering of the electrons the collision processes cause a change in the electron energy. Since the mean energy of the electrons is distinctly larger than that of the heavy particles, usually the electrons lose energy when undergoing elastic or inelastic collisions.

The velocity distribution of the electrons finally established in special plasma conditions is the result of a complex interplay between the field action on the electrons and the various binary collision processes of the electrons.

In order to gain a better understanding of basic aspects of the collision interaction, the most significant electron collision processes, namely elastic and exciting collisions, are briefly considered in the following [2, 3]. In the course of each binary collision the electron undergoes a change in its kinetic energy and momentum. Since the mean energy of the electrons is much higher than that of the heavy particles, the latter can be considered to be at rest before each collision. The energy losses of the electrons in elastic and exciting collisions can be deduced from the energy and momentum conservation laws of the respective binary collision process. The main contributions to these losses are given by the expressions

$$\frac{m_e}{2}v^2 - \frac{m_e}{2}(v^\star)^2 = \frac{m_e}{2}v^2 \, 2\frac{m_e}{M}\left(1 - \cos\theta\right), \quad \frac{m_e}{2}v^2 - \frac{m_e}{2}(v^\star)^2 = E_{in}^* - E_{in}, \quad (2.1)$$

where m_e and M denote the mass of the electron and the heavy particle, respectively, v and $v^\star$ are the absolute values of the electron velocities before and after the collision event, θ is the scattering angle related to the elastic collisions, and $E_{in}^* - E_{in}$ represents the increase in the internal energy of the heavy particle in the excitation process.

The expressions in Eq. 2.1 reflect important properties of the elastic and exciting electron collision processes already mentioned in the introduction. The first expression of Eq. 2.1 indicates that the energy loss in each elastic collision is proportional to the electron energy $m_e v^2/2$ before the collision and to the very small mass ratio m_e/M. Thus, the poor energetic contact between the electrons and the heavy particles in an elastic collision becomes immediately obvious. However, according to the second expression of Eq. 2.1, in each exciting collision the electron loses the energy $E_{in}^* - E_{in}$ necessary to excite the heavy particle from its lower energy level E_{in} to its higher level E_{in}^*.

The energy and momentum conservation laws related to the elastic and exciting collisions represent together four scalar equations connecting the two vectorial velocities of both colliding particles before the collision event with the corresponding ones after the event. If the initial velocities are given, ultimately the components of the two velocities after the collision event, i.e., six scalar quantities, have to be determined in order to describe the result of the binary collision. Thus, when using the energy and momentum conservation laws two scalar quantities remain undetermined. The remaining lack of knowledge on the collision process can be eliminated by using additional information on the electron scattering process as being involved in the differential scattering cross section $\sigma(m_e v^2/2, \cos\theta)$ of the corresponding collision process [1, 2, 3].

Differential cross sections calculated by averaging over the solid angle $\sin\theta\, d\theta\, d\phi$ are the basis for determining the appropriate total cross sections $Q(m_e v^2/2)$, as detailed below.

With respect to the binary inelastic collision processes of the electrons with the heavy particles other important types should be mentioned [2, 3]. Usually these collision processes are subdivided into conservative and nonconservative processes with respect to the change in the number of electrons in the course of the collision event.

Other important conservative collision processes are the dissociation of molecules and the de-excitation of excited atoms or molecules. In each dissociation process the colliding electron loses the dissociation energy of the molecule and at least two heavy particle fragments occur due to the dissociation process. In each de-exciting collision process the colliding electron receives the excitation energy from the excited heavy particle. Thus, in de-exciting collisions with an excited atom the electron is transferred to a region of much larger energy by one collision only.

Important nonconservative collision processes with respect to the kinetics of the electrons are ionization and attachment. While in the first process after each collision event two electrons occur, in the second process the colliding electron is lost and a negative ion is generated by the attachment of an electron to the neutral heavy particle. In the course of the ionization event the ionization energy has to be covered by the initially available electron energy and the remaining electron energy is distributed among both electrons. In the case of attachment, the electron itself and its initial energy disappear from the electron component of the plasma.

2.2 Kinetic treatment of the electrons

2.2.1 Velocity distribution and macroscopic properties

The distribution of the electrons with respect to their velocity space $\vec{v}$ at the coordinate space position $\vec{x}$ and at time t is described by the velocity distribution function $F(\vec{v},\vec{x},t)$ [1]. The contribution $dn(\vec{x},t)$ of the velocity space interval $d\vec{v}$ around $\vec{v}$ to the electron density $n(\vec{x},t)$ is then determined by the expression $dn(\vec{x},t) = F(\vec{v},\vec{x},t)\, d\vec{v}$ with $d\vec{v} \equiv dv_x\, dv_y\, dv_z$.

If the velocity distribution $F(\vec{v},\vec{x},t)$ is known, important macroscopic properties of the electrons can be calculated by appropriate velocity space averaging over the distribution. To give a few examples, the density $n(\vec{x},t)$ and the vectorial particle current density $\vec{j}(\vec{x},t)$ of the electrons are given by the averages [1]

$$n(\vec{x},t) = \int F(\vec{v},\vec{x},t)\, d\vec{v}, \quad \vec{j}(\vec{x},t) = \int \vec{v}\, F(\vec{v},\vec{x},t)\, d\vec{v}. \tag{2.2}$$

In the preceding relations the usual normalization of the velocity distribution $F(\vec{v},\vec{x},t)$ with respect to the electron density $n(\vec{x},t)$ has been used. In special conditions, for example, in steady-state or in time-dependent conditions with the inclusion of conservative electron collision processes only, the electron density becomes a constant and can easily be separated from the velocity distribution. Especially in these cases a normalization of the velocity distribution on one electron is often used.

2.2.2 Kinetic equation of the electrons

The various microphysical interactions such as the field action and the various binary collision processes, in which the electrons are involved in an weakly ionized plasma, lead to a complex redistribution of the electrons in phase space, i.e., their combined coordinate and velocity space.

Owing to the short range interaction in binary electron collisions the appropriate phase space balance equation for the velocity distribution $F(\vec{v}, \vec{x}, t)$ of the electrons is given by the Boltzmann equation [1, 2, 3]

$$\frac{\partial F}{\partial t} + \vec{v} \cdot \nabla_{\vec{x}} F - \frac{e_0}{m_e} \vec{E} \cdot \nabla_{\vec{v}} F = C^{el}(F) + \sum_l C_l^{cs}(F) \,. \tag{2.3}$$

Here $-e_0$ denotes the charge of the electron, $\vec{E}(\vec{x}, t)$ is the electric field and $C^{el}(F)$ and $C_l^{cs}(F)$ are the collision integrals for elastic collisions and important conservative (cs) inelastic collisions, i.e., the lth excitation or dissociation process of the electrons in collisions with the ground state atoms or molecules of the gas. For simplicity only the most essential electron collision processes have been taken into account in the kinetic equation 2.3. When applied to a specific plasma, other collision processes, for example the excitation and de-excitation of excited atoms or molecules and the ionization and attachment of ground state and excited atoms or molecules, may be of importance, and corresponding collision integrals have to be added to the right-hand side of Eq. 2.3.

In order to avoid the more complex treatment of ionizing collisions in the kinetic approach, but still taking the energy dissipation in these collisions approximately into account, the ionization is often treated in the same way as an excitation process neglecting the appearance of an additional electron after the ionization event.

The kinetic equation 2.3 is of very complex nature and covers a large area of special electron kinetic problems. Consequently, there does not seem to be any way of finding some kind of *general* solution of this equation which can be specified afterwards to the plasma conditions of special interest.

As a consequence, for different plasma conditions, e.g., steady-state, time-dependent or space-dependent problems, different methods of solution and numerical techniques have been developed and applied. In addition, the specific structure of the electric field acting on the electrons is of particular importance for the establishment of a special symmetry in the velocity distribution and thus for a specific simplification of the method of solution.

Therefore, the objective of this paper can only be to give a brief introduction into the study of the kinetics of the electrons under different plasma conditions and to illustrate some typical aspects of the kinetics.

2.2.3 Treatment of the kinetic equation

In order to find an approximate solution of the kinetic equation an orthogonal expansion of the velocity distribution with respect to the direction $\vec{v}/v$ of the velocity $\vec{v}$ is commonly used in the treatment of the kinetic equation. Depending on the structure of the electric field and the expected inhomogeneity of the plasma, either a reduced expansion with respect to only one angular coordinate or a more complex expansion with respect to both angular coordinates of $\vec{v}/v$ has to be used.

If the electric field and, in addition, a possible inhomogeneity in the plasma are parallel to a fixed space direction, for example the $\vec{e}_z$ direction of the coordinate space, the velocity distribution becomes symmetrical around this direction. Then, for example, for steady-state, time-dependent or space-dependent conditions, the velocity distribution has the reduced dependence $F(U, v_z/v)$, $F(U, v_z/v, t)$ or $F(U, v_z/v, z)$ and can be given the corresponding expansion [2, 3]

$$F\left(U, \frac{v_z}{v}\right) = C \sum_{n=0}^{\infty} f_n(U) P_n\left(\frac{v_z}{v}\right),$$

$$F\left(U, \frac{v_z}{v}, t\right) = C \sum_{n=0}^{\infty} f_n(U, t) P_n\left(\frac{v_z}{v}\right), \tag{2.4}$$

$$F\left(U, \frac{v_z}{v}, z\right) = C \sum_{n=0}^{\infty} f_n(U, z) P_n\left(\frac{v_z}{v}\right)$$

in Legendre polynomials $P_n(v_z/v)$ with $U = m_e v^2/2$ being the kinetic energy, $v = |\vec{v}|$ and the constant $C \equiv (1/2\pi)(m_e/2)^{3/2}$.

In these expansions the dependence of the velocity distribution on the direction $\vec{v}/v$ is fixed by the Legendre polynomials $P_n(v_z/v)$. Thus, averages with respect to the angular space $\vec{v}/v$ over the respective velocity distribution F and appropriate weight functions can already be calculated. For example, the angular space averages occurring in the macroscopic quantities 2.2 are given by

$$dn = \int_{\vec{v}/v} F \, d\vec{v} = U^{1/2} f_0 \, dU,$$

$$d\vec{j} = \int_{\vec{v}/v} \vec{v} F \, d\vec{v} = \frac{1}{3}\left(\frac{2}{m_e}\right)^{1/2} U f_1 \, dU \, \vec{e}_z \tag{2.5}$$

as a result of the orthogonality of the Legendre polynomials. This means that the contributions of the interval dU of the kinetic energy to the density n and the particle current density $\vec{j}$ are, apart from scalar factors, completely determined by the two lowest coefficients f_0 and f_1 of the respective expansion 2.4. Furthermore, expression 2.5 already indicates that the lowest coefficient f_0 represents the isotropic part of the velocity distribution and all other terms of each expansion 2.4 are contributions to the anisotropy of the velocity distribution. The relations 2.5 additionally show that, in a strict sense, the expression $U^{1/2} f_0$ represents the energy distribution of the electrons and $(1/3)(2/m_e)^{1/2} U f_1$ the energetic distribution of their particle current density. The latter possesses a component in the $\vec{e}_z$ direction only.

Substitution of each of the expansions 2.4 into the correspondingly specified kinetic equation 2.3 leads, after several intermediate rearrangements, to an analogous expansion in Legendre polynomials of the entire kinetic equation, and because of the orthogonality of the polynomials ultimately to a corresponding hierarchy of equations [2, 3]. This equation system includes the corresponding expansion coefficients f_n and its approximate solution finally yields these coefficients and thus the velocity distribution.

In deriving the equation system it is commonly assumed with respect to the collision integrals that the atoms or molecules are at rest before the collision events. Furthermore, each collision integral is additionally expanded with respect to the mass ratio m_e/M and only the leading term in m_e/M of each collision integral is taken into account in each coefficient of the Legendre polynomial expansion of the kinetic equation. Often the various inelastic collision processes of the electrons are assumed to occur with isotropic scattering, i.e., the corresponding differential collision cross sections are considered to be dependent only on the kinetic energy of the electrons.

The resultant infinite system of equations has to be truncated in order to obtain a closed system and its approximate solution. It has been found in recent years that a restriction of each of the expansions 2.4 to its two lowest terms, i.e., the so-called two-term approximation, leads to an unexpectedly good approximation for the velocity distribution under many steady-state, time-dependent as well as space-dependent plasma conditions. This two-term approximation already allows a detailed microphysical analysis of various electron kinetic problems related to each of these plasma conditions, which will be demonstrated in the following.

2.2.4 Macroscopic properties of the electrons

Owing to the orthogonality of the expansions 2.4 all essential macroscopic quantities of the electrons can be represented by energy space averages over the lowest two expansion coefficients f_0 and f_1. Similarly to the averaging in the expressions 2.5, this is a consequence of the integration over the angle space $\vec{v}/v$.

Thus, the electron density n, the mean energy density u_m (i.e., the mean electron energy times its density) and the particle and energy current density $\vec{j} = j_z \vec{e}_z$ and $\vec{j}_e = j_{ez} \vec{e}_z$ are given by the expressions

$$n = \int_0^\infty U^{1/2} f_0 \, dU \,, \qquad u_m = \int_0^\infty U^{3/2} f_0 \, dU \,, \tag{2.6}$$

$$j_z = \frac{1}{3}\sqrt{\frac{2}{m_e}} \int_0^\infty U f_1 \, dU \,, \quad j_{ez} = \frac{1}{3}\sqrt{\frac{2}{m_e}} \int_0^\infty U^2 f_1 \, dU \,. \tag{2.7}$$

The particle and energy current density $\vec{j}$ and $\vec{j}_e$ possess a z-component due solely to the rotational symmetry of the velocity distribution F around the direction of $\vec{e}_z$. This is a consequence of the assumption that the field action and a possible inhomogeneity only occur parallel to this direction.

With the electric field $\vec{E} = E \, \vec{e}_z$ the power and momentum gain from the electric field P^f and I^f are given by

$$P^f = -j_z \, e_0 E \,, \qquad I^f = -n \frac{e_0}{m_e} E \,. \tag{2.8}$$

The power losses P^{el}, P_l^{cs} by elastic collisions and by the lth excitation or dissociation process and the corresponding momentum losses I^{el}, I_l^{cs} may be written as

$$P^{el} = 2\frac{m_e}{M}\sqrt{2/m_e}\int_0^\infty U^2 N Q^d(U) f_0\, dU, \tag{2.9}$$

$$P_l^{cs} = U_l^{cs}\sqrt{2/m_e}\int_0^\infty U N Q_l^{cs}(U) f_0\, dU, \tag{2.10}$$

$$I^{el} = \frac{2}{3m_e}\int_0^\infty U^{3/2} N Q^d(U) f_1\, dU, \tag{2.11}$$

$$I_l^{cs} = \frac{2}{3m_e}\int_0^\infty U^{3/2} N Q_l^{cs}(U) f_1\, dU\,, \tag{2.12}$$

where N is the gas density. Furthermore, the total power and momentum losses P^c and I^c in all collision processes are given by

$$P^c = P^{el} + \sum_l P_l^{cs}\,,\qquad I^c = I^{el} + \sum_l I_l^{cs}. \tag{2.13}$$

Finally, the mean collision frequency ν_l^{cs} and the corresponding rate coefficient k_l^{cs} of the lth excitation or dissociation process are represented by the averages

$$\nu_l^{cs} = \sqrt{2/m_e}\int_0^\infty U N Q_l^{cs}(U) f_0\, dU\,/n\,,\qquad k_l^{cs} = \nu_l^{cs}/N\,. \tag{2.14}$$

According to the elastic and inelastic collisions of the electrons considered in Eq. 2.3 the cross section $Q^d(U)$ for momentum transfer in elastic collisions and the total cross section $Q_l^{cs}(U)$ of the lth excitation or dissociation process with the corresponding excitation or dissociation energy U_l^{cs} are involved in the respective collisional power and momentum losses, expressions 2.9 and 2.10 and expressions 2.11 and 2.12, and the related mean collision frequencies 2.14. These cross sections are obtained by averaging the differential cross sections $\sigma^{el}(U, \cos\theta)$ and $\sigma_l^{cs}(U, \cos\theta)$ together with appropriate weight factors over the solid angle of scattering according to

$$Q^d(U) = \int \sigma^{el}(U, \cos\theta)(1 - \cos\theta)\sin\theta\, d\theta\, d\phi\,,$$

$$Q_l^{cs}(U) = \int \sigma_l^{cs}(U, \cos\theta)\sin\theta\, d\theta\, d\phi\,. \tag{2.15}$$

If, for example the isotropic and anisotropic distribution f_0 and f_1 have been determined by solving the adequate equation system of the two-term approximation related to a specific kinetic problem, the steady-state values, the temporal evolution or the spatial alteration of the macroscopic quantities can be calculated by appropriate energy space averaging over these distribution functions according to the corresponding representations 2.6 to 2.14.

Furthermore, appropriate energy space averaging over the Boltzmann equation 2.3 or equations obtained from the Boltzmann equation by using the appropriate expansion 2.4 yields the consistent macroscopic balance equations of the electrons. In particular, the particle, power and momentum balance equation of the electrons can be derived in this way. As detailed below, valuable information about the physics involved in the kinetic treatment of a specific problem can already be obtained when considering the consistent macroscopic balance equations of the electrons.

2.3   Kinetics in time- and space-independent plasmas

Kinetic studies of plasmas under steady-state conditions represent the conventional area of electron kinetics. Such studies have been made on many atomic and molecular gases and on mixtures of such gases. In addition to the basic electron collision processes, such as elastic collisions and exciting and dissociating collisions with ground state atoms and molecules, exciting and de-exciting electron collision processes with excited atoms and molecules and, at higher electron densities, partly the Coulomb interaction between electrons have been taken into account.

These investigations are largely performed on the basis of the two-term approximation allowing for anisotropic scattering in elastic collisions, whereas mainly isotropic scattering is assumed for the inelastic collision processes.

2.3.1   Basic equations and consistent macroscopic balances

Within the framework of the two-term approximation with $\vec{E} = E\,\vec{e}_z$ the kinetic treatment of the electrons in steady-state plasmas is based on the equation system [2, 4]

$$-e_0 E \frac{d}{dU}\left[\frac{U}{3}f_1(U)\right] - \frac{d}{dU}\left[2\frac{m_e}{M}U^2 NQ^d(U)f_0(U)\right] + \sum_l UNQ_l^{cs}(U)f_0(U)$$

$$- \sum_l (U + U_l^{cs})NQ_l^{cs}(U + U_l^{cs})f_0(U + U_l^{cs}) = 0\,,$$

$$-e_0 E \frac{d}{dU}f_0(U) + [\,NQ^d(U) + \sum_l NQ_l^{cs}(U)\,]f_1(U) = 0 \qquad (2.16)$$

for determining the isotropic and anisotropic distribution $f_0(U)$ and $f_1(U)$.

Under steady-state conditions the electron density n can be separated from the velocity distribution according to $F(U, v_z/v) = n\,\hat{F}(U, v_z/v)$ and, consequently, in the two-term approximation the same separation can be applied to the isotropic and anisotropic distributions, i.e., $f_i(U) = n\,\hat{f}_i(U)$ with $i = 0$ and 1. Substitution of this relation into the representation 2.6 of the density leads to the normalization condition

$$\int_0^\infty U^{1/2}\hat{f}_0(U)\,dU = 1 \qquad (2.17)$$

for the per one electron normalized isotropic distribution $\hat{f}_0(U)$.

By eliminating $f_1(U)$ from the first equation of system 2.16 by using the second equation of 2.16, the equation

$$-\frac{(e_0 E)^2}{3}\frac{d}{dU}\left[\frac{U}{NQ^d(U) + \sum_l NQ_l^{cs}(U)}\frac{d}{dU}\hat{f}_0(U)\right] - \frac{d}{dU}\left[2\frac{m_e}{M}U^2 NQ^d(U)\hat{f}_0(U)\right]$$

$$+ \sum_l UNQ_l^{cs}(U)\hat{f}_0(U) - \sum_l (U + U_l^{cs})NQ_l^{cs}(U + U_l^{cs})\hat{f}_0(U + U_l^{cs}) = 0 \qquad (2.18)$$

is obtained for the normalized isotropic distribution $\hat{f}_0(U)$.

This equation and the condition 2.17 are commonly used to determine the normalized isotropic distribution. Consideration of equation 2.18 shows that various quantities of the collision processes and few plasma parameters are involved in its coefficients and naturally make an immediate impact on its solution. The atomic data of the various collision processes are the momentum transfer cross section $Q^d(U)$, the total cross sections $Q_l^{cs}(U)$, the corresponding excitation or dissociation energies U_l^{cs} of the ground state atoms or molecules and the mass ratio m_e/M. With regard to the plasma parameters, the electric field strength E and the density N of the atoms or molecules occur, however, only in the form of the reduced field strength E/N. All these quantities have to be known for a specific weakly ionized plasma in order to determine the isotropic distribution $\hat{f}_0(U)$ by solving equation 2.18.

If the isotropic distribution $\hat{f}_0(U)$ has been obtained from Eq. 2.18 the normalized anisotropic distribution $\hat{f}_1(U)$ can easily be determined from the second of equation 2.16. By means of both the normalized distributions the steady-state values of all the important macroscopic quantities can be calculated, up to the electron density n as a common factor, using the representations 2.6 to 2.14 and replacing the distributions $f_i(U)$ by $n\,\hat{f}_i(U)$ with $i = 0$ and 1 in all integrals.

Furthermore, in steady-state plasmas the consistent power and momentum balances mentioned above are given by the equations

$$\frac{P^f}{n} = \frac{P^c}{n} \quad \text{and} \quad \frac{I^f}{n} = \frac{I^c}{n}. \tag{2.19}$$

According to the power balance the mean power gain from the electric field is compensated for by the mean power loss in collisions, and this happens for any given gas and its specific atomic or molecular data and for any reduced field strength E/N. An analogous compensation occurs in the momentum balance between the mean momentum gain from the field and the mean momentum loss in collisions.

Equation 2.18 is a linear ordinary second-order differential equation with the additional terms $\hat{f}_0(U + U_l^{cs})$ involving the shifted energy arguments $U + U_l^{cs}$. These terms are caused by the inelastic electron collision processes with the corresponding energy losses $U_l^{cs} > 0$ in the collision events. The solution of Eq. 2.18 is required in an appropriate energy range $0 \le U \le U^\infty$, with the upper limit U^∞ chosen in such a way that $\hat{f}_0(U)$ becomes negligibly small for energies larger than U^∞.

In order to determine the desired solution of the second order equation, two boundary conditions have to be imposed on the solution. For the latter, the normalization by Eq. 2.17 and the condition $\hat{f}_0(U \ge U^\infty) = 0$ can be used. Various solution techniques for equation 2.18 have been developed in the past. A very efficient solution technique is obtained when performing a discretization of the second order differential equation 2.18 at all internal points of an equidistant energy grid, using a finite difference approach with second-order-correct difference analogues.

The discretization of the ordinary differential equation 2.18 and of the aforementioned boundary conditions leads finally to a complete linear system of equations. An efficient resolution of this system becomes possible when iteratively treating all the terms with shifted energy arguments.

A particular advantage of this solution approach is that other collision processes and even the nonlinear Coulomb interaction between the electrons can be included and can be success-fully treated after corresponding extensions of the solution technique.

2.3.2 Illustration of distribution functions and macroscopic quantities

To demonstrate the behaviour of electron kinetic quantities in steady-state conditions, the weakly ionized plasma in krypton is considered in the following as a typical representative of an atomic gas plasma.

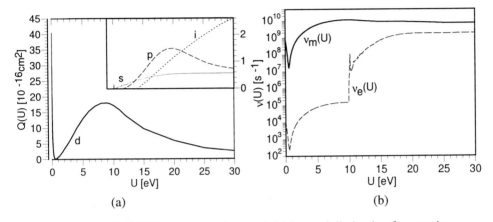

Figure 2.1: (a) Collision cross sections and (b) lumped dissipation frequencies.

Using two different scales for the cross section values, the important inelastic collision cross sections of krypton are shown in Fig. 2.1 (a) together with the respective cross section $Q^d(U)$ for momentum transfer in elastic collisions, denoted by d. The individual collision cross sections have been taken from Ref. [5]. The relevant total cross sections $Q^s(U)$, $Q^p(U)$ and $Q^i(U)$ for the respective lumped excitation of the s- and p-states and for the ionization from the ground state are denoted by s, p and i. The corresponding energy losses in these inelastic processes are $U^s = 9.9$ eV, $U^p = 11.3$ eV and $U^i = 14.0$ eV.

Despite the lumped description of some individual excitation processes by one total cross section, the characteristic features of the different collision processes and their consequences for the various electron kinetic properties are preserved to a large extent.

The ionization process has been taken into account. However, as mentioned above, this process is treated as an excitation process, conserving the electron number in this inelastic collision process.

In order to fix the plasma parameters involved in the kinetic equation 2.18 and in the representations 2.6 to 2.14 of the macroscopic quantities, the gas density $N = 3.54 \times 10^{16}$ cm^{-3}, i.e., a density which corresponds to 1 Torr pressure at 0°C gas temperature, is used henceforth in all examples. Equation 2.18 has been solved for a larger range of field strengths E using the mass ratio m_e/M for krypton and the set of collision cross sections presented in Fig. 2.1 (a).

The resultant energy dependence of the isotropic distribution $\hat{f}_0(U)$ in a krypton plasma is shown at field strengths E between 0.2 and 10 V/cm in Fig. 2.2 (a). The structural change of the distributions is the result of the competing action of the electric field and of the elastic and inelastic collisions.

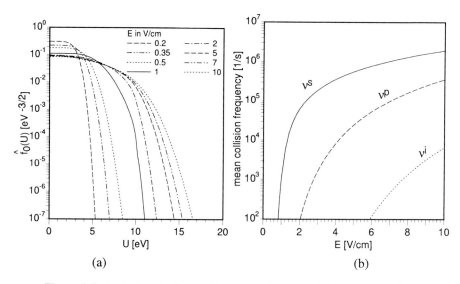

Figure 2.2: (a) Isotropic distribution and (b) mean collision frequencies.

At field strengths lower than about 1 V/cm elastic collisions take place with only a small energy loss in each collision event. An increase in the field strength causes a monotonic growth of the isotropic distribution at higher energies. However, if the electron population markedly overcomes the energy threshold U^s of the lowest excitation process, a larger structural change in the energy dependence of the isotropic distribution becomes obvious. This is mainly the result of the growing occurrence of inelastic collisions.

In each exciting collision the colliding electron suffers the large energy loss U^s. As a consequence this electron is backscattered into the low energy region of the isotropic distribution, which leads ultimately to an efficient depopulation of the isotropic distribution at higher energies. Thus, particularly at medium electric field strengths a pronounced nonequilibrium behaviour of the isotropic distribution is usually observed or, in other words, large deviations from the well-known Maxwell distribution, i.e., a straight line in a semi-logarithmic plot, is usually found. An additional increase in the field increasingly causes a smoothing of the isotropic distribution due to the stronger field action, which slowly reduces its nonequilibrium behaviour.

The mean collision frequencies ν_l^{cs} (or the corresponding rate coefficients k_l^{cs}) which are related to the various inelastic electron collision processes are of particular importance with respect to the application of electron kinetic quantities. According to the formula 2.14 these mean collision frequencies are determined by the isotropic distribution. Using the normalized isotropic distributions given in Fig. 2.2 (a), the resultant evolution with growing field strength of the various mean collision frequencies in the krypton plasma is presented in Fig. 2.2 (b).

Very different evolutions with field strength and very different magnitudes of the various mean collision frequencies can be observed. At medium field strengths a sensitive increase in the mean collision frequencies ν^s and ν^p for the excitation of the s- and p-states occurs because of the sensitive dependence of these frequencies on the high energy tail of the isotropic distribution.

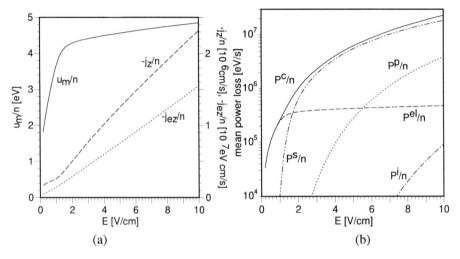

Figure 2.3: (a) Transport properties and (b) mean power losses of the electrons.

Furthermore, using these isotropic distributions together with their corresponding anisotropic distributions, some transport properties and mean power losses in the different collision processes have been determined by means of expressions 2.6 to 2.13 and represented as a function of the field strength in Fig. 2.3. The mean energy u_m/n and the magnitudes of the reduced particle and energy current densities j_z/n and j_{ez}/n of the electrons given in Fig. 2.3 (a) naturally increase with the field strength because of the correlated increase of the field action on each electron. However, it can be seen that the growth of the mean energy is quite different from that of the current densities. The mean energy in the atomic gas plasma reaches large values of some eV even at lower field strengths. The large structural change in the behaviour of the mean energy with growing field strength around $E = 1$ V/cm is caused by the transition from the sole action of elastic collisions to the increasing impact of inelastic collisions. Owing to the large energy loss in each inelastic collision event, a noticeably smaller increase in the mean energy as a function of the field is obtained at higher field strengths.

This interpretation is confirmed to a large extent by the representation of the mean power losses in the various collision processes given in Fig. 2.3 (b). It can easily be observed from the loss in elastic collisions P^{el}/n and the total loss in collisions P^c/n that in the steady-state krypton plasma the dominant contribution to the power loss changes around $E = 1$ V/cm from that caused by elastic to that caused by inelastic collisions. With respect to the latter, Fig. 2.3 (b) additionally shows that the excitation of the s-levels represents the dominant power loss channel among the three inelastic power losses P^s/n, P^p/n and P^i/n in the range of field strengths considered. This behaviour also can be expected from the behaviour and magnitude of the corresponding total collision cross sections shown in Fig. 2.1 (a).

So far the kinetics of the electrons in steady-state was considered for plasmas in a pure gas. However, in many applications mixtures of several gases occur and the kinetic treatment of the electrons has to include all the important electron collision processes for each component of the mixture. Owing to the above-mentioned short range of electron-heavy particle interactions, all these processes are considered to occur independently of each other. As a consequence of this concept the collision integrals related to all mixture components have to be summarized on the right-hand side of the Boltzmann equation 2.3. Then, for plasmas in the steady-state instead of Eq. 2.18 an extended equation for the determination of the isotropic and anisotropic distribution $\hat{f}_0(U)$ and $\hat{f}_1(U)$ in the mixture plasma is obtained. In this equation the quantities $Q^d(U), Q_l^{cs}(U), U_l^{cs}, m_e/M$ and N in particular are replaced by the corresponding quantities $Q_k^d(U), Q_{kl}^{cs}(U), U_{kl}^{cs}, m_e/M_k$ and N_k related to the kth mixture component.

A similar extension has to be made if, in addition to electron collisions with ground state atoms, electron collisions with excited atoms have to be considered [4].

2.4   Electron kinetics in time-dependent plasmas

According to the relevant power and momentum balance equations 2.19 the electron kinetics in steady-state plasmas is characterized by the condition that at any instant the power and momentum input from the electric field is dissipated by elastic and inelastic electron collisions into the translational and internal energy of the gas particles.

This instantaneous and complete compensation of the respective gain from the field and the loss in collisions does not usually occur in time-dependent plasmas, and often the collisional dissipation follows with a delay which can be quite pronounced, for example, the temporally varying action of a time-dependent field. Thus, the temporal response of the electrons to certain disturbances in the initial value of their velocity distribution or to rapid changes of the electric field becomes more complicated and the study of kinetic problems related to time-dependent plasmas naturally becomes more complex and sophisticated. Despite this extended interplay between the action of the binary electron collisions and the action of the electric field, the electron kinetics in time-dependent plasmas can also be treated, in many cases with good accuracy, on the basis of the time-dependent two-term approximation.

2.4.1   Basic equations for the distribution components

The kinetic treatment in the two-term approximation of electrons in a time-dependent plasma is based upon the equation system [6, 7, 8, 9, 10]

$$\left(\frac{m_e}{2}\right)^{1/2} U^{1/2} \frac{\partial}{\partial t} f_0 - e_0 E(t) \frac{\partial}{\partial U}\left(\frac{U}{3} f_1\right) - \frac{\partial}{\partial U}\left(2\frac{m_e}{M} U^2 N Q^d(U) f_0\right)$$

$$+ \sum_l U N Q_l^{cs}(U) f_0 - \sum_l (U + U_l^{cs}) N Q_l^{cs}(U + U_l^{cs}) f_0(U + U_l^{cs}, t) = 0 , \qquad (2.20)$$

$$\left(\frac{m_e}{2}\right)^{1/2} U^{1/2} \frac{\partial}{\partial t} f_1 - e_0 E(t) U \frac{\partial}{\partial U} f_0 + U\left(N Q^d(U) + \sum_l N Q_l^{cs}(U)\right) f_1 = 0 .$$

The system describes the temporal evolution of the isotropic and anisotropic distribution $f_0(U, t)$ and $f_1(U, t)$.

Thus, in a strict sense, now a system of two first-order partial differential equations with additional terms $f_0(U + U_l^{cs}, t)$ of shifted energy arguments has to be solved.

In order to obtain a simpler structure of this mathematical problem the system has often been reduced in the past by neglecting the first term in the second equation 2.20, i.e., the derivative of $f_1(U, t)$ with respect to time [6, 7, 8]. When accepting this additional approximation the anisotropic distribution can be eliminated by means of the second equation 2.20 and ultimately a partial differential equation of the second order with additional terms of shifted energy arguments is obtained. However, in recent years techniques have been developed [9, 10] to solve numerically the equation system 2.20 without additional reductions or simplifications. This modern approach is used as the basis of the following explanations concerning the time-dependent two-term treatment.

The system of partial differential equations 2.20 of the first order has usually to be treated as an initial boundary-value problem in an appropriate energy region $0 \leq U \leq U^\infty$ and for times $t \geq 0$, where the time represents the evolution direction of the kinetic problem.

Initial values for each of the distributions $f_0(U, t)$ and $f_1(U, t)$, which are suitable for the problem under consideration, have to be fixed, for example at $t = 0$. Appropriate boundary conditions for the system are given by the requirements $f_0(U \geq U^\infty, t) = 0$ and $f_1(0, t) = 0$.

2.4.2 Balance equations and dissipation frequencies

The balance equations for conservation of particle number, power and momentum of the electrons in a time-dependent plasmas are

$$\frac{d}{dt} n(t) = 0, \quad \frac{d}{dt} u_m(t) = P^f(t) - P^c(t), \quad \frac{d}{dt} j_z(t) = I^f(t) - I^c(t), \quad (2.21)$$

and the total power and momentum loss in collisions 2.13 can be rewritten as

$$P^c(t) = \int_0^\infty \nu_e(U) \, U^{3/2} f_0(U, t) \, dU \, ,$$

$$I^c(t) = \int_0^\infty \nu_m(U) \frac{1}{3} \sqrt{2/m_e} \, U f_1(U, t) \, dU \, , \quad (2.22)$$

where the energy-dependent lumped frequencies $\nu_e(U)$ and $\nu_m(U)$ for power and momentum dissipation in collisions have been introduced by the expressions

$$\nu_e(U) = \left(\frac{2}{m_e} \right)^{\frac{1}{2}} U^{\frac{1}{2}} \left(2 \frac{m_e}{M} N Q^d(U) + \sum_l N Q_l^{cs}(U) \frac{U_l^{cs}}{U} \right), \quad (2.23)$$

$$\nu_m(U) = \left(\frac{2}{m_e} \right)^{\frac{1}{2}} U^{\frac{1}{2}} \left(N Q^d(U) + \sum_l N Q_l^{cs}(U) \right). \quad (2.24)$$

Since only conservative collisions are considered in Eq. 2.20 for simplicity, the consistent particle balance given in 2.21 simply says that the electron density n is independent of time.

Because of the linear dependence of all terms of the basic equation system 2.20 on the isotropic or anisotropic distribution, instead of the latter, the per one electron normalized distributions $f_0(U,t)/n$ and $f_1(U,t)/n$ can be introduced into this system. If nonconservative collisions are considered in addition, the electron density $n(t)$ becomes time-dependent and such a normalization will no longer be possible.

As can be seen from the power and momentum balance presented in 2.21 a temporal evolution of the mean energy density $u_m(t)$ and the particle current density $j_z(t)$ is initiated if the instantaneous compensation of the respective gain from the field and the corresponding total loss in collisions is disturbed for any reason. Generally, the electron component tries to reduce these disturbances by collisional dissipation and to re-establish the compensated state in both balances.

The rapidity of the collisional dissipation of power and momentum finally determines whether the compensation in both balances and thus the establishment of the steady-state occurs almost immediately or with a noticeable temporal delay after the occurrence of a disturbance.

The expressions 2.22 and 2.6 for the total power loss in collisions $P^c(t)$ and of the mean energy density $u_m(t)$ clearly indicate that the rapidity of the dissipation of the kinetic energy per volume unit $U^{3/2} f_0(U,t)\, dU$ contained in the energy interval dU is determined by the lumped energy dissipation frequency $\nu_e(U)$. In an analogous manner the expressions 2.22, 2.7 and 2.5 of the total momentum loss $I^c(t)$ in collisions and the particle current density $j_z(t)$ show that the rapidity of the dissipation of the contribution $(1/3)\sqrt{2/m_e}\,U f_1(U,t)\, dU$ of the energy interval dU to the particle current density, and thus to the momentum of the electrons per volume unit, is determined by the lumped momentum dissipation frequency $\nu_m(U)$.

These energy-resolved dissipation frequencies are very important for characterizing the rapidity of the response of the electron component in different regions of their energy space to disturbances, for example of the established steady-state.

These dissipation frequencies can be determined as soon as the atomic data of the important electron collision processes and the gas density are known.

To give an example, Fig. 2.1 (b) represents both these dissipation frequencies for krypton with a gas density $N = 3.54 \times 10^{16}$ cm^{-3} calculated using the atomic data of this gas given in Fig. 2.1 (a). It can be observed from this figure that the momentum dissipation frequency exceeds the energy dissipation frequency by at least one order of magnitude. This means that momentum dissipation occurs much faster than energy dissipation. The energy dissipation frequency shows a complicated energy dependence. According to Eq. 2.1 the energy loss in an elastic collision event is proportional to the mass ratio m_e/M and is thus very small compared with that in an inelastic collision event.

As a consequence, the lumped energy dissipation frequency assumes very small values in the energy region where only elastic collisions contribute, but far larger values at those energies where inelastic collisions occur.

Thus, the efficiency of the collisional energy dissipation depends considerably on the electron population and its temporal evolution in different parts of the energy region.

The numerical solution of a specific initial boundary-value problem based on the equation system 2.20 can be performed [10] by applying a finite difference method to an equidistant grid in energy U and time t. A discrete form of the equation system 2.20 is obtained using second-order-correct centered difference analogues on the rectangular grid for both distributions $f_0(U,t)/n$ and $f_1(U,t)/n$ and for their partial derivatives of first order.

2.4.3 Temporal relaxation of the electrons

To illustrate the relaxation of the electrons [6], the temporal evolution of their velocity distribution under the action of a time-independent field and of the important electron collision processes has been calculated by solving equation system 2.20. The solution procedure started at $t = 0$ from a Gaussian distribution as initial value of the isotropic distribution and from a vanishing anisotropic distribution, i.e., from

$$f_0(U,0)/n = c \, \exp\left(-(U - U_c)^2/U_w^2\right), \quad f_1(U,0)/n = 0, \quad (2.25)$$

with the center energy U_c, the energy width U_w and the factor c used to normalize the initial value of the isotropic distribution on one electron according to $\int_0^\infty U^{1/2} f_0(U,0)\, dU/n = 1$. The following results on the relaxation in krypton have been determined for two field strengths at $N = 3.54 \cdot 10^{16}\ \mathrm{cm}^{-3}$ and with $U_c = 8\,\mathrm{eV}$ and $U_w = 2\,\mathrm{eV}$. Fig. 2.4 shows the temporal

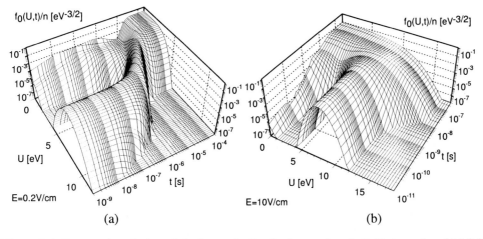

Figure 2.4: Temporal evolution of the isotropic distribution at electric field strengths of (a) 0.2 V/cm and (b) 10 V/cm.

evolution of the isotropic distribution $f_0(U,t)/n$ on a logarithmic time scale for the field strengths (a) 0.2 V/cm and (b) 10 V/cm. The evolution starts at $t = 0$ from an isotropic distribution with a single peak around 8 eV and the course of the relaxation finally leads to the establishment of the corresponding steady-state distributions. Because of the logarithmic time scale the plot starts somewhat after the onset of the relaxation process. In this time region a distribution peak at low energies is already created by the backscattering of electrons which have undergone inelastic collisions at the very beginning of the relaxation. During the relaxation process quite different evolutions occur with strong depletion of the electron population at higher energies under low field action in (a) or with strong enhancement of the electron population at higher energies under the action of a larger field in (b), and very different steady-state distributions are ultimately established. With respect to the total relaxation time of the isotropic distribution, i.e., the time needed to reach a steady-state, very different times of approximately 10^{-4} s (a) and approximately 10^{-7} s (b) are found. It is therefore clear that the relaxation time for the establishment of the steady-state in a time-independent electric field can differ by several orders of magnitude and drastically depends on the field strength and thus on the collisional energy dissipation efficiency of the electrons. The pronounced variation of this efficiency can also be expected from the strong field dependence of the total energy loss in collisions P^c/n given for the steady-state plasma in Fig. 2.3 (b).

Establishment of the steady-state and the corresponding relaxation times can be fairly accurately evaluated when considering the temporal behaviour of the gain from the field and the total loss in collisions due to the power and momentum balance during the relaxation process.

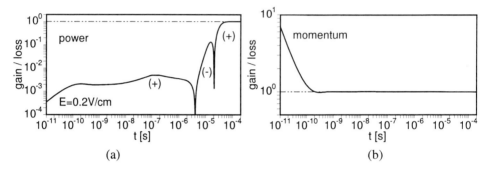

Figure 2.5: Temporal evolution of (a) power and (b) momentum gain/loss ratio.

For the same conditions as considered in Fig. 2.4 (a), the power gain/loss ratio $P^f(t)/P^c(t)$ is displayed in Fig. 2.5 (a) and the momentum gain to loss ratio $I^f(t)/I^c(t)$ in Fig. 2.5 (b). The plot of both gain to loss ratios shows that these ratios undergo a quite different evolution involving changes of several orders of magnitude up to the value of unity, i.e., complete compensation of the respective gain and loss is established. However, the graph of the momentum gain to loss ratio shows clearly that the evolution of this quantity towards unity takes place much faster. The almost complete compensation of the momentum gain and loss is reached much earlier, at about 5×10^{-10} s, and this is in agreement with the magnitude of the lumped frequency $\nu_m(U)$ for momentum dissipation by collisions as presented above in Fig. 2.1 (b).

Consideration of the consistent power and momentum balance in Eq. 2.21 clearly shows that the establishment of the steady-state simultaneously requires complete compensation of the respective gain and loss terms and its continuation with increasing time in both these balances. This means that the relaxation process and the corresponding relaxation time is mainly determined by the much slower establishment of the power balance, which, according to Fig. 2.5 (a), requires about 10^{-4} s.

2.5 Electron kinetics in space-dependent plasmas

Similarly to time-dependent plasmas the electron component in space-dependent plasmas is usually characterized by the property that its local power and momentum gain from an electric field cannot be dissipated at almost the same position by elastic and inelastic electron collisions into translational and/or internal energy of the gas particles. Thus, the spatial response of the electrons, for instance, to a disturbance of their velocity distribution is of a complex nature [11, 12, 13, 14]. Despite the complicated character of the electron response, in many cases the electron kinetics of inhomogeneous plasmas can also be analyzed with good accuracy by using the space-dependent two-term approximation including the impact of the important binary electron collision processes with the gas particles and the action of an electric field.

2.5.1 Basic equations and consistent balances

Within the framework of the two-term approximation the kinetic treatment of the electrons in space-dependent plasmas is based on the following system of equations

$$\frac{\partial}{\partial z}\left(\frac{U}{3}f_1\right) - e_0 E(z)\frac{\partial}{\partial U}\left(\frac{U}{3}f_1\right) - \frac{\partial}{\partial U}\left(2\frac{m_e}{M}U^2 NQ^d(U)f_0\right)$$
$$+ \sum_l UNQ_l^{cs}(U)f_0 - \sum_l (U + U_l^{cs})NQ_l^{cs}(U + U_l^{cs})f_0(U + U_l^{cs}, z) = 0\,,$$

$$U\frac{\partial}{\partial z}f_0 - e_0 E(z)U\frac{\partial}{\partial U}f_0 + U\left(NQ^d(U) + \sum_l NQ_l^{cs}(U)\right)f_1 = 0 \qquad (2.26)$$

which describe the evolution of the isotropic and anisotropic distribution $f_0(U, z)$ and $f_1(U, z)$ in the coordinate and energy space of the electrons. The second of equations 2.26 can easily be resolved with respect to the anisotropic distribution and the result can be used to eliminate $f_1(U, z)$ from the first of equations 2.26. A second-order partial differential equation for the isotropic distribution with additional terms $f_0(U + U_l^{cs}, z)$ of the shifted energy arguments $U + U_l^{cs}$ is then obtained.

This differential equation contains cross derivative terms, which cause its direct numerical treatment to become difficult. The standard form of equation system 2.26 which is suitable for numerical solution is obtained by replacing the kinetic energy U by the total energy ε according to $\varepsilon = U + W(z)$, where $W(z) = e_0 \int_0^z E(\tilde{z})d\tilde{z}$ is the potential energy of the electrons in the electric field. This transformation and the elimination of the anisotropic distribution coupled with the further abbreviation $U_{\varepsilon z} \equiv \varepsilon - W(z)$ leads ultimately to the parabolic equation [15, 16, 17]

$$\frac{\partial}{\partial z}\left(\frac{1}{3}\frac{U_{\varepsilon z}}{NQ^d(U_{\varepsilon z}) + \sum_l NQ_l^{cs}(U_{\varepsilon z})}\frac{\partial}{\partial z}\bar{f}_0\right) + \frac{\partial}{\partial \varepsilon}\left(2\frac{m_e}{M}U_{\varepsilon z}^2 NQ^d(U_{\varepsilon z})\bar{f}_0\right)$$
$$- \sum_l U_{\varepsilon z}NQ_l^{cs}(U_{\varepsilon z})\bar{f}_0 + \sum_l (U_{\varepsilon z} + U_l^{cs})NQ_l^{cs}(U_{\varepsilon z} + U_l^{cs})\bar{f}_0(\varepsilon + U_l^{cs}, z) = 0 \quad (2.27)$$

for the transformed isotropic distribution $\bar{f}_0(\varepsilon, z) = f_0(U_{\varepsilon z}, z)$, including the additional terms

$\bar{f}_0(\epsilon + U_l^{cs}, z)$ with the shifted energy arguments $\varepsilon + U_l^{cs}$. Furthermore, the equation

$$\bar{f}_1(\varepsilon, z) = -\frac{1}{NQ^d(U_{\varepsilon z}) + \sum_l NQ_l^{cs}(U_{\varepsilon z})} \frac{\partial}{\partial z} \bar{f}_0(\varepsilon, z) \tag{2.28}$$

for determining the transformed anisotropic distribution $\bar{f}_1(\varepsilon, z) = f_1(U_{\varepsilon z}, z)$, using the solution $\bar{f}_0(\varepsilon, z)$ of Eq. 2.27, is obtained.

The parabolic equation 2.27 describes the evolution of the isotropic distribution and must be solved as an initial boundary-value problem on a nonrectangular solution region whose boundaries are partly determined by the spatial course of the electric field and thus by the specific kinetic problem under consideration. The parabolic problem must be solved with appropriate initial and boundary conditions, which are partly described below.

In space-dependent plasmas the self-consistent particle, power and momentum balance equations of the electrons are given by [15, 17]

$$\frac{d}{dz} j_z(z) = 0,$$

$$\frac{d}{dz} j_{ez}(z) = P^f(z) - P^c(z), \tag{2.29}$$

$$\frac{d}{dz}\left(\frac{2}{3m_e} u_m(z)\right) = I^f(z) - I^c(z),$$

where all macroscopic quantities occurring in these balances are given by the expressions 2.6 to 2.13.

Since only conservative inelastic collision processes are taken into account in the basic kinetic equations 2.26, the self-consistent particle balance given in Eqs. 2.29 states that the particle current density of the electrons j_z becomes independent of the z coordinate. Furthermore, the power and momentum balances given in Eqs. 2.29 show that an incomplete compensation of the respective gain from the electric field and the corresponding loss in collisions enforces the occurrence of sources or sinks in these balances. According to the power balance a spatial evolution of the energy current density $j_{ez}(z)$ and according to the momentum balance a spatial evolution of the mean energy density $u_m(z)$ of the electrons are then initiated. Thus, the magnitude of these sources or sinks compared, for example, with the respective gain from the electric field makes an assessment of the degree of the nonlocal or nonhydrodynamic behaviour of the electron component possible. In contrast to plasmas in the steady-state and to time-dependent plasmas the electron density $n(z)$ in space-dependent plasmas always depends on the z coordinate and this already occurs if conservative inelastic collisions only are considered. Consequencely, separating the electron density from the isotropic and anisotropic distribution no longer makes sense.

2.5.2 Spatial relaxation of the electrons

Essential aspects of the spatial relaxation of the electrons in collision-dominated plasmas can be revealed when studying under the influence of a space-independent electric field the evolution of the electrons whose velocity distribution has been disturbed at a certain position in space [15, 16]. Ultimately a uniform state becomes established sufficiently far from this position in the field acceleration direction of the electrons. Such relaxation problems can be analyzed on the basis of the parabolic equation 2.27 determining the isotropic distribution when adapting the initial boundary-value problem to the relaxation model.

In order to cause a disturbance in the velocity distribution, the anisotropic distribution has been fixed at the position $z = 0$ by a Gaussian-like distribution, i.e., by

$$f_1(U,0) = cU \exp\left(-(U - U_c)^2/U_w^2\right),\tag{2.30}$$

with center energy U_c, energy width U_w and constant of proportionality c. The latter is used to normalize the boundary-value of the anisotropic distribution on the space-independent value of j_z according to the particle balance given in Eqs. 2.29.

A numerical solution of the parabolic equation 2.27 as an initial boundary-value problem can be obtained by using a finite difference approach according to the well-known Crank-Nicholson scheme for parabolic equations [17].

Results for the spatial relaxation of the electron component in the krypton plasma, obtained by this approach for a gas density $N = 3.54 \times 10^{16}$ cm^{-3} and parameter values $U_c = 5$ eV and $U_w = 2$ eV in the boundary condition 2.30 for $f_1(U,0)$, are presented below. If negative values are chosen for the uniform electric field E, the electron acceleration and thus the spatial evolution of the various kinetic properties of the electrons takes place in the positive z-direction. For the two field strengths $E = -2$ V/cm and -5 V/cm Fig. 2.6 illustrates the

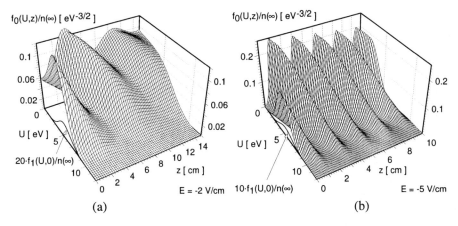

Figure 2.6: Spatial relaxation of the isotropic distribution at (a) –2 V/cm and (b) –5 V/cm.

initial part of the spatial relaxation of the normalized isotropic distribution $f_0(U,z)/n(\infty)$ together with the corresponding boundary-value of the anisotropic distribution that initiates the relaxation process. $n(\infty)$ denotes the electron density established at large values of z. The figure clearly shows that the relaxation behaviour of the isotropic distribution is very different for both field strengths. Whereas at the higher field value (b) a weakly damped, distinctly periodic evolution with a short period length dominates, the spatial relaxation at the lower field (a) occurs in a less pronounced way and with larger periodic length. Thus, the detailed relaxation behaviour of the isotropic distribution is controlled to a great extent by the field acting upon the plasma.

These results indicate that the more or less damped periodic evolution in the relaxation process is a typical feature of the spatial electron relaxation at medium electric field strengths.

This means that despite the continuous power dissipation by collisions a damped periodic response of the electrons occurs during the relaxation process of the electrons. The basic processes involved in this periodic response are as follows. Ultimately driven in the positive z direction by the uniform electric field the electrons gain potential energy from the electric field and the electron population at higher energies increases monotonically. If an appreciable number of electrons overcome the energy threshold of the lowest excitation process then these electrons will undergo inelastic collisions.

As a result of these events the electrons lose the excitation energy, i.e., with respect to krypton the excitation energy U^s of the lumped s-states, and are backscattered to the region of low energies, where they are subject to the field action again and gain potential energy from the field. Continuous repetition of this basic sequence leads to a periodic response of the electrons in the relaxation process. If this basic mechanism dominates the inherent periodic length λ satisfies the simple relation $e_0 |E| \lambda = U^s$. The spatial relaxation process will now

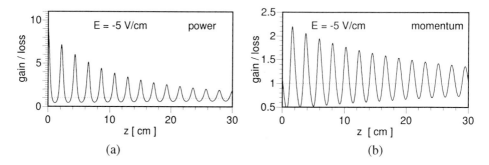

Figure 2.7: (a) Power and (b) momentum gain to loss ratio.

be considered with respect to the local compensation of the gain and loss in the power and momentum balance equations 2.29. Fig. 2.7 shows the spatial evolution of the respective gain to loss ratio of the power (a) and momentum (b) balance at a field strength of -5 V/cm. The figure shows clearly that the deviations from the local compensation of gain and loss are generally large and extend over a large spatial range. This is a reflection of large relaxation lengths and of a pronounced nonlocal behaviour of the electron component.

A comparison of the ratio of the power balance with that of the momentum balance shows that the deviations from local compensation are generally somewhat more pronounced in the power balance. Contrary to the temporal relaxation, in the spatial relaxation process the establishment of a uniform state occurs, at fixed field strength, on the same spatial scale in both balances. This means that the coupling between the isotropic and anisotropic distributions is much closer in the spatial than in the temporal relaxation process.

The relaxation of the isotropic distribution, shown for two field strengths in Fig. 2.6, and the corresponding evolution of the gain to loss ratios of the power and momentum balance, displayed in Fig. 2.7, indicate that the relaxation length depends sensitively on the field strength. In order to show the dependence of the relaxation length on the magnitude of the uniform field, the relaxation process has been studied for several field strengths keeping all other parameters unchanged.

Fig. 2.8 shows the spatial relaxation behaviour of the normalized density $n(z)/n(\infty)$ (a) and of the mean energy $u_m(z)/n(z)$ (b) of the electrons at field values between –1 V/cm and –15 V/cm. From both figures a drastic change in the relaxation behaviour and in the involved relaxation length can be seen. This is caused by the change of the dominant relaxation mechanism with increasing field strength. At medium fields, i.e., around –5 V/cm, the gain in potential energy and the energy loss in the lowest excitation process, i.e., the excitation of the lumped s-states, dominate. These are the two basic processes that cause a weakly damped periodic relaxation of the electrons and very large relaxation lengths. In the limiting case where only these two power transfer channels are active, an undamped, purely periodic spatial evolution without any spatial relaxation is excited. The occurrence of other power loss channels or

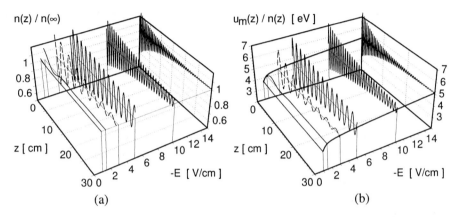

Figure 2.8: Spatial relaxation of (a) density and (b) mean energy.

additional ones causes damping of the spatial evolution to take place. Thus, with decreasing field the power loss in elastic collisions monotonically increases and eventually becomes the dominant power loss process. In this case the basic mechanism for the excitation of the periodic response is increasingly lost and a strongly damped, aperiodic relaxation process with a markedly reduced relaxation length occurs.

However, if the field strength rises above the range of medium field values, additional power loss channels become active and of comparable importance as the lowest power loss channel due to the excitation of higher electronic states and eventually because of ionization. Consequently, the electrons undergo different inelastic collision processes and are backscattered into the region of low energies according to markedly different energy losses. This backscattering by different energy losses causes a strong increase in the damping and a marked diminution of the relaxation length.

Furthermore, the increase in the field strength leads to a monotonic decrease in the period length, since a comparably large gain in potential energy already occurs over a shorter distance.

2.6 Concluding remarks

An attempt has been made to give an introduction into the kinetic treatment of the electron component in steady-state, time-dependent and space-dependent plasmas on the basis of the electron Boltzmann equation, and to illustrate the large variety in electron kinetics in non-thermal weakly ionized plasmas by means of a few selected examples.

In this chapter particular emphasis has been laid on a uniform basis for the electron kinetics under different plasma conditions. The main points in this context concern the consistent treatment (i) of the isotropic and anisotropic contributions to the velocity distribution, (ii) of the relations between these contributions and the transport properties, the collisional energy and momentum transfer rates and the rate coefficients and (iii) of the macroscopic particle, power and momentum balances.

Furthermore, special attention has been paid to presenting the basic equations for the kinetic treatment, briefly explaining their mathematical structure, giving a few hints on appropriate boundary and/or initial conditions and indicating the main aspects of a suitable solution approach.

Such complete studies of electron kinetic problems enables one to reveal essential non-equilibrium properties of the electron component and to reach a deeper understanding of the interplay between the various microphysical processes involved in the kinetics of the electrons. This point in particular has been illustrated by a few examples relating to the temporal and spatial relaxation of the electrons.

Kinetic problems within the same framework of the time-dependent two-term approximation have been treated in quite different time-dependent fields [6, 9]. In this context, the temporal evolution of the kinetic properties initiated by jump-like and continuous field transitions between steady-states, pulse-like field alterations, decaying electric fields and the periodic evolution [9] caused by the action of *rf* electric fields should also be mentioned.

In these studies a similar variety of electron collision processes has been treated as for steady-state kinetics.

By using the space-dependent two-term approximation the spatial evolution of electron kinetic quantities have been analyzed in very different space-dependent electric fields [14, 18], and even in highly modulated, spatially periodic fields occurring in moving and standing striations. Nonconservative inelastic electron collisions can also be included in these studies and cause a spatial evolution of the particle current density in the consistent particle balance.

Future aspects of the study of electron kinetics based upon the electron Boltzmann equation certainly concern extensions to cover spatially two- and even three-dimensional kinetic problems, coupled space- and time-dependent problems, more complex field structures and more sophisticated boundary conditions. First attempts in this direction have already been published [19, 20, 21] or are just underway.

An other important point, closely connected with the electron kinetics, concerns the self-consistent treatment of the electron kinetics, of the particle and/or power balance equations of heavy particles (as ions and excited atoms) and of the Maxwell equations (or a reduced set) to obtain a more complete description of all the important plasma components as well as of the internal electric field. The self-consistent treatment of such equation systems is usually very difficult and based on an iterative solution approach of the various types of equations involved [8, 18, 20]. In order to integrate an adequate electron kinetic equation into such an approach, an effective solution procedure for the equation is of particular importance, though remarkable progress has been achieved in recent years with respect to the speed of computation.

Acknowledgments

I would like to thank my colleagues Dr. Detlef Loffhagen and Dr. Florian Sigeneger for their kind assistance in preparing the figures of this chapter.

References

[1] E. A. Desloge, Statistical Physics, Holt, Rinehart and Winston, Inc.: New York (1966) 273

[2] I.P. Shkarofsky, T.W. Johnston, M.P. Bachynski, The Particle Kinetics of Plasmas, Addison-Wesley Publishing Company: Reading, Massachusetts (1966) 46, 70, 119, 161, 243

[3] V.E. Golant, A.P. Zhilinsky, I.E. Sakharov, Fundamentals of Plasma Physics, John Wiley & Sons, Inc.: New York (1980) 16, 108, 181

[4] R.B. Winkler, J. Wilhelm, R. Winkler, Contrib. Plasma Phys. **22**, 401 (1982)

[5] M. Hayashi, Electron collision cross sections, in Plasma Material Science Handbook, Editor, Japan Society for the Promotion of Science, appendix-3, pp. 748-766. Ohmsha Ltd.: Tokyo (1992)

[6] J. Wilhelm, R. Winkler, Journal de Physique, Colloque C7, **40**, 251 (1979)

[7] R. Winkler, M.W. Wuttke, Appl. Phys. B **54**, 1 (1992)

[8] D. Loffhagen, R. Winkler, J. Comput. Phys. **112**, 91 (1994)

[9] R. Winkler, in Microwave Discharges - Fundamentals and Applications, Editors C.M. Ferreira and M. Moisan, NATO ASI Series, Series B: Physics Vol. 302, Plenum Press: New York (1993) 339

[10] R. Winkler, G.L. Braglia, J. Wilhelm, Contrib. Plasma Phys. **35**, 179 (1995)

[11] L.D. Tsendin, Plasma Sources Sci. Technol. **4**, 200 (1995)

[12] V.I. Kolobov, V.A. Godyak, IEEE Trans. Plasma Sci. **23**, 503 (1995)

[13] R. Winkler, F. Sigeneger, D. Uhrlandt, Pure & Appl. Chem. **68**, 1065 (1996)

[14] R. Winkler, G. Petrov, F. Sigeneger, D. Uhrlandt, Plasma Sources Sci. Technol. **6**, 118 (1997)

[15] F. Sigeneger, R. Winkler, Plasma Chem. Plasma Process. **17**, 1 (1997)

[16] F. Sigeneger, R. Winkler, Plasma Chem. Plasma Process. **17**, 281 (1997)

[17] F. Sigeneger, R. Winkler, Contrib. Plasma Phys. **36**, 551 (1996) 551

[18] D. Uhrlandt, R. Winkler, J. Phys. D: Appl. Phys. **29**, 115 (1996)

[19] P.M. Meijer, The Electron Dynamics of RF Discharges, Dissertation, Rijks-University: Utrecht (1991)

[20] Y. Yang, H. Wu, J. Appl. Phys. **80**, 3699 (1996)

[21] S.C. Arndt, D. Uhrlandt, R. Winkler, Plasma Chem. Plasma Process. **21**, 175 (2001)

3 Elementary collision processes in plasmas

Kurt H. Becker

Department of Physics and Engineering Physics, Stevens Institute of Technology, Hoboken, NJ 07030, U.S.A.

3.1 Introduction

Plasmas are complex systems which consist of various groups of mutually interacting particles. These include in the simplest cases neutral gas atoms and molecules in their ground states and in excited states, electrons, and positive and negative ions. In principle, one would need to understand and describe all possible interactions between these particles in order to model the properties of the plasma and to predict its behavior. In addition, one would have to include photon interactions with all plasma constituents. In many instances, however, two-body binary interactions are the only processes of relevance and only a subset of all possible collisional interactions are important. This book deals primarily with low-temperature, non-equilibrium plasmas which are weakly ionized. Electron collisions play a particularly important role in these plasmas. Heavy particle collisions are less important in the bulk of the plasma, but they do play an important role in the sheath regions. The main power input to the plasma is provided by an externally applied electric field acting on the charged particles. Because of their small masses, electrons are only poorly coupled to the heavy particles in the plasma. Electrons cannot transfer a large part of their kinetic energy to the heavy particles in elastic collisions. As a consequence, electrons reach a much higher average kinetic energy compared to the heavy particles (neutrals and ions). In plasmas relevant to the subject area of this book, typical mean electron energies may range from 0.5 eV to 5 eV, whereas the energy of the heavy particles (0.025 eV–0.05 eV) corresponds to temperatures in the range of 1–2 times the room temperature (300 K–600 K). Typical charge carrier densities are 10^8/cm^3 to 10^{12}/cm^3. There is an appreciable number of electrons in such plasmas which are capable of overcoming the threshold energies above which the various inelastic electron collision processes can occur. In each inelastic collision process the electron loses at least the threshold energy for this process and is thus transferred from the region of high kinetic energy to the low energy region. Inelastic collisions are a very effective way to lower the mean energy of the electrons in a plasma.

The list of specific target atoms and molecules of interest in various plasmas is comprehensive and ranges from atomic hydrogen, oxygen and nitrogen to very polyatomic molecules such as Si-organic and metal-organic compounds. Many species in between these two extreme categories are relevant in a variety of diverse plasma-assisted processes: atoms such as Al, Si and the halogens, simple diatomics such as H_2, Cl_2, HBr, polyatomics such as all partially and fully halogenated methane compounds, BCl_3, SF_6 and SiH_4, free radicals such as CF_x,

SiH$_x$ (x=1–3) to name just a few. The remainder of this article will focus mainly on electron collision processes as they relate to an understanding of low-temperature plasmas.

The relevance of electron collision processes in low-temperature plasmas is largely determined by the overlap of the electron energy distribution function (EEDF) with the respective electron impact cross section. This overlap, in turn, is determined by two factors, by the threshold and shape of the respective electron impact cross section and by the shape of the EEDF. Inelastic electron collision processes have threshold energies ranging from a few tenth of an electron-volt for vibrational excitation of molecules and dissociative attachment processes to 5–15 eV for dissociation, electronic excitation, and ionization processes. The higher threshold energies for dissociation, electronic excitation, and ionization means that the effectiveness of these processes for the plasma properties is crucially dependent on the shape of the high-energy tail of the EEDF. The high-energy tail of the EEDF, in turn, is determined to a large extent - assuming a constant reduced electric field strength - by low-energy electron collisions such as vibrational excitation of the plasma constituents and dissociative attachment which are among the most effective energy-loss mechanisms for the plasma electrons. In cases where the cross sections for vibrational excitation and/or dissociative attachment are small, these energy loss channels are weak and, as a consequence, there is a pronounced high-energy tail of the EEDF. This usually results in a situation where single electron dissociation and ionization processes are very important. Molecules with large vibrational excitation cross sections and/or large dissociative attachment cross sections, on the other hand, provide efficient energy-loss mechanisms for the plasma electrons. As a consequence, the high energy tail of the EEDF is significantly reduced in such cases. Consequently, single electron dissociation, electronic excitation, and ionization are less significant fundamental collision processes in the plasma under these circumstances.

We shall limit the discussion in what follows to selected electron impact-induced processes involving atomic and molecular targets. Ionization and dissociation processes will be discussed in detail as the two most important *high-energy* electron collision processes relevant to low-temperature plasma. The discussion of excitation processes will include a review of recent developments in the area of collisions with targets in excited/metastable states which is a very important, albeit often neglected, *low-energy* loss mechanism for the plasma electrons.

3.2   Summary of electron impact-induced collision processes with atoms

In atomic plasmas (i.e., plasmas where the main constituents are atoms rather than molecules) the most relevant primary inelastic electron collision processes are electron impact excitation and ionization of the atoms. Unlike photo-excitation processes which are governed by electric dipole selection rules, electron impact can populate essentially all excited states of an atom via optically allowed, optically forbidden, and spin-forbidden processes. Electron impact excitation can also populate long-lived metastable atomic states, which are subsequently involved in secondary collision processes in a plasma via, e.g., step-wise excitation and ionization which can be very important reaction channels in non-equilibrium plasmas.

Electron impact excitation of atomic states has been studied extensively both by experimentalists and theorists since the 1930s with increasing levels of sophistication. To date, there appears to be reasonably good agreement between experiment and theory for the excitation of the lowest lying resonance levels of simple atoms such as the alkalis, the rare gases, and some atoms such as hydrogen, nitrogen, and oxygen. The most common experimental technique for the measurement of excitation cross sections is electron-energy loss spectroscopy [1]. As an alternative, optical techniques for the excitation of those states that subsequently decay radiatively can also be used [2]. Optical techniques measure a photoemission cross section from which an excitation cross section can be obtained when proper corrections are made for cascading and branching. Detailed reviews of both techniques can be found in the literature [1, 2].

Atomic excitation cross sections from the ground state of the atom exhibit a characteristic energy dependence which is determined by the nature of the excitation process [3], optically allowed, optically forbidden, and spin forbidden. Optically allowed excitation processes (e.g., excitation of the helium ^{1}P states from the ^{1}S ground state), which are governed by the selection rules $\Delta L = \pm 1$ and $\Delta S = 0$ for the orbital angular momentum L and the spin S, result in cross sections which peak at an energy of about 3–5 times the threshold energy. The cross sections decline gradually towards higher impact energies with a high-energy behavior proportional to $\ln(E)/E$, where E is the impact energy. Optically forbidden excitation processes (e.g., excitation of the helium ^{1}S and ^{1}D states), which violate the ΔL selection rule (but still satisfy the ΔS selection rule), are characterized by a maximum in the cross section at lower impact energies followed by steeper decline described by a $1/E$-dependence at high energies. Lastly, the cross sections for excitation of, e.g., the triplet states of helium from the ground state, a process which violates the ΔS selection rule, reach their maximum at energies just a few electronvolts above threshold and decline sharply with increasing impact energy as E^{-3}.

Recent advances in experimental techniques such as the use of spin-polarized electrons and atoms and the correlated detection of two or more post-collision products (scattered electrons, recoiling atoms, emitted photons) using coincidence techniques has allowed very detailed studies of atomic excitation processes at the level of quantum mechanical scattering amplitudes and their phase differences. This allows very stringent tests of theoretical models and it is the most fundamental level at which one can study atomic excitation processes [4]. The level of agreement between these sophisticated experiments and the results of state-of-the-art quantum mechanical calculations is quite good from the excitation of the lowest excited levels from the ground state for many simple atoms such as the alkalis, the noble gases, the earth alkalis, and even for a heavy atom such as mercury. As an example, figure 3.1 shows the so-called alignment angle γ as a function of the electron scattering angle for excitation of the 5s[3/2]$_{J=1}$ state in krypton by 30 eV electrons. The alignment angle γ denotes the orientation of the charge cloud of the electron impact excited Kr state relative to the direction of the incident electrons. The experimental results of Murray et al. [5] are compared with two theoretical calculations using a distorted-wave Born approximation (DWBA) [6] and a first-order many-body theory (FOMBT) [7]. It is apparent that the experimental data are in excellent agreement with the DWBA results, but are very different from the FOMBT predictions. This demonstrates how the results of these sophisticated experiments can be used as stringent tests of theoretical models.

The status of experiments and calculations is much less satisfactory as far the excitation of higher-lying atomic energy levels from the ground state is concerned.

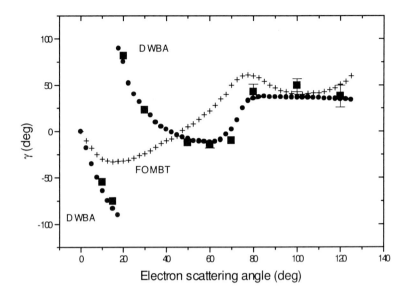

Figure 3.1: Alignment angle γ for excitation of the $5s[3/2]_{J=1}$ state in Kr by 30 eV electrons as a function of electron scattering angle (after Murray et al. [5]), DWBA •, FOMBT +, experiments ■ (see text).

In a stationary (steady-state) plasma the production rate of ions equals the loss rate of ions. Electron impact-induced ionization processes of the heavy neutral particles in a plasma (atoms and molecules in their ground state or in excited states) are an important ion formation process. In many plasmas electron impact ionization is the dominant ion formation process [8]. The experimental determination of an ionization cross section requires an ion source in which collisions occur between a well-characterized beam of electrons and the target atoms under single collision conditions. The resulting product ions are extracted from the ion source and are subsequently detected by a suitable ion detector. The determination of the total ionization cross section requires a careful measurement of the number density of targets, the current of the ionizing electron beam, the length over which the electron beam and the target gas interact, and the number of product ions (ion current). Partial ionization cross sections in the case of molecules can be obtained, if the detection of the ion current is restricted to a particular product ion. In this case, a mass selective device has to be employed, e.g., a mass spectrometer. Since it is often difficult to determine all the quantities that determine an absolute ionization cross section with sufficient precision, it has become common practice to utilize normalization procedures. This typically involves gases of well-known ionization cross sections such as Ar or Kr and requires the simultaneous measurement of the known ionization cross section and the unknown cross section for the target under study under essentially identical experimental conditions.

To date, absolute single ionization cross sections have been measured for less than 40

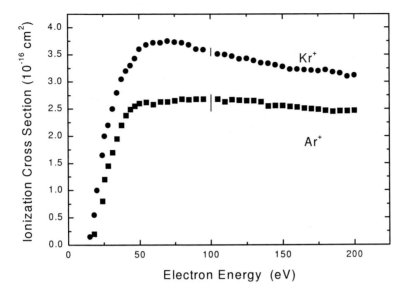

Figure 3.2: Absolute single ionization cross sections for Ar and Kr as a function of electron energy. The vertical bars in both curves at 100 eV represent the spread among recent measurements of these cross sections [10].

atoms, cross sections for multiple ionization for even fewer atoms [9]. Many of these cross sections have only been measured once using one particular experimental technique. This is unsatisfactory in view of the fact that measurements of absolute cross sections are tedious, cumbersome, difficult, and prone to systematic errors whose presence and importance sometimes becomes evident only after the same absolute cross section has been re-measured using a different experimental technique. The most reliable cross section measurements are those for the rare gases, where total and single ionization cross sections are known with an accuracy of 10% or better [8, 9]. This is shown in figure 3.2, which compares the results of the most recent experimental determinations of the single ionization cross sections of the rare gases Ar and Kr.

More atomic ionization cross sections have been measured by the fast-neutral-beam technique than by any other technique [11, 12]. A schematic diagram of the fast-beam apparatus used in our group is shown in figure 3.3. A direct current (dc) discharge through a suitably chosen target gas biased at typically 2–3 kV serves as the primary ion source. The primary ions are mass selected in a Wien filter and a fraction of them is neutralized by near-resonant charge transfer in a charge-transfer cell filled with an appropriately chosen gas for resonant or near-resonant charge transfer. Residual ions are removed from the target gas beam by electrostatic deflection and most species in Rydberg states are quenched in a region of high electric field. The neutral beam is subsequently crossed at right angles by a well-characterized electron beam (5–200 eV beam energy, 0.5 eV full width at half maximum (FWHM) energy spread, 0.03–0.4 mA beam current). The product ions are focused into the entrance plane

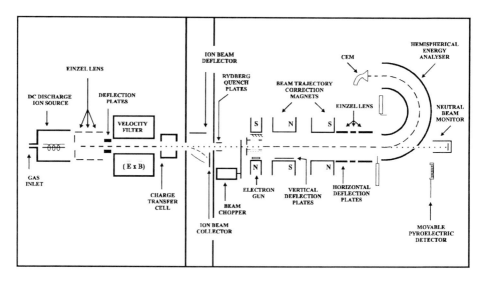

Figure 3.3: Schematic diagram of the fast-neutral-beam apparatus used for ionization cross section measurements at Stevens Institute of Technology.

of an electrostatic hemispherical analyzer which separates ions of different charge-to-mass ratios (i.e. singly-charged from multiply-charged ions and parent from fragment ions in the case of a molecular target). The ions leaving the analyzer are detected by a channel electron multiplier (CEM). The neutral beam density in the interaction region can be determined from a measurement of the energy deposited by the fast neutral beam into a pyroelectric crystal whose response is first calibrated by a well-characterized ion beam. As an alternative, the well-established Kr or Ar absolute ionization cross sections (known to better than 5%) can be used to calibrate the pyroelectric crystal. The calibrated detector, in turn, is then used to determine the flux of the neutral target beam in absolute terms.

Rigorous calculations of ionization cross sections are much more difficult compared to excitation cross section calculations because of the inherent complexity of the ionization process (a three-body Coulomb problem with a multitude of final states). Semi-classical and semi-empirical approaches are quite common [13]. Recently, Kim and co-workers have introduced the so-called BED (binary encounter dipole) method which proved quite successful in the calculation of atomic ionization cross sections [14]. The BED model combines the binary-encounter theory and the Bethe theory and expresses the total single ionization cross section in terms of the binding energies of the occupied orbitals, the average kinetic energy of the bound electrons in their orbitals, the electron occupation number, and the continuum dipole oscillator strength. The semi-classical Deutsch-Märk (DM) formalism [15] is another approach which proved to work well for calculating ionization cross sections for many atoms.

There has been much less activity regarding experiments and calculations dealing with excitation (and ionization) processes out of excited atomic states such as, e.g., the metastable levels of the rare gases. Yet secondary electron collision processes involving atoms, particularly metastable rare gas atoms, are important processes in many plasmas due to the large cross sections for collision processes out of these metastable states and due to their long life-times. For the experimentalist, the main obstacles in studying electron collision processes out of excited/metastable atomic levels are the preparation and characterization of the excited state target. On the theoretical side, first attempts have been made to treat the excitation of excited targets [16].

A survey of the experimental data base on electron collisions with excited atoms reveals the following findings:

1. Ionization: data are available for the excitation of metastable rare gases, metastable H and O, and laser-excited Na and Ba.

2. Excitation: data are available for metastable He (see below) and laser-excited Na, Rb, Cs, and Ba.

Electronic excitation of excited target species such as metastable rare gas atoms is also a collision process of fundamental importance. Metastable rare gas atoms, e.g., He (2^3S) atoms represent targets which are different from the spherically symmetric, tightly bound helium ground state. Lin, Anderson and collaborators at the University of Wisconsin have carried out extensive experimental investigations of the electron impact excitation out of the He metastable states [17, 18]. In a first generation of experiments [17], these authors used a hollow-cathode discharge to produce a mixture of metastable and ground state He atoms. Collisions with electrons whose energy was kept below the onset energy of the excitation of ground state He atoms present in the beam resulted in the collisionally induced population of higher-lying levels in He which subsequently decay radiatively. The detection of the emitted radiation allows the determination of the cross sections for exciting the higher levels out of the metastable level. Once the energy of the electron beam exceeds the energy necessary to excite the same levels from the ground state, this experimental technique breaks down due to the preponderance of ground state atoms in the target beam. A set of low-energy cross sections measured from excitation of the higher-lying 3^3S, 3^3P, and 3^3D states is shown in figure 3.4. Two observations are noteworthy: (i) the measured cross sections are larger than the cross sections for exciting the same levels from the He ground state by 1–2 orders of magnitude and (ii) the conventional rules of thumb regarding the cross section shapes for electron impact excitation of ground state atoms (see previous discussion) appear to be violated. Specifically, the following observations should be noted:

(i) for the n = 3, 4, 5 levels, the following pattern emerges:

    - sharply peaked excitation cross sections for the n^3S levels

    - less sharply peaked cross sections for exciting the n^3P levels

    - broad excitation cross sections for the n^3D levels

(ii) the 3^3D cross section is 4 times larger than the 3^3P cross section, even though the excitation of the 3^3P state is optically allowed whereas the excitation of the 3^3D is not.

Very recently the same authors have succeeded in obtaining cross section data for the excitation of metastable He atoms up to impact energies of 1000 eV in a second generation apparatus which is based on a fast-beam concept - metastable He atoms are selectively produced by appropriate charge transfer thus minimizing the number of ground state He atoms

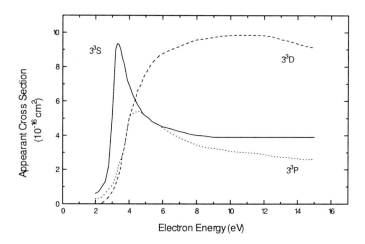

Figure 3.4: Absolute cross sections for excitation of various He 3^3L states from the metastable He 2^3S state by electron impact at low electron energies.

in the target beam [18]. Other rare gases such as Ne and Ar are currently being studied by the Wisconsin group. An extended review of electron interactions with excited atoms and molecules is given by Christophorou and Olthoff [19].

3.3 Electron impact-induced collision processes with molecules

Electron collisions with molecular targets are much more complex processes in many ways compared to electron-atom collisions. The additional degrees of freedom of the target (vibrational and rotational motion of the nuclei) and the fact that the target consists of several atoms increase the number of possible inelastic electron collision processes drastically. Rotational and vibrational excitation becomes possible in addition to electronic excitation [20]. Dissociative processes essentially double the number of possible electron-molecule interactions, since dissociation can be combined with just about any other inelastic process (e.g., dissociative excitation, dissociative ionization, and dissociative electron attachment).

A detailed discussion in this article will be largely limited to dissociation and ionization as the two most important *high-energy* electron-molecule collision processes relevant to plasmas. Other electron-molecule collisions will only be dealt with in a very general way. The reader is referred to reviews and pertinent references in the literature regarding the other processes [20, 21].

From an experimental point of view, it is very difficult to distinguish rotational excitation of a molecule from elastic scattering, since the rotational spacing in molecules is very small and the energy that is exchanged is smaller than that which can be resolved in most experiments (about 0.001 eV). Swarm experiments can provide some insight into the rotational excitation of simple molecules [20, 22]. Vibrational excitation can be a very effective energy loss mechanism for the plasma electrons. Vibrational excitation of molecules has been studied extensively (see, e.g., refs. [20, 21]), although it still requires a high level of experimental sophistication, since the vibrational energy level spacing in most molecules is comparable to the best energy resolution that can be routinely obtained in electron scattering experiments (about 0.05 eV or less). Electron attachment and dissociative attachment to molecules is an area that has received much attention in the past and a broad data base of cross sections and/or attachment coefficients for a large number of molecules can be found in the literature. Both swarm techniques and beam techniques have been successfully employed in experimental studies. The experiments are challenging, since many attachment and dissociative attachment processes have cross sections which are sharply peaked at very low electron energies, often at or near zero energy [22].

In addition to being an important ion formation process in plasmas containing molecules in the feedgas mixture, electron impact ionization of molecules also serves as the *trigger* for many plasma chemical reactions through the formation of reactive fragment ions via dissociative ionization of inert parent molecules. For instance, carbon-tetrafluoride, CF_4, the main constituent of fluorocarbon plasmas used for semiconductor etching is a very inert molecule. The only reason why fluorocarbon plasmas can be used in plasma processing applications is due to the fact that electron impact dissociation and dissociative ionization of CF_4 produce a myriad of reactive neutral and ionic CF_x (x=1–3) and F radicals which are responsible for the plasma chemistry in fluorocarbon plasmas.

CF_4 is one of many polyatomic molecules which does not have a stable positive parent ion. CF_4^+ undergoes rapid metastable decay on a time scale of microseconds. Other molecules in that category are silane, SiH_4 and sulfur-hexafluoride, SF_6. The dissociative ionization of CF_4 by electron impact has served as a test case for the reliability of partial ionization cross section measurements using various experimental techniques. Some of the fragment ions are produced with a substantial amount of excess kinetic energy (up to several electronvolts per ion in the case of F^+). As pointed out by Märk and co-workers [23], in many experimental arrangements used in ionization cross section measurements the efficiency of ion extraction from the ion source and the efficiency of their transport through the mass selective device (e.g., mass spectrometer) to the detector is critically dependent on the kinetic energy of the ions (ion discrimination). Energetic fragment ions are generally detected much less efficiently than thermal or near-thermal ions. For many years, there was poor agreement between different partial ionization cross section measurements carried out in different apparati using different experimental techniques. These problems have now been recognized and reliable partial ionization cross section measurements require a careful analysis of possible ion discrimination effects. This is usually accomplished by a combination of ion trajectory modeling calculations and in-situ experimental studies of the ion beam profile for each fragment ion [23].

Figure 3.5 shows a summary of the some recent total CF_4 ionization cross section measurements. There is good agreement between the experimental data of ref. [23] (solid triangles), ref. [24] (open circles), ref. [25] (solid line), and ref. [26] (open squares) within the combined error margins of the various measurements. The dashed line is the result of a binary encounter Bethe (BEB) calculation of Kim and co-workers, a simplified version of the

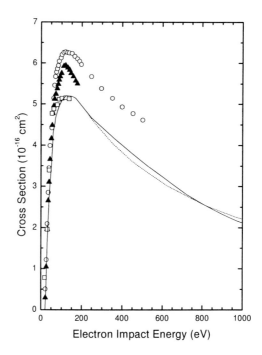

Figure 3.5: Selected electron impact ionization cross sections of CF_4 (see text for details).

previously mentioned BED method, which is particularly suitable for molecules. The BEB calculation agrees best with the experimental results of ref. [25]. For clarity of presentation, the results of other calculations based on the DM-formalism and on a modified additivity rule [27] have been omitted. Both calculations agree well with the BEB calculation and also with the experimental data [27].

There are two reasons why ionization processes involving free radicals are as important in plasmas as the electron impact ionization of the stable parent molecules [28]. Free radicals are abundant constituents in molecular plasmas and they are, in fact, the reactive species which make plasma-assisted processing a viable technology. Threshold ionization mass spectroscopy (TIMS) has evolved as a powerful plasma diagnostics technique [29, 30] which enables the absolute determination of radical species concentrations in plasmas. TIMS requires a quantitative knowledge of radical ionization cross sections. The fast-beam apparatus used by our group is particularly suited for measurement of absolute ionization and dissociative ionization cross sections of free radicals and other transient species, since the primary ion source is capable of generating well-characterized, mass-selected beams of a large number of free radicals which serve as a target for subsequent electron impact ionization studies. Fig. 3.6 shows the absolute cross sections for the electron impact ionization and dissociative ionization of the CF_3 free radical [31] which was carried out as part of a comprehensive series of ionization cross section measurements of halogen-containing radicals. As was the case with

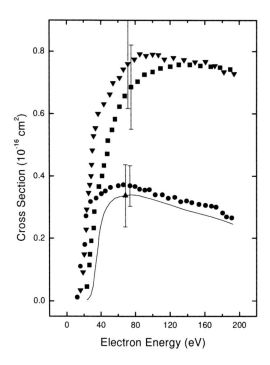

Figure 3.6: Electron impact ionization of the CF_3 free radical; circles (CF_3^+), squares (CF^+), inverted triangles (CF_2^+), solid line (F^+).

many complex molecules and free radicals, the ionization of CF_3 is dominated by the formation of fragment ions via a dissociative ionization process, whereas the parent ionization cross section is comparatively small.

Rigorous calculations of molecular ionization cross sections are even more scarce than atomic ionization cross section calculations. As a consequence, the calculation of absolute electron impact ionization cross sections for molecules have relied largely on empirical and semi-empirical methods and on simplistic additivity rules due to the complexity of more rigorous calculations for these processes and these targets [13]. In the past few years, several new developments have emerged:

(1) several modifications of conventional additivity rules [32]–[34] which attempt to account for molecular bonding in different ways [27, 35]

(2) the Deutsch-Märk (DM) formalism which combines a Gryzinski-type energy dependence of the cross section with quantum mechanically calculated molecular structure information [36, 37]

(3) a binary encounter Bethe (BEB) theory [38]

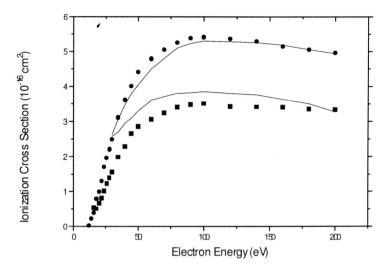

Figure 3.7: Measured and calculated (solid lines) total single ionization cross sections for the SF$_5$ (circles) and SF$_3$ (squares) free radicals.

(4) a theory based on the calculation of the maximum in the electron impact ionization cross section as a function of the electron-molecule approach geometry and subsequent averaging over all orientations [39, 40].

The modified additivity rule introduced by Deutsch et al. [27] attempts to account for the effects of molecular bonding by introducing empirically determined weighting factors which depend on the atomic orbital radii and the electron occupation numbers of the various atomic orbitals. A detailed comparison with a large number of existing molecular ionization cross section data demonstrated that this semi-empirical approach yields a level of agreement with experimental data which is as good as and in some cases even better than the results of more rigorous and more complex calculations. As an example, figure 3.7 shows a comparison between measured total single ionization cross sections for the SF$_5$ and SF$_3$ free radicals and calculated cross sections using the modified additivity rule. The agreement is excellent for both targets over the entire range of impact energies.

Conventional ionization cross section measurements do not distinguish between different final states of the ion that is produced. In almost all cases, this lack of selectivity regarding the final state of the ionization process is unimportant. Atomic and molecular ionization cross sections have absolute maximum values of $1–50 \times 10^{-16}$ cm^2. Typical cross sections for the formation of ions in excited states, on the other hand, are 2 orders of magnitude smaller. This means that in almost all cases, the formation of ions in the ground state is by far the dominant ionization process. A notable exception to this is molecular nitrogen, N$_2$, where ions in the first (A $^2\Pi_2$) and in the second (B $^2\Sigma_g^+$) excited state are formed with appreciable cross sections in addition to N$_2^+$ (X $^2\Sigma_g^+$) ground state ions. Since the total cross section for the formation of N_2^+ ions is well-known from experiment (to within about 10%), one would hope to be able to deduce the $N_2^+(X)$ cross section, which is difficult to measure, by subtracting the measured $N_2^+(A)$ and $N_2^+(B)$ cross section from the total N_2^+ cross section. While the N_2^+ B-state cross section has been measured by several groups and appears to be known with an accuracy of better than 20%, the same is not true for the N_2^+ A-state cross section. The lifetime of the A-state is too long to allow precise photoemission cross section measurements of the A → X transition, nor do electron-energy loss measurements seem to have yielded a reliable cross section value. The margin of uncertainty in the A-state cross section is about a factor of 2 which precludes a reliable estimate of the N_2^+ X-state cross section from these data.

Recently laser-induced fluorescence techniques have been successfully used in an absolute measurement of the cross section for formation of N_2^+ ions in the ground state. This was done in a triple-beam apparatus in which N_2^+ ground-state ions were formed by electron impact on N_2 molecules. A tunable dye-laser is then used to pump a specific rotational-vibrational state of the N_2^+ X → B transition and the subsequent fluorescence of the N_2^+ B → X transition is recorded [42]. After proper normalization of the laser-induced fluorescence signal one can obtain an absolute cross section for the formation of N_2^+ ions in the ground state. The cross section obtained by this method is shown in figure 3.8.

Electron impact dissociation processes can be divided into two categories, dissociative excitation and dissociation into neutral ground-state fragments. Dissociative excitation refers to

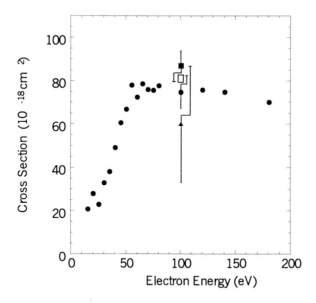

Figure 3.8: Absolute cross section for the formation of $N_2^+(X)$ ground-state ions following electron impact on N_2. The measured data of Abramzon et al. [42] for impact energies from 20 eV to 200 eV are compared with 2 estimated cross section values at 100 eV (for further detail, see Ref. [42]).

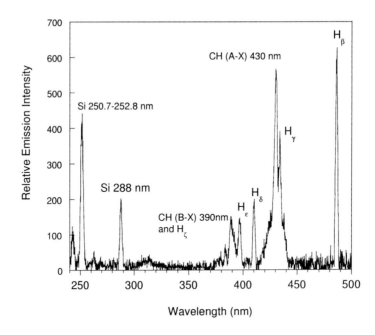

Figure 3.9: Optical emission spectrum produced by 100 eV electron impact on TMS (tetramethylsilane). The main emission features have been labeled (see text for furthur detail).

dissociation processes which leave the fragment(s) in an excited state. The subsequent emission associated with the radiative decay of the excited species is then recorded. Dissociative excitation processes can be studied with relatively little experimental effort by monitoring and analyzing these optical emissions. There is a reasonably broad data base on the dissociative excitation of many molecules relevant to low-temperature plasma processing, particularly for H_2, O_2, N_2, and many halogen-containing molecules as discussed in detail in a recent review [43]. Even some of the more complex polyatomic molecules such as the Si-organic molecules HMDSO, TMS, and TEOS, which are widely used in deposition plasmas, have been studied recently [44]–[46]. Figure 3.9 shows the optical emission spectrum in the 250–500 nm region recorded with a FWHM resolution of 2 nm produced by 100 eV controlled single electron impact on tetramethylsilane (TMS). The CH A $\rightarrow$ X and B $\rightarrow$ X band systems, the H Balmer lines, and the Si lines at 251 nm and 288 nm are readily observed. The HMDSO emission spectrum looks very similar to the TMS spectrum, while there are no indications of the Si lines in the TEOS spectrum [44, 45]. This can be understood in terms of the molecular structure of the three target molecules.

In addition to a detailed analysis of the emission spectra in terms of the identification of the emitting species of the various features, one can measure absolute emission cross sections of the spectral features and their appearance energies. The results of these studies provide insight into the collision-induced break-up mechanism of the complex species. This provides invalu-

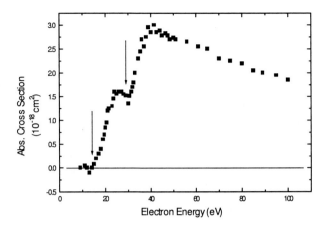

Figure 3.10: Absolute emission cross section of the BCl 272.4 nm band system following dissociaitive electron impact excitation of BCl_3 as a function of electron energy. The two onsets at 14 eV and 29 eV have been marked by arrows.

able information for the interpretation of data from plasma-induced emission spectroscopic studies. When combined with mass spectrometric data, photoemission cross sections can elucidate the dominant plasma processes that govern the deposition process. In some cases, the measured emission cross sections exhibit complicated structures which indicate that several break-up mechanisms with different minimum energies contribute to the emission of the same spectral feature. This is shown in figure 3.10 which displays the measured emission cross section of the BCl $A^1\Pi \rightarrow X^1\Sigma^+$ band system at 272.4 nm following dissociative electron impact excitation of boron-trichloride, BCl_3 as a function of electron energy [47]. Two onsets are apparent at about 14 eV and 29 eV. A similar double-onset structure was also observed for the emission cross sections of 2 atomic boron lines following dissociative excitation of BCl_3. This was interpreted as evidence for the presence of two channels contributing to the formation of BCl fragments in the excited state via the following mechanisms:

(i) $BCl_3 + e^- \rightarrow BCl(A^1\Pi) + 2\,Cl$ with a minimum energy of 13.1 eV,

(ii) $BCl_3 + e^- \rightarrow BCl(A^1\Pi) + Cl + Cl^+$ with a minimum energy of 26.0 eV.

Mechanism (i) was labeled the *neutral* channel, since all fragments produced by the dissociative excitation process are neutrals, while process (ii) was labeled the *ionic* channel because one of the dissociation fragments is a positively charged chlorine ion. Gilbert et al. [47] discussed the relevance of the existence of these two channels to the actual properties of processing plasmas containing BCl_3 in the feed gas mixture.

The break-up of a molecule into two or more neutral ground-state fragments caused by the impact of a single electron is one of the most fundamental collisional interactions between electrons and molecules. Dissociation into neutral ground-state species is also an important process from an application-oriented view-point, e.g., in low-temperature processing plasma, where these processes are responsible for the formation of a multitude of reactive species from the often rather inert parent molecules. It is, therefore, astonishing that these processes have not received much attention by the collision physics community until recently. The reason for the lack of research activities relating to the study of electron impact dissociation processes by theorists and experimentalists alike is due to the serious difficulties associated with the investigation of these processes compared to other electron collision processes. Neutral dissociation fragments in the ground state carry neither a charge nor any excess energy that could easily be exploited for the quantitative detection of these species in experimental studies. Complex target molecules and processes such as dissociation and ionization with a multitude of final states are inherently difficult to handle in ab initio, fully quantum mechanical theoretical calculations. The need to understand electron impact dissociation processes into neutral ground-state fragments in low-temperature plasmas stimulated innovative and novel approaches by the collision physics community to study these processes.

A brief review of the experimental techniques used to study electron impact dissociation into neutral ground-state fragments reveals the following:

- The chemical getter technique pioneered by Corrigan [48] and by Winters and collaborators [49] used a chemical getter to trap the dissociation products. The drawbacks of this approach are a very limited range of target molecules which can be studied and a lack of selectivity as far as the dissociation products are concerned (i.e. only total dissociation cross sections can be obtained)

- The two-electron-beam technique in which the first electron beam dissociates the target molecule and a second electron beam which is spatially separated from the first one probes the dissociation fragments. The main drawback of this technique is the problem of absolute calibration.

Both techniques were developed about 25 years ago and have rarely been used recently because of the many problems associated with their application.

Recently, Cosby and collaborators developed a new approach using a fast neutral beam in conjunction with a correlated product detection scheme [50]. This techniques enables the measurement of absolute electron impact dissociation cross sections for diatomic molecules or polyatomic molecules whose dissociation is dominated by a two-fragment break-up process with relative ease and has been successfully applied to molecules such as CO and Cl_2. The method is obviously inadequate for the investigation of dissociation processes where more than two fragments are formed.

Sugai and co-workers [51] developed another variant of the previously described two-electron-beam technique. They were able to overcome many of the problems associated with earlier approaches by using the technique of threshold ionization mass spectroscopy to probe the dissociation fragments produced by the interaction of the first electron beam with the target gas. The absolute calibration of the dissociation cross sections obtained by this technique requires a knowledge of the dissociative ionization cross sections of the parent molecules (which are available in most cases) and of the fragment (radical) ionization cross sections (which are available only in a few cases). Molecules relevant to low-temperature processing plasmas for which absolute dissociation cross sections have been obtained by this technique include SiF_4, CH_4, CHF_3, SiH_4 and CF_4. The most recent CF_4 data of Sugai and co-workers

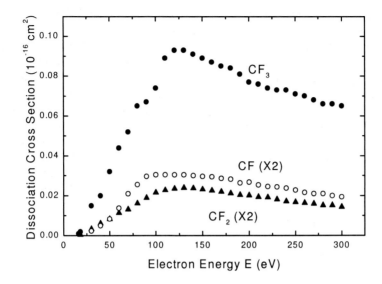

Figure 3.11: Absolute cross sections for the formation of CF_3, CF_2, and CF free radicals following controlled electron impact dissociation of CF_4.

[52] are shown in figure 3.11. In some cases such as SF_6 and SiH_4 only relative neutral dissociation cross sections could be reported due to a lack of the necessary radical ionization cross sections. It should be noted, though, that the cross sections shown in Fig. 3.11 are not undisputed. They are inconsistent with cross section values for the neutral dissociation of CF_4 obtained indirectly in two different ways [53, 54].

Very recently, Moore and co-workers [55] introduced another experimental technique to measure neutral dissociation cross sections. The method employs a unique detection scheme for the dissociation fragments. The fragments produced by neutral dissociation of stable molecules are allowed to react with the element tellurium to produce volatile tellurides, which are themselves stable and can readily be detected by conventional mass spectrometric methods.

In a different approach, McConkey and collaborators modified the two-electron beam technique mentioned above by replacing the second (probing) electron beam with a tunable dye-laser beam and used laser-induced fluorescence (LIF) to detect the electron impact produced neutral ground-state dissociation products [56]. In this approach, the laser beam probes the dissociation fragments directly in the interaction region of the electron beam and the gas beam. Processes studied so far include OH formation from H_2O and CN formation from HCN. The main problem with this techniques lies in obtaining a reliable absolute calibration. The limited wavelength range covered by currently available tunable dye-lasers is another limitation (no tunable vacuum ultraviolet lasers are routinely available as yet!). However, one can argue that this approach is perhaps the most versatile and most promising technique for the future in view of the rapid developments in laser technology.

Becker and co-workers [57] have used a variant of the LIF detection technique, which is

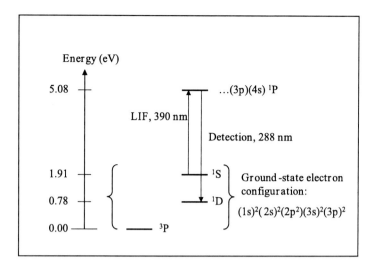

Figure 3.12: Partial energy level diagram of the Si atoms showing the optical pumping and detection scheme employed in the LIF detection of $Si(^1S)$ atoms following electron impact dissociation of SiH_4.

essentially an extension of the technique used by these authors previously in their absolute measurement of the $N_2^+(X)$ final-state specific ionization cross section, in neutral dissociation studies of the silane (SiH_4) molecule using the LIF detection scheme shown in Fig. 3.12. Si atoms in the ^1S state of the ground-state electron configuration resulting from the electron impact dissociation of SiH_4 were detected by pumping the $(3p)^2\ ^1S \rightarrow (3p)(4s)\ ^1P$ transition around 390 nm and recording the subsequent $(3p)(4s)^1P \rightarrow (3p)^2\ ^1D$ fluorescence at 288 nm. LIF spectra in Si were recorded for impact energies from 20 eV to 120 eV. The LIF spectra recorded at different electron energies yield the absolute cross section for the formation of $Si(^1S)$ following neutral dissociation of SiH_4 as a function of energy after proper normalization [42]. Fig. 3.13 shows our measured absolute $Si(^1S)$ cross section for the formation of $Si(^1S)$ following electron impact dissociation of SiH_4 from 20 eV to 100 eV. Please note that the two data points at 40 eV represent the result of two LIF measurements carried out under very different experimental conditions. The shape of the cross section curve displayed in Fig. 3.13 was reproducible to better than 6% over the energy range studied here. Our measured cross section rises rapidly from threshold to a plateau around 30 eV and peaks in the energy range from 50–70 eV with a maximum value of about $4 \times 10^{-17}\ cm^2$. The cross section declines fairly rapidly with increasing impact energy. The shape of our cross section is rather similar to the shapes of various photoemission cross sections of Si and SiH_4 reported by Perrin and Aarts [58] all of which displayed a prominent structure in the low-energy regime around 30 eV followed by a maximum around 60 eV and rapid decline towards higher impact energies.

It is interesting to compare our final-state specific $Si(^1S)$ neutral dissociation cross section

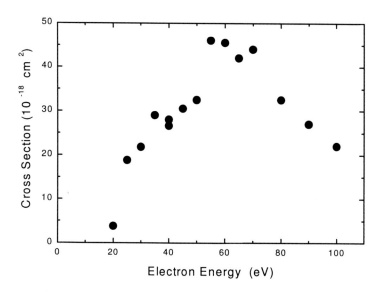

Figure 3.13: Absolute cross section for the formation of Si (^{1}S) atoms following electron impact dissociation of SiH$_4$. The two data points at 40 eV were obtained under very different operating conditions.

from SiH$_4$ with the total neutral dissociation cross section of SiH$_4$ obtained by Perrin et al. [59] using a method based on the kinetic analysis of SiH$_4$ dissociation in a constant-flow plasma reactor. These authors determined a maximum total neutral SiH$_4$ dissociation cross section of 1.2×10^{-15} cm^2 at 60 eV with a cross section shape rather similar to our measured Si(^{1}S) cross section shape. This results in a branching ratio of about 3.3×10^{-2} for the formation of Si(^{1}S) atoms. We can further compare our final-state specific Si(^{1}S) neutral dissociation cross section from SiH$_4$ with the photoemission cross section of SiH$_4$ leading to the formation of excited Si(^{1}P) atoms which subsequently radiate according to the process $(3p)(4s)^1P \rightarrow (3p)^2 \ ^1D$ [23]. These authors determined a cross section of 2.2×10^{-18} cm^2 at 60 eV which is much smaller (by a factor of 20) than our value for the final-state specific Si(^{1}S) ground-state dissociation cross section. This supports the notion that the formation of dissociation fragments by electron impact is usually dominated by processes leading to fragments in the ground state [8, 43].

On the theoretical side, first attempts are under way to treat electron impact dissociation processes in the framework of quantum mechanical theories. The application of variational methods such as the complex Kohn method or the Schwinger method (in conjunction with the use of massively parallel computers) has produced first results for the dissociation of Cl$_2$, BCl$_3$ and NF$_3$. In the case of Cl$_2$ and NF$_3$, the results of the calculations seem to agree rather well with experimental results [60].

3.4   Concluding remarks

We hope to have demonstrated in this chapter how collision physics, a discipline which has
its home in basic science and low-temperature plasma processing, a very application-oriented
technology-driven field, can benefit from each other. Fundamental research focusing on electron collisions, ion collisions and photon interactions with technologically relevant targets
provides invaluable collision and spectroscopic data for plasma modeling and for CAD of
plasmas processes and plasma reactors, helps identify key plasma processes and bottlenecks
in reaction mechanisms, enables the development of new and more sophisticated plasma diagnostics techniques, and provides the necessary tools for a more quantitative interpretation
of optical and mass spectrometric plasma diagnostics studies. At the same time, challenges,
unresolved questions and problems facing the plasma processing community have stimulated
novel research in collision physics. The need to understand the complex plasma processes at
a microscopic level have initiated new experimental work involving complex target molecules
and led to the development of new experimental techniques to generate, study and detect unstable, transient, reactive and/or corrosive species. Theory has been encouraged to treat complex target species and processes with many final states such as ionization and dissociation.
The development of new and improved scaling laws and semi-empirical methods has been
advanced by the need to compare data for a large number of target species and for classes of
targets of similar structure. Computational methods have benefited from ever more powerful
and faster computers and the implementation of highly parallel computers.

Acknowledgments

I am grateful to my collaborators over the past few years who made significant contributions
to many aspects of the topics described in this article. Thanks are also due to many colleagues,
too numerous to list, for many helpful and stimulating discussions and for making some of
their recent and/or unpublished data available to me. The financial support of our research
in the subject area of this article by the U.S. National Science Foundation (NSF), the U.S.
Department of Energy (DOE), and the U.S. National Aeronautics and Space Administration
(NASA) is gratefully acknowledged.

References

[1] S. Trajmar and D.F. Register, in *Electron-Molecule Collisions*, editors: I. Shimamura
     and K. Takanyanagi, Plenum Press, New York (1984), p. 427-494

[2] A.R. Filippeli, C.C. Lin, L.W. Anderson, and J.W. McConkey, in *Advances in Atomic,
     Molecular, and Optical Physics*, editor: M. Inokuti, Academic Press, New York (1994),
     Vol. 33, p. 1-62

[3] H.S.W. Massey and E.H.S. Burhop, *Electronic and Ionic Impact Phenomena*, Oxford
     University Press, Oxford (1969)

[4] K. Becker, A. Crowe, and J.W. McConkey, J. Phys. B **25**, 3885 (1992)

[5] P.D. Murray, S.F. Gough, P.A. Neill , and A. Crowe, J. Phys. B **23**, 2137 (1990)

[6] K. Bartschat and D.H. Madison, J. Phys. B **20**, 5839 (1987)

[7] G.D. Meneses, F.J. da Paixao, and N.T. Padial, Phys.Rev. A **32**, 156 (1985)

[8] K. Becker and V. Tarnovsky, Plasma Sources Sci. Technol. **4**, 307 (1995)

[9] V. Tarnovsky and K. Becker, Invited Papers, XVIII ICPEAC, Aarhus, Denmark (1993), editors: T. Anderson, B. Fastrup, F. Folkmann, H. Knudsen, and N. Anderson, AIP Press, New York, 1993, p. 234-48

[10] V. Tarnovsky and K. Becker, CUNY Internal Report (1991), unpublished

[11] R.C. Wetzel, F.A. Biaocchi, T.R. Hayes, and R.S. Freund, Phys. Rev. A **35**, 559 (1987)

[12] V. Tarnovsky and K. Becker, Z. Phys. D **22**, 603 (1992)

[13] S.M. Younger and T.D. Märk, in *Electron Impact Ionization*, editors: T.D. Märk and G.H.Dunn, Springer Verlag, Vienna (1985)

[14] Y.-K. Kim and M.E. Rudd, Phys. Rev. A **50**, 3954 (1994)

[15] H. Deutsch and T.D. Märk, Int. J. Mass Spectrom. Ion Proc. **79**, R1 (1987)

[16] see, e.g., E.J. Mansky, Bull. Am. Phys. Soc. **40**, 1305 (1995)

[17] R.B. Lockwod, L.W. Anderson, and C.C. Lin, Z. Phys. D **24**, 155 (1992)

[18] C.C. Lin and L.W. Anderson, private communication (1996, 1997)

[19] L.G. Christophorou and J.K. Olthoff, Adv. At. Molec. Opt. Phys. **44**, 155 (2000)

[20] *Electron-Molecule Collisions* editors: I. Shimamura and K. Takayanagi, Plenum Press, New York (1984)

[21] *Electron-Molecule Interactions and their Applications*, editor: L.G. Christophorou, Academic Press, New York (1984)

[22] L.G. Christophorou, D.L. McCorkle, and A.A. Christodoulides, in *Electron-Molecules Interactions and their Applications*, editor: L.G. Christophorou, Academic Press, New York (1984), p. 478-617

[23] H.U. Poll, C. Winkler, D. Margreiter, V. Grill, and T.D. Märk, Int. J. Mass Spectrom. Ion Proc. **112**, 1, (1992)

[24] R.A. Bonham, Jpn. J. Appl. Phys. **33**, 4157 (1994)

[25] M.V.V.S. Rao and S.K. Srivastava, in Contributed Papers, XX. ICPEAC, Vienna (1997), editors: F. Aumayr, G. Betz, and HP. Winter, MO150

[26] H. Nishimura, Proc. 8th Symposium on Plsama Processing, K. Tachibana, Ed., Nagoya, Japan (1991), p. 333

[27] H. Deutsch, K. Becker, and T.D. Märk, Int. J. Mass Spectrom. Ion Proc. **167/168**, 503 (1997)

[28] K. Becker, in *Electron Collisions with Molecules, Clusters, and Surfaces*, H. Ehrhardt and L.A. Morgan (editors), Plenum Press, New York (1994), p. 127-140

[29] R. Robertson, D. Hils, H. Chatham, and A. Gallagher, Appl. Phys. Lett. **43**, 544 (1983)

[30] H. Kojima, H. Toyoda, and H. Sugai, Appl. Phys. Lett. **55**, 1292 (1989)

[31] V. Tarnovsky, P. Kurunczi, D. Rogozhnikov, and K. Becker, Int. J. Mass Spectrom. Ion Proc. **128**, 181 (1993)

[32] J.W. Otvos and D.P. Stevenson, J. Am. Chem. Soc. **78**, 546 (1956)

[33] W.L. Fitch and A.D. Sauter, Analyt. Chem. **55**, 832 (1983)

[34] H. Deutsch and M. Schmidt, Contr. Plasma Phys. **25**, 475 (1985)

[35] M. Bobeldijk, W.J. van der Zande, and P.G. Kistemaker, Chem. Phys. **179**, 125 (1994)

[36] H. Deutsch, T.D. Märk, V. Tarnovsky, K. Becker, C. Cornelissen, L. Cespiva, and V. Bonacic-Koutecky, Int. J. Mass Spectrom. Ion Proc. **137**, 77 (1994)

[37] H. Deutsch. C. Cormelissen, L. Cespiva, V. Bonacic-Koutecky, D. Margreiter, and T.D. Märk, Int. J. Mass Spectrom. Ion Proc. **129**, 43 (1993)

[38] W. Hwang, Y.-K. Kim, and M.E. Rudd, J. Chem. Phys. **104**, 2965 (1996)

[39] C. Vallence, P.W. Harland, and R.G.A.R. MacLagan, J. Phys. Chem. **100**, 15021 (1996)

[40] P.W. Harland and C. Vallence, Int. J. Mass Spectrom. Ion Proc. **171**, 173 (1997)

[41] N. Abramzon, R.B. Siegel, and K. Becker, Int. J. Mass. Spectrom. **188**, 147 (1999)

[42] N. Abramzon, R.B. Siegel, and K. Becker, J. Phys. B **32**, L247 (1999)

[43] K. Becker, Comm. At. Mol. Phys. **30**, 261 (1994)

[44] M. Ducrepin, J. Dike, R. Siegel, V. Tarnovsky, and K. Becker, J. Appl. Phys. **73**, 7203 (1993)

[45] P. Kurunczi, A. Koharian, K. Becker, and K.E. Martus, Contr. Plasma Phys. **36**, 723 (1996)

[46] P. Kurunczi, K. Becker, and K E. Martus, Can. J. Phys. **76**, 153 (1998)

[47] P.G. Gilbert, R.B. Siegel, and K. Becker, Phys. Rev. A **41**, 5594 (1990)

[48] S.J.B. Corrigan, J. Chem. Phys. **43**, 4381 (1965)

[49] H.F. Winters, J. Chem. Phys. **63**, 3462 (1975)

[50] P.C. Cosby, J. Chem. Phys. **98**, 6813 (1993)

[51] T. Nakano, H. Toyoda, and H. Sugai, Jap. J. Appl. Phys. **30**, 2908 & 2912 (1991)

[52] H. Sugai, H. Toyoda, T. Nakano, and M. Goto, Contr. Plasma Phys. **35**, 415 (1995)

[53] R.A. Bonham, private communication (1996)

[54] L.G. Christophorou, J.K. Olthoff, and M.V.V.S. Rao, J. Chem. Phys. Ref. Data **25**, 1341 (1996)

[55] S. Motlagh and J.H. Moore, J. Chem. Phys. **109**, 432 (1998)

[56] M. Darrach and J. W. McConkey, J. Chem. Phys. **95**, 754 (1991)

[57] N. Abramzon, K. E. Martus, and K. Becker, J. Chem. Phys. **113**, 2250 (2000)

[58] J. Perrin and J. F. M. Aarts, Chem. Phys. **80**, 351 (1983)

[59] J. Perrin, J. P. M. Schmidt, G. de Rosy, B. Drevillion, J. Huc, and A. Loret, Chem. Phys. **73**, 383 (1982)

[60] T. N. Rescigno, Bull. Am Phys. Soc. **50**, 1382 (1994); ibid. **52**, 329 (1995)

4 Fundamental processes in plasma-surface interactions

Rainer Hippler

Institut für Physik, Ernst-Moritz-Arndt-Universität Greifswald, Domstraße 10a, D-17487 Greifswald, Germany

4.1 Introduction

The interaction between a plasma and its surrounding walls plays an important role in almost all kinds of plasmas including low temperature plasmas for technical applications and high temperature plasmas in fusion research. Many of the underlying fundamental processes have not been fully examined yet and details of the interactions of the plasma particle with solid surfaces (substrate, walls) are very often unknown. The main constituents of a plasma, i.e., electrons, ions, neutrals, radicals, and metastables, all interact differently with a solid, giving rise to a large variety of different effects. Interest arises due to the fundamental importance of such processes, and because of the technological applications for which these processes are important. For example, plasma etching due to heavy particle impact on surfaces plays an important role in the manufacturing of solid state devices and computers (see chapter 18). On the other hand, the erosion of the surrounding plasma wall by energetic particle impact is one of the major problems in fusion devices, giving rise to an unwanted cooling of the fusion plasma. In this chapter we shall review some of the basic mechanisms of the interaction of sufficiently energetic electrons, atoms or ions with solid surfaces. Interaction processes include scattering of the impinging atomic particle from the surface, deposition on and implantation into the surface, emission of secondary particles, and modification and erosion of the surface.

4.2 Theoretical considerations

The basic theoretical concepts describing the interaction of a sufficiently massive and energetic particle with a surface are the *binary collision (BC) model* and the *molecular* or *classical dynamics (MD) model*.

4.2.1 Binary collision model

The binary collision model is applicable as long as classical trajectories are justified and quantum mechanical trajectory and interaction effects are negligible. This is frequently the case for impinging atoms and ions but generally not the case for incident electrons. The interaction of an impinging particle with the surface is considered via individual binary collisions with atoms constituting the surface of the solid, and only one binary collision event at each instant is considered. The total interaction of the ion with the atoms which constitute the solid as well as the interactions among the target atoms then fall into a sequence of binary collisions. The binary collision itself may be treated either fully classically, quantum mechanically or within the semi-classical approximation. In many cases of practical relevance to plasma physical applications, a classical description suffices. In that case, a relatively simple and straightforward relationship exists between the scattering angle and the impact parameter and the kinetics of the collisions which is further complicated by the interaction potential between the colliding partners which is often not precisely known.

The interaction of an ion or atom with another atom leads to an exchange of momentum and energy. Conservation of momentum and energy requires that the total momentum and the total energy are the same before and after each collision, i.e., for a target atom initially at rest,

$$E_0 = E_1 + E_2 + \Delta E_{in} , \tag{4.1}$$

where E_0 is the kinetic energy of the projectile prior to the collision, E_1 and E_2 are the kinetic energies of projectile and target atom, respectively, after the collision, and ΔE_{in} is the inelastic energy transfer. Similarly, conservation of momentum requires

$$m_1 \vec{v}_0 = m_1 \vec{v}_1 + m_2 \vec{v}_2 , \tag{4.2}$$

where m_1 and m_2 are the mass of projectile and target atom, respectively, $\vec{v}_0$ is the projectile velocity prior to the collision, and $\vec{v}_1$ and $\vec{v}_2$ are projectile and target atom velocity after the collision. Eq. 4.2 may be rewritten as

$$\begin{aligned} m_1 v_0 &= m_1 v_1 \cos\theta_1 + m_2 v_2 \cos\theta_2 , \\ 0 &= m_1 v_1 \sin\theta_1 + m_2 v_2 \sin\theta_2 , \end{aligned} \tag{4.3}$$

where θ_1 and θ_2 are the angles of the scattered projectile and the recoiling target atom, respectively, with respect to the incident direction.

4.2.1.1 Scattering angle and energy transfer

The kinetics for one collision partner at rest prior to the collision are depicted in Fig. 4.1. The exact amounts of transferred momentum and energy depend on the details of the collision, for example, the impact parameter b and scattering angle θ_s. In the laboratory system, the energies E_1 and E_2 of projectile and target atom after the collision are given by

$$\begin{aligned} E_1 &= \frac{E_0}{(1+A)^2} \left(\cos\theta_1 \pm \sqrt{A^2 f^2 - \sin^2\theta_1} \right)^2 , \\ E_2 &= \frac{E_0 A}{(1+A)^2} \left(\cos\theta_2 \pm \sqrt{f^2 - \sin^2\theta_2} \right)^2 , \end{aligned} \tag{4.4}$$

where $A = m_2/m_1$ is the mass ratio and $f^2 = 1 - (1 + A^{-1})\Delta E_{in}/E_0$. Simpler relations are obtained in the center-of-mass system. The center-of-mass scattering angle θ_{cm} is related

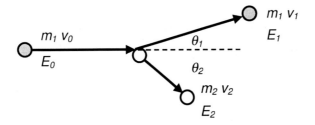

Figure 4.1: Binary collision kinematics.

to the scattering angles θ_1 and θ_2 by

$$\tan \theta_1 = \frac{Af \sin \theta_{cm}}{1 + Af \cos \theta_{cm}} ,$$

$$\tan \theta_2 = \frac{\sin \theta_{cm}}{1 - f \cos \theta_{cm}} . \tag{4.5}$$

For the kinetic energies we obtain

$$E_1 = \frac{E_0}{(1 + A)^2} \left((1 + Af)^2 - 4Af \sin^2 \frac{\theta_{cm}}{2} \right) ,$$

$$E_2 = \frac{E_0 A}{(1 + A)^2} \left((1 - f)^2 + 4f \sin^2 \frac{\theta_{cm}}{2} \right) . \tag{4.6}$$

The maximum energy transferred from the projectile to a target atom $E_{max}(\Delta E_{in})$ is obtained at $\theta_{cm} = 90^\circ$ and given by

$$E_{max}(\Delta E_{in}) = \gamma \frac{(1 + f)^2}{4} E_0 \tag{4.7}$$

where $\gamma = 4m_1 m_2 / (m_1 + m_2)^2$ and

$$E_{max}(0) = \gamma E_0 \tag{4.8}$$

for $\Delta E_{in} = 0$ ($f = 1$).

The scattering angles not only depend on the collision parameters, for example, the impact parameter b, but also on the interaction potential $V(r)$. Particularly at low relative velocities of the colliding partners, a proper choice of the interaction potential is crucial. In the center-of-mass system, the scattering angle θ_{cm} may be obtained from

$$\theta_{cm} = \pi - 2b \int_{R_0}^{\infty} \frac{dr}{r^2 \sqrt{1 - V(r)/E_{cm} - (b/r)^2}} , \tag{4.9}$$

where r is the internuclear separation, R_0 is the minimum distance during the collision, obtained from

$$1 - V(R_0)/E_{cm} - (b/R_0)^2 = 0 \tag{4.10}$$

with

$$E_{cm} = \frac{1}{2}\frac{m_1 m_2}{m_1 + m_2} v_0^2 \tag{4.11}$$

is the center-of-mass energy.

4.2.1.2 Stopping power

Projectile ions impinging on the surface which hit the surface without being reflected or back-scattered are either deposited on the surface or implanted in the solid. The depth distribution of the implanted projectile ions is governed by the energy loss of the projectile inside the substrate. The specific energy loss per distance or *stopping power* S is expressed as

$$S(E) = \frac{dE}{dx}; \tag{4.12}$$

this is related to a *stopping cross section* $S_x(E)$ by

$$S(E) = N S_x(E), \tag{4.13}$$

where N is the atomic density in the solid. The concept of a stopping cross section is based on several approximations, e.g., the continuous-slowing-down approximation and the assumption of straight-line trajectories (e.g., Refs. [1, 2]). The maximum depth R_m of the implanted ions is obtained from

$$\int_0^{R_m} S(E)dx = E_0 \tag{4.14}$$

where E_0 is the kinetic energy of the incident projectile, $E = E(x)$ is the kinetic energy of the projectile inside the solid, and x the distance from the surface. Again, Eq. 4.14 only holds in the limit of the continuous-slowing-down approximation which becomes questionable for energy losses ΔE comparable to E_0, which is the case at low kinetic energies.

Both nuclear and electronic interactions contribute to the total stopping power.

4.2.1.2.1 Nuclear stopping power
The nuclear stopping power results from the direct interaction (scattering) of two nuclei colliding with each other. This interaction leads to a relatively large momentum and energy transfer from the projectile to the substrate atoms, and, thereby, to a large displacement of the atoms involved. The nuclear stopping power may be calculated from the Rutherford scattering cross section, i.e., for the scattering of two unscreened nuclei with projectile charge $Z_1 e$ and target atom charge $Z_2 e$ as in [3]

$$d\sigma(E, E_0) = \pi \frac{m_1}{m_2} (Z_1 Z_2 e^2)^2 \frac{dE}{E^2 E_0}; \quad \text{for } 0 \leq E \leq E_{max}. \tag{4.15}$$

Eq. 4.15 holds for an unscreened Coulomb potential and is thus justified for sufficiently large projectile energies, i.e., for $\epsilon \geq 1$. Here the *reduced energy* ϵ is defined as

$$\epsilon = \frac{E_{cm}}{E_c(a)} \tag{4.16}$$

where $E_c(a) = Z_1 Z_2 e^2 / a$, and a is the so-called screening length (see below). At lower projectile energies ($\epsilon \leq 1$), the colliding nuclei penetrate less deeply into their respective Coulomb fields, and the screening of the repulsive nuclear potential by surrounding electrons becomes important. Assuming an interatomic potential $V(r)$ of the type

$$V(r) \propto r^{-1/m} ,$$ (4.17)

the energy loss cross section may be expressed as [3]

$$d\sigma(E, E_0) \cong C_m \frac{dE}{E_0^m E^{1+m}} ; \quad \text{for } 0 \leq E \leq E_{max}$$ (4.18)

where

$$C_m = \frac{\pi}{2} \lambda_m a^2 \left(\frac{m_1}{m_2} \right)^m \left(\frac{2 Z_1 Z_2 e^2}{a} \right)^{2m} ,$$ (4.19)

and λ_m is a dimensionless factor which varies slowly from $\lambda_0 = 24$ at low energies where $m = 0$ to $\lambda_1 = \frac{1}{2}$ at high energies where $m = 1$ [3]. The nuclear stopping cross section $S_n(E_0)$ is obtained from

$$S_n(E_0) = \int E d\sigma(E, E_0) ,$$ (4.20)

which yields

$$S_n(E_0) = \frac{1}{1-m} C_m \gamma^{1-m} E_0^{1-2m} .$$ (4.21)

The stopping cross section thus rises approximately linearly as a function of incident energy from low energies ($m = 0$) and, after reaching a maximum, decreases as E_0^{-1} at large energies where $m = 1$ (Fig. 4.2). In compact form, the nuclear stopping cross section $S_n(E_0)$ may be expressed as [3]

$$S_n(E_0) = 4\pi a Z_1 Z_2 e^2 \frac{m_1}{m_1 + m_2} s_n(\epsilon)$$ (4.22)

where the so-called *reduced stopping power* $s_n(\epsilon)$ is a universal function which is largely independent of the mass and charge of both projectile and target. For the reduced nuclear stopping power $s_n(\epsilon)$ an approximate analytical expression has been given by Bohdansky [4, 5],

$$s_n(\epsilon) = \frac{3.441\sqrt{\epsilon} \ln(\epsilon + 2.718)}{1 + 6.355\sqrt{\epsilon} + \epsilon(6.882\sqrt{\epsilon} - 1.708)}$$ (4.23)

where ϵ is defined by Eq. 4.16. The reduced nuclear stopping power calculated for three different interaction potentials is shown in Fig. 4.2. It is quite apparent that particularly at low reduced energies significant deviations may arise for different choices of the interaction potential. The reduced nuclear stopping power calculated with the help of Eq. 4.23 is also shown.

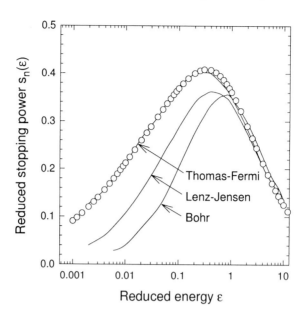

Figure 4.2: Reduced nuclear stopping power *vs.* reduced energy ϵ calculated for three different interaction potentials (solid lines; after Sigmund [3]). Also shown are the predictions from Eq. 4.23 (open circles).

4.2.1.2.2 Electronic stopping power

The electronic stopping power results from the interaction of the impinging ion or atom with the electrons inside the solid. The electronic stopping cross section S_e may be calculated from quantum mechanical theory. Bethe's [6] result is based on the first Born approximation and may be expressed as [1, 7]

$$S_e = \frac{8\pi a_0^2 R^2 z^2}{T} \left[Z \ln(4T/I_0) \right] , \tag{4.24}$$

where $T = E_0 \times m_e/m_p$, E_0 is the incident energy, m_e and m_p are the electron and projectile mass, respectively, z the projectile charge, Z the number of target atom electrons, a_0 the Bohr radius, R the Rydberg energy, and I_0 the mean excitation energy. Although Eq. 4.24 is based on atomic scattering theory, it has certain merits, particularly for large incident velocities, in that it describes the interaction of fast particles with solids as well. Figure 4.3 displays the stopping cross section (in units of eV cm^2) for protons incident on solid (amorphous) carbon and for a mean excitation energy $I_0 = 78$ eV calculated with the help of Eq. 4.24. In comparison with other estimates for the stopping cross section [8, 9], Eq. 4.24 appears reasonable for large energies $T \gg I_0$ which are, however, of little relevance to low-temperature plasma physics. The apparent failure at low energies largely reflects the use of a mean excitation energy I_0 which is a reasonable approximation at high energies but ignores the electronic shell structure of atoms and solids. Typical excitation energies (e.g., electron-hole pair production, excitation of plasmons) can be as small as a few eV particularly in solid materials. Such small energy losses are thus not properly accounted for if a mean excitation energy is used. So-

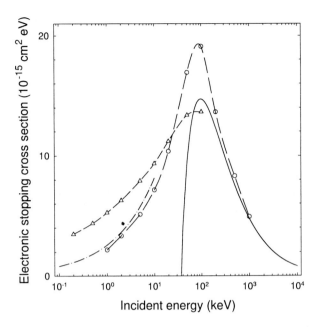

Figure 4.3: The electronic stopping cross section *vs.* incident energy for protons on amorphous carbon: Solid line: Bethe theory (Eq. 4.24), dash-dotted line: Lindhard and Scharff (Eq. 4.25) [11], o: Kaneko [8], and △: Ziegler et al. [9].

called shell corrections as well as other corrections have been introduced in order to account for this deficiency. Better agreement is thus expected from theories which additionally take the shell structure or, alternatively, the dielectric response of the electrons inside a solid into account (e.g., Refs. [7, 10], and references therein). An analytical formula based on the dielectric response model that is applicable at low particle velocities was given by Lindhard and Scharff [11]

$$S_e = 8\pi\sqrt{2}a_0\hbar\frac{Z_1^{7/6}Z_2}{(Z_1^{2/3} + Z_2^{2/3})^{3/2}}\sqrt{\frac{E_0}{m_p}}\; ; \tag{4.25}$$

the results from Eq. 4.25 are also shown in Fig. 4.3.

While electronic stopping may give rise to a significant excitation and ionisation of electrons including emission of secondary electrons, thereby causing an additional stopping of fast moving projectiles, the momentum and energy transfer to the substrate nuclei nevertheless remains relatively small. The major contribution to the sputtering yield therefore stems from nuclear stopping while contributions from electronic stopping are generally weak and comparatively small. Exceptions may occur at very high projectile energies where the nuclear stopping is small, and with highly-charged ions at low velocities due to the large potential energy then carried by the projectile (potential sputtering, e.g., Ref. [12]).

4.2.1.3 Sputtering yield

As an application of the binary collision model we present some results for the sputtering of atoms by ion impact. Sputtering unlike (thermal) evaporation is a collision process by which an atom formerly bound to a solid becomes liberated; the energy needed to liberate the atoms is provided by binary collisions either with the projectile (*direct knock-on*) or with other target atoms which received their kinetic energy through a sequence of collisions (*collision cascade*) originating from the projectile. The number of atoms involved in such a collision cascade may vary widely, depending, for example, on the kinetic energy of the projectile and on the projectile/target atom combination.

The sputtering yield $Y(E_0)$ is defined as the number of sputtered atoms per incident projectile ion,

$$Y(E_0) = \frac{\text{number of sputtered atoms}}{\text{number of projectile ions}} \, ; \tag{4.26}$$

this is a function of the projectile energy E_0 and also depends on collision parameters such as projectile mass m_1, target atom mass m_2, projectile and target nuclear charge $Z_1 e$ and $Z_2 e$ (where e is the elementary charge), and the angle of ion incidence θ_{inc}.

Within the framework of the collision cascade model, the sputtering yield is calculated from the energy deposited in a certain depth of the target or substrate, which is related to the nuclear stopping power $S_n(E_0)$, and from the transport of the moving substrate atoms to the surface. The sputtering yield Y may be expressed as [4]

$$Y(E_0) = Q \, S_n(E_0) \, g(E_0/E_{th}) \, , \tag{4.27}$$

where Q is the so-called yield factor, and $g(E_0/E_{th})$ is a semi-empirical correction factor which accounts for threshold effects,

$$g(E_o/E_{th}) = \left(1 - (E_{th}/E_0)^{2/3}\right) \left(1 - E_{th}/E_0\right)^2 . \tag{4.28}$$

The threshold energy E_{th}, which is a function of the mass ratio m_2/m_1, may be calculated from an analytical expression that was obtained from a fit to experimental and theoretical sputtering yield data [4, 13, 14],

$$\frac{E_{th}}{U_0} = 7 \left(\frac{m_2}{m_1}\right)^{-0.54} + 0.15 \left(\frac{m_2}{m_1}\right)^{1.12} . \tag{4.29}$$

Following Sigmund [3], the yield factor Q can be expressed as

$$Q = 0.01175 \, a_0^{-2} \times \frac{\alpha}{U_0} \tag{4.30}$$

where U_0 is the surface binding energy of the atoms, and a_0 the Bohr radius. The dimensionless factor $\alpha = \alpha(E_0, m_1/m_2, \theta_{inc})$ depends weakly on the incident energy E_0 but shows a pronounced dependence on the mass ratio m_2/m_1, and on θ_{inc}, the angle of incidence relative to the surface normal. Figure 4.4 shows the dependence of α as a function of the mass

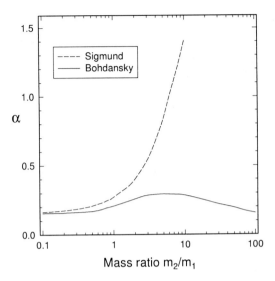

Figure 4.4: The dimensionless parameter α *vs.* mass ratio m_2/m_1 according to Sigmund [3] and Bohdansky [4].

ratio m_2/m_1 as calculated by Sigmund [3] and according to Bohdansky (e.g., Ref. [14]). Sigmund's original calculation shows a rather steep dependence of α on m_2/m_1, while Bohdansky's result, being close to $\alpha \approx 0.17$, is in better agreement with experiment. The reason for this apparent deficiency of Sigmund's calculation is believed to be due to an overestimation of the energy deposited inside the solid by light ions.

4.2.1.4 Computer simulations based on the binary collision model

The binary collision approximation (BCA) model was the first to be used in computer simulations of ion-solid interactions [15]. The usefulness of computer simulations was further demonstrated by Robinson and Oen [16] during their discovery of the channeling effect. Computer simulations based on the BCA model in essence fall into two categories, whether they assume a crystalline structure of the solid or, as in calculations based on the TRIM code a randomized or structureless target.

In simulations assuming a crystalline structure, the collision sequence is deterministic once the impact point of the projectile at the surface and its direction into the solid are given. A list of target atom positions is then required that can be constructed as in standard solid state theory with the help of three primitive translation vectors starting from a basis of one or more atoms. A major and elaborate task in such simulations is therefore the calculation of a list of nearest neighbors for each collision sequence and the search procedure to find the next collision partner.

These problems are to a large extent circumvented by simulations which employ a randomly structured solid target. The next collision partner can then be found by a random selection process which is why these simulations are sometimes called Monte Carlo programmes. A rather popular code is the *TRIM (TRansport of Ions in Matter)* programme [17] which comes in various versions and modifications, e.g., TRIM.SP [18] and TRIDYN [19]. A discussion of the various simulation programmes may be found in Eckstein's book [13].

4.2.2 Molecular dynamics model

In the *classical (CD)* or *molecular dynamics (MD) model* the movement of all atoms inside a solid is studied as a function of time. The model thus takes interactions with all neighboring atoms into account. The starting point is Newton's equation for the motion of a single atom i which is governed by the interaction forces from all target atoms in the neighborhood,

$$m_i \frac{d^2 \vec{x}_i(t)}{dt^2} = \sum_{j=1}^{N} \vec{F}_{ij} = \vec{F}_i(\vec{x}_i(t)), \tag{4.31}$$

where N is the number of atoms taken into consideration. The forces $\vec{F}_{ij}$ can be calculated from the interaction potentials assumed in the calculations. Once all forces interacting with atom i have been computed, Eq. 4.31 may be integrated numerically to yield the new position $\vec{x}_i(t + \Delta t)$. It is apparent that a good algorithm is required to achieve simultaneously good computational accuracy and a sufficient computational speed. A common algorithm is the so-called Verlet [20] algorithm, but other schemes are in use as well [13, 21]. The Verlet algorithm is based on a Taylor expansion of $\vec{x}_i(t + \Delta t)$,

$$\vec{x}_i(t + \Delta t) = \vec{x}_i(t) + \vec{v}_i(t)\Delta t + \frac{\vec{F}_i(t)}{2m_i}(\Delta t)^2 + \frac{1}{3}\frac{d^3 \vec{x}_i(t)}{dt^3}(\Delta t)^3 + O((\Delta t)^4) \tag{4.32}$$

and of $\vec{x}_i(t - \Delta t)$,

$$\vec{x}_i(t - \Delta t) = \vec{x}_i(t) - \vec{v}_i(t)\Delta t + \frac{\vec{F}_i(t)}{2m_i}(\Delta t)^2 - \frac{1}{3}\frac{d^3 \vec{x}_i(t)}{dt^3}(\Delta t)^3 + O((\Delta t)^4). \tag{4.33}$$

Combining these two equations, one obtains the new position $\vec{x}_i(t + \Delta t)$ from the two former positions $\vec{x}_i(t)$ and $\vec{x}_i(t - \Delta t)$,

$$\vec{x}_i(t + \Delta t) = 2\vec{x}_i(t) - \vec{x}_i(t - \Delta t) + \frac{\vec{F}_i(t)}{m_i}(\Delta t)^2, \tag{4.34}$$

which is accurate to within $O((\Delta t)^4)$. Obviously, making the time step Δt sufficiently small enhances the computational accuracy at the expense of long computing times. The choice of the step size is thus a compromise between these two limitations; typical time steps are in the range of several fs, and several hundred time steps may be required for a full calculation. The molecular dynamics approach has become quite popular for scattering at low projectile velocities where the interaction can be limited to a small number of atoms and where small cell sizes (several 100 to a few 1000) may suffice.

4.2.3 Scattering potentials

Various alternatives for the interaction potential $V(r)$ between projectile and target atom or between two colliding target atoms can be found in the literature. In most cases, the interaction potentials are only approximately known, however.

4.2.3.1 Repulsive potentials

The interaction between two colliding atoms, particularly for small internuclear separations, is dominated by the Coulomb force between the two positively charged nuclei, which is repulsive by nature. Surrounding electrons modify and partly screen the repulsive Coulomb potential, particularly at large internuclear distances, and therefore realistic potentials drop off more quickly than pure Coulomb potentials.

4.2.3.1.1 Screened Coulomb potentials A simple *ansatz* is to use screened Coulomb potentials

$$V_B(r) = \frac{Z_1 Z_2 e^2}{r} f_s(r/a) \tag{4.35}$$

where $f_s(r/a)$ is a screening function and a a (typical) screening length. The simplest screened Coulomb (Bohr) potential uses a single exponential for the screening function

$$f_B(r) = \exp(-r/a) . \tag{4.36}$$

For the screening parameter, Firsov [22] proposed

$$a_F = \frac{0.8853 \, a_0}{(Z_1^{1/2} + Z_2^{1/2})^{2/3}} , \tag{4.37}$$

where a_0 is the Bohr radius. This choice of screening function ignores the shell structure of realistic atoms which are composed of many electron shells. Hence a more realistic *ansatz* is to use screened Coulomb potentials which (partly) take the shell structure into account. The first one to come to mind is the Molière potential which is obtained using Eq. 4.35 with a modified screening function f_M,

$$f_M(r) = 0.35 \exp(-0.3r/a_F) + 0.55 \exp(-1.2r/a_F) + 0.10 \exp(-6.0r/a_F) . \tag{4.38}$$

Fig. 4.5 compares these screening functions with each other. As a major difference we note that the Bohr potential drops off rather quickly while the Molière screening function shows a significantly weaker dependence on the internuclear separation. There are other potentials that are in frequent use; for a discussion see, e.g., Ref. [13].

4.2.3.2 Attractive potentials

So far we have discussed interaction potentials which are purely repulsive. These potentials, hence, do not take into account that realistic interactions, particularly for atomic combinations

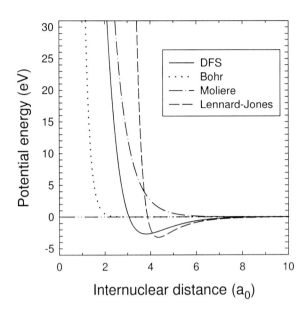

Figure 4.5: A comparison of different interatomic Si-Si potentials (solid line, Dirac-Fock-Slater (DFS) potential, Eq. 4.41; dotted line, Bohr potential, Eq. 4.36; dash-dotted line, Molière potential, Eq. 4.38; dashed line, Lennard-Jones potential, Eq. 4.40).

which form stable molecules or solids, may have to become attractive around intermediate separations. Such potentials exhibit a potential well D at a certain internuclear distance r_0. A frequently applied potential has been proposed by Morse [23]

$$V(r) = D \exp[-2\alpha_M (r - r_0)] - 2D \exp[-\alpha_M (r - r_0)], \qquad (4.39)$$

where α_M is a constant that determines the slope of the potential and, hence, the zero-point. Also in use is another attractive potential introduced by Lennard-Jones [24]. A popular version of the Lennard-Jones potential is the so-called 6-12 potential,

$$V(r) = D \left(\frac{r_0}{r}\right)^{12} - 2D \left(\frac{r_0}{r}\right)^6, \qquad (4.40)$$

where the r^{-6} dependence is due to the dipole-dipole interaction leading to van der Waals forces which govern the binding of van der Waals complexes but not necessarily that of other compounds. No physical justification is given for the r^{-12} term except that it has to drop off more quickly than the first term to yield a (partly) attractive potential.

It is to be noted that both the Lennard-Jones and Morse potentials while providing a realistic description of the attractive part of the interaction become insufficient at small internuclear separations where purely repulsive potentials are more adequate. While more sophisticated potentials calculated, for example, by employing the Dirac-Fock-Slater (DFS) method [25] have recently become available, such potentials are generally more complicated and only available in numerical form and thus not very convenient for the calculations of interest here. For Si-Si collisions, the interaction potential calculated with the DFS method may be approximated by Eq. 4.35 with a screening function given by

$$
\begin{aligned}
f_{DFS}^{Si-Si}(r/a_F) &= \left(1 - 0.005713(\frac{r}{a_F})^2\right) \times \left[0.35\exp(-0.28\frac{r}{a_F})\right. \\
&+ 0.55\exp(-1.2\frac{r}{a_F}) + 0.1\exp(-6.0\frac{r}{a_F}) \\
&- \left. 0.002327(\frac{r}{a_F})\exp(-0.15(\frac{r}{a_F} - 5.757)^2)\right] .
\end{aligned}
\tag{4.41}
$$

Fig. 4.5 compares the Dirac-Fock-Slater potential for Si-Si collisions with the corresponding Bohr and Molière potentials. Whereas for low internuclear separations the DFS and Molière potentials are almost indistinguishable from each other, significant differences occur for medium and large internuclear separations where both potentials are weak. Nevertheless, trajectories calculated with these two potentials differ significantly for large impact parameters where the DFS potential becomes attractive.

4.2.3.2.1 Embedded atom potential   In the embedded atom model the energy U_i of each atom inside a metal is calculated from the energy needed to embed the atom at a given locality in the local electron density provided by the surrounding atoms [26],

$$
U_i = -F(\rho_i) + \frac{1}{2}\sum_j V_{ij}
\tag{4.42}
$$

where V_{ij} describes the core-core repulsion between atoms i and j. The electron density ρ_i may be approximated by a superposition of atomic densities $\phi_{ij}(r_{ij})$ which depend on the interatomic distance between atom i and j,

$$
\rho_i = \sum_j \phi_{ij}(r_{ij})
\tag{4.43}
$$

where

$$
\phi_{ij}(r) = Z_i(r)Z_j(r)/r
\tag{4.44}
$$

with effective charges $Z_i(r)$ and $Z_j(r)$. The embedding function F is chosen to fit the bulk properties of the solid. A rather simple form is $F(\rho) \propto \rho^{1/2}$ but various other and more accurate alternatives for the embedding function have been proposed; for a discussion see, e.g., Ref. [21].

4.3   Scattering of ions at surfaces

4.3.1   Implantation of ions

The interaction of energetic (fast) ions or atoms with surfaces leads to the deposition of the incident atomic or molecular particles on the surface and to implantation in the solid. The range

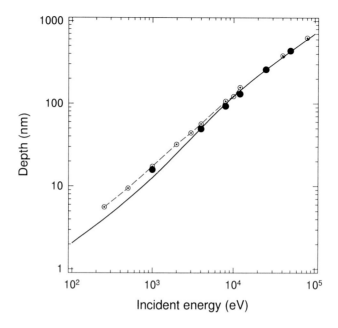

Figure 4.6: Mean projected range vs. incident energy for He^+ incident on silicon. The experimental results ($\bullet$) are compared with mean ranges obtained from analytical theory (solid line) and from TRIM calculations (dashed line) (after Ref. [13]).

of the incident projectiles in the solids is largely determined by the incident kinetic energy and increases with increasing projectile energy. Since the slowing-down of particles in matter in general does not proceed along straight-lines, several quantities such as the mean projected range (i.e., projected along the incident direction) and spread (i.e., the transverse distribution) have been introduced (e.g., Ref. [13]). Experimental results for the mean projected range of He^+ ions in silicon compared to theoretical results obtained from analytical theory and from TRIM calculations [13] are displayed in Figure 4.6. In general, the agreement between experiment and theoretical predictions is rather good.

4.3.2 Backscattering of ions

Particularly at low incident energies, the incident ion has a fair chance of escaping deposition or implantation and being backscattered instead. The particle reflection coefficient R is defined as the ratio of the number of reflected particles relative to the number of incident particles (ions),

$$R(E_0, \theta_{inc}) = \frac{\text{number of reflected particles}}{\text{number of incident particles}}. \tag{4.45}$$

Calculated reflection coefficients may be as large as unity for low incident energies in the region of several 10 eV. For these low velocities the repulsive part of the interaction potential

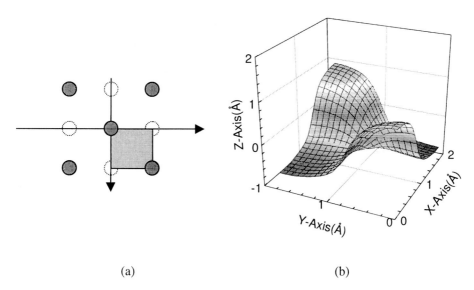

(a)   (b)

Figure 4.7: (a) Top view of a Ni(100) surface. The (shaded) impact area is indicated.
(b) Penetration depth vs. impact point for 30 eV argon atoms on a Ni(100) surface (after
Ref. [27]).

is strong enough to prevent the ion from entering the solid. Moreover, the scattering of ions at
and from surfaces is strongly influenced by the particular choice of interaction potential [27].
As an example, Fig. 4.7 displays the penetration depth distribution of 30 eV argon atoms
incident on a Ni(100) surface as a function of the position on the surface (XY–plane) onto
which the incident atom is directed. The origin ($X = Y = Z = 0$) of the plot corresponds
to the position of a nickel atom. The molecular dynamics calculations were performed by
employing a purely repulsive Lenz-Jensen potential for the Ar-Ni interaction [27]. At these
low incident energies, the majority of ions penetrate at most a few Å into the solid. It can be
seen that the deepest penetration is achieved in the middle between two surface atoms, while
for head-on collision the argon ions are reflected well above the surface. The penetration depth
critically depends on the chosen interaction potential [27].

The calculated particle reflection coefficients for nickel by H^+ and Ne^+ ion bombardment
versus incident energy using the TRIDYN code [19] are displayed in Fig. 4.8. The reflection
coefficient is relatively large at low incident energies amounting to about 50% at 100 eV for
incident H^+ and about 30% for Ne^+ ions. The reflection coefficient decreases monotonically
towards larger energies [13]. This behaviour reflects the fact that the average scattering angle
decreases with incident energy and increasing projectile mass.

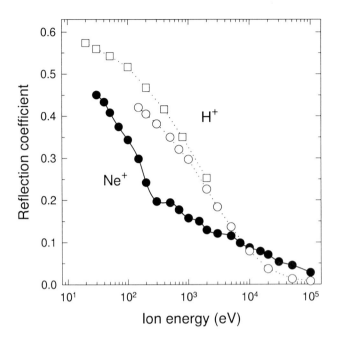

Figure 4.8: Reflection coefficient of nickel for H^+ (open symbols) and Ne^+ (closed symbols) ion impact *vs.* incident energy (after Ref. [13]).

4.4   Physical sputtering

4.4.1   Projectile energy dependence

Measured sputtering yields for nickel by H^+, Ne^+ and Ni^+ ion bombardment versus incident energy are displayed in Fig. 4.9. Also shown are theoretical calculations based on the TRIDYN code [19] and the Bohdansky [4] formula. It is rather obvious that the sputtering yields by the heavier Ne^+ and Ni^+ projectiles are quite different both in magnitude and with respect to the position of the maximum compared to H^+ impact. There are several mechanisms which contribute to the total sputtering yield. According to Eckstein [13], one may distinguish between *primary knock-on (PK)* atoms which receive kinetic energy and momentum directly from the projectile, and *secondary knock-on (SK)* atoms which receive energy and momentum from other fast moving target atoms. A further distinction arises from the actual direction of the projectile at the time of the interaction, i.e., whether the projectile is moving *inwards (projectile-in)* or *outwards (projectile-out)* with respect to the target surface. At low incident energies the projectile-out contributions dominate, largely because there is little energy to transfer and the chance of the projectile becoming back-scattered is relatively large. Here, primary knock-on is by far the most important process. At higher energies, the fraction of back-scattered projectiles decreases as E_0^{-2} while the total energy transferred to the target atoms increases. This increases the number of projectile-in events at the expense

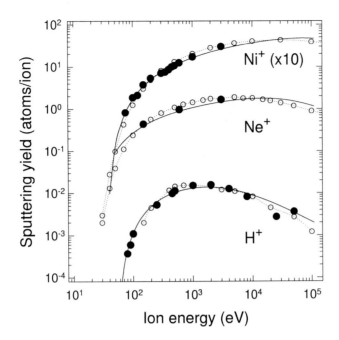

Figure 4.9: Sputtering yield of nickel for H^+, Ne^+, and Ni^+ ion impact *vs.* incident energy (after Ref. [13]).

of the projectile-out processes. In this case, secondary knock-on through the formation of a collision cascade dominates the sputtering events.

4.4.2 Energy distribution of sputtered particles

Within the framework of the collision cascade model, the energy distribution $N(E)$ of sputtered atoms is proportional to

$$\frac{dY(E_0, E_s)}{dE_s} \propto \frac{E_s}{(E_s + U_0)^{3-2m}}, \tag{4.46}$$

where E_s is the kinetic energy of the sputtered target atoms. In the derivation of Eq. 4.46, a potential of the type $V(r) \propto r^{-1/m}$ (Eq. 4.17) has been assumed. For $m = 0$, Eq. 4.46 is the same as the so-called Thompson formula [28]. At low energies, threshold effects due to the possible maximum energy transfer which modify the energy dependence need to be taken into account [29],

$$\frac{dY(E_0, E_s)}{dE_s} \propto \frac{E_s(1 - \sqrt{(E_s + U_0)/E_{max}})}{(E_s + U_0)^{3-2m}}, \tag{4.47}$$

where E_{max} is given by Eqs. 4.7 or 4.8. Calculated energy distributions of sputtered nickel atoms following 50–1000 eV Ar bombardment of a Ni surface are shown in Fig. 4.10, together

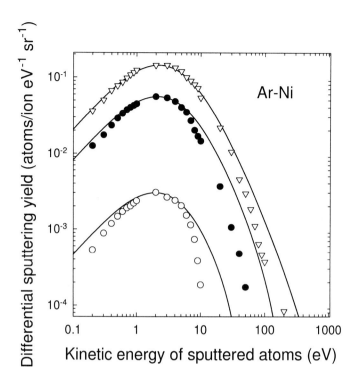

Figure 4.10: Calculated energy distributions of sputtered nickel atoms following 50 eV ($\circ$), 200 eV ($\bullet$), and 1000 eV ($\triangle$) Ar bombardment of a Ni surface (after Ref. [13]). Also shown are the predictions according to Eq. 4.47 (solid lines).

with predictions according to Eq. 4.47. It should be noted that the energy distributions peak around $U_0/2$. As expected, the influence of threshold effects is more pronounced for low incident energies.

Experimentally, the energy distribution of sputtered target atoms are frequently determined by optical means. In that case, the Doppler shift and Doppler broadening of emitted spectral lines of excited atoms are investigated (e.g., Ref. [29]). Figure 4.11 shows the velocity distribution of sputtered Fe atoms in the ground and in the metastable state following 10 keV Ar^+ bombardment of iron [30]. Also shown is the velocity distribution predicted by the Thompson formula (Eq. 4.47 with $m = 0$). Taking the surface binding energy as a fit parameter, the predicted velocity distributions are in reasonable agreement with experiment for both the ground state and the metastable atoms. Nevertheless, the velocity distributions of ground state and metastable atoms differ significantly from each other. The observed shift towards larger velocities of the metastable atom distribution may be caused by de-excitation mechanisms which prevent excited or metastable atoms from escaping intact from the surface if their velocity is too slow. Among the possible mechanisms we mention resonance tunneling of electrons from a sputtered excited atom back to the surface [31, 32]. It should be mentioned here that reso-

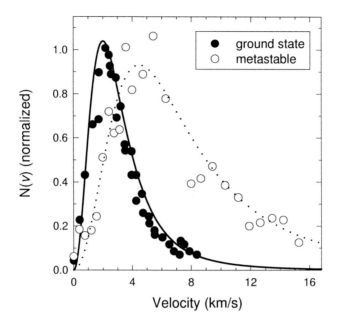

Figure 4.11: Velocity distribution of sputtered iron atoms in the ground state (•) and in the metastable $a^5 F_5$ state (○) following 10 keV Ar^+ bombardment of Fe [30]. The solid and dashed lines are corresponding fits according to Eq. 4.47 (see text).

nance tunneling may be prevented by oxidation of a metal or semi-conductor surface leading to the formation of a band gap in the solid. The resonance tunneling model may be incorporated into the Thompson formula by adding an exponential escape factor which depends on the velocity component $v_\perp$ perpendicular to the surface,

$$\frac{dY(E_0, E)}{dE} \propto \frac{E}{(E + U_0)^{3-2m}} \exp\left(-\frac{A}{av_\perp}\right), \quad (4.48)$$

allowing only those atoms to escape from the surface without being ionised if their velocity is sufficiently great. Here A and a are constants which depend on the transition probability for such a non-radiative transition and the interaction distance. Another striking feature observed in such experiments is the very large Doppler broadening of spectral lines from excited Al^{2+} ions. In this case the broadening of the ionic line is found to be almost double that of the neutral atomic line. Moreover, when the ionic line profile is fitted to Eq. 4.48, with U_0 taken as a free parameter, one obtains a rather high value of $U_0 = 500$ eV [33, 34]. These observations imply that an appreciable fraction of the excited Al^{2+} ions has kinetic energies of more than 1 keV. This is considerably larger than the kinetic energies with which the excited neutral Al atoms leave the surface and may be taken as evidence of different production mechanisms for excited Al atoms and Al^{2+} ions. Presumably, the Al^{2+} ions are produced in close encounters and, hence, in more violent encounters than the neutral atoms which are believed to result from the sputtering cascade.

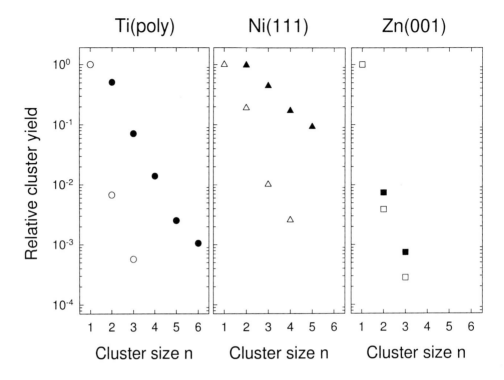

Figure 4.12: Relative yield of neutral (open symbols) and positively charged (closed symbols) Ti_n ($\circ$), Ti_n^+ ($\bullet$), Ni_n ($\triangle$), Ni_n^+ ($\blacktriangle$), Zn_n ($\square$), and Zn_n^+ ($\blacksquare$) clusters *vs.* clusters size n following 5 keV Ar^+ impact of polycrystalline Ti, Ni(111), and Zn(001) surfaces (after Ref. [35]).

4.4.3 Sputtering of clusters

Beside the emission of atomic particles the emission of small agglomerates or clusters also occurs during sputtering events. Cluster sizes of up to $n = 200$ have been observed. In most cases small cluster sizes dominate, however. Apart from the formation process is the occurrence of clusters also influenced by cluster stability which may give rise to oscillations as a function of cluster size. Fig. 4.12 shows some cluster size distributions obtained during the bombardment of Ti, Ni, and Zn targets with 5 keV Ar^+ ions [36] (after Ref. [35]). In these experiments, the atomic species ($n = 1$) dominate, and the size distributions, with few exceptions, are more or less monotonically falling functions. The small number of higher Zn clusters is attributed to the poor stability of those in comparison to, e.g., Ni or Cu clusters.

4.4.4 Potential sputtering employing highly-charged ions

So far, sputtering has been considered as a kinetic process in which momentum and kinetic energy is transferred from the projectile either directly or indirectly via a collision cascade to the target atoms. It was realised some time ago that the potential energy of, e.g., highly-

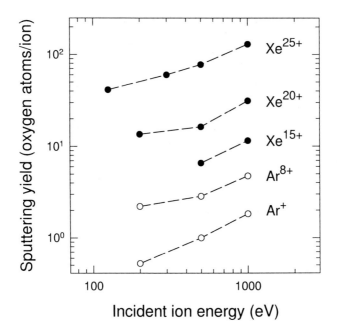

Figure 4.13: Sputtering yield of SiO_2 bombarded by highly-charged Ar^{q+} (open symbols) and Xe^{q+} ions (closed symbols) *vs.* incident energy and for different incident charge states q (after Ref. [37]).

charged ions brought into collision with a solid may also lead to the ejection of atoms, electrons and photons from the surface (e.g., Ref. [12]). This so-called *potential* sputtering has been recently discovered during the interaction of highly charged Ar^{q+} and Xe^{q+} ions (q is the incident charge state) with insulating LiF and SiO_2 surfaces [37]. In such sputtering events more than a 100 atoms may be removed per incident highly charged ion (Fig. 4.13). The physical mechanism leading to such a high sputtering yield is not yet clear, however. Two distinct potential sputtering models are discussed in Ref. [37]. One is the so-called *Coulomb explosion* model, while the other is called the *defect-mediated sputtering* model. The Coulomb explosion model is based on rapid electron depletion at the surface by the approaching highly charged ion and the breaking up of molecular bonds among the surface atoms. This model has attracted a fair amount of interest in the past. However, no firm evidence for enhanced sputtering due to the Coulomb explosion process has been demonstrated yet. The *defect-mediated sputtering* model is based on valence band excitation by the approaching highly charged ion and the formation of *self-trapped excitons*. This mechanism would work for SiO_2 surfaces but not for Si and could thus provide a novel cleaning procedure for Si surfaces [37].

4.5   Electron emission

The interaction of sufficiently energetic particles (e.g., neutrals, ions, electrons) from the plasma with a solid surface may give rise to the emission of electrons. The processes leading to the ejection of electrons may be subdivided, as for the sputtering process, into *kinetic* and *potential* processes. A *kinetic* process is due to the exchange of kinetic energy and momentum between the incident particle and the electrons inside the solid. The other process is *potential emission* of an electron due to a surplus of potential energy of an incoming particle. Potential emission results from Auger-type transitions at the surface involving conduction band electrons that are transferred to high-lying vacant states of the projectile. The potential emission process is, hence, restricted to particles which carry a large amount of potential energy into a collision and thus to atomic particles which are either (singly or multiply) ionized or excited.

Electron emission by the kinetic process is considered to be a three-stage process [38]:

- projectile interaction with electrons within the solid creating freely mobile excited electrons,

- migration of some of the excited electrons to the surface of the solid, where

- some of the electrons escape through the potential barrier at the surface into the vacuum.

As a consequence, the electron emission coefficient δ may be expressed as

$$\delta = D_e(E, x, \cos\theta)\,\Lambda\,, \tag{4.49}$$

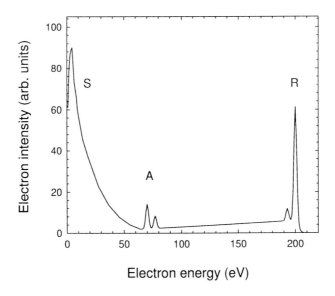

Figure 4.14: Typical energy distribution of electrons emerging from a solid following electron bombardment.

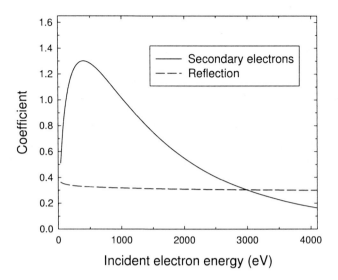

Figure 4.15: Calculated secondary electron coefficient δ (solid line, Eq. 4.53) and reflection coefficient η (dashed line, Eq. 4.50) of electrons incident on iron *vs.* incident electron energy.

where D_e corresponds to the energy deposited in electronic excitation in an interval $[x, x+dx]$ inside the solid, and x is the distance from the surface. D_e is, hence, very much dependent on the projectile parameters, in particular, the incident energy E_0 and the angle of incidence θ with respect to the surface normal. The parameter Λ describes the motion of the excited electrons within the solid and the escape of these electrons into the vacuum. As such, Λ is a material-dependent factor and largely independent of the parameters of the projectile.

The potential ejection process involves near-resonant electron transfer from the solid to high-lying states of the projectile, followed by a non-radiative (Auger type) de-excitation process. The surplus energy is transferred to another electron that can escape into the vacuum. The potential ejection process works particularly well for highly-charged ions and, since it does not require the exchange of kinetic energy, operates even at low incident energies where it is only weakly dependent on the projectile energy.

4.5.1 Emission of electrons by electron impact

A typical energy spectrum of electrons ejected from a surface during electron impact is shown in Fig. 4.14. The most prominent features are peaks at low and high energies due to the emission of so-called secondary electrons (S) and back-scattered or reflected primary electrons (R), respectively. The structures just below the main back-scattered electron peak are due to reflected primary electrons that have suffered specific inelastic energy losses inside the solid. The peaks labeled A are due to ionization of inner atomic shells and the subsequent filling of these vacancies through emission of Auger electrons.

It is obvious from Fig. 4.14 that the majority of back-scattered electrons have energies

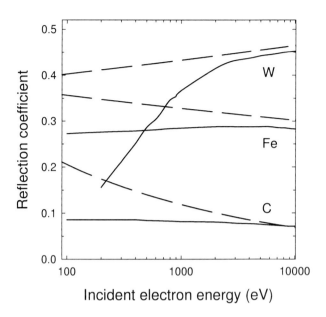

Figure 4.16: Electron reflection coefficient of carbon, iron, and tungsten versus incident electron energy. Solid lines: experimental data from Ref. [38]; dashed lines according to Eq. 4.50.

equal to or somewhat below the primary electron energy, while most secondary electrons have received kinetic energies of only a few electron volts. In between, both distributions overlap and there is no way of distinguishing between back-scattered primary electron and ejected secondary electrons. It has thus been considered convenient to count all electrons just below 50 eV as secondary electrons and those above as back-scattered primary electrons [38]. The energy dependence of the electron coefficient and of the reflection coefficient on the incident electron energy is displayed in Fig. 4.15. A pronounced energy dependence is observed for the electron emission coefficient with a material-dependent maximum around several 100 eV. A much weaker energy dependence is observed for the reflection coefficient, which is largely independent of the incident energy over a wide range.

4.5.1.1 Reflection of electrons from surfaces

Electron reflection coefficients for three different materials (carbon, iron, tungsten) are displayed in Fig. 4.16. As mentioned above, the electron reflection coefficient η is rather weakly dependent on the incident energy. Above 4 keV, the measured energy dependencies are fairly well reproduced by an empirical fit formula due to Hunger and Kuchler [39],

$$\eta = E_0^{m(Z)} \exp(C(Z)), \qquad (4.50)$$

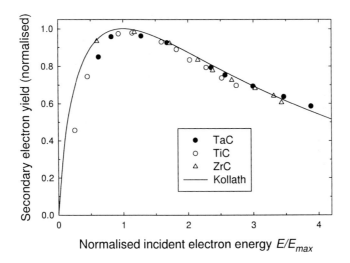

Figure 4.17: Yield of electrons following electron bombardment of TaC ($\bullet$), TiC ($\circ$), and ZrC ($\triangle$), from the work of Thomas and Pattison [40], after Ref. [38]). The solid line shows the prediction of Kollath [41] given by Eq. 4.53.

where

$$m(Z) = 0.1382 - 0.9211 Z^{-0.5}, \tag{4.51}$$

$$\exp(C(Z)) = 0.1904 - 0.2236 \ln Z + 0.1292 \ln^2 Z - 0.01491 \ln^3 Z, \tag{4.52}$$

and Z is the atomic number of the investigated material.

4.5.1.2 Emission of secondary electrons by electron impact

The secondary electron emission coefficient δ for impact of electrons is shown in scaled form for various composite materials [40] in Fig. 4.17. As has been noted previously, the secondary emission coefficient when suitably scaled follows a universal curve that, according to Kollath [41], may be expressed as

$$\frac{\delta}{\delta_{max}} = 2.72^2 \frac{E_0}{E_{max}} \exp\left(-2\left(\frac{E_0}{E_{max}}\right)^{0.5}\right), \tag{4.53}$$

where δ_{max} is the maximum of the electron coefficient, and E_{max} the incident electron energy at which the maximum occurs. It is noted from the comparison in Fig. 4.17 that the agreement between experimental results and Eq. 4.53 is excellent in general. Typical values for δ_{max} lie in the range 0.5 (for carbon) to 1.4 (for tungsten), while typical values for E_{max} vary between 200 eV (for beryllium) and 650 eV (for tungsten), c.f., Ref. [38]. Tabulated values for δ_{max} and E_{max} may be found in Refs. [38, 42].

4.5.2   Emission of electrons by ion impact

Secondary electron emission by ion impact is governed by the kinetic emission process for sufficiently large incident velocities. A typical energy dependence of the secondary electron yield during the bombardment of Al and Au targets by H^+ ions is shown in Fig. 4.18 over a wide range of incident energies. The maximum of the secondary electron yield occurs around 100 keV where, depending on the material bombarded, about 1 to 3 electrons per incident ion are emitted.

The secondary electron yield decrease towards low (and high) incident energies. Fig. 4.19 displays the secondary electron yield at low incident energies and for different (singly and multiply charged) projectiles. Large differences of several orders of magnitudes are already noted among the singly charged H^+, He^+, Ar^+, and Xe^+ ions. Furthermore, the data show a tendency to become constant for very low incident energies rather than to display a threshold energy below which the electron yield drops to zero. This behaviour, which is most pronounced for He^+ ions, cannot be explained by kinetic emission of secondary electrons. The threshold velocity v_{th} of incident ions below which no electron emission by the kinetic process should occur can be estimated from the maximum possible energy transfer in an ion-electron

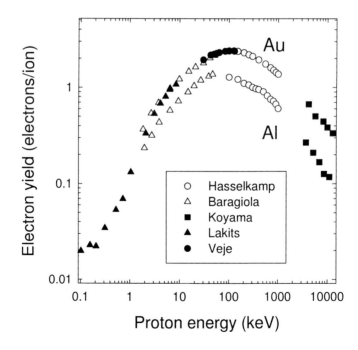

Figure 4.18: Secondary electron yield of H^+ ions on gold (upper curves) and aluminum surfaces (lower curves). Experimental data are from Lakits et al. [43] (▲), Baragiola et al. [44] (△), Veje [45] (●), Hasselkamp et al. [46] (○) and Koyama et al. [47] (■).

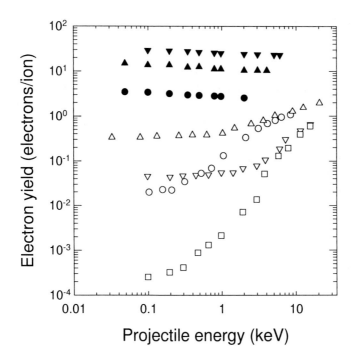

Figure 4.19: Secondary electron yield of singly-charged H^+ (○), He^+ (△), Ar^+ (▽) and Xe^+ (□) ions and multiply-charged Ar^{5+} (●), Ar^{10+} (▲) and Ar^{15+} (▼) ions on gold. Data for singly-charged ions (open symbols) are from Lakits et al. [43], those for multiply-charged ions (closed symbols) from Kurz et al. [48].

collision as [43]

$$v_{th} = -\frac{v_F}{2} + \left(\frac{v_F^2}{4} + \frac{W_\phi}{2m_e}\right)^{\frac{1}{2}}, \tag{4.54}$$

where v_F is the Fermi velocity of (free) conduction band electrons of the solid, m_e the electron mass, and W_ϕ the work function. Eq. 4.54 when applied to polycrystalline gold yields a threshold energy value of ≈ 300 eV/amu, i.e., threshold energies of 1.2 keV, 12 keV, and 38 keV for He, Ar, and Xe projectiles, respectively. The observed secondary electron yields below the respective threshold energy are thus attributed to potential emission. This idea is further corroborated by observations that show that the electron yield increases with increasing charge state of the projectile [48] (Fig. 4.19), while the electron yields of neutral projectiles, e.g., He, Ar, or Xe, that do not carry potential energy are significantly reduced or even orders of magnitude smaller than comparable yields for the corresponding singly-charged ions [49].

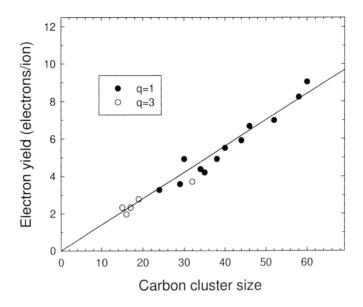

Figure 4.20: Secondary electron yield of gold bombarded by singly-charged (○) and triply-charged (●) carbon cluster ions C_n^{q+} *vs.* cluster size n. The impact velocity was 1.2×10^5 m/s (after Ref. [50]).

4.5.3 Emission of electrons by cluster impact

Studies of the interaction of large clusters with surfaces have recently become feasible. For example, Winter et al. [50] studied electron emission during the impact of C_n^{q+} ions ($n \leqslant 60$, $q = 1-5$) on polycrystalline gold surfaces. The results are displayed in Fig. 4.20 as a function of cluster size. It is noted that the electron yield rises approximately linearly with cluster size n, this being attributed to the kinetic emission of electrons, since simultaneously the kinetic energy of the cluster ion was increased. No difference between impact of differently charged clusters was observed. The apparent suppression of potential electron emission is explained by the complex structure of the impinging clusters and their high electron mobility. As such, electrons on the carbon cluster move apart rather quickly staying as far away from each other as possible. This reduces the probability for autoionisation and, hence, leads to a drastic suppression of the potential emission process.

4.6 Chemical effects

4.6.1 Chemical sputtering and plasma etching

Unlike physical sputtering, where the sputtered particles receive sufficient kinetic energy to overcome the surface binding energy via collisions with other energetic particles, surface erosion by chemical effects relies on chemical reactions by which the chemical bonds of sur-

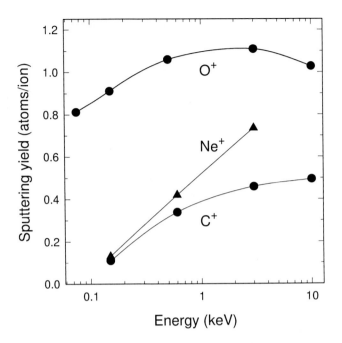

Figure 4.21: Sputtering yield of carbon by low-energy C^+, O^+, and Ne^+ ions versus incident energy (after Ref. [51]).

face atoms with their neighbors are broken. Chemical effects are thus rather sensitive to the chemical affinity (reactivity) of the agent-surface combination and may show a pronounced temperature dependence. For this reason a general picture of chemical erosion processes is much more difficult to achieve and no simple picture has emerged yet. However, chemical sputtering is generally considered to be a multi-step process finally leading to the formation of a volatile molecule which escapes into the gaseous phase. Since the formation of these molecules may be exothermic, no minimum energy transfer is required for chemical sputtering. Fig. 4.21 shows the sputtering yield of carbon by low-energy C^+, O^+ and Ne^+ ions. Whereas the sputtering yield for C^+ and Ne^+ ions has a threshold at about 100 eV incident energy, the threshold energy appears to be much lower or even zero for O^+ ion bombardment. Whereas C^+ and Ne^+ ions are subject to physical sputtering, O^+ ions forming stable molecules with C (e.g., CO, CO_2) in addition give rise to chemical sputtering, particularly at low energies where physical sputtering is weak. This example also demonstrates the selectivity of chemical sputtering and its dependence on a particular agent-substrate combination. As a further significance of chemical sputtering we mention the pronounced temperature dependence of the sputtering yield that is frequently observed. For example, Roth et al. [51] measured the temperature dependence of the sputtering yield of Si by 300 eV H^+ ions and the formation of SiH_4 molecules with a mass spectrometer. The measurements indicate that chemical formation of volatile SiH_4 molecules is the major erosion process. Further evidence for chemical sputtering is inferred from measurements of the kinetic energy of the sputtered

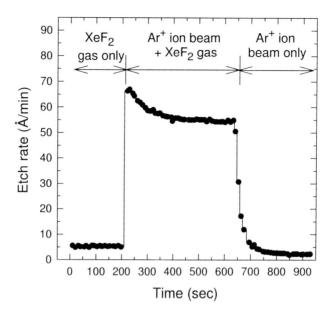

Figure 4.22: Etch rate of polycrystalline silicon under the influence of XeF_2 gas, 450 eV Ar^+ ion impact, and under the combined influence (after Ref. [52]).

particles. While the kinetic energy distribution of particles ejected by physical sputtering is well described by Eq. 4.46, the energy distribution of atoms due to chemical sputtering should be controlled by thermal desorption and thus be in accordance with

$$\frac{dY(E_0, E)}{dE} \propto E \exp\left(-\frac{E}{kT}\right) . \tag{4.55}$$

The measured kinetic energy distribution of the sputtered molecules was indeed significantly different, i.e., shifted to lower energies, compared to that expected for pure physical sputtering. Nevertheless, at the higher kinetic energies above about 1 eV the results followed the E^{-2} dependence which is consistent with the physical sputtering picture provided that a surface binding energy as low as 0.3 eV and, hence, significantly lower than the regular binding energy was assumed. Towards lower energies there have been indications for an additional thermal desorption consistent with chemical sputtering.

Chemical and chemically enhanced sputtering have important applications, for example, in dry etching in computer chip production or plasma cleaning (see chapter 18). In order to achieve the small structures required for the most advanced computer technologies, a high directionality together with a good etching rate is required. Fig. 4.22 shows results of Coburn [52] which compare chemical etching of silicon by XeF_2 molecules, sputtering of Si by 450 eV Ar^+ ions, and the combined effect, the latter resembling the situation in a real plasma where chemically active molecules and energetic ions simultaneously interact with surfaces. The comparison shows that neither chemical etching by XeF_2 nor physical sputtering by Ar^+

ions provides high erosion rates, in contrast to the combined interaction of XeF$_2$ and energetic Ar$^+$ ions giving $\approx$ 10 times larger rates with the simultaneous benefit of a high directionality. This synergetic effect most probably has to do with the activation of chemical reactions due to Ar$^+$ impact which enable the formation of volatile SiF$_4$ molecules.

References

[1] M. Inokuti, Int. J. Quantum Chem. **57**, 173 (1996)

[2] International Commission on Radiation Units and Measurements, ICRU Report No. 49, Bethesda, Maryland (1993)

[3] P. Sigmund, in *Sputtering by Particle Bombardment I*, R. Behrisch, Ed., Topics in Applied Physics 47, Heidelberg: Springer-Verlag, p. 9 (1981)

[4] J. Bohdansky, Nucl. Instr. Meth. Phys. Res. **B2**, 587 (1984)

[5] W. Eckstein, J. Bohdansky, J. Roth, in *Atomic and Plasma-Material Interaction Data for Fusion*, Nuclear Fusion, Vol. 1, Vienna: International Atomic Energy Agency, p. 51 (1991)

[6] H. Bethe, Ann. Physik **5**, 325 (1930)

[7] H. Paul, M.J. Berger, H. Bichsel, in *Atomic and Molecular Data needed in Radiotherapy*, Vienna: International Atomic Energy Agency IAEA-TECDOC (1994), Chapter 7

[8] T. Kaneko, At. Data Nucl. Data Tabl. **53**, 271 (1993)

[9] J.F. Ziegler, J.P. Biersack, U. Littmark, *The Stopping and Ranges of Ions in Matter*, Vol. 1, New York: Plenum (1985)

[10] R.F. Egerton, *Electron Energy-loss Spectroscopy in the Electron Microscope*, New York: Plenum (1996)

[11] J. Lindhard, V. M. Scharff, Phys. Rev. **124**, 128 (1961)

[12] A. Arnau, F. Aumayr, P.M. Echenique, M. Grether, W. Heiland, J. Limburg, R. Morgenstern, P. Roncin, S. Schippers, R. Schuch, N. Stolterfoht, P. Varga, T.J.M. Zouros, HP. Winter, Surf. Sci. Rep. **27**, 113 (1997)

[13] W. Eckstein, *Computer Simulations of Ion-Solid Interactions*, Springer Series in Materials Science 10, Heidelberg: Springer-Verlag (1991)

[14] W. Eckstein, G. Garcia-Rosales, J. Roth, W. Ottenberger, *Sputtering Data*, Max-Planck-Institut für Plasmaphysik: Garching, Report IPP 9/82 (1983)

[15] M.M. Bredov, I.G. Lang, M.N. Okuneva, Sov. Phys. - Tech. Phys. **3**, 228 (1958)

[16] M.T. Robinson, O.S. Oen, Phys. Rev. **132**, 2385 (1963)

[17] J.P. Biersack, L.G. Haggmark, Nucl. Instr. Meth. **174**, 257 (1980)

[18] W. Möller, W. Eckstein, Nucl. Instr. Meth. **B2**, 814 (1984)

[19] J.P. Biersack, W. Eckstein, Appl. Phys. **34**, 73 (1984)

[20] L. Verlet, Phys. Rev. **159**, 98 (1967)

[21] D. Frenkel, B. Smit, *Understanding Molecular Simulation*, Academic Press: San Diego (1996)

[22] O.B. Firsov, Sov.–Phys. JETP **6**, 534 (1958)

[23] P.M. Morse, Phys. Rev. **34**, 57 (1929)

[24] J.E. Lennard, I. Jones, Proc. Roy. Soc. **A 106**, 441, 463 (1924)

[25] W. Eckstein, S. Hackel, D. Heinemann, B. Fricke, Z. Physik **D 24**, 171 (1992)

[26] M.S. Daw, M.I. Baskes, Phys. Rev. Letters **50**, 1285 (1983); Phys. Rev. **B 29**, 6443 (1984)

[27] Z.B. Güvenç, Y. Hundur, R. Hippler, Nucl. Instr. Meth. Phys. Res. **B 164–165**, 854 (2000)

[28] M.W. Thompson, Phil. Mag. **18**, 377 (1968)

[29] G. Betz, K. Wien, Int. J. Mass Spectr. Ion Processes **140**, 1 (1994)

[30] B. Schweer, H.L. Bay, Appl. Phys. **A 29**, 53 (1982)

[31] H. D. Hagstrum, Phys. Rev. **96**, 336 (1954)

[32] D. Ghose, R. Hippler, in *Luminescence of Solids* (D.R. Vij, Ed.), Plenum: New York, p. 189 (1998)

[33] S. Reinke, D. Rahmann, R. Hippler, Vacuum **42**, 807 (1991)

[34] R. Hippler, S. Reinke, Nucl. Instr. Meth. **B68**, 413 (1992)

[35] W.O. Hofer, in *Sputtering by Particle Bombardment III* (R. Behrisch, K. Wittmark, Eds.), Topics in Appl. Phys. **64**, 1991), p. 15

[36] H. Gnaser, W.O. Hofer, Appl. Phys. **A 48**, 261 (1989)

[37] M. Sporn, G. Libiseller, T. Neidhart, M. Schmid, F. Aumayr, HP. Winter, P. Varga, M. Grether, D. Niemann, N. Stolterfoht, Phys. Rev. Letters **79**, 945 (1997)

[38] E.W. Thomas, in *Atomic and Plasma-Material Interaction Data for Fusion*, Vol 1, IAEA: Vienna (1991), p.93

[39] H.J. Hunger, L. Kuchler, phys. stat. sol. **A 56**, K45 (1979)

[40] E.B. Pattison, S. Thomas, J. Phys. **D 3**, 349 (1970)

[41] R. Kollath, in *Handbuch der Physik* (S. Flügge, Ed.), Band XXI, Springer: Berlin (1956), p. 232

[42] *CRC Handbook of Chemistry and Physics*, 80th Edition, D.R. Lide, Ed., CRC Press: Boca Raton (1999)

[43] G. Lakits, F. Aumayr, M. Heim, HP. Winter, Phys. Rev. **A 42**, 5780 (1990)

[44] R.A. Baragiola, E.V. Alonso, J. Ferròn, A. Oliva-Florio, Surf. Sci. **90**, 240 (1979)

[45] E. Veje. Nucl. Instr. Meth. **194**, 433 (1982)

[46] D. Hasselkamp, K.G. Lang, N. Stiller, A. Scharmann, Nucl. Instr. Meth. **180**, 349 (1981)

[47] A. Koyama, T. Shikata, H. Sakairi, Jpn. J. Appl. Phys. **20**, 65 (1981)

[48] H. Kurz, F. Aumayr, C. Lemell, K. Töglhofer, HP. Winter, Phys. Rev. **A 48**, 2182 (1993)

[49] Z. Šroubek, Phys. Rev. Letters **78**, 3209 (1997)

[50] HP. Winter, M. Vana, G. Betz, F. Aumayr, H. Drexel, P. Scheier, T.D. Märk, Phys. Rev. **A 56**, 3007 (1997)

[51] J. Roth, J. Bohdansky, W. Ottenberger, IPP-Report 9/26, Garching (1979)

[52] J.W. Coburn, Appl. Phys. **A 59**, 451 (1994)

5 Modeling of plasma-wall interaction

Holger Kersten

Institut für Physik, Ernst-Moritz-Arndt-Universität Greifswald, Domstraße 10a,
D-17487 Greifswald, Germany

5.1 Introduction

At present, plasma-wall interaction is one of the fastest growing branches in plasma physics and has got a leading position in the field of applied surface science. But its basic question includes the mastering of an old problem: the contact of different states of matter. The investigation of plasma-wall interaction plays an essential role in low temperature, low-pressure plasma processing as etching, deposition, or modification of surfaces as well as in fusion research.

The characterization of plasmas in contact with solid surfaces is incomparably much more complicated than that of unbounded plasmas. It requires the consideration of marked anisotropies and inhomogeneities, deviations from quasi-neutrality, exchange of energy and matter, and may result in strongly non-linear equations containing many unknown quantities. The aim of understanding of plasma-surface interaction is preferably an extensive description of the parameters which characterize the plasma and the solid surface (substrate) by taking into consideration the particle and energy balances of the involved species.

The complex processes at etching or deposition of thin surface layers can be described as macroscopic expressions between the several rates and the corresponding experimental conditions which are realized in the plasma reactor. The experimental process parameters are gas mixture, gas flow, pressure, discharge power, reactor geometry, temperature etc. On the other hand, the overall rates involve a variety of elementary processes, which have to be described by models on a micro-physical scale.

Elementary processes in plasmas – in particular during the interaction with walls – are rather complicate, and often the experimental separation of the single phases of the involved mechanisms is very difficult. Depending on nature and intensity of the particle fluxes from the gas phase (plasma), on the energy distribution of the incoming particles, and on the thermal conditions as well as on the electrical potentials at the substrate surface different mechanisms are dominating. For example, at physical sputtering (PS) one has to investigate mainly direct interactions between energetic ions and solids. Even in this case an exact description is quite difficult but it becomes nearly impossible for surface film reactions which proceed via molecular intermediate stages. Examples for such reactions are ion beam assisted etching (IBAE), plasma enhanced chemical vapor deposition (PECVD), and plasma polymerization (PP). The mentioned intermediate stages may include adsorption, surface diffusion, activation, reaction, and desorption of the different species.

5.2 Characterization of the elementary mechanisms in low-temperature plasma processing

Particles from the gas phase, which collide with a solid surface, can undergo several interactions [1]. They can be simply reflected at the wall or they can be adsorbed by forming a bond to surface atoms. After adsorption the particles may diffuse along the surface or into the solid (bulk diffusion). Furthermore, they can react with already adsorbed species or with surface atoms of the solid material in order to form product molecules. Finally, depending on the binding energy of the particles, they may either desorb back again into the gas phase or stick by forming a growing film.

5.2.1 Adsorption, desorption, diffusion

Many surface processes proceed via an adsorption precursor as a first step of the interaction between gas/plasma and solid. If a particle originating from the gas phase looses sufficiently enough energy, it can be bound at the surface.

The bonds are of electromagnetic nature. In the case of rather weak physisorption the bonds are mediated by electrostatic forces as dipole-dipole-interaction or dispersion forces, respectively. The released adsorption heat is in the order of $E_{ads} < 0.5$ eV. Typical examples for physisorption are the adsorption of noble gases on metals at very low temperatures by van der Waals interaction (example Ar on Zr: $E_{ads} = 0.24$ eV or -25 kJ/mole) or the adsorption of nitrogen molecules on most surfaces. The other binding type which is called chemisorption is mainly caused by valence forces of the exchanging electronic orbitals of adsorbed particles and substrate atoms. In this case, the incoming particles form a strong chemical bond to the surface ($E_{ads} > 0.5$ eV) as for example in the adsorption of CO on metals (example CO on Pd: $E_{ads} = 1.43$ eV or -147kJ/mole) or the adsorption of oxygen on most surfaces. The incoming molecules can also dissociate by approaching the surface and the fragments are adsorbed. This process of dissociative chemisorption occurs, for example, at passivation of metals or silicon by hydrogen molecules. In Fig.5.1 some typical mechanisms for the $Si/SiH_4/H_2$-system are illustrated schematically.

The potential for adsorption above a real substrate surface consists usually of a physisorbed precursor and a chemisorbed state. A qualitative potential diagram for the adsorption of a diatomic molecule is shown in Fig.5.2. Depending on the energy of the incoming particle, it gets first trapped in the weak precursor and it either chemisorbs by overcoming the activation barrier between both states or it desorbs back into the gas phase. If the activation barrier between the physisorbed and chemisorbed state is above zero, the process is an activated dissociative chemisorption. The rate of adsorption R_{ads} of a species A on a surface B is determined by the particle flux j_A and the sticking coefficient γ_A of A on B:

$$R_{ads} = dn_{ads}/dt = \gamma_A j_A f(\Theta) . \tag{5.1}$$

The sticking coefficient γ_A describes the probability of the incoming particle to loose its kinetic energy and get trapped in the potential well [2]. For example, the sticking coefficient of NO on Pt is close to unity (~ 0.9), while the sticking coefficient of molecular hydrogen on a wall is rather low due to the inefficient energy transfer in this collision. This is also the reason for the long residence time of hydrogen in a plasma reactor volume. The particle flux j_A is

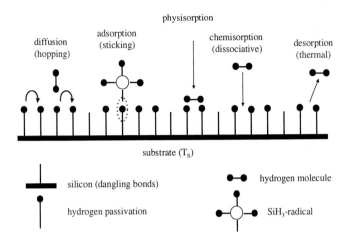

Figure 5.1: Several elementary gas-surface processes during treatment of silicon by a SiH$_4$/H$_2$ plasma.

related to the partial gas pressure p_A of the species A and its gas temperature T_A according to:

$$j_A = \frac{p_A}{\sqrt{2\pi m_A k T_A}} \, ,$$
(5.2)

where m_A is the particle mass, and k the Boltzmann constant. Especially in chemisorption, the sticking coefficient depends on different quantities as impact angle and kinetic energy of the striking particle, respectively, and is usually calculated by molecular dynamics. Furthermore, the adsorption process is characterized by the so called *degree of coverage*: $\Theta = n_{ads}/n_0$, which represents the ratio of the surface density of adsorbed particles (n_{ads}) to the total density of available surface sites (n_0). In general, the coverage is a function $f(\Theta)$ of surface morphology, vacant sites etc. In the most simple case of physisorption the function $f(\Theta)$ reflects the fraction of vacant sites, i.e., $f(\Theta) = 1 - \Theta$.

During their residence on the substrate the physisorbed particles may diffuse and come into positions of stronger bonds. The migration of atoms or molecules along the surface is one of the most important elementary steps of gas-surface interaction, reactive or non-reactive. The main reason for surface diffusion is due to the crystal structure and chemical composition of the substrate. Depending on the structure of the crystal face, diffusion coefficients may vary by orders of magnitude. Diffusion rates parallel to steps are greater than diffusion rates perpendicular to them. Both diffusion along the surface and diffusion into the near-surface layers of the solid is possible. In this case one speaks of absorption or bulk diffusion, respectively.

The adsorbed particles may leave the surface after a certain residence time τ_{des} which is a statistical quantity. The energy of desorption E_{des} is delivered by the lattice in the case of thermal desorption or by collisions with energetic particles as it is the case in collisional

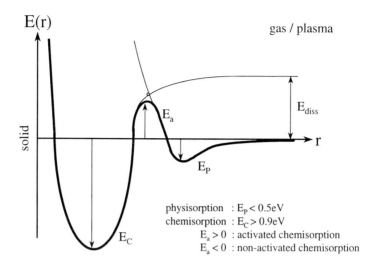

Figure 5.2: One-dimensional potential energy diagram showing the possible transition from molecular physisorption to dissociative chemisorption. E_p and E_c are the desorption energies from a physisorbed and chemisorbed, respectively, state.

desorption. The residence time in thermal desorption is defined by the Frenkel equation, where the surface temperature T_S becomes important:

$$\tau_{des} = \tau_0 \, \exp(E_{des}/kT_s) \,. \tag{5.3}$$

The pre-factor τ_0 is the reciprocal characteristic vibration frequency in the order of 10^{-12} s. During collisional induced desorption (e.g. by photons, electrons or ions) the residence time depends on the flux of these energetic particles j_X and a characteristic cross section σ_X:

$$\tau_{des} = (\sigma_X j_X)^{-1} \,. \tag{5.4}$$

Very similar to adsorption, desorption into the gas phase may be described by a desorption rate R_{des} which can be expressed as

$$R_{des} = dn_{ads}/dt = -n_{ads}/\tau_{des} \,. \tag{5.5}$$

In equilibrium, there is a balance between the elementary processes of adsorption and desorption. Solutions of the adsorption-desorption equilibrium result in adsorption isotherms which reflect the coverage-pressure dependence at a constant substrate temperature. In the most simple case, where every surface site is equivalent and only single monolayer adsorption occurs without any adsorbate-adsorbate interactions, one obtains the rate equation

$$dn_{ads}/dt = R_{ads} + R_{des} = \gamma_A j_A(1 - \Theta) - n_{ads}/\tau_{des} = 0 \,. \tag{5.6}$$

In this case one gets the so called Langmuir isotherme:

$$\Theta = \frac{n_{ads}}{n_0} = \left(1 + \frac{n_0}{\gamma_A j_A \tau_{des}}\right)^{-1} . \tag{5.7}$$

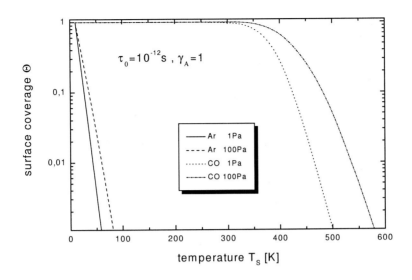

Figure 5.3: The initial coverage of Ar on Zr ($E_{des} = 0.24\,\text{eV}$) and CO on Pd ($E_{des} = 1.43\,\text{eV}$) in dependence on the surface temperature. The surface density is assumed to $n_0 = 6 \times 10^{14}$ cm^{-2}.

Since the balance depends essentially on the substrate temperature, Fig. 5.3 shows corresponding Langmuir-isobares $\Theta = \Theta(T_S)$ as functions of temperature for different process pressures and gases.

5.2.2 Reflection, activation, mixing

The plasma particles striking the solid surface exhibit a wide energy spectrum. In addition to low-energetic neutrals (atoms, molecules, radicals) there exist also particles with higher energy ($E \gg 1$ eV), which can be distributed on the different excitation modes.

If the energy of an incoming particle is too high for entering an adsorption process, or if the energy is too small for sputtering or implantation, respectively, or if it finds no suitable surface site, the particle will be simply reflected at the surface. Due to the energy transfer at such elastic scattering the substrate is heated until a thermal equilibrium between gas phase and solid surface is reached. A characteristic quantity for the energy exchange is the accommodation coefficient α:

$$\alpha = \frac{E_{in} - E_f}{E_{in} - E_S}. \tag{5.8}$$

E_{in} is the energy of the incident particle, E_f marks the energy of the scattered particle and $E_S = kT_S$ the thermal energy of the surface [1]. Scattering experiments of inert gases from a graphite surface have shown that especially for hot surfaces the energy transfer is incomplete. For diatomic or polyatomic molecules, in addition to changes of kinetic energy (translation),

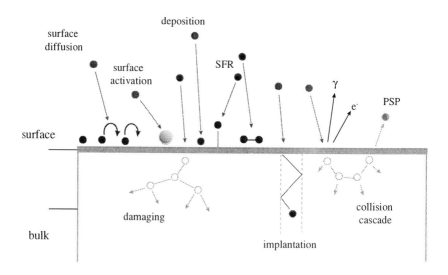

Figure 5.4: Processes which are enhanced by energetic particle bombardment. SFR and PSP denote surface film reaction and physical sputtering, respectively.

the transfer involves de-excitation or excitation of vibrational and rotational modes during collision with the surface.

On the other hand, an energetic impact can induce a collision cascade in the lattice of the surface layer. The bindings to the neighbour atoms are broken due to the cascades, and the generated dangling bonds are preferential positions for surface reactions. The reactivity of the surface atoms can be extraordinarily increased by such an activation which is also called *mixing* (Fig.5.4).

5.2.3 Physical sputtering and implantation

An important example for the influence of high-energetic charge carriers (ions) on surfaces is the physical sputtering (PS). Physical sputtering is the erosion of solid atoms, which received sufficiently high momentum and energy in order to overcome their binding forces in the lattice and which become separated particles in the gas phase, see Fig. 5.4 and 5.5. Sputtering of metal targets has been investigated to a large extent [3, 4]. In this case one distinguishes three different regimes: single-knock-on-sputtering, linear cascades, and spikes. The sputter rate R_{PS} depends on the flux of incident ions j_x and a characteristic surface cross section σ_{PS}:

$$R_{PS} = \frac{dn_B}{dt} = -n_B < \sigma_{PS} > j_x = - < Y_B > j_x \, . \tag{5.9}$$

The sputtering yield Y, which is the ratio between released particles to incident particles (see chapter 4), is determined by the discharge conditions and the target properties. The discharge

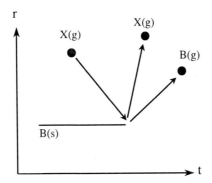

Figure 5.5: Interaction graph of a physical sputtering process $B(s) + X(g) \rightarrow B(g) + X(g)$. B denotes a bulk atom, X an incident energetic particle, while s or g denote a bound (solid) or free (gaseous) atom.

conditions as the electric field (e.g., cathode fall in direct current glow discharges, or sheath width in radiofrequency discharges) and the pressure (e.g., mean free path, collisions in the gas phase) influence the energy of the incident ions, while the chemical composition and the crystal structure determine the bindings of the target material.

Furthermore, the angle of incidence and the mass ratio of the ions hitting the target surface as well as the binding energy of the target atoms have an important influence. In connection with the energetic particle bombardment also implantation should be mentioned. Besides mixing and bulk diffusion it is an important method for surface modification and is widely used.

The mathematical treatment of ion induced processes at surfaces is mainly based on collision and scatter theories. There exists a lot of powerful computer codes (e.g., TRIM, TRIDYN [5]) for the computation of sputtering processes (see chapter 4).

5.2.4   Chemical reactions at the surface

An important component of plasma-wall-interaction, which has an extraordinary meaning for the erosion and deposition of thin films, is the chemical component. The chemical processes are due to two generic types: Eley-Rideal process, Langmuir-Hinshelwood process [6, 7].

First of all, in analogy to physical sputtering, one can introduce sputtering under involvement of chemical reactions. This channel is called chemical sputtering (CS): incident particles (A) react immediately (without adsorption) with the surface atoms of the solid (B) or adsorbate under formation of a product molecule AB (Eley-Rideal mechanism). The product molecule may desorb instantly (direct chemical sputtering) or it remains for a certain residence time (delayed chemical sputtering) at the surface. Fig. 5.6 shows the interaction graphs of chemical sputtering. The rate R_{CS} for chemical sputtering is proportional to the flow rate j_A (see Eq. 5.2) of incoming chemically active particles to the surface:

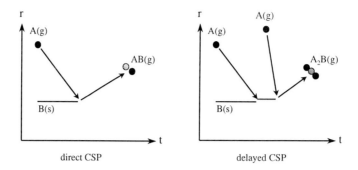

Figure 5.6: Interaction graphs of a chemical sputtering process (CSP) $B(s) + nA(g) \rightarrow A_nB(g)$, where B denotes a bulk atom, A is an incident reactive particle, while s or g denote a bound (solid) or free (gaseous) atom.

$$R_{CS} = \frac{dn_B}{dt} = \varepsilon_{CS} j_A = -n_B \sigma_{CS} j_A, \qquad (5.10)$$

where ε_{CS} denotes the coefficient of gasification which describes the surface reaction probability with the incident species. Generally, the surface reaction probability can be written as the product of a cross section and the surface site density. The cross section is the invariant property of the surface for the reaction under consideration and may be visualized as an area of the surface surrounding the active site upon which the reaction can occur. According to the mechanisms of activated chemical reactions, ε_{CS} contains the probability for the appearance of relative particle energies above a certain limit E_C which results in the well-known

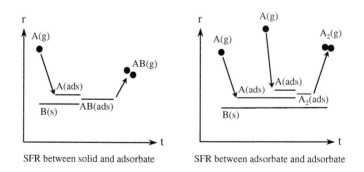

Figure 5.7: Interaction graphs of surface film reactions. Gasification of surface or adsorbate molecules by a surface film reaction (SFR) process may proceed via intermediate steps, for example, via (a) $B(s) + A(ads) \rightarrow AB(ads)$, or (b) $A(ads) + A(ads) + B \rightarrow A_2(ads) + B$. B denotes a bulk atom, A is an incident reactive particle, AB and A_2 are product molecules, while s, ads or g denote a bound (solid), adsorbed or free (gaseous) atom or molecule, respectively.

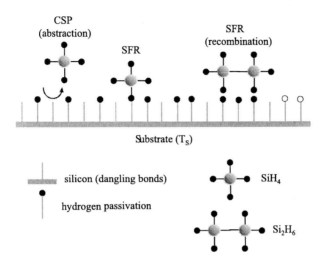

Figure 5.8: Chemical surface processes during treatment of silicon by a SiH$_4$/H$_2$ plasma.

exponential term of an Arrhenius expression:

$$\varepsilon_{CS} = \varepsilon_0 \exp(-E_C/kT_S) \,. \tag{5.11}$$

In contrast to chemical sputtering, the other basic reaction mechanism is surface film reaction (SFR). In this case the adsorption of the arriving particle from the gas phase is a characteristic step, before the chemical reaction takes place within a certain residence time (Fig. 5.7). These reactions are also called recombinations of the Langmuir-Hinshelwood-type. Now the reaction rate R_{SFR} depends primarily on the degree of coverage with adsorbed particles A:

$$R_{SFR} = k_{SFR} n_A n_0 = k_{SFR} n_0^2 \Theta \,. \tag{5.12}$$

At full coverage ($\Theta = 1$) the temperature dependence of R_{SFR} is governed by $k_{SFR}(T_S)$ which shows again an Arrhenius-type behaviour:

$$k_{SFR} = k_0 \exp(-E_{SFR}/kT_S) \,. \tag{5.13}$$

However, with imperfect coverage more complex functions $R_{SFR}(T_S)$ appear.

Chemical sputtering and surface film reaction are the two limiting cases of the chemical plasma-wall-interactions in etching or deposition, respectively. There exist different transitions in between. In practice and on principle, a sharp distinction will be difficult. The different types of chemical reactions at a surface during plasma processing are summarized schematically in Fig. 5.8.

5.3   Modeling of etching and deposition processes

Because of the high level complexity of plasma-wall-interaction, in technological applications approximative methods which permit to some extent generalizations have been favoured. Simultaneously, the interest in the as much as possible exact handling of simplest models has considerably increased.

The sophistication of the phenomena at plasma-wall-interaction is emphasized by the following items:

- combination of chemistry of neutral particles and charge carrier effects,

- simultaneous appearance of erosion and deposition, sometimes even the formation of dust,

- complexity of the plasma and surface structures.

Originating from these points, a lot of problems arise in the optimization of plasma processes and they are often the reason for a discrepancy between empirical use in technological practice and exact physical description.

5.3.1 Particle balance

An essential access to model the processes is the treatment of particle balances of the involved species on the basis of elementary mechanisms [8]. Since a variety of different particles are involved in the surface processes, one has – in principle – to establish a balance for each species. The several species originate on the one hand from the gas/plasma and on the other hand from interactions on the surface. The particle balances of the different particle sorts are linked to each other. However, such a balance equation system can only be solved for very simple and idealized systems. Alternatively, one has to simplify the system by restrictions.

In the following, a simple and idealized example will be discussed in more detail. It will be supposed the etching of a substrate B by means of surface film reaction (SFR) and chemical sputtering (CS). The gas/plasma should only contain the reactive etching component A which forms by reaction with B the product AB, as illustrated in Fig. 5.9. For those particles which originate from the bulk (B) the removal rate equals the etch rate R, and for those particles which arrive from the gas phase (A) the impinging rate must be equal to the removal or conversion rate, respectively:

$$\frac{dn_B}{dt} = -k^B_{SFR}n_An_B - \sigma^B_{CS}n_Bj_A = R$$

$$\frac{dn_{AB}}{dt} = k^B_{SFR}n_An_B + \sigma^B_{CS}n_Bj_A - \frac{n_{AB}}{\tau^{AB}_{des}} = 0 \qquad (5.14)$$

$$\frac{dn_A}{dt} = \gamma_Aj_A\Theta_B - \frac{n_A}{\tau^A_{des}} - k^B_{SFR}n_An_B = 0.$$

Under the assumption that at each B only either one A or one AB can be adsorbed one obtains for the coverage:

$$\Theta_A + \Theta_B + \Theta_{AB} = 1, \qquad (5.15)$$

where $\Theta_A = n_A/n_0$, $\Theta_B = n_B/n_0$, and $\Theta_{AB} = n_{AB}/n_0$. Combining and solving the equations 5.14 and 5.15 one obtains the normalized etch rate R/n_0:

$$\frac{R}{n_0} = \left(\tau^{AB}_{des} + \frac{(\tau^A_{des}\gamma_Aj_A)^{-1} + n_0k^B_{SFR}(\gamma_Aj_A)^{-1} + n_0^{-1}}{k^B_{SFR} + \sigma^B_{CS}\gamma^{-1}_A(\tau^A_{des} + n_0k^B_{SFR})^{-1}}\right)^{-1}. \qquad (5.16)$$

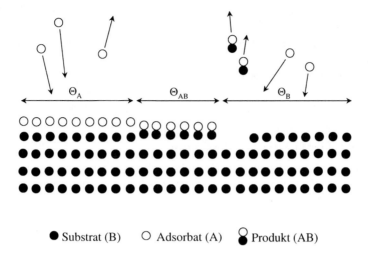

Figure 5.9: A supposed situation for modeling of reactive surface etching.

This expression that is obtained for such a simple example, becomes much more difficult in real systems with more involved components and processes as considered here. Therefore, it is of importance to study some limiting cases of equation 5.16. For instance, at a very low adsorption of A (e.g., $\Theta_A \ll 1$), as it is the case for high substrate temperatures where $\tau_{des}^A \approx 0$, or for low sticking coefficients $\gamma_A \approx 0$, the etch rate becomes

$$\frac{R}{n_0} = \left(\tau_{des}^{AB} + \frac{1}{j_A \sigma_{CS}^B} \right)^{-1}. \tag{5.17}$$

In this case the etch rate R is mainly determined by the chemical sputtering of B and the desorption of the products, respectively. If the formed product molecule AB desorbs immediately, the overall rate depends only on σ_{CS}. Vice versa, in the case of a nearly complete coverage with A, i.e., $\Theta_A \approx 1$, chemical sputtering can be neglected and the surface film reaction dominates:

$$\frac{R}{n_0} = \left(\tau_{des}^{AB} + \frac{n_0}{\gamma_A j_A} + \frac{1}{n_0 k_{SFR}^B} \right)^{-1}. \tag{5.18}$$

At very small flux densities of reactive particles j_A the etching process is transport-limited, whereas at large $\gamma_A j_A$ the process is limited by the surface reactions.

In modeling the global etch mechanism the basic processes mentioned above must be combined and have to be completed by volume processes in the plasma which supply the various fluxes of particles to the substrate surface [9].

5.3.2 Temperature dependence of the processes

As already mentioned, the central question in modeling of the global surface modification processes is the description of relevant rates R in dependence on the process parameters. Among these parameters the substrate temperature T_S is of special interest. A detailed investigation of the temperature dependence of the different rates $R(T_S)$ and surface properties can provide a lot of information on the kinetics and the interaction of the elementary processes of plasma-wall interaction because the different rate coefficients depend on T_S [8].

An interesting question in modeling of $R(T_S)$ is the interpretation of non-monotonous curves. In the case of plasma etching exist mainly three mechanisms for this behaviour:

- The etching or deposition process, respectively, may proceed via different reaction channels which exhibit different temperature dependencies (e.g. surface film reaction, chemical sputtering).

- During the process product molecules are formed which may react or desorb at different temperatures.

- Different sensitivity of the elementary processes against temperature-dependent surface structures of the bulk material.

For example, the appearance of a maximum in the $R(T_S)$ dependence is quite general for etching via surface film reactions. At low temperatures the concentration of reactants in the surface film is nearly saturated (full coverage). In this region where $\Theta \approx 1$, the etch rate increases with increasing temperature according to the Arrhenius rate of a chemical reaction with constant reactants concentration. But at higher temperatures normally the concentration of the surface film decreases due to faster desorption of adsorbed particles. Reaching a maximum the decreasing supply of reactants results, finally, in a decreasing etch rate with further increasing temperature. The appearance of a minimum in the etch rate R can be modeled by the superposition of two or more processes of gasification, e.g., of one mechanism with a maximum at low temperatures together with another mechanism which increases monotonously and dominates at higher temperatures. A possible combination of such mechanisms might be SFR and CS. In the first reaction channel etching occurs by conversion from a precursor weakly bound to the surface, and the latter etching occurs directly by the reactants arriving from the gas phase. A characteristic example for the combination of both reaction channels is the dry etching of silicon by fluorine-containing compounds. A non-monotonous $R(T_S)$-curve due to the different desorption behaviour of the formed products has been observed, for instance, at etching of GaAs by chlorine [10]. Starting at low temperature the etch rate increases because of large surface coverage of reactive species and increasing chemical reaction rates for the formation of $GaCl_3$ and $AsCl_3$. At higher temperature the fast decreasing coverage of the origin species (Cl_2) and of the products (e.g., $GaCl_3$) vanishes and the formation of trichlorides stops. However, at these high temperatures a desorption of the very fast formed layer of monochlorides (AsCl, GaCl) starts. The situation for the total reaction probability ε_T which is directly related to the etch rate $R(\varepsilon_T) = (1/2)R/j_{Cl2}$ is illustrated in Fig. 5.10. The comparison of the measurements (points) with a simple model on the basic assumptions mentioned above yield typical quantities of the etch process like activation energies, residence times, etc.

Summarizing one can conclude, that the several elementary mechanisms of adsorption, diffusion and desorption as well as the chemical reactions (CS, SFR) depend sensitively on the surface temperature T_S of the substrate, while the mechanisms of activation, physical sputtering and implantation (at least in the relevant range) are almost independent of T_S.

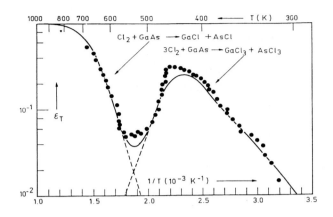

Figure 5.10: Total reaction probability of Cl_2 on GaAs in dependence on the substrate temperature.

5.3.3 Energy balance of the substrate during low-temperature plasma processing

On the other hand one can also obtain information on the energy influx J_{in} from the plasma towards the substrate by evaluation of the substrate temperature T_S. For the energy balance during plasma-wall interaction one has to include the internal heat sources (energetic particle bombardment, radiation, surface reactions etc.) as well as the external sources (heating or cooling of the substrate) [11]. In the following the energy balance in plasma-wall interaction will be discussed shortly:

When a solid comes into contact with a plasma, energy transfer takes place. The substrate is heated and, after a certain time, it may reach a thermal equilibrium. This steady state is determined by a balance of energy gain Q_{in} from the plasma processes and energy losses Q_{out} due to conduction and radiation. The general power balance at the substrate is given by:

$$Q_{in} = \dot{H}_S + Q_{out} , \qquad (5.19)$$

where $\dot{H}_S = C_S \frac{dT_S}{dt}$ denotes the enthalpy of the substrate. It should be mentioned that the flux Q_{in} is the surface integral of the related energy flux density J_{in} over the substrate surface A_S:

$$Q_{in} = \int_{A_S} J_{in} dA . \qquad (5.20)$$

In general, the total energy influx J_{in} is the sum of the fluxes due to the kinetic energy of electrons, J_e, and of ions, J_i, the energy which is released when a positive ion recombines at the surface, J_{rec}, the energy which is supplied when a gas phase species associate with another gas phase species at the surface, J_{ass}, and the energy which is released if chemical reactions occur between gas phase species and the surface, J_{chem}:

$$J_{in} = J_e + J_i + J_{rec} + J_{ass} + J_{chem} . \qquad (5.21)$$

Contributions due to plasma radiation can mostly be neglected because the environment (gas, walls) is nearly at room temperature.

A certain part of energy is delivered by charge carriers to the substrate surface. In general, the mean kinetic ion energy is determined by the ion energy distribution function (IEDF). At elevated pressures the energy distribution of the ions arriving at the substrate is affected by collisions in the sheath in front of the substrate. But at low pressures as it is commonly the case the maximum ion energy is determined mainly by the free fall energy $e_0 V_{bias}$ where V_{bias} is the potential drop across the sheath in front of the substrate which corresponds to the sum of the plasma potential V_{pl} and the substrate potential V_S:

$$V_{bias} = V_{pl} - V_S . \tag{5.22}$$

It should be emphasized that the simple expression of Eq. 5.22 for the mean kinetic energy of the ions striking the substrate is applicable in most cases of plasma processing. Only if the IEDF for the ions near the substrate is much more complex the assumption of $e_0 V_{bias}$ for the kinetic energy is no longer justified.

In addition to the directed kinetic energy of the ions which originate from acceleration in the electrical field in front of the substrate, the ions have also thermal energy. However, this part can be neglected if the ions are nearly at room temperature. Hence, the contribution of the ions, which can be expressed as a product of the ion flux density j_i and their mean energy E_i, is

$$
\begin{aligned}
J_i = j_i E_i &= j_i e_0 (V_{pl} - V_{fl}) = n_e v_{amb} e_0 (V_{pl} - V_{fl}) \\
&= 0.6 n_e \sqrt{\frac{kT_e}{m_i}} \cdot e_0 (V_{pl} - V_{fl}) ,
\end{aligned}
\tag{5.23}
$$

where the ambipolar diffusion $n_e v_{amb}$ has been approximated by the Bohm flux:

$$j_i = 0.6 n_e \sqrt{\frac{kT_e}{m_i}} . \tag{5.24}$$

The substrate has been assumed to be at floating potential, i.e., $V_S = V_{fl}$. Under low pressure conditions ($p < 10\ Pa$) the Bohm equation is applicable, because in the sheath almost no collisions occur. The Bohm equation yields the ion flux j_i by knowing the electron density n_e which equals the ion density n_i at the sheath edge.

In addition to kinetic energy, ions transfer a part of their potential energy when striking a surface. For metallic substrates the neutralization of ions is caused by long-range interactions and may be accompanied by the emission of secondary electrons. The released recombination energy E_{rec} is given by:

$$E_{rec} = E_{ion} - \Phi - \gamma_i \Phi , \tag{5.25}$$

and the resulting contribution J_{rec} to the energy influx to the substrate due to the recombination is

$$J_{rec} = j_i E_{rec} , \tag{5.26}$$

where E_{ion} is the ionization potential of the incident ion, Φ the work function of the metal, and γ_i the yield of secondary electrons. Data for E_{ion}, Φ and γ_i, respectively, may be taken from the literature (see also chapter 4).

In general, the cooling effect due to sputtering of substrate material can be ignored because the energy $e_0 V_{bias}$ of the impinging ions is always smaller than 100 eV and the sputtering yield Y of most materials in this low energy range is rather small ($Y \leq 0.1$) as well as the mean energy of the few sputtered particles (≈ 5 eV). Therefore, the flux of sputtered surface atoms which may contribute to an energy loss of the substrate is negligible.

The electrons have to overcome the bias voltage V_{bias} in front of the substrate in order to reach the substrate surface and to transfer their energy. The kinetic energy of the plasma electrons arises from the integration over the electron energy distribution function (EEDF) from V_{bias} up to infinity which yields in case of an Maxwellian electron energy distribution (EEDF) the energetic influx J_e as

$$J_e = n_e \sqrt{\frac{kT_e}{2\pi m_e}} \exp\left\{ -\frac{e_0(V_{pl} - V_{fl})}{kT_e} \right\} 2kT_e .$$
(5.27)

An analysis of the charged plasma components by using the general equations 5.21–5.27 listed above will in principle yield the part of surface heating caused by positive ions and electrons. Because several heat sources act together, e.g., radiation, chemical reactions, neutrals, and charge carriers, it is possible to separate the contribution of the charge carriers by variation of the bias potential.

In a process plasma containing reactive species (e.g., N_2, O_2 etc.) in addition to energy transfer by charge carriers the thermal balance of a substrate may be influenced also by atomic recombination (association) and exothermic reactions. Evidence for substrate heating by exothermic reactions on the processed surface has been reported, for example, with respect to plasma etching of silicon with fluorine containing compounds [12], and during plasma cleaning of contaminated metal surfaces [13]. The percentage of the recombination energy which is used for surface heating varies with the chemical composition of the surface and of the plasma.

For a reactive oxygen plasma, the energy influx J_{ass} by atom recombination, i.e., the formation of oxygen molecules, is described by:

$$J_{ass} = j_O \Gamma_O E_{diss} = \Gamma_O \frac{1}{2} n_O \sqrt{\frac{8kT_g}{\pi m_O}} E_{diss} ,$$
(5.28)

where Γ_O is the association probability of O-atoms on the substrate surface, T_g is the gas temperature (room temperature), n_O is the density of O-atoms, m_O is the mass of O-atoms, and E_{diss} is the dissociation energy of O_2 molecules.

The contribution J_{chem} due to exothermic chemical reaction (i.e. surface oxidation) can be calculated by the product of the average energy E_{react} released per reaction with the flux density $j_{react} = j_O$ of the reactive oxygen radicals:

$$J_{chem} = j_{react} E_{react} .$$
(5.29)

The radical flux density j_O which has to be known in Eqs. 5.28 and 5.29 is mainly determined by electron impact dissociation taking place in the negative glow of the discharge. If the growth rate of the oxide which is formed during the plasma process is known, one can J_{chem} estimate easily from the growth rate R_{dep}, the mass density ρ of the layer, and the average specific oxidation enthalpy gain h_{ox}:

$$J_{chem} = R_{dep} \rho h_{ox} .$$
(5.30)

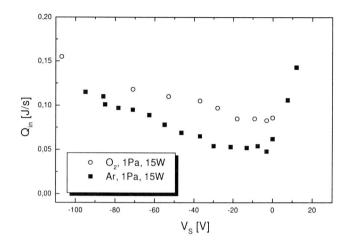

Figure 5.11: Measured integral energy influx from a weak rf plasma towards a metal substrate ($A_S = 18$ cm^2) for argon (■) and oxygen (○), as a function of the substrate voltage.

Fig.5.11 shows an example for energy influxes Q_{in} measured for a weak radiofrequency (rf) plasma in argon and oxygen, respectively. The discharge power was kept constant at 15 W and the gas pressure at 1 Pa, while the substrate voltage (V_S) was varied. Comparison of the measured total energy influx with the different energetic contributions calculated on the basis of the equations mentioned above gives insight into the involved processes and dominant mechanisms which determine the thermal balance of a substrate during plasma treatment: For negative substrate voltages only the ions dominate, while the contributions of electrons and recombination become important for positive voltages. At floating potential the energy influx from the plasma and, hence, the substrate heating shows a minimum. In principle, both gases (argon and oxygen) exhibit a similar qualitative behaviour. However, the measured energy influx in the case of an oxygen plasma is higher than for an argon plasma. This observation is due to the processes of oxygen atom recombination (association) and surface oxidation which supply, in addition, chemical reaction heat.

The determination of the energy influx in practical applications of plasma discharge devices is important when it is necessary to know the surface temperature of an object immersed in the plasma. For example, in plasma processing it may be important to know the substrate temperature in order to avoid overheating of the substrate during processing.

Also the morphology and micro-structure of the plasma-treated surfaces may be influenced by the thermal conditions at the substrate. It is well known that density and roughness of deposited films show a correlation with the substrate temperature and with the energy influx. This observation is illustrated in the pictures of Fig. 5.12, where thin aluminium films have been sputtered onto small iron powder particles. The layers deposited on the powder particles exhibit a rough, cauliflower-like structure while films sputtered on the large substrates are comparatively smooth. An essential reason for the observed differences in surface roughness might be the different substrate temperatures which are reached during the sputter process.

Figure 5.12: SEM micrograph of Al coated iron particles.

Although under comparable deposition conditions the energy flux towards a floating powder particle is the same as towards a flat substrate, the resulting equilibrium temperature may be quite different. Due to a much better heat conduction along the substrate holder a Si wafer do not reach as high temperatures as microscopic powder particles which are mainly cooled by radiation. Hence, the rather high deposition temperature of the powder particles resulting from the thermal power during the sputter process causes an increased grain size and rough microstructure.

References

[1] G.A. Somorjai, *Introduction to Surface Chemistry and Catalysis*, Wiley & Sons, Chichester (1994).

[2] H. Doshita, K. Ohtani, A. Namiki, J. Vac. Sci. Techn. A 16, 265 (1998).

[3] R. Behrisch, Ed., *Sputtering by Particle Bombardment*, Springer, Berlin, Vol. I and II (1981, 1983).

[4] R. Behrisch, K. Wittmaack, Eds., *Sputtering by Particle Bombardment*, Springer, Berlin, Vol. III (1991).

[5] W. Eckstein, J.P. Biersack, Z. Physik B 63, 471 (1986).

[6] M.P. Bonzel, R. Ku, Surf. Sci. 33, 91 (1972).

[7] D.D. Eley, Chem. Ind. 1, 12 (1976).

[8] H. Deutsch, H. Kersten, A. Rutscher, Contrib. Plasma Phys. 29, 263 (1989).

[9] R.A. Haefer, *Oberflächen- und Dünnschichttechnologie*, Springer, Berlin (1987, 1991).

[10] M. Balooch, D.R. Olander, W.J. Siekhaus, J. Vac. Sci. Techn. B4, 794 (1986).

[11] H. Kersten, H. Deutsch, G.M.W. Kroesen, R. Hippler, *The Energy Balance at Substrate Surfaces during Plasma Processing*, Vacuum, in print (2001).

[12] A. Durandet, O. Joubert, J. Pelletier, M. Pichot, J. Appl. Phys. 67, 3862 (1990).

[13] H. Kersten, H. Deutsch, J.F. Behnke, Vacuum 48, 123 (1997).

6   Langmuir probe diagnostics of low-temperature plasmas

Sigismund Pfau[†] *and Milan Tichý*[‡]

[†] Institut für Physik, Ernst-Moritz-Arndt-Universität Greifswald, Domstraße 10a,
D-17487 Greifswald, Germany
[‡] Charles University in Prague, Faculty of Mathematics and Physics, Department of
Electronics and Vacuum Physics, V Holešovičkách 2, 18000 Praha 8, Czech Republic

6.1   Introduction

6.1.1   Probe shapes and probe characteristics

Measurements with electric probes belong to the oldest as well as to the most often used procedures of the low-temperature plasma diagnostics. The method has been developed by Langmuir and his co-workers in the twenties [1]. Since then it has been subject of many extensions and of further developments in order to extend its applicability to problems with more general conditions as those presumed by Langmuir. Such investigations proceeded continuously and the research on extension of applicability of Langmuir probe diagnostics continues in the present time, too.

The method of the Langmuir probe measurements is based on the estimation of the current-voltage characteristics – the so-called *probe characteristics* – of a circuit consisting of two metallic electrodes that are both immersed into the plasma under study. Two cases are of interest:

(a) the surface areas of both the electrodes being in contact with plasma differ by several orders of magnitude, and

(b) the surface areas of both the electrodes being in contact with plasma are very small in comparison with the dimensions of the vessel containing plasma and approximately equal to each other.

Case (a) is called *the single probe method*, case (b) *the double probe method*. Most of this text is devoted, in accordance with the frequency of use of either method, to the single probe method; the double probe method is discussed in section 6.10.

The Langmuir probe is usually constructed in simple geometric shapes: spherical probe, cylindrical probe and planar (flat) probe, see Fig. 6.1. When constructing the probe we have to take into account that not only the active metallic part, i.e., the collecting surface of the probe, has to be small in comparison with the characteristic dimensions of the plasma vessel. The isolated parts of the probe must fulfill the same condition since often just these passive parts substantially influence the plasma around the probe .

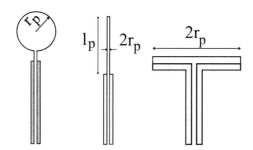

Figure 6.1: Typical shapes of Langmuir probes .

A simple experimental set-up for measurements of probe characteristics is shown in Fig. 6.2. Apart from the direct current (*dc*) high-voltage power supply that feeds the glow discharge and the stabilizing resistor Z_a (the *dc* supply can also work in the constant-current mode) the experimental system consists of the following parts:

1. a *dc* voltage source for compensation of the potential difference between the reference electrode (anode) and the probe,

2. a sawtooth- or staircase-like voltage generator,

3. a current measuring instrument, and

4. a computer that stores the measured current and voltage data, controls the measurement procedure and processes the acquired data.

Let us assume that the probe potential φ_p differs from the plasma (space) potential φ_s at the place where the probe is located by $U_p = \varphi_p - \varphi_s \neq 0$. In such a case the electric field that

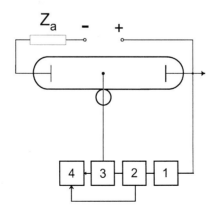

Figure 6.2: Typical probe circuit. 1 *dc* bias voltage, 2 sawtooth or staircase U_p generator, 3 current–voltage converter, 4 $I_p - U_p$ data acquisition (computer).

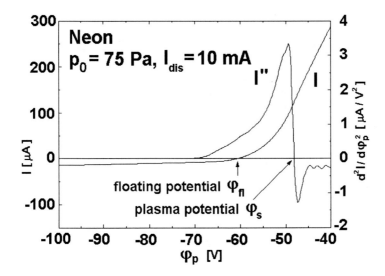

Figure 6.3: Typical course of the probe characteristics $I(\varphi_p)$ (left scale) with its second derivative $I'' \equiv d^2I/d\varphi_p^2$ (right scale).

arises between the plasma and the probe surface accelerates the charged particles with one sign and repels those with the opposite sign. Since the charged particles have thermal energy that randomizes their movement this process finishes with the creation of a space charge sheath around the probe which shields the plasma from the electric field of the probe. A simplified physical model of such a probe assumes that the space charge sheath has a finite thickness. This means that the influence of the electric field extends only up to a certain distance from the probe and that the plasma farther away from the probe is perfectly shielded. The thickness of the probe sheath $(r_p - r_s)$ is of the order of the so-called Debye length (λ_D). The Debye length for the electrons can be obtained from the well known formula (see also chapter 1)

$$\lambda_D = \sqrt{\frac{\varepsilon_0 k_B T_e}{q_0^2 n_e}}, \tag{6.1}$$

where T_e is the electron temperature, n_e the concentration of the electrons, ε_0 the permittivity of the vacuum, k_B the Boltzmann constant, and q_0 the elementary charge. The thickness of the probe sheath at moderate U_p actually is several times λ_D and increases with increasing U_p. If the potential of the probe surface is changed with respect to the space potential the probe current changes too, and finally the probe characteristics will be obtained. A typical course of the probe characteristics is shown in Fig. 6.3.

Three regions can be clearly distinguished in the course of the probe characteristics. Assuming that no negative ions exist in the plasma the probe current consists of the electron and the positive ion current. The three regions of the probe characteristics are characterized by:

I. $U_p \leq 2U_{fl}$ $\rightarrow$ $|\,I_{pi}/I_{pe}\,|\;\gg\;1$ Positive ion acceleration region

II. $2U_{fl} < U_p < 0$ $\rightarrow$ $|\,I_{pi}/I_{pe}\,|\;\approx\;1$ Transition region

III. $0 \leq U_p$ $\rightarrow$ $|\,I_{pi}/I_{pe}\,|\;\ll\;1$ Electron acceleration region

In the ion acceleration region (sometimes called ion saturation region since the ion current is close to saturation) an almost pure positive ion current flows to the probe. At the floating potential the electron and the positive ion current compensate each other and the total probe current equals to zero. For positive probe voltages the electron current dominates the probe current; the probe operates in the electron acceleration region. If the probe immersed into plasma is not connected to the outer circuit, and therefore it cannot carry the current, then the probe is negatively charged with respect to the plasma potential. This takes place since the electrons (even at the same temperature like positive ions) have a much higher thermal velocity due to their much lower mass (typically for argon is the square root of the mass ratio around 300). The more negative potential of the probe, in consequence, attracts more positive ions, and this process continues until the electron and the positive ion current components equal each other; the floating potential is established on the probe.

The contact of the plasma and the probe surface, i.e. the contact of two basic states of aggregation of matter causes the creation of a largely inhomogeneous transition region close to the probe surface which is very difficult to describe. Characteristic for this transition region are the space charges and high electric fields. The theoretical description of this region is coupled with substantial difficulties.

With the help of a theoretical description of the relation between the probe current and the probe voltage it is possible to determine the basic parameters of the surrounding plasma. In such a way we can determine the density of the electrons n_e and of the positive ions n_i, the electron temperature T_e or the electron velocity (energy) distribution function $f_e(\vec{w}, \vec{r})$ as well as the space potential at the place where the probe is located. The theory of the probe method will actually be confronted with all the problems connected with the contact between plasma and solid surface. The importance of the probe theory therefore reaches far behind the borders of probe diagnostics.

6.1.2   The working regimes of the Langmuir probe

In order to characterize the different regimes of operation of the Langmuir probe we introduce the following parameters:

- the characteristic dimension r_p of the probe,

- the mean free path λ_ν of a particular charged particle $\nu = e, i$ (e, electron; i, ion),

- the Debye screening length λ_D (Eq. 6.1), and

- the plasma *anisothermicity parameter* $\tau = T_e/T_i$, i.e. the ratio of the electron temperature to the ion temperature.

The following part restricts at first to the presentation of the theoretical foundations of the *classic* Langmuir *collisionless* single probe theory. The adjective *collisionless* represents the assumption posed by Langmuir that the mean free path of charged particles is much greater than the characteristic dimension of the probe: $\lambda_\nu \gg r_p$. Assuming this three working regimes of the probe we have to distinguish:

1. $\lambda_\nu \gg r_p \gg \lambda_D$ collisionless movement of charge carriers in the thin sheath,

2. $\lambda_\nu \gg \lambda_D \gg r_p$ collisionless movement of charge carriers in the thick sheath,

3. $\lambda_D \gg \lambda_\nu \gg r_p$ collision determined description of the probe current.

The case of the so-called *continuum probe* working regime, when the condition $\lambda_\nu \ll r_p$ is fulfilled, shall not be discussed here.

6.1.3   Advantages and disadvantages of probe diagnostics

The most important advantages of probe diagnostics in comparison to other diagnostic methods can be found in the following attributes:

- The technical complexity of the probe method implementation is comparatively small. Also, the necessary experimental set-up for the probe measurements is comparatively simple.

- From the probe data it is possible to determine the time and spatial distribution of quite a large variety of quantities characterizing the plasma under study (i.e., n_e, n_i, T_e, f_e, φ_s, $\vec{E}$, etc.).

The spatial resolution of the probe method is of the order of the magnitude of the Debye length λ_D. The temporal resolution may be estimated from the plasma frequency ω of a charged particle with mass m

$$\omega = \sqrt{\frac{q_0^2 n_i}{\varepsilon_0 m}}, \tag{6.2}$$

it is approximately given by $2\pi/\omega_i$, where $\omega = \omega_i$ is the plasma frequency for an ion of mass $m = m_i$.

The disadvantages of the probe method are the following:

- The probe data processing method (aiming at the determination of the plasma parameters) depends on the parameters being measured.

- The plasma surrounding the probe is disturbed by the drain of the charge carriers to the probe.

- The presence of the probe in the plasma can initiate inhomogeneities in the plasma.

- The application of the probe method is difficult or even impossible in a plasma containing fluctuations, oscillations and waves.

- It is very difficult to assess the effect of the secondary- and photo-emission of electrons from the probe surface as well as the effect of reflection of charge carriers from the probe surface.

- The heavy particles can cause the creation/destruction of thin films on the probe surface during the course of measurements which change the properties of the probe surface, this making the interpretation of the probe data difficult or impossible.

- The fact that the probe is in direct contact with the plasma under study restricts its use substantially to the (non-thermal or thermal) low-temperature plasma. The use of probes for the diagnostics of plasma in high-temperature plasma devices such as tokamaks is restricted to peripheral parts of the plasma, the so-called scrape-off layer (SOL).

6.2   The Langmuir single probe method

6.2.1   Theoretical foundations of the Langmuir probe method

The range of the conventional Langmuir probe theory covers the conditions of the collision-less movement of charge carriers in the space charge sheath around the probe. Further it is assumed that the sheath boundary is well-defined and that behind this boundary the plasma is completely undisturbed by the presence of the probe. This means that the electric field caused by the difference between the potential of the probe and the plasma potential at the place where the probe is located is limited to the volume inside the probe sheath boundary and does not penetrate it.

Assuming that the velocity distribution function f_ν of a charged particle ν ($\nu = e, i$) at the surface area of the probe sheath is known, we can obtain the following expression for the number of charged particles passing the surface area element dA_s of the sheath boundary per unit of time:

$$dI_{\nu s} = dA_s dj_{\nu s} = dA_s q_\nu n_\nu w_z f_\nu(w_x, w_y, w_z) dw_x dw_y dw_z \,. \tag{6.3}$$

Here q_ν is the charge and n_ν the density of the charged particle ν, $j_{\nu s}$ its current density at the sheath boundary, and w_z, w_x and w_y are the velocity vector components perpendicular respectively parallel to the surface area element dA_s. If the distribution function f_ν is isotropic and the surface A_s is not concave, then the contribution of each surface element is independent of its orientation in space. In such a case the total current can be obtained by simple integration over the surface elements dA_s.

For a certain probe potential the fraction of particles that fall on the probe surface without passing it can be determined by the choice of proper integration limits. In general we have

$$I_{p\nu} = A_s q_\nu n_\nu \int_{w_{x1}}^{w_{x2}} \int_{w_{y1}}^{w_{y2}} \int_{w_{z1}}^{w_{z2}} w_z f_\nu(w_x, w_y, w_z) \, dw_x dw_y dw_z \,. \tag{6.4}$$

In the case of the spherical probe it is advantageous to transform the cartesian velocity coordinates w_x, w_y, w_z into spherical coordinates w, ϑ, φ,

$$I_{p\nu} = \frac{A_s q_\nu n_\nu}{4} \int_{w_1}^{w_2} 4\pi w^3 \, f_\nu(w) \left[\sin^2 \vartheta_2 - \sin^2 \vartheta_1\right] dw \,. \tag{6.5}$$

6.2.2 Probe characteristics – example of the spherical probe

6.2.2.1 Probe current at $q_\nu U_p \geq 0$

As an example for the calculation of the current to a Langmuir probe the case of a spherical probe shall be examined. Fig. 6.4 enlightens the movement of a charged particle in the retarding field (i.e., for $q_\nu U_p \geq 0$) of a spherical probe. where ϑ is the incident angle of

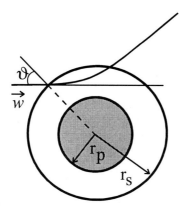

Figure 6.4: Model trajectory.

the particle with respect to the surface normal, $\vec{w}$ its incident velocity into the space charge sheath, and r_p and r_s are the probe and sheath radii, respectively. It is assumed that a particle that enters the space charge sheath moves collision-free in the central field of the probe. Its movement is then governed by two laws of conservation, namely that of energy and that of angular momentum. For the particles that fall on the probe surface the condition

$$\sin \vartheta \leq \frac{r_p}{r_s} \left(1 - \frac{2q_\nu U_p}{m_\nu w^2}\right)^{\frac{1}{2}} \tag{6.6}$$

holds, from which the following limits of integration can be derived

$$\vartheta_1 = 0 \quad \text{and} \quad \sin \vartheta_2 = \frac{r_p}{r_s}\left(1 - \frac{2q_\nu U_p}{m_\nu w^2}\right)^{\frac{1}{2}} \geq 0 \tag{6.7}$$

and

$$w_1 = \left(\frac{2q_\nu U_p}{m_\nu}\right)^{\frac{1}{2}} \quad \text{and} \quad w_2 = \infty \,. \tag{6.8}$$

Hence the probe current $I_{p\nu}$ of the particle ν is obtained by the integral

$$I_{p\nu} = \frac{A_s q_\nu n_\nu}{4} \int_{\sqrt{\frac{2q_\nu U_p}{m_\nu}}}^{\infty} 4\pi w^3 f_\nu(w) \left(\frac{r_p}{r_s}\right)^2 \left[1 - \frac{2q_\nu U_p}{m_\nu w^2}\right] dw. \tag{6.9}$$

Assuming that the velocity distribution of the charged particle ν is Maxwellian this relation after a few manipulations can be simplified to

$$I_{p\nu} = \frac{A_p q_\nu n_\nu}{4} \overline{w}_\nu \exp\left(-\frac{q_\nu U_p}{k_B T_\nu}\right), \tag{6.10}$$

where

$$\overline{w}_\nu = \left(\frac{8k_B T_\nu}{\pi m_\nu}\right)^{\frac{1}{2}} \tag{6.11}$$

and $A_p = 4\pi r_p^2$. The same equation is obtained for the planar probe and for the cylindrical probe (in these cases A_p denotes the surface area of the collecting probe plane or cylinder, respectively).

The temperature T_ν of the charged particles ν is obtained from the derivative (slope) of the probe characteristics in the respective retarding voltage range on a semi-logarithmic scale, i.e., as

$$T_\nu = -\frac{q_\nu}{k_B} \frac{d}{dU_p} \ln\left(\frac{I_{p\nu}}{I_{p\nu 0}}\right). \tag{6.12}$$

The charged particle density n_ν can be estimated from the probe current $I_{p\nu 0}$ at the space potential ($U_p = 0$), provided the space potential is known and that it is possible to separate the current contribution of the particle ν from the total probe current. The corresponding relation obtained from Eq. 6.10 is

$$I_{p\nu 0} = A_p q_\nu \frac{n_\nu}{4} \overline{w}_\nu \tag{6.13}$$

from which n_ν is obtained as

$$n_\nu = \frac{4 I_{p\nu 0}}{A_p q_\nu \overline{w}_\nu}. \tag{6.14}$$

We have to keep in mind, however, that we only are able to measure the total probe current. In the transition region II (see section 6.1.1) between the positive ion acceleration region I and the electron acceleration region III both, electron and positive ion current, are comparable to each other. In order to separate these contributions from each other one can use several procedures, e.g., one can extrapolate the *saturated* positive ion current to the space potential in a linear manner and subtract this positive ion current component from the total probe current. From this consideration it also follows that only for electrons it is possible to estimate the particle temperature T_ν from the relation 6.12, but not for ions except under very special operating conditions.

6.2.2.2 Probe current at $q_\nu U_p \leq 0$

In a manner similar to the procedure above we can derive, for $q_\nu U_p \leq 0$ and with the help of the energy and angular momentum conservation laws, the integration limits for ϑ and w for particles which fall on the probe surface at a given probe potential. In analogy to Eqs. 6.7 and 6.8 we get

$$\vartheta_1 = 0 \quad \text{and} \quad \sin \vartheta_2 = \frac{r_p}{r_s} \left(1 - \frac{2 q_\nu U_p}{m_\nu w^2} \right)^{\frac{1}{2}} \leq 1 . \tag{6.15}$$

With the collected particle velocities now in the range $w_1 = 0$ to $w_2 = \infty$ we can distinguish 2 cases: for $0 \leq w \leq w^*$ we get $\sin \vartheta_2 = 1$, while for $w \geq w^*$ expression 6.15 holds. By substituting these limits into equation 6.5 for the probe current we obtain

$$I_{p\nu} = \frac{A_s n_\nu q_\nu}{4} \left\{ \int_0^{w^*} 4\pi w^3 f_\nu(w) dw + \int_{w^*}^\infty 4\pi w^3 f_\nu(w) \frac{r_p^2}{r_s^2} \left(1 - \frac{2 q_\nu U_p}{m_\nu w^2} \right)^2 dw \right\} , \tag{6.16}$$

where

$$w^* = \left\{ -2 q_\nu U_p / m_\nu \left[r_s^2 / r_p^2 - 1 \right] \right\}^{\frac{1}{2}} . \tag{6.17}$$

For the velocity interval $0 \leq w \leq w^*$ the probe acts as a collector with an effective collecting area that is larger than the true surface area A_p of the probe. For charged particle velocities $w \geq w^*$ a part of the charged particles which entered the probe sheath passes by the probe and is not collected by the probe (so-called orbital motion limited current, OML). Assuming a Maxwellian velocity distribution $f_\nu(w)$ we obtain from (6.16)

$$I_{p\nu} = \frac{n_\nu}{4} q_\nu \overline{w}_\nu A_s \left[1 - \left(1 - \frac{r_p^2}{r_s^2} \right) \exp \left(\frac{q_\nu U_p}{k_B T_\nu \left(\frac{r_s^2}{r_p^2} - 1 \right)} \right) \right] . \tag{6.18}$$

On condition that

$$-\frac{k_B T_\nu}{q_\nu U_p} \left(\frac{r_s^2}{r_p^2} - 1 \right) \gg 1 \tag{6.19}$$

we get from the above relation

$$I_{p\nu}^{(s)} = q_\nu \frac{n_\nu}{4} \overline{w}_\nu A_p \left\{ 1 - \frac{q_\nu U_p}{k_B T_\nu} \right\} \tag{6.20}$$

for the spherical probe, and

$$I_{p\nu}^{(c)} = q_\nu \frac{n_\nu}{4} \overline{w}_\nu A_p \left\{ 1 - \frac{q_\nu U_p}{k_B T_\nu} \right\}^{\frac{1}{2}} \tag{6.21}$$

for the cylindrical probe. For the planar probe we obtain a current that is independent of the probe potential,

$$I_{p\nu}^{(p)} = q_\nu \frac{n_\nu}{4} \overline{w}_\nu A_p . \tag{6.22}$$

While the evaluation of the relations 6.20–6.22 provides usable results for electrons, this does not hold for ions. A more detailed analysis of the transition region between the undisturbed non-thermal plasma and the region of the space charge sheath was performed by Bohm [2]. He showed that the sheath screens the electric field from the probe imperfectly, i.e., a *rest* potential $k_B T_e/(2q_0)$ (Bohm criterion) remains. This *rest* potential is screened by the so-called pre-sheath that spans from the probe sheath to infinite distance. For this reason the positive ions entering the probe sheath do not have room temperature as in an undisturbed plasma, but are accelerated in the pre-sheath, their temperature being determined by the temperature of the electrons.

6.3 General theories of the current to a Langmuir probe

6.3.1 Starting system of equations

The general theoretical description of the current to a Langmuir probe requires the simultaneous solution of the following fundamental equations

- the Poisson equation
- the collision-free Boltzmann (Vlasov) equation and
- the continuity equation

with regard to the respective boundary conditions at the probe surface and in the plasma. At a larger distance from the probe such solution must be identical to the solution which corresponds to the undisturbed plasma. At the probe surface the boundary conditions are determined by complex processes of the plasma-wall interaction. The starting system of equations reads:

$$\mathrm{div}\vec{E} \; = \; \frac{q_0}{\varepsilon}(n_i - n_e) \, , \tag{6.23}$$

$$\vec{w}_\nu \mathrm{grad} f_\nu \; + \; \frac{q_\nu}{m_\nu}\vec{E}\,\mathrm{grad}_w f_\nu \; = \; 0 \, , \tag{6.24}$$

$$\mathrm{div}(\vec{j}_e \, + \, \vec{j}_i) \; = \; 0 \tag{6.25}$$

with the normalizing conditions

$$\int_{-\infty}^{+\infty}\int_{-\infty}^{+\infty}\int_{-\infty}^{+\infty} f_\nu(w_x, w_y, w_z) dw_x dw_y dw_z \; = \; n_\nu \, ; \qquad \nu = i, e \, . \tag{6.26}$$

The simultaneous solution of this set of equations presents a very complicated problem which can be solved only numerically and by using substantial computational effort. If the method of solution takes into account the redistribution of charges induced by the change of the probe potential and consequently of the probe current, such a solution is called self-consistent. Such a self-consistent solution of this integro-differential system of equations was firstly stated by Bernstein and Rabinowitz under the assumption of a monoenergetic ion energy distribution [3]. This solution was developed further by Laframboise who assumed a Maxwellian ion energy distribution [4]. A simplified model for the calculation of the ion current was developed by Allen, Boyd and Reynolds [5]. Since this model is one of the most often used theories for the interpretation of the probe data it shall be explained in more detail in the following section.

6.3.2 The *cold ion* model by Allen, Boyd and Reynolds $(T_i/T_e = 0)$

The starting system of equations of this model [5] consists of the Poisson equation, the simplified ion energy balance, the Boltzmann distribution of the electron density and the continuity equation. Assuming spherical symmetry such a system of equations can be written as:

$$\text{Poisson equation}: \quad \frac{1}{r^2}\frac{d}{dr}\left(r^2\frac{dU}{dr}\right) = -\frac{q_0}{\varepsilon}(n_i - n_e), \quad (6.27)$$

$$\text{ion energy balance}: \quad \frac{1}{2}m_i w_r^2 = -q_0 U, \quad (6.28)$$

$$\text{Boltzmann distribution}: \quad n_e = n_{e0}\exp\left(\frac{q_0 U}{k_B T_e}\right), \quad (6.29)$$

$$\text{continuity equation}: \quad \frac{1}{r^2}\frac{d}{dr}(r^2 j_{ir}) = 0 \quad \text{with} \quad 4\pi r^2 j_{ir} = I_{pi}. \quad (6.30)$$

It is usual to introduce dimension-less variables, i.e., variables that are normalized with respect to *reasonable* or *representative* values of each variable. Allen, Boyd and Reynolds [5] introduced the normalized variables as follows:

$$\eta = \frac{q_0 U}{k_B T_e} \quad ; \quad \eta_p = \frac{q_0 U_p}{k_B T_e}; \quad (6.31)$$

$$\xi = \frac{r}{\lambda_D} \quad ; \quad \xi_p = \frac{r_p}{\lambda_D}; \quad (6.32)$$

$$i_i = \frac{I_{pi}}{I_{n_i}} \quad ; \quad I_{n_i} = 4\pi r_p^2 n_{e0} q_0 \left\{\frac{k_B T_e}{2\pi m_i}\right\}^{\frac{1}{2}}. \quad (6.33)$$

Using these variables the Poisson equation 6.27 reads

$$\frac{d}{d\xi}\left(\xi^2\frac{d\eta}{d\xi}\right) = \xi^2\exp(\eta) - \xi_p\frac{i_i}{2\sqrt{-\pi\eta}}. \quad (6.34)$$

The solution of this equation proceeds in such a way that for large ξ it converges to the so-called plasma solution. The plasma solution is obtained by setting the left-hand side of the equation 6.34 equal to zero. For $\eta \to -\frac{1}{2}$ the derivative increases to infinity, $\frac{d\eta}{d\xi} \to \infty$. This result is known as the Bohm sheath criterion [2]. From the solution of the differential equation

$$\eta = \eta(\xi, \xi_p, i_i) \quad (6.35)$$

follows the mutual relation between the three quantities η_p, ξ_p and i_i. The results are available in the form of tables, graphical representations and analytical approximations. In the following section the method of estimation of the positive ion density from the ion probe current on the basis of the described cold ion model by Allen, Boyd and Reynolds shall shortly be outlined. Since the normalized ion current i_i depends on ξ_p which in turn depends on the ion density to be estimated the first possibility is the iterative procedure. For the selected pair of values I_{pi} and η_p (in other words the electron temperature has to be determined first) on the measured probe characteristics we have to estimate (the first approximation of) $\xi_p^{(1)}$. From the dependence $i_i = i_i(\xi_p, \eta_p)$ then it is possible to estimate the normalized current $i_i^{(1)}$. Since I_{pi} is already known it is possible to determine the $I_{n_i}^{(1)}$ from equation 6.33 and hence the first

approximation of the ion density $n_i^{(1)}$. From this density value the second order approximation of $\xi_p^{(2)}$ is calculated, etc., until the approximations $n_i^{(k-1)}$ and $n_i^{(k)}$ do not differ *too much*
from each other. The convergence of this procedure is unclear and depends very much on the
first choice of $\xi_p^{(1)}$.

The necessity of the iterative procedure was removed by Sonin by introducing the so-called
Sonin plot [6]. This plot relates the artificially created quantity $\xi_p^2 i_i$ to i_i at a pre-selected value
of η_p, usually $\eta_p = \eta_{fl} - 10$ [21] or $\eta_p = -15$ [18]. Since ξ_p^2 is proportional to n_i and i_i
is inversely proportional to n_i the quantity $\xi_p^2 i_i$ does not depend on n_i and hence excludes the
necessity of the iterative procedure

$$\xi_p{}^2 i_i = \frac{1}{\varepsilon} \left\{ \frac{m_i}{8\pi q_0} \right\}^{\frac{1}{2}} \left(\frac{q_0}{k_B T_e} \right)^{\frac{3}{2}} I_{p_i} . \tag{6.36}$$

Analytical approximations of the relation $i_i = i_i(\xi_p, \eta_p)$ are available for the spherical as well
as for the cylindrical probe. For example, the approximation for the cylindrical probe can be
written as [7]

$$i_i(\eta_p) = a(\eta_p/b)^\alpha , \tag{6.37}$$

where

$$a = (\xi_p + 0.6)^{0.05} + 0.04 , \quad b = 0.09 \left(\exp(-\xi_p^{-1}) + 0.08 \right) , \tag{6.38}$$

$$\alpha = (\xi_p + 3.1)^{-0.6} . \tag{6.39}$$

The radial motion theory has been further developed and tested. The relevant publications are
[8] (application of the radial motion theory in a radiofrequency (*rf*) discharge in argon), [9]
(review on OML theories, recent experimental work), [10] (new modern aspects together with
recent developments in experimental techniques), [11] (extension for double probe), and [12]
(small body floating in a plasma). Further the radial motion theory has been extended to the
interesting case of the so-called standard electron energy distributions [13], i.e.,

$$f^*(u_p) = \text{const.} \times \varepsilon_p^{-3/2} \exp\left(-\frac{u_p^k}{k\varepsilon_p^k} \right) \quad \text{with } k \geq 1 . \tag{6.40}$$

Here u_p represents the voltage equivalent of the electron energy (see the paragraph below) and
ε_p is the so-called effective electron temperature. For $k = 1$ this distribution represents the
Maxwellian, for $k = 2$ the Druyvesteynian, and for $k = 4$ the Davydov distribution (see also
chapter 1). The case of finite ion temperature values, i.e. of arbitrary ratios T_i/T_e has been
treated in Ref. [14]. This theory includes the classical ABR theory (developed for $T_i/T_e \to 0$)
as a special case.

6.4 The Druyvesteyn method for estimation of the electron energy distribution function (EEDF)

One of a few procedures that permit the direct experimental determination of the electron
energy distribution function EEDF in a plasma is based on the probe measurements. In order

to explain this procedure the investigation starts at the probe characteristics in the electron retarding region given by equation 6.9. In order to express the variables in terms of energy the quantities q_ν, U_p and w are substituted by $-q_\nu = q_0$, $-U_p = u_p$ and $w = (2q_0u/m_e)^{1/2}$. Then by differentiating equation 6.9 with respect to u at the point $u = u_p$ the well-known Druyvesteyn relation follows [15]:

$$\frac{d^2 I_{p_e}}{du_p^2} = \left(\frac{q_0}{2}\right)^{\frac{3}{2}} m_e^{-\frac{1}{2}} n_e A_p \overline{f}(u_p) \,, \tag{6.41}$$

where $\overline{f}(u_p)$ fulfills the normalizing condition

$$\int_0^\infty \overline{f}(u_p) \, u_p^{\frac{1}{2}} \, du_p = 1 \,. \tag{6.42}$$

Eq. 6.41 enables us to estimate the electron density n_e and the mean electron energy $q_0 u_m$ using the relations

$$n_e = \left(\frac{2}{q_0}\right)^{\frac{3}{2}} \frac{m_e^{\frac{1}{2}}}{A_p} \int_0^\infty u_p^{\frac{1}{2}} \frac{d^2 I_{p_e}}{du_p^2} du_p \tag{6.43}$$

and

$$q_0 u_m = \frac{q_0 \int_0^\infty u_p^{\frac{3}{2}} \frac{d^2 I_{p_e}}{du_p^2} \, du_p}{\int_0^\infty u_p^{\frac{1}{2}} \frac{d^2 I_{p_e}}{du_p^2} \, du_p} \,. \tag{6.44}$$

It is very important to fix the origin of the energy scale for the EEDF, in other words to determine the space (plasma) potential at the probe position. In accordance with the works of Luijendijk and van Eck [16] and Herrmann and Klagge [17] the space potential is most accurately determined as the probe voltage at the zero-cross of the second derivative of the total probe current. Other methods, such as the *method of tangents*, estimation from the position of the maximum of the second derivative [25] or from the probe voltage corresponding to the floating potential [18] etc. are less accurate and not commonly used.

Errors in the determination of the plasma potential by means of this method are related to the phenomena at the probe surface. Among those the change of the work function over the probe surface due to impurities on the probe surface, reflection and secondary emission of electrons, bombardment of the probe surface by metastable atoms and photoemission may be mentioned [19, 20]. The described phenomena round the knee of the probe characteristics in the vicinity of the plasma potential and thus the zero-crossing point of the second derivative does not exactly correspond to the plasma potential. In a case in which the collisions play an important role in the collection of charged particles by the probe the effect of rounding the knee of the probe characteristics also occurs. Klagge and Tichý [21] have specified the conditions under which this effect can be neglected (cylindrical probe), i.e., for

$$K_e(K_e + 1) \left(K_e + \ln\left(\frac{l_p}{r_p}\right) \right)^{-1} > 1 \,, \tag{6.45}$$

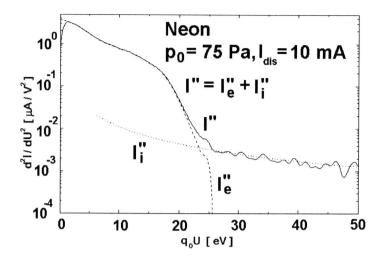

Figure 6.5: Sample of the measured second derivative I''. The relation between the magnitudes of the I_e'' and I_i'' is shown.

where $K_e = \lambda_e / r_p$ denotes the Knudsen number for electrons.

The second derivative of the electron probe current with respect to the probe potential is usually replaced by the second derivative of the total probe current. This approximation is valid only under the assumption that the ion probe current at higher probe potentials is constant or is a linear function of the probe potential. However, the commonly measured dependence of the positive ion current in a low-temperature non-thermal plasma can be fitted in a better way by the so-called double-logarithmic approximation introduced by Nuhn and Peter [22], i.e. by the relation (for normalized current) $I_i = I_{i0}(1 + \eta_p)^\kappa$, where I_{i0} and κ are parameters to be determined from the fit to the experimentally measured dependence. The second derivative of such dependence does not vanish, and sometimes it is a useful practice [23] to correct the second derivative of the total probe current for the second derivative of the positive ion current. An example of the measured second derivative of the total probe current and its decomposition to the second derivative of the electron and of the ion probe current is shown in Fig. 6.5.

The second derivative, necessary for the estimation of the EEDF, either can be directly measured (on-line methods) or computed numerically from the experimental data (off-line methods). Many experimental set-ups for on-line measurement of the second derivative of the total probe current have been developed. They are based on

- making use of the non-linearity of the probe characteristics, i.e. the relation between the *curvature* of the characteristics and second harmonic generation [24], mixing [16], detection of the modulated signal [25] etc.,

- direct analog differentiators using operational amplifiers and sawtooth-like probe voltage [26],

- analog difference amplifiers and stepwise-like probe voltage [27].

The off-line differentiating methods are either based on algorithms of the numerical analysis [28] or on a direct numeric solution of the integral equation 6.9 [29] or on the non-recursive digital filtering of a dependence given as a set of data. The non-recursive digital filter, also called finite impulse response (FIR) filter is applied to a given set of data. It converts the selected data ordinate y_l to a value that is given by a linear combination of itself and of ordinates from its adjacent neighbourhood ($\pm m$ points). The resulting value characterizes the given data in the selected interval around the data point y_l, i.e. it represents the derivative of nth order ($0 \leq n \leq 2m$) at the abscissa corresponding to y_l [30, 31]:

$$y_l^{(n)} = \frac{1}{h^n} \sum_{i=-m}^{m} c_i y_{l+i} .$$

(6.46)

Equation 6.46 requires the given data to be equally spaced by h on the abscissa axis. If this condition is not fulfilled the linear interpolation between adjacent points can be used to make the data equally spaced prior to the calculation of $y_l^{(n)}$. It is seen that the value of $y_l^{(n)}$ cannot be obtained at the first m and the last m data points in the given data set. For $n = 0$ Eq. 6.46 represents a weighted smoothing operation. The usual numeric procedures for the differentiation of the experimental data can also be regarded as digital filters since they use formulae analogous to Eq. 6.46.

The most common procedure for differentiation of noisy experimental data is the *sliding polynomial approximation*. This represents the approximation of the selected odd number $M = 2m + 1$ ($m = 1, 2, \ldots$) adjacent data points around selected x_l by a second order polynomial $p(x) = a_0 + a_1 x + a_2 x^2$ (for $x_l - mh \leq x \leq x_l + mh$). Hence the second derivative at x_l is $y_l^{(2)}(x_l) = 2a_2$. Since the coefficient a_2 is a weighted sum of the ordinates y_i (for $l - m \leq i \leq l + m$) this method may be regarded as digital filtering where the weights are the filter coefficients $c_i/(h^2)$. Tables of such coefficients are given by Savitzky and Golay [32] and formulae for the evaluation of these filter coefficients in Ref. [33]. Computer programs that make use of this procedure have already been constructed and reported Ref. [34]. However, the variability and the noise-suppression are not always sufficient in the probe diagnostics [35, 36].

Better signal-to-noise-ratio can be obtained if the elements of the digital filter theory are applied. This method consists in a calculation of the second derivative by the method described in the above paragraph and in a subsequent application of the digital filtering on the resulting second derivative data set. The derivative is calculated e.g. by using the 3 or 5 point method, i.e. according to the formulae

$$y_l^{(2)} = \frac{1}{h^2}(y_{l-1} - 2y_l + y_{l+1}) \qquad \text{or}$$

(6.47)

$$y_l^{(2)} = \frac{1}{12h^2}(-y_{l-2} + 16y_{l-1} - 30y_l + 16y_{l+1} - y_{l+2}) .$$

(6.48)

The derivatives calculated in this manner feature the quadratic increase of the high noise frequencies in their frequency spectrum, see e.g., Ref. [36]. In order to suppress this noise increase a smoothing filter with an attenuation increasing at higher frequencies steeper than quadratic has to be applied. Such a filter may be obtained by the window function based filter design [36]. The simplest of such filters is the Hanning filter. It has a beginning stopband attenuation of 44 dB. A design procedure for a smoothing filter with strong increasing noise suppression having a variable beginning stopband attenuation between 27 dB and 80 dB is given in Ref. [36].

Further discussion on the numerical differentiation of probe characteristics is given in Ref. [37]. Here the numerical filtering of the probe data by means of B–spline approximation is applied. Errors that can arise due to the numerical differentiation are treated in Ref. [38]. A further novel smoothing method using the so-called Hayden numerical filtering is described in Ref. [39]. Generally, one has to be careful in the choice of the degree of smoothness in any of the methods mentioned above. Secondary maxima or local deficits of the EEDF due to elementary processes characteristic of the investigated plasma that create or loose electrons in a certain range of energies should not be suppressed. When in doubt whether a particular irregularity on the EEDF is due to the plasma processes or due to noise the best recommendation is to process a set of data with different kinds of differentiating methods and compare the results with theoretical expectations.

In all procedures for the estimation of the second derivative of the probe characteristics the second derivative is estimated not from the infinitesimally small vicinity of the point where the derivative is estimated, but from the finite voltage (energy) interval around this point. This leads to a distortion of the estimated second derivative in comparison with the theoretical one. It has been shown that the estimated second derivative $\overline{I}''(u_p)$ is related to the theoretical one $I''(u_p)$ via convolution with so-called apparatus function $T(u_p - u'_p)$ [40]:

$$\overline{I}''(u_p) = \int\limits_{-\infty}^{+\infty} I''(u'_p)T(u_p - u'_p)du'_p .$$

(6.49)

A typical course of the apparatus function (case of the second harmonic procedure [41]) is

$$T(u_p - u'_p) = \frac{8}{3\pi} \left(1 - \left(\frac{u_p - u'_p}{a}\right)^2\right)^{\frac{3}{2}} ,$$

(6.50)

where a is the amplitude of the applied fundamental harmonic. The function $T(u_p - u'_p)$ has non-zero values only for $-1 \leq (u_p - u'_p)/a \leq 1$. Practice has shown that in order to get EEDF's that are close to reality it is a better way to enhance the signal-to-noise ratio of the measured signal from the probe than to try to process numerically the data with a bad signal-to-noise ratio. In any case the deconvolution is not necessary if the peak-to-peak amplitude of the differentiating signal in volts (in the on-line as well as in the off-line case) is less than half the electron temperature expressed in electron-volts, i.e., in our example $2a < k_B T_e/(2q_0)$ [7].

On-line methods have the advantage that the experimenter is at once aware what the signal-to-noise ratio is like and can take measures to increase it if necessary. On the other hand, on-line methods usually do not yield absolute values (unless calibrated which requires additional efforts). However, even if only relative data of the second derivative of the probe characteristics are available it is possible to estimate the electron temperature and/or the electron mean energy.

On the other hand the off-line numeric methods of differentiation yield the absolute values of the second derivative of the probe characteristics. Also, since the experimenter records *merely* the probe characteristics that have a much higher amplitude than the signal of the second derivative, it may seem that usually it takes less effort to set up the experimental system. Yet the noise that may not be visible at first sight on the probe characteristics data can substantially influence the second derivative results. This is easy to comprehend from the fact that the amplitude of the second derivative is usually 2-3 orders of magnitude lower than the amplitude of the probe characteristics. Therefore, as a rule the numeric procedures for differentiating the probe characteristics require probe data of comparatively very good signal-to-noise ratio.

6.5   Probe diagnostics of anisotropic plasmas

All discharge plasmas are to a larger or lesser degree anisotropic. It is possible to make the direction-resolved diagnostics of the electron velocity distribution function (EVDF) with the aid of a planar probe. The normal to the probe plane must be adjustable in different angles θ to the preferred direction (e.g. current flow). It is possible to make the analysis of the probe data acquired in different angles to the preferred direction under the assumption that the EVDF is symmetrical along the preferred direction. In such a case one can expand the EVDF into a series of spherical (Legendre) polynomials with the angle θ as a parameter. In general the expansion has the form

$$f(w_x, w_y, w_z) = \sum_k f_k(w) P_k(\cos\theta) . \tag{6.51}$$

The first three coefficients of the expansion are given by $P_0 = 1$, $P_1 = \cos\theta$, and $P_2 = \frac{3}{2}(\cos^2\theta - 1)$. In these relations θ represents the angle between the preferred direction and the instantaneous velocity $\vec{w}$.

Substituting the expansion 6.51 of the EVDF into the general formula 6.4 for the probe current we obtain

$$I_{p_e} = -q_0 n_e \int dA_s \int_{w_{x1}}^{w_{x2}} \int_{w_{y1}}^{w_{y2}} \int_{w_{z1}}^{w_{z2}} w_z \sum_k f_k(w) P_k(\cos\theta) dw_x dw_z dw_z . \tag{6.52}$$

Due to the rotational symmetry of the problem in question all surface area components dA_s having the same orientation to the anisotropy component of the EVDF should carry the same current. Therefore we can integrate over dA_s to get A_s. Then equation 6.52 can be written in spherical coordinates

$$I_{p_e} = -q_0 n_e A_s \int_{w_1^*}^{\infty} \int_0^{2\pi} \int_0^{\arccos\theta^*} w^3 \sum_k f_k(w) P_k(\cos\theta(\vartheta, \varphi, \Phi)) \cos\vartheta d\vartheta d\varphi dw \tag{6.53}$$

where Φ is the angle between the surface normal and the anisotropy component, and w_1^* and θ^* are the integration limits for the electrons that fall on the probe surface. Differentiating this expression twice with respect to u_p we obtain

$$I_{p_e\Phi}'' = I_{p_e}''(f_0, f_1, f_2, \Phi) . \tag{6.54}$$

By measuring the second derivative of the probe characteristics with respect to the probe voltage in three different orientations $(0°, 90°, 180°)$ relative to the preferred direction we obtain a set of three equations from which we can determine the first three components in the expansion. The exact formulae were developed by Mesentsev et al. who obtained the following relations [42]:

1. for the electron density:

$$n_e = \left(\frac{1}{3}\right)(2m_e)^{1/2}q_0^{-3/2}A_p^{-1}\int_0^{\infty}u_p^{1/2}(I''_{p_e 0} + 4I''_{p_e 90} + I''_{p_e 180})du_p, \qquad (6.55)$$

2. for the isotropic part of the EVDF:

$$f_0(q_0u_p) = \left(\frac{1}{3}\right)(2m_e)^{1/2}q_0^{-5/2}n_e^{-1}A_p^{-1}u_p^{1/2}(I''_{p_e 0} + 4I''_{p_e 90} + I''_{p_e 180}), \qquad (6.56)$$

3. for the first-order anisotropy of the EVDF:

$$f_1(q_0u_p) = (2m_e)^{1/2}q_0^{-5/2}n_e^{-1}A_p^{-1}u_p^{1/2}g_1(q_0u_p), \qquad (6.57)$$

$$\text{where} \quad g_1(q_0u_p) = G_1 + (2q_0u_p)^{-1}\int_{q_0u_p}^{\infty}G_1 d\varepsilon,$$

$$G_1 = (I''_{p_e 0} - I''_{p_e 180})$$

$$\text{and} \quad \varepsilon = q_0u_p,$$

4. for the second-order anisotropy of the EVDF:

$$f_2(q_0u_p) = \left(\frac{2}{3}\right)(2m_e)^{1/2}q_0^{-5/2}n_e^{-1}A_p^{-1}u_p^{1/2}g_2(q_0u_p), \qquad (6.58)$$

$$\text{where} \quad g_2(q_0u_p) = G_2 + \left(\frac{3}{2}\right)(q_0u_p)^{-3/2}\int_{q_0u_p}^{\infty}\varepsilon^{1/2}G_2 d\varepsilon$$

$$\text{and} \quad G_2 = (I''_{p_e 0} - 2I''_{p_e 90} + I''_{p_e 180}).$$

6.6 Probe diagnostics under non-collision-free conditions

The collisionless OML model has been found satisfactory by experimenters for the electron collection by the probe over a comparatively wide range of experimental conditions. However, the positive ion current was underestimated in this model, and this leads to an overestimation of the positive ion density with respect to the electron density when determined from the same probe characteristics even at conditions when the plasma could be assumed as quasi-neutral. This has been found in non-thermal discharge plasmas, see e.g., Ref. [43], as well as in thermal afterglow plasmas [44]. Moreover the Bernstein–Rabinowitz [3] and the Laframboise [4] theory predict that the ion saturation current for the orbital motion limited conditions (OML) should be independent of the Debye number ξ_p, if the probe potential is constant. A comparison of the theoretical and experimental results has shown that in contrary to the

Bernstein–Rabinowitz [3] and Laframboise [4] theory the ion saturation current depends on the Debye number ξ_p at orbital motion limited (OML) conditions and that its magnitude is larger than the theoretically predicted one.

Experimental results obtained by several authors [44, 45, 46] have been summarized by Smith and Plumb [44]. They demonstrated that the ion density n_i determined by the ion current I_i may be expressed in terms of the electron density n_e determined by the electron current I_e by the empirical relationship $n_i/n_e \approx (1 + 0.07\sqrt{m_i/m_H})$, where m_H is the proton mass. From the quasi-neutrality of the plasma follows, however, that the electron density n_e should be equal to the positive ion density n_i assuming that no negative ions are present within the plasma. A difference between n_i and n_e has been observed in Ref. [44] even in the flowing afterglow plasma where the condition for the stable space charge sheath $T_i > T_e/2$ (see, e.g., Ref. [47]) is satisfied. The difference between n_i and n_e cannot be explained by an end-effect phenomenon, i.e. transition to a spherical-like sheath configuration [48]. The experimental results obtained in Refs. [44, 45, 46] therefore indicate an inaccuracy in the assumptions of the OML model of the positive ion collection for the used experimental conditions.

It is believed that these discrepancies can be explained, at least at certain simple experimental conditions, if the collisions of positive ions with neutrals in the space charge sheath around the probe are taken into account. Monte Carlo simulations have shown that already a small number of collisions in the sheath can destroy the orbital motion of an ion [49]. Further discussions are therefore devoted to probe theories that take account of the ion-neutral collisions in the sheath around the Langmuir probe. For the characterization of the pressure the Knudsen number $K_i = \lambda_i/r_p$ has been introduced.

The collection of positive ions by the probe under the influence of ion-neutral collisions has been studied for many years; examples of earlier papers on this subject are Refs. [50, 51, 52, 53, 54, 55, 56]. The theories that treat the problem of the positive ion collection by a Langmuir probe under the influence of collisions of ions with neutrals in the probe sheath can be divided into kinetic and continuum theories; such theoretical models are usually called *collision models* or *collision theories*. The kinetic theories aim at the description of the influence of the individual collision processes to the ion collection; the continuum theories use the hydrodynamical description. The kinetic approach can be regarded as more general since it describes the collisionless conditions too while the continuum theories require the supposition of many collisions. The kinetic theories discussed in this review present the effect of collisions in form of a correction to a collisionless ion collection model. Two collisionless models are used: Laframboise's model [4] that includes the classical Langmuir's orbital motion limited model (OML) [1] and the radial motion cold ion approximation model, called ABR-Chen [5, 57]. We shall not present the thin sheath collision models such as *continuum plus free fall theory* [58, 59] since they require higher plasma densities than usually used in plasma-aided technologies.

The collision probe theory of the positive ion as well as of the electron current to a cylindrical (spherical) probe has been presented by Chou, Talbot and Willis [60, 61]. It describes the Langmuir probe characteristics at arbitrary Knudsen numbers $0 \leq K_i = \lambda_i/r_p \leq \infty$. The result of this theory is a reduction of the ion as well as of the electron probe current due to elastic scattering of these particles caused by collisions with neutrals in the probe sheath. The influence of collisions is calculated as a correction F_T to Laframboise's collisionless ion current i_i^L. This correction was in a shape more suitable for practical calculations presented by Talbot and Chou [60] and by Kirchhoff, Peterson and Talbot [62].

It is to be noted that there were also other kinetic models for the description of the charged particle collection by a Langmuir probe. Bienkowski and Chang [63] found the solution of the Poisson and Boltzmann equation with the collision term for the limiting case $\xi_p \rightarrow \infty$ while Wasserstrom et al. [64] only in the limiting case $\eta \rightarrow 0$. Both limits can be, however, derived from the Talbot and Chou theory with the same results. Substantially simpler than the Talbot and Chou theory is the procedure employed by Self and Shih [65] that modified the ABR radial model (for spherical probes) [5] by introduction of the *friction term*. The results are presented as dependencies of the normalized ion current at a certain probe voltage on the ξ_p and the ion-neutral collision frequency ν_{in}. For cylindrical probes the same procedure is applied to the ABR-Chen [5, 57] radial model in Ref. [66].

Klagge and Tichý [21] employing the procedure of Ref. [60] carried out a set of numerical calculations of the positive ion current as a function of the Knudsen number K_i at the normalized potential $\eta = \eta_{fl} - 10$. The difference to the original Talbot and Chou paper [60] consisted in the fact that they used the ABR-Chen [5, 57] collisionless current i_i^A in place of i_i^L in the calculation of F_T. The reason was that the model i_i^A describes with reasonable accuracy the experimental results at conditions close to collision-free. The authors used their own analytical approximation for the numerical Chen [57] currents, see section 6.3.2. The exact step-by-step procedure described in Ref. [21] enables to set up a program that directly calculates the correction F_T.

Zakrzewski and Kopiczynski [67] have introduced another model in which, contrasting the Talbot and Chou theory, elastic collisions of ions with neutral particles have two consequences: the destruction of the orbital motion of ions and the elastic scattering of positive ions. The destruction of the orbital motion can lead to an increase of the positive ion current, too. The effect of the orbital motion destruction predominates for lower pressures when the mean free path of positive ions is bigger or comparable to the sheath thickness. The elastic scattering of the positive ions causes a monotonous decrease of the positive ion current and dominates for higher pressures. As a result the current peak appears at a pressure when on the average one collision of an ion with a neutral in the space charge sheath occurs. Similar as in the ABR-Chen and Laframboise theory and in the Talbot et al. theory [60, 61], the ion current in the Zakrzewski and Kopiczynski theory is normalized by Eq. 6.33. The resulting normalized dimensionless ion current i_i to a cylindrical probe then is $i_i = \gamma_1 \gamma_2 i_i^L$, where γ_1 describes the rate of increase of ion current due to destruction of orbital motion, and γ_2 corresponds to the rate of reduction of ion current due to scattering. The correction factor F_Z is given by $F_Z = (\gamma_1 \gamma_2)^{-1}$. Zakrzewski and Kopiczynski have derived the coefficient γ_1 under the assumption that at the orbital motion-limited conditions the positive ion current collected by a perfectly absorbing probe is saturated and described by the Laframboise theory. This physical argument can be expressed analytically, assuming that the dimensionless ion current at the sheath edge i_i^A is given approximately by the Allen et al. theory [5]. In the collisionless limit only the current i_i^L predicted by Laframboise reaches the probe surface. The current $(i_i^A - i_i^L)$ leaves the sheath due to the orbital motion.

When an orbiting positive ion undergoes a collision with a neutral particle in the space charge sheath it looses energy and is attracted to the probe. An ion on the average makes $X_i = S/\lambda_i$ collisions in the sheath, if we denote the thickness of the sheath by S and the ion mean free path by λ_i. According to Zakrzewski and Kopiczynski the rate of increase of the positive ion current due to the destruction of the orbital motion by elastic collision is $\gamma_1 = 1 + \left(i_i^A/i_i^L - 1\right)\frac{S}{\lambda_i}$. Kopiczynski [68] determined the thickness of the sheath S on the basis of numerical calculations carried out by Basu and Sen [69]. Zakrzewski and Kopiczynski estimated the corresponding reduction rate of the ion current due to the elastic scattering γ_2 according to Schulz and Brown [70] and Jakubowski [71]. It should be noticed that for a given probe potential γ_1 and γ_2 are functions of ξ_p and K_i.

For a normalized probe potential $\eta_p = -15$ Kopiczynski [68] carried out an extensive numerical calculation of the dependence of the quantity $i_i\xi_p^2$ on the ion Knudsen number K_i, with the Debye number ξ_p as a parameter. David [72] and David et al. [18] extended these calculations towards lower Knudsen numbers K_i occurring in a medium pressure discharge. The comparison of several of the already mentioned theories of the ion current collection by the probe in the transition pressure regime has been done by David [72].

The results of Talbot and Chou and those of Zakrzewski and Kopiczynski are not in contradiction with each other within the region $2 \leq \xi_p \leq 3$. For lower ξ_p ($\xi_p < 2$) the ion current calculated by using the Zakrzewski and Kopiczynski theory [67, 68] exhibits a well pronounced maximum at $X_i \cong 1$. At higher values of $X_i \gg 1$ the ion current decreases with decreasing K_i more rapidly than the ion current obtained from the theory [60, 61] in which no current peak is observed.

Since the Chou and Talbot theory [60, 61] does not take into account the effect of the probe current increase due to the orbital motion destruction caused by ion collisions with neutrals within the probe sheath, the theory developed by Zakrzewski and Kopiczynski describes the ion collection by a Langmuir probe at OML conditions in the region where not all the ions suffer a collision with a neutral particle within the probe sheath ($X_i < 1$) more precisely than the Talbot and Chou theory. On the other side for a larger number of ion collisions within the sheath ($X_i \gg 1$) the theory developed by Zakrzewski and Kopiczynski is not applicable at OML conditions since it uses for the evaluation of the factor F_Z the formulae which have been derived in [71] under the assumption of only a few ion collisions within the probe sheath. In order to extend the validity of the theory developed by Zakrzewski and Kopiczynski [67, 68] for an arbitrary K_i it was suggested in Ref. [73] to apply the Talbot and Chou theory [60, 61] for the determination of the factor (or its equivalent) describing the effect of scattering of ions due to their collisions with neutrals within the probe sheath. The corresponding expression following the model used by Zakrzewski and Kopiczynski is $i_i = \gamma_1\gamma_2^*i_i^L$, where γ_2^* is the coefficient describing the effect of the ions scattering due to collisions with neutrals within the probe sheath determined from the Talbot and Chou model [60]. The advantage of the new model [73] of the ion collection by a cylindrical Langmuir probe mainly consists in the fact that it is valid for any K_i, i.e. for $0 \leq K_i < \infty$, as long as the condition of the OML regime of the probe is fulfilled ($\xi_p \leq 3$). In contrast to the calculations of the γ_2 factor which have been made by Zakrzewski and Kopiczynski the correction factor γ_2^* depends on the ratio of the electron to ion temperature $\tau = (T_e/T_i)$. It was suggested in Ref. [73] to call this new model the *modified* Talbot and Chou model since it refines the *classical* kinetic Chou and Talbot theory [60, 61]. A comparison of the theory [73] with the Chen-Talbot [21] and with the Zakrzewski-Kopiczynski [67, 68] theory has been made by Kudrna [74]. He calculated also the dependence of the directly measurable quantity $\xi_p^2i_i$ on K_i with ξ_p as a parameter. Samples of his results are presented in Fig. 6.6. Note that in both figures the ion current

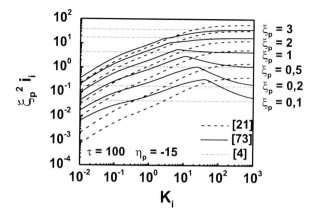

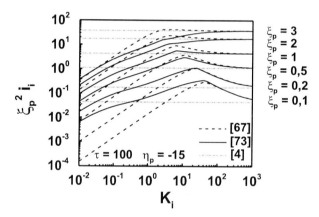

Figure 6.6: (a) Comparison of the Chen-Talbot positive ion current model [21] with the modified Chou-Talbot model [73]. For comparison also the collisionless current model of Laframboise [4] is shown. (b) Comparison of the Zakrzewski-Kopiczynski positive ion current model [67] with the modified Chou-Talbot model [73]. The collisionless current model by Laframboise [4] is also shown.

calculated according to Ref. [73] decreases at lower K_i slower than predicted by the theories of Refs. [21, 67, 68].

Experimental assessment of the applicability of all mentioned collision theories is presented in Ref. [75]. The measurements have been done in a low-pressure flowing afterglow system with a Langmuir probe and downstream mass analysis. Several kinds of ions have been used for this study. Fig. 6.7 presents typical results that have been obtained with rare gas ions Ar^+ in a He carrier gas. The plot shows the numerical values of the ratio of the ion

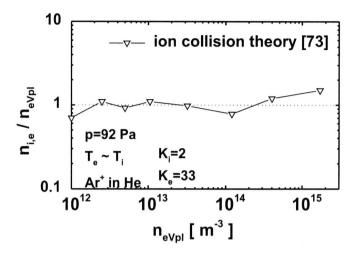

Figure 6.7: The ratio of the ion density evaluated using the theory of Ref. [73] to the electron number density *vs.* electron number density in a flowing afterglow discharge [75].

density according to the theory of Ref. [73] to the electron density n_{eVpl} estimated from the electron current at the plasma potential *vs.* n_{eVpl}.

The present state of knowledge about the influence of positive ion-neutral collisions in the probe sheath does not, however, allow to name a universally applicable theory that would give satisfactory results in a broad range of experimental conditions; for afterglow conditions the theory of Ref. [73] yields fairly consistent results, for an active plasma (also when the supposition of the Maxwellian EEDF is not quite satisfied) the ABR-Chen theory with the collision correction [21] gives the best agreement with the experiment. More general results might bring the Monte-Carlo simulation of the ion motion in the probe vicinity. First such steps have been made in Ref. [49].

6.7 Langmuir probe in a magnetized plasma

In many technological applications of the low-temperature plasma, for example deposition of thin films or etching of microstructures, a magnetic field is used that confines the plasma and increases the grow/etch rate. The magnetic field in these systems can be either non-homogeneous (usually created by permanent magnets; one example is the planar unbalanced magnetron) or almost homogeneous fields (created by coils) with a strength usually not reaching too high values. When using the Langmuir probe as a diagnostic tool in these systems the question arises up to what limit of the ratio B/p (magnetic field strength to the working pressure; this is a similarity parameter in magnetized plasma) it is possible to use the conventional methods for the evaluation of the basic plasma parameters such as charged particle density and the electron energy distribution function (EEDF) from the Langmuir probe data.

The discussion of the influence of the magnetic field on the probe measurement in the collision-free case can be, in accord with Ref. [76], divided in 4 categories in dependence on the parameter $\beta = r_p/r_L$ (r_L is the radius of the cyclotron motion of a charged particle): (i) At $\beta \ll 1$, weak B, the influence is small and can be neglected (rare case of technologically interesting applications), (ii) at $\beta \approx 1$, B is still weak, but it is sometimes necessary to introduce small corrections (frequently the case in plasma-aided technologies), (iii) at $\beta > 1$, B is strong, but it is still possible to interpret at least a part of the Langmuir probe characteristics (case of tokamak edge plasma), and (iv) at $\beta \gg 1$, i.e., at very strong B, the probe characteristics are no longer interpretable. From the definition of the parameter β it is evident that β decreases with increasing charged particle velocity and mass; therefore the faster electrons and the ions are less influenced by the magnetic field during their collection by the probe. In other words, the part of the probe characteristics that shows the least distortion due to the magnetic field is that for a very negative probe with respect to the plasma, see, e.g., Refs. [77, 78]. The changes in the probe characteristics therefore are most apparent in the region close to the plasma potential when the probe is applied in a magnetized plasma. The directional movement of charged particles along the magnetic field lines reduces the diffusion of particles in the direction across the field lines. Considering the electron motion which is mostly influenced by the magnetic field, we see that the diffusion coefficient across the field $D_\perp$ reduces to $D_\perp = D_0/(1+\Omega_e^2\tau_{en}^2)$ with $D_0 = v_{th}^2\tau_{en}$, where $\Omega_e = q_0 B/m_e$, $\tau_{en} = 1/\nu_{en}$ and v_{th} are the electron cyclotron frequency, electron-neutral collision time and electron thermal velocity, respectively [79]. If $\Omega_e\tau_{en} \gg 1$ (i.e., a magnetized plasma at low pressures) which holds for most technologically used plasmas, then the expression for $D_\perp$ reduces to $D_\perp \approx v_{th}^2/\Omega_e^2\tau_{en} = r_L^2/\tau_{en}$. In other words in the direction perpendicular to the magnetic field the effective mean free path of electrons is roughly equal to the radius of the cyclotron motion. If the probe draws too much current at probe potentials close to the plasma potential the electrons are absorbed by the probe more rapidly than they can be supplied by diffusion from the distant regions where they are produced. Another effect concerns the change of the effective collecting area of the probe, since the charged particles flow to the probe mostly from the direction of the field lines reducing hence the original probe area to double of its projection to the field direction. Finally, at higher magnetic fields, e.g., in tokamaks, the probe is connected to its reference electrode only by the *current tube*, which reduces the ratio of the reference-electrode-probe surface collecting areas. All three effects then lead to *blurring of the knee* of the probe characteristics near the space potential as seen in Fig. 6.8. When for the data analysis from such an affected probe the conventional methods are used the resulting plasma number density is underestimated, the plasma potential shifted towards probe retarding voltages and the electron temperature deduced from the slope in the electron retarding regime is overestimated. Note that this effect not only depends on the magnetic field strength, but rather on the ratio Ω_e/ν_{en}, or on B/p. The degree of anisotropy of the problem (and hence of the influence to the probe measurements) will therefore depend also on B/p and not on B itself. Thus the assessment of the error caused by the effect of the magnetic field to the probe data and consequently to the accuracy of the estimated plasma parameters is most interesting. Experimental assessment of this effect in weak-to-medium magnetic fields is presented in Refs. [80, 81, 82]. The work is based on the assumption that the influence of the magnetic field on the positive ion collection by the probe is negligible in the range of pressures and of magnetic fields employed. Thus the comparison of the positive ion density estimated from the ion accelerating region of the probe characteristics with the electron density estimated from

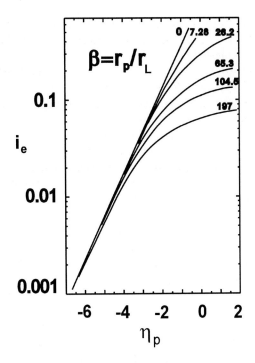

Figure 6.8: The normalized electron probe current *vs.* η_p in dependence on the parameter β for a spherical probe (from Ref. [77]).

the electron current at the plasma potential (by a conventional method, i.e. without any correction to the magnetic field) at lower and higher B/p should give an idea of the magnitude of this influence. A sample of experimental results is given in Fig. 6.9. The figure is arranged in a similar manner as Fig. 6.7. It is seen that at higher B/p the electron density estimated from the electron current at the space potential is systematically underestimated, while at low B/p there is an agreement between the estimated values of the ion and electron density; the difference gives a measure for the error due to the magnetic field.

Anisotropy of the plasma in a magnetic field has been studied, e.g., in Refs. [79, 83, 84]. Aikawa [79] used for the determination of plasma anisotropy in a magnetic field two Langmuir probes with their collecting surfaces perpendicular and parallel to the magnetic field.

Ref. [85] deals with the radial distribution of the EEDF in the positive column of a glow discharge in neon in a magnetic field collinear with the electric field. The authors used the *conventional* second harmonics method to obtain the second derivative of the probe characteristics from which they calculated the electron density and the electron mean energy. They found that even at comparatively low values of B/p of the order 10^{-3} T/Pa the magnetic field influences the radial distribution of the mean electron energy in the sense that higher B/p causes the increase of the mean electron energy at larger radii.

Arslanbekov et al. [86] discuss the EEDF measurements by a Langmuir probe at elevated

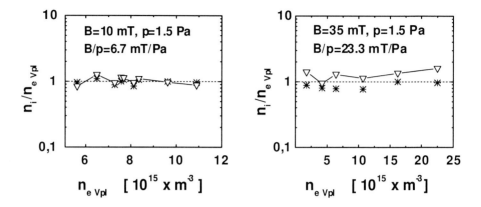

Figure 6.9: Comparison of the relative positive ion ($-\nabla-$, theory Ref. [21]) and electron densities (∗, OML theory [1, 4]) estimated from the same probe data at different magnetic fields. *dc* cylindrical magnetron discharge, working gas argon. Reference electron density from the electron current at plasma potential.

pressures and in a magnetic field. They create an analogy between the effect of an increased pressure and the effect of the magnetic field on the probe characteristics. However in the analysis of the influence of the magnetic field on the cylindrical probe in a plasma the case of a thin sheath is assumed and hence this work is not directly applicable to the low temperature discharge plasma of technological interest since in such a case the sheath thickness is usually large or comparable to the probe radius. Measurements of the EEDF in the cylindrical magnetron discharge have recently been published in Refs. [87] and [88].

6.8   Space and time resolved Langmuir probe method

6.8.1   Space resolved Langmuir probe measurements

In the case that the plasma is spatially inhomogeneous it is possible to use a probe that is movable along the path of interest, e.g., radially in case of a study of the radial plasma density and electron temperature distribution in a cylindrical discharge tube. The spatial resolution is given by λ_D or by the probe dimension, whichever is bigger. Therefore it is important that the probe dimension should be smaller than the characteristic length of the investigated spatial change. This means that along the probe dimension the plasma parameters should not change considerably. Since the minimum dimension of the Langmuir probe is limited mainly by the dimension of its holder and therefore cannot be easily made smaller than some tenths of a mm, it is almost impossible to study reliably by probes spatial changes of plasma parameters, where the characteristic length is smaller than a few mm (e.g., in narrow plasma jets). Much better spatial resolution can, however, be achieved, if the plasma parameters do not change in all directions in the same manner. In case of cylindrical symmetry, e.g., where the characteristic dimension of change in longitudinal direction is much larger than in the radial one, a thin cylindrical Langmuir probe positioned perpendicularly to the radial direction can yield a spatial resolution down to some hundred microns.

Most common mechanisms for the movement of the probe in the plasma vessel are based on the *screw-and-nut* principle. The thread, however, should be made sufficiently lose in order to enable the outgasing of the system by baking without the danger of seizing the mechanism afterwards (stub acme screw thread). The transfer of the rotational movement into the plasma vessel is made either magnetically or by using the bellow type rotational transfer (cats tail). The simplest mechanism for moving the probe is a ferromagnetic slug inside the diamagnetic stainless steel or glass tube that is moved by an external (electro-)magnet. The electric lead is made either by a spiral-like wire permanently connected to the probe and the vacuum feedthrough or by a sliding contact.

6.8.2 Time resolved Langmuir probe measurements

Often it is necessary to investigate a plasma that changes in time. Famous are the studies of the variations of the EEDF in ionization waves (moving striations) [89, 90] or in a stationary afterglow [44]. In addition there are many studies using probes for studying plasma in single shot systems, such as tokamaks [91]. Last but not least there are many technologically important applications, where plasmas are generated by means of alternating current (*ac*) or radiofrequency (*rf*) power.

The approach to the problem differs depending on the possibility to make the studied changes of plasma parameters periodic in time with a reasonably short period (at least several Hz). Examples of systems, where the periodicity is realizable, are the mentioned studies of stationary afterglow or of ionization waves, but also *ac* or *rf* generated plasmas. Since the Langmuir probe technique relies upon the formation of a space charge sheath around the probe, its time resolution cannot be higher than the ion plasma frequency, $\omega_{pi} = \sqrt{\frac{q_0^2 n_i}{\varepsilon_0 m_i}}$. If the frequency of plasma variations (e.g., in case of *rf* generated plasmas) is higher than ω_{pi}, only measurements of time-averaged values of plasma parameters are possible by the Langmuir probe. Another interesting frequency is the electron plasma frequency, $\omega_{pe} = \sqrt{\frac{q_0^2 n_e}{\varepsilon_0 m_e}}$. With $n_i \approx n_e = 10^{16}$ m^{-3} we get $\omega_{pe} \approx 1$ GHz, and $\omega_{pi} \approx 3.16$ MHz (for argon). In the assessment of the different working regimes of the probe diagnostics in a time-dependent discharge plasma we restrict ourselves to the case that the plasma is excited by the periodically changing electric field. For the discussion of the relations in such a discharge plasma the detailed knowledge of the complex time response of the electron velocity distribution function in dependence on the frequency of the applied electric field is of highest importance. For weakly ionized plasmas where the Coulomb electron-electron collisions can be neglected the calculations already have been made [92]. The results presented in this work show that the time response of the electron kinetics is given by the ratio of the field frequency to the collision frequencies for the electron momentum (ν_i) and the electron energy (ν_ε) dissipation. Systematic investigations of the case when the Coulomb interactions cannot be neglected are to our knowledge not yet available. In the following discussion we therefore shall use the simplified discussion as it is known from classic books on probe technique [93].

There are 5 regions of operation for Langmuir probes as far as the frequency of plasma generation or plasma changes is concerned:

- $\omega \ll \omega_{pi}$ The ions and electrons are in equilibrium with the superimposed periodically varying electric field. The steady ion as well as the electron current increases due to the *rectification effect* of the non-linear probe characteristics. This incremental increase is independent of the frequency. Also harmonics of ion and electron currents are generated due to the non-linearity of the probe characteristics. Time resolved measurements of plasma parameters are possible. This range of frequencies is typical for low frequency plasma oscillations such as ionization waves, periodically switched discharge etc.

- $\omega \approx \omega_{pi}$ The electrons are in equilibrium with the oscillating electric field. The period of oscillations is comparable to the transit time of ions through the space charge sheath and therefore a small resonance peak in the incremental increase in the steady ion current is observed. Since the resonance effects are difficult to assess it is not recommendable to make probe measurements in this frequency range. This frequency range can be entered when studying the frequency dependence of PECVD technologies, i.e., when the plasma is generated not by a single frequency power *rf* generator, but by a small power signal generator followed by a power amplifier.

- $\omega_{pi} < \omega \ll \omega_{pe}$ The electrons are in equilibrium with the oscillating electric field whereas the ions are not. The incremental increase of electron current remains the same as in the previous two cases, the one of the ion current vanishes. Harmonics of the electron current are generated in the same manner as in the previous cases, those of the ion current are zero. Probe measurements of time-averaged values of plasma parameters are possible provided that the *rf* voltage component between the probe and the plasma is removed. This frequency range is typical for single frequency *rf* discharges (13.56 MHz, 27.12 MHz).

- $\omega \approx \omega_{pe}$ The period of oscillations is comparable to an electron's transit time through the sheath. Therefore a small resonance peak in the incremental increase in the steady electron current is observed. Since the resonance effects are difficult to assess it is not recommendable to do probe measurements in this frequency range. This range of frequencies is not typical for any plasma of technological interest.

- $\omega > \omega_{pe}$ Neither ions nor electrons are able to respond to the oscillating electric field. There is neither incremental increase nor harmonics generation of the steady probe current. The probe measurements can be done in a normal way, i.e., as in a *dc* plasma, provided that the *rf* pick-up does not interfere with sensitive current measuring input of the probe set-up. This range of frequencies is typical for magnetron generated microwave plasmas (2.45 GHz).

The case of periodic changes with $\omega < \omega_{pi}$, where the time resolved measurements can be done, the case of $\omega_{pi} < \omega \ll \omega_{pe}$, where it is possible to measure the time-averaged values of plasma parameters and their relevance to single shot experiments, will be discussed separately.

6.8.2.1 Time resolved probe measurements in periodically changing plasmas at $\omega < \omega_{pi}$

Earlier studies made use of an electronic gate which allowed the passing of the signal from the probe to the signal processing system only for a short time period that was synchronized with the plasma process under study. Gates were constructed differently, either the probe was attached to the experiment permanently [89, 90] and the probe current was sampled or the gate connected directly the bias to the probe [94, 95], i.e., for the remainder of the time period the probe was floating. The first arrangement avoids the transitional effects that arise by application of the square-wave-like voltage to the probe (for discussion on this effect see, e.g., Refs. [96, 97, 98, 99, 100]), the other one avoids the over-current to the probe if there exists a large change of plasma potential along the studied time period, e.g., when measuring in the periodically switched discharge.

Most of up-to-date equipment uses the multichannel scaler (either factory made, or in the shape of an add-on card to a personal computer) in order to record the probe current continuously and in synchronism with the plasma periodicity. The modern devices have a time resolution around 1-2 μs per channel. By recording many time dependencies of the probe current at different probe voltages a 3-D matrix of current values at different voltages and times can be formed in a computer memory from which it is easy to construct the probe characteristics in dependence on time by swapping the parameter (voltage) and variable (time) [101, 102]. In order to avoid the destruction of the probe due to the excess current care must be taken to protect the probe if a large change of the plasma potential along the studied time period can be expected [103, 104]. Time-resolved probe measurements in time varying plasma with the frequency 100 kHz [105], and even up to 400 kHz [106], have been reported.

6.8.2.2 Probe measurements of time-averaged plasma parameters at $\omega_{pi} < \omega \ll \omega_{pe}$

There is a permanent interest to study the time-averaged plasma parameters in plasmas generated by an *rf* single frequency generator that operates most often at 13.56 MHz. The incremental increase of the electron probe current in this frequency range degrades the probe characteristics in a sizeable manner making the probe measurements of even just simple plasma parameters such as the floating potential quite useless. However, if a provision is made to remove the *ac* bias voltage component which arises between the plasma and the probe in *rf* generated plasmas the additional electron incremental current can be made negligibly small and decent probe measurements are possible. To remove the *ac* component between the plasma and the probe means to *ac* decouple the probe from the *dc* current measuring circuit and let the probe swing in phase with the plasma potential oscillations. Decoupling is usually done for the generator frequency and its second and eventually also third harmonics by using the parallel LC resonant circuits tuned to the appropriate frequencies (see Fig. 6.10). These decoupling circuits must be located in closest vicinity to the probe in order to avoid any parasitic capacitance of the probe lead to ground. A closer analysis [26] showed that at 13.56 MHz and $T_e \approx 1$eV any parasitic capacitance of the probe against ground bigger than approximately 10^{-2} pF can significantly distort the measured probe characteristics. The tuning of the filters is often influenced by stray capacitances when the probe is inserted into a metallic plasma reactor. For this reason some authors tried to place the actual tuned filter outside the vacuum

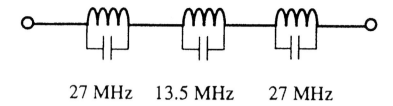

27 MHz   13.5 MHz   27 MHz

Figure 6.10: Passive LC filters [26].

chamber and call their probe a *tuned probe* [107].

Instead of the tuned filter the use of the low-pass broadband filter is also possible. Such filters may be constructed of RC elements, i.e. of the resistor in series with the probe together with the parasitic capacitance [108]. In this case the voltage drop on the series resistor must be accounted for. Attempts of using LC broadband filters also have been made [109]. This method uses the resistive impedance of ferrite core chokes. In other words, the choke wound on a low cut-off frequency ferrite core presents high resistance at high frequency while keeping negligible resistance for *dc* current. Therefore it is not necessary to take the voltage drop on the choke(s) into account.

A similar effect as with the decoupling can be achieved by driving the probe by the same voltage waveform as is the waveform of the plasma potential at the probe position. This is a technique used to minimize the capacitance of the inner conductor against ground in a tri-axial cable when the internal shield is driven by the amplified voltage of the inner conductor [110]. The realization of this technique is, however, not quite simple, since the plasma potential waveform is not exactly known. Two clones of this technique exist, the drive by the voltage from an external *rf* generator [111] and by the voltage from an additional probe that is positioned close to the measuring one [112, 113]. This additional probe is sometimes isolated from the plasma by a glass tube and is called *capacitive probe* [114]. The *ac* voltage between the plasma and the probe can be minimized by minimizing the plasma potential changes at the probe position, i.e. by symmetric drive of the parallel-plane-electrodes plasma reactor and by positioning the probe in the plane of symmetry between the planar electrodes [26]. A comparison of the passive and the active compensation are presented in Ref. [115].

An interesting modification of the probe method in *rf* plasma presents Ref. [116]. Here the authors measure in the 13.56 MHz *rf* discharge the time-dependence of the *ac* current to the probe in dependence on the *dc* probe bias. From this total current they eliminate the electron component by determining and subsequently subtracting the ion current component and the current component corresponding to the displacement current. From the electron current component they were able to calculate the second derivative with respect to the probe voltage and obtained therefore the EEDF in dependence on the time over the *rf* period. They detected the changes in the tail of the EEDF along the *rf* period as predicted by Winkler et al. [92]. Ref. [116] presents therefore the only (up to our knowledge) time-resolved measurements of the EEDF with the time resolution better than that given by $1/\omega_{pi}$.

6.8.2.3   Time resolved probe measurements in single-shot experiments

In single-shot systems, e.g., in the Langmuir probe studies of the scrape-off layer (plasma in the limiter shadow) of tokamaks the whole probe characteristics have to be acquired during a time which is much shorter than the characteristic time of the change of plasma parameters. The probe voltage is generated either as a sawtooth-like voltage or (simpler) as a sinusoidal voltage with an appropriate period. Care must be taken to measure only the *true* probe current and not the complex (capacitive, displacement) current components that arise due to parasitic capacitances of the probe. If the capacitive current cannot be directly avoided (e.g. due to the indispensable long probe leads) the bridge probe circuit is recommended. The frequency of the probe bias change must fulfill the condition $\omega < \omega_{pi}$. The plasma density in the limiter shadow of tokamaks is generally much higher than 10^{16} m^{-3} therefore also the ω_{pi} is higher and the admissible bias frequency ranges up to several hundreds of kHz. During the stationary phase of a tokamak discharge which lasts up to several seconds it is therefore possible to record several probe characteristics [117].

6.9   Probe diagnostic of chemically active plasmas

During the deposition of conducting and non-conducting layers in a plasma environment the probe diagnostic becomes more complicated. The reason for this consists in the fact that the similar layer as is intended to cover the substrate is deposited both on the probe and on the reference electrode. Generally, the probe surface contamination changes the work function resulting in a deformation of the probe characteristics and/or in hysteresis [118, 119]. If the layer is conducting it helps to centre the probe wire in the insulating holder in such a way that it does not touch the holder walls (see Fig. 6.11). In the case of a non-conducting layer several

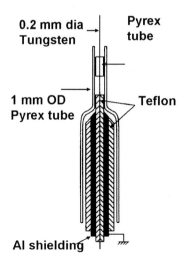

Figure 6.11: Langmuir probe for the use in chemically active plasmas (from Ref. [39]).

approaches have been developed.

- The film is regarded as a series resistance to the probe. Such model has been developed, e.g., in Refs. [120, 121]. This model is simple but its application can be questionable in some cases.

- The film from the probe surface is removed either by periodic ion bombardment of the probe surface and measuring with a short sweep time [122, 123] or by its direct [7, 121, 124] or indirect [125, 126] continuous heating. The heating has the advantage that it prevents the formation of the non-conducting layers, however, the probe construction is more complicated. Further, the heated probe can be usually operated also as the emissive probe that enables easy estimation of the plasma potential [127]. The directly heated probe can also be heated periodically: only prior to the actual measurement. This eliminates the uncertainty of the probe bias due to the voltage drop along the probe wire during heating.

- If the conductive reactor wall is used as a reference for a single probe in plasma-chemical environment, the layer on the reference electrode can complicate the measurements, too [128, 129]. If this happens the application of the so-called triple probe method may help [130, 131]. This method uses a floating system of 3 probes, and hence is independent of a reference electrode. The three probes can also be heated thus preventing the layer deposition (see Fig. 6.12). An application of such a probe system is described in Ref. [132].

- A probe that works on another principle is used. One example is the *rf* probe. It makes use of the change of the plasma impedance with the plasma parameters (density, collision frequency). The plasma impedance can be measured directly by means of a vector impedance meter [133]. Alternatively, assuming some simplifications, the conductive part of the plasma impedance can be measured by a bridge circuit [134]. Further, the plasma oscillation probe uses an interesting principle [135]. A beam of electrons injected into a plasma by a suitably biased hot filament generates plasma oscillations with the plasma frequency ω_{pe} (see section 6.8.2). This frequency can be detected by a small wire-antenna (probe) and measured by a high-frequency spectrum analyzer. Since it is directly

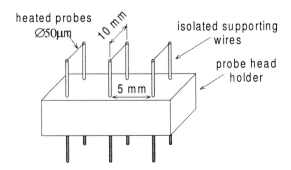

Figure 6.12: Triple heated probe.

coupled to the electron density the knowledge of the plasma frequency means that the plasma density is known, too. Finally, let us mention a novel electrostatic probe method for ion flux measurements [136]. A flat probe is biased by a square-wave-modulated *rf* voltage via a capacitor. During each *rf* voltage pulse the negative *dc* self-bias is generated on the probe. This *dc* bias is then discharged by means of a positive ion flux coming from the plasma. Since this ion current is almost independent of the probe voltage the change of the probe voltage is almost linear, and from its time-derivative it is possible to calculate the ion flux. This method is almost independent of the thickness of the layer on the probe; however, for the method to work properly it is necessary to have a large flat probe (in Ref. [136] the diameter of the probe was 50 mm).

Further discussion on applications of Langmuir probes in a plasma-chemical environment can be found, e.g., in Refs. [137, 138].

6.10   Double probe technique

When a voltage is applied between two small electrodes immersed in plasma a current flows, and the current-voltage characteristics resemble the ion current part of the single probe characteristics in both polarities of the applied voltage (see Fig. 6.13). In fact the single probe technique obeys the same laws, only one of the electrodes has a much larger surface area than the other one. Therefore we speak of a double probe method only, if the surface areas of both electrodes are small and not very different from each other. The advantage of the double probe method is that it does not need the large reference electrode and therefore can be applied in electrode-less (*rf* generated) plasma. The double probe technique is usually applied to measure the electric field in a discharge plasma, but it can also be applied to determine the electron temperature and, at certain experimental conditions, the plasma density.

A usual arrangement in the double probe technique places both probes close to each other so that it is possible to assume that the plasma parameters are the same at both probe positions. Here the words *close* and *small* relate to the characteristic dimension of the change of plasma parameters as discussed in chapter 6.8. If, moreover, both the probes have the same shape

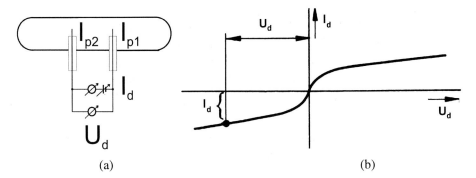

Figure 6.13: (a) Basic double probe circuit. (b) Typical double probe characteristics.

and the same surface collecting area, the resulting current-voltage characteristic is centre-symmetrical with respect to the point of zero current.

Just the double probe theory that was originally presented by Johnson and Malter will shortly be described [139]. The suppositions of this theory are the same as that of the Langmuir theory; mainly that the charged particles do not undertake collisions neither with themselves nor with the neutrals in the space charge sheath around the probes. The double probe technique for the electron temperature determination in the middle and higher pressure plasma when the collisions of charged particles with neutrals start to be important has been studied for various plasma conditions by Bradley and Matthews [140], Kirchhoff et al. [141], Klagge and Tichý [21], and Chang and Laframboise [142]. Since the total current drawn by the double probe system from the plasma is zero (the system is floating) it is evident that the current, carried to one probe by particles of a certain charge must be compensated by the current of the second probe due to carriers of the opposite charge. Denoting the first and second probe by the subscripts 1 and 2, respectively, we can write down the following equations based on Kirchhoff's laws:

$$I_{pe1} + I_{pi1} = -(I_{pe2} + I_{pi2}) = I_d \qquad \text{1}^{\text{st}} \text{ Kirchoff law} \qquad (6.59)$$

$$U_1 = U_2 + U_d \qquad \text{2}^{\text{nd}} \text{ Kirchhoff law} \qquad (6.60)$$

where

$$I_{pe1} = I_{pe1}(U_1) \quad \text{and} \quad I_{pi1} = I_{pi1}(U_1)$$

$$I_{pe2} = I_{pe2}(U_2) \quad \text{and} \quad I_{pi2} = I_{pi2}(U_2).$$

Here $U_{1,2}$ denotes the probe voltage with respect to the space potential, and U_d is the voltage between the two probes. In case of a Maxwellian EVDF it is possible to use the following expression for the electron probe current to the first probe:

$$I_{pe1} = I_{pe}(U_1 = 0) \exp\left(-\frac{q_0 U_1}{k_B T_e}\right); \qquad (6.61)$$

a similar expression is valid for the second probe. The current in the double probe circuit then is given by:

$$I_d(U_d) = I_{pi1}(U_1) + I_{pe1}(U_1 = 0) \exp\left(-\frac{q_0 U_1}{k_B T_e}\right)$$

$$-I_d(U_d) = I_{pi2}(U_2) + I_{pe2}(U_2 = 0) \exp\left(-\frac{q_0 U_2}{k_B T_e}\right). \qquad (6.62)$$

If we differentiate this double probe characteristics with respect to U_d and take into account that $(dU_d/dU)_{fl} = 2$, then we obtain the expression for the electron temperature firstly derived by Johnson and Malter [139]:

$$\frac{k_B T_e}{q_0} = \left[2\left(\frac{dI_d}{dU_d}\right)_{fl} - \frac{1}{2}\left(\frac{dI_{pi1}}{dU} + \frac{dI_{pi2}}{dU}\right)_{fl}\right]^{-1} \times \frac{2(I_{pi1})_{fl}(I_{pi2})_{fl}}{(I_{pi1})_{fl} + (I_{pi2})_{fl}}. \qquad (6.63)$$

Here $(I_{pi1,2})_{fl}$ denotes the extrapolated dependence of the ion current on $U_d = 0$. A modification of this procedure is based on the estimation of the so-called Γ-function, $\Gamma = I_{e1}/I_{e2}$, where the electron temperature T_e is determined from the slope of the dependence $\ln(\Gamma)$ *vs.* U_d

$$\ln(\Gamma) = -\frac{q_0 U_d}{k_B T_e} + \ln\left(\frac{A_1}{A_2}\right),$$ (6.64)

where A_1 and A_2 are the surface collecting areas of the first and second probe, respectively. The great advantage of this method is its insensitivity to the effect of collisions in the space charge sheath around the probe. It has been shown that for $K_i \to 0$ the error of this method does not exceed 15% [21]. Therefore (for the purpose of estimating T_e) it is sometimes advantageous to convert the single probe characteristics to those of the double probe [18]. The procedure is based on the determination of the voltage difference U_{d1} between the two points on the single probe characteristics which correspond to the same absolute value of the chosen probe current I_{d1} but with opposite signs; by choosing I_{d1} from 0 to the maximum ion probe current we obtain the dependence I_{d1} *vs.* U_{d1} that corresponds to one half of the double probe characteristics. The second half is centre-symmetrical with respect to the origin of U_d.

The double probe technique can also be used for the plasma density estimation provided that the model of the positive ion collection by the probe which is suitable for the particular experimental conditions can be applied. A fairly general model has been derived in Ref. [143].

Acknowledgements

The authors wish to thank Dr. Petr Špatenka for his substantial contribution to section 6.9. Part of this work was supported by the Deutsche Forschungsgemeinschaft (DFG) in Sonderforschungsbereich (SFB) 198 *Kinetik partiell ionisierter Plasmen*, by the Grant Agency of the Czech Republic, Grants No. GACR-202/00/1689, GACR-202/98/P078 and GACR-202/98/0116, and by the Grant Agency of Charles University, Grants No. GAUK-075/98 and 171/2000.

References

[1] I. Langmuir, H.M. Mott-Smith, Gen. Elec. Rev. **26** (1923) 731.

[2] D. Bohm, E.H.S. Burhop, H.S.W. Massey, The Characteristics of Electrical Discharges in Magnetic Fields, A. Guthrie, R.K. Wakerling (Eds.), McGraw-Hill: New York (1949), p.77.

[3] I.B. Bernstein, J.N. Rabinowitz, Phys. Fluids **2** (1959) 112.

[4] J.G. Laframboise, Univ. of Toronto, Institute for Aerospace Studies (UTIAS), Report **100** (1966).

[5] J.E. Allen, R.L.F. Boyd, P. Reynolds, Proc. Phys. Soc. B **70** (1957) 297.

[6] A.A. Sonin, AIAA Journal **4** (1966) 1588.

[7] S. Klagge, PhD Thesis, University of Greifswald, Germany (1988).

[8] B.M. Annaratone, M.W. Allen, J.E. Allen, J. Phys. D: Appl. Phys. **25** (1992) 417.

[9] J.E. Allen, Physica Scripta **4** (1992) 497.

[10] J.E. Allen, Plasma Sources Sci. Technol. **4** (1995) 234.

[11] C.M.C. Nairn, B.M. Annaratone, J.E. Allen, Plasma Sources Sci. Technol. **4** (1995) 416.

[12] J.E. Allen, B.M. Annaratone, U. de Angelis, J. Plasma Phys. **63** (2000) 299.

[13] J. Rohmann, S. Klagge, Contrib. Plasma Phys. **33** (1993) 111.

[14] J.I. Fernández Palop, J. Ballesteros, V. Colomer, M.A. Hernández, J. Phys. D: Appl. Phys. **29** (1996) 2832.

[15] M.J. Druyvesteyn, Z. Physik **64** (1930) 781.

[16] S.C.M. Luijendijk, J. Van Eck, Physica **36** (1967) 49.

[17] D. Herrmann, S. Klagge, Beitr. Plasmaphysik **12** (1972) 17.

[18] P. David, M. Šícha, M. Tichý, T. Kopiczynski, Z. Zakrzewski, Contrib. Plasma Phys. **30** (1990) 167.

[19] S. Aisberg, 3rd Symp. Engineering Aspects of Magnetohydrodynamics (1963), p.89.

[20] W. Seifert, D. Johanning, H.R. Lehmann, N. Bankov, Beitr. Plasmaphysik **26** (1986) 237.

[21] S. Klagge, M. Tichý, Czech. J. Phys. B **35** (1985) 988.

[22] B. Nuhn, G. Peter, Proc. XIII. ICPIG Berlin (1977), p. 999.

[23] M. Šícha, Czech. J. Phys. B **29** (1979) 640.

[24] G.R. Branner, E.M. Friar, G. Medicus, Rev. Sci. Instr. **34** (1963) 231.

[25] N.A. Vorobjeva, J.M. Kagan, V.M. Milenin, Zhurnal tekhnicheskoi fiziki (J. Tech. Phys. USSR) **34** (1964) 2079.

[26] V.A. Godyak, R.B. Piejak, B.M. Alexandrovich, Plasma Sources Sci. Technol. **1** (1992) 36.

[27] B. Saggau, Z. Angew. Physik **32** (1972) 324.

[28] W.H. Press, S.A. Teukolsky, W.T. Wetterling, B.P. Flannery, Numerical Recipes in C (The art of scientific computing), Cambridge University Press (1988, 1992).

[29] L.M. Volkova, A.M. Devyatov, G.A. Kralkina, N.N. Sedov, M.A. Sherif, Vestnik Mosk. Univ. (fizika, astronomija) **16** (1957) 502.

[30] A.V. Oppenheim, R.W. Schafer, Discrete-Time Signal Processing, Prentice-Hall (1989).

[31] J.F. Kaiser, W.A. Reed, Rev. Sci. Instr. **48** (1977) 1447.

[32] A. Savitzky, M.J.E. Golay, Anal. Chem. **36** (1964) 1627.

[33] H. Madden, Anal. Chem. **50** (1978) 1383.

[34] M. Tichý, P. Kudrna, J.F. Behnke, C. Csambal, S. Klagge, Journal de Physique IV **7** (1997) C4-397.

[35] M. Hannemann, Greifswalder Physikalische Hefte **6** (1983) 145; PhD Thesis, University of Greifswald, Germany (1991).

[36] M. Hannemann, INP-Report VII, Institute for Low-Temperature Plasma Physics, Greifswald (1995).

[37] D. Trunec, Contrib. Plasma Phys. **32** (1992) 523.

[38] M. Tichý, M. Šícha, J. Bok, S. Pfau, Czech. J. Phys. B **37** (1987) 179.

[39] J.I. Fernández Palop, J. Ballesteros, V. Colomer, M.A. Hernández, Rev. Sci. Instr. **66** (1995) 4625.

[40] H. Amemyia, J.Appl.Phys. **15** (1976) 1767.

[41] V.I. Demidov, N.B. Kolokolov, Zhurnal tekhnicheskoi fiziki (J. Tech. Phys. USSR) **51** (1981) 888.

[42] A.P. Mesentsev, A.S. Mustafaev, V.L. Fedorov, Zhurnal tekhnicheskoi fiziki (J. Tech. Phys. USSR) **55** (1985) 544; V.L. Fedorov, Zhurnal tekhnicheskoi fiziki (J. Tech. Phys. USSR) **55** (1985) 926.

[43] M.B. Hopkins, W.B. Graham, Rev. Sci. Instr. **57** (1986) 2210.

[44] D. Smith, I.C. Plumb, J. Phys. D: Appl. Phys. **6** (1973) 196.

[45] L. Tonks, I. Langmuir, Phys. Rev. **34** (1929) 876.

[46] J.F. Shaeffer, PhD Thesis, University of St. Louis (1971).

[47] F.F. Chen, in Plasma Diagnostics Techniques, R.H. Huddelstone, S.L. Leonard (Eds.), Academic Press: New York, London (1965).

[48] J. Virmont, Proc. IX. ICPIG, Bucharest, Contrib. papers (1969), p. 609.

[49] D.Trunec, P. Španěl, D. Smith, Contrib. Plasma Phys. **35** (1995) 203.

[50] C.V. Goodall, D. Smith, Plasma Physics **10** (1968) 249.

[51] C.V. Goodall, B. Polychronopoulos, Planet. Space Sci. **22** (1974) 1585.

[52] M.G. Dunn, J.A. Lordi, AIAA Journal **7** (1969) 1458.

[53] M.G. Dunn, J.A. Lordi, AIAA Journal **8** (1970) 1077.

[54] R.A. Johnson, P.C.T. DeBoer, AIAA Journal **10** (1972) 664.

[55] J.A. Thornton, AIAA Journal **9** (1971) 342.

[56] V.M. Zakharova, J.M. Kagan, K.S. Mustafin, V.I. Perel, Sov. Phys. Tech. Phys. **5** (1960) 411.

[57] F.F. Chen, Plasma Phys. **7** (1965) 47.

[58] J.D. Swift, Proc. Phys. Soc. **79** (1969) 697.

[59] J.F. Waymouth, Phys. Fluids **7** (1964) 1843.

[60] L. Talbot, Y. S. Chou, Rarefied Gas Dynamics, Academic Press (1966), p. 1723.

[61] Y.S. Chou, L. Talbot, D.R. Willis, Phys. Fluids **9** (1966) 2150.

[62] R.H. Kirchhoff, E.W. Peterson, L. Talbot, AIAA Journal **9** (1971) 1686.

[63] K.G. Bienkowski, K.W. Chang, Phys. Fluids **11** (1968) 784.

[64] E. Wasserstrom, C.H. Su, R.F. Probstein, Phys. Fluids **8** (1965) 56.

[65] S.A. Self, C.H. Shih, Phys. Fluids **11** (1968) 1532.

[66] C.H. Shih, E. Levi, AIAA Journal **9** (1971) 1673, AIAA Journal **9** (1971) 2417.

[67] Z. Zakrzewski, T. Kopiczynski, Plasma Physics **16** (1974) 1195.

[68] T. Kopiczynski, PhD Thesis, IMP PAN, Institute of Fluid-Flow Machines, Polish Academy of Sciences, Gdansk, Poland (1977).

[69] J. Basu, C. Sen, Japan J. Appl. Phys. **12** (1973) 1081.

[70] G.J. Schulz, S.C. Brown, Phys. Rev. **98** (1955) 1642.

[71] A.K. Jakubowski, AIAA Journal **8** (1972) 988.

[72] P. David, CSc Thesis, Faculty of Mathematics and Physics, Charles University Prague (1985).

[73] M. Tichý, M. Šícha, P. David, T. David, Contrib. Plasma Phys. **34** (1994) 59.

[74] P. Kudrna, PhD thesis, Charles University Prague, Czech Republic (1996)

[75] O. Chudáček, P. Kudrna, J. Glosík, M. Šícha, M. Tichý, Contrib. Plasma Phys. **35** (1995) 503.

[76] P.M. Chung, L. Talbot, K.J. Touryan, Electrical Probes in Stationary and Flowing Plasmas, Theory and Application, Springer-Verlag: Berlin, Heidelberg, New York (1975).

[77] J.R. Sanmartin, Phys. Fluids **13** (1970) 103.

[78] J.G. Laframboise, J. Rubinstein, Phys. Fluids **19** (1976) 1900.

[79] H. Aikawa, J. Phys. Soc. Japan **40** (1976) 1741.

[80] E. Passoth, P. Kudrna, C. Csambal, J.F. Behnke, M. Tichý, V. Helbig, J. Phys. D: Appl. Phys. **30** (1997) 1763.

[81] P. Kudrna, E. Passoth, Contrib. Plasma Phys. **37** (1997) 417.

[82] P. Kudrna, O. Chudáček, M. Šíicha, M. Tichý, E. Passoth, J.F. Behnke, Proc. XXII. ICPIG, Hoboken, K.H. Becker, W.E. Carr, E.E. Kunhardt (Eds.), Contributed Papers, Vol. 2 (1995), p. 185.

[83] J.C. Blancod, K.S. Golovanievski, T.P. Kravchov, Proceedings X. ICPIG, Oxford (1971), p. 25.

[84] J.G. Brown, A.B. Compher, K.W. Ehlers, D.F. Hopkins, W.B. Kunkel, P.S. Rostler, Plasma Phys. **13** (1971) 47.

[85] Z. Zakrzewski, T. Kopiczynski, M. Lubanski, Czech. J. Phys. B **30** (1980) 1167.

[86] R.R. Arslanbekov, N.A. Khromov, A.A. Kudryavtsev, Plasma Sources Sci. Technol. **3** (1994) 528.

[87] J.F. Behnke, E. Passoth, C. Csambal, M. Tichý, P. Kudrna, D. Trunec, A. Brablec, Czech. J. Phys. B **49** (1999) 483.

[88] E. Passoth, J.F. Behnke, C. Csambal, M. Tichý, P. Kudrna, Ju.B. Golubovskii, I.A. Porokhova, J. Phys. D: Appl. Phys. **32** (1999) 2655.

[89] S.W. Rayment, N.D. Twiddy, Brit. J. Appl. Phys. **2** (1969) 1747.

[90] M. Šícha, V. Veselý, V. Řezáčová, M. Tichý, M. Drouet, Z. Zakrzewski, Czech. J. Phys B **21** (1971) 62.

[91] J. Dvořáček, L. Kryška, J. Stöckel, F. Žáček, M. Šícha, M. Tichý, V. Veselý, Czech. J. Phys. B **40** (1990) 678.

[92] R. Winkler, J. Wilhelm, A. Hess, Annalen der Physik **42** (1985) 537.

[93] J.D. Swift, M.J.R. Schwar, Electrical Probes for Plasma Diagnostics, Iliffe Books: London (1970).

[94] A.B. Blagoev, N.B. Kolokolov, V.M. Milenin, Zhurnal tekhnicheskoi fiziki (J. Tech. Phys. USSR) **42** (1972) 1701.

[95] A.B. Blagoev, J.M. Kagan, N.B. Kolokolov, R.I. Lyaguschenko, Zhurnal tekhnicheskoi fiziki (J. Tech. Phys. USSR) **44** (1974) 333.

[96] D. Kamke, H.J. Rose, Z. Physik **145** (1956) 83.

[97] P.R. Smy, J.R. Greig, J. Phys. D **1** (1968)3 51.

[98] R.W. Carlson, T. Okuda, H.J. Oskam, Physica **30** (1964) 182; T. Okuda, R.W. Carlson, H.J. Oskam, Physica **30** (1964) 193; H.J. Oskam, R.W. Carlson, T. Okuda, Physica **30** (1964) 375.

[99] K.-D. Weltmann, PhD thesis, University of Greifswald, Germany (1993).

[100] T. Bräuer, PhD thesis, University of Greifswald, Germany (1997).

[101] A.G. Dean, D. Smith, I. Plumb, J. Phys. E **5** (1972) 776.

[102] B.M. Wunderer, J. Phys. E **8** (1975) 938.

[103] T. Bräuer, S. Gortschakow, D. Loffhagen, S. Pfau and R. Winkler, J. Phys. D: Appl. Phys. **30** (1997) 3223.

[104] T. Bindemann, M. Tichý, J.F. Behnke, H. Deutsch, Rev. Sci. Instr. **69** (1998) 2037.

[105] M.B. Hopkins, C.A. Anderson, W.G. Graham, Europhysics Letters **8** (1989) 141.

[106] K.R. Stalder, C.A. Anderson, A.A. Mullan, W.G. Graham, J. Appl. Phys. **72** (1992) 1290.

[107] Ajit P. Paranpje, James P. McVittie, Sidney A. Self, J. Appl. Phys. **67** (1990) 6718.

[108] U. Flender, B.H. Nguyen Thi, K. Wiesemann, N.A. Khromov, N.B. Kolokolov, Plasma Sources Sci. Technol. **5** (1996) 61.

[109] P. Špatenka, V. Brunnhofer, Meas. Sci. Technol. **7** (1996) 1065.

[110] U. Tietze, Ch. Schenk, Halbleiter-Schaltungstechnik, Springer-Verlag: Berlin, Heidelberg, New York (1980).

[111] N.St.J. Braithwaite, N.M.P. Benjamin, J.E. Allen, J. Phys. E **20** (1987) 1046.

[112] P.A. Chatterton, J.A. Rees, W.L. Wu, K. Al-Assadi, Vacuum **42** (1991) 489.

[113] W. Kasper, Proc. XXII. ICPIG, Hoboken, K.H. Becker, W.E. Carr, E.E. Kunhardt (Eds.), Contributed Papers, Vol. 2 (1995), p. 171.

[114] M. Sugawara, J. Ichihashi, Y. Kobayashi, Proc. XVIII. ICPIG, Swansea, W. Terry Williams (Ed.), Vol. 4 (1987), p. 830.

[115] B.M. Annaratone, N.St.J. Braithwaite, Meas. Sci. Technol. **2** (1991) 795.

[116] R.B. Turkot, Jr., D.N. Ruzic, J. Appl. Phys. **73** (1993) 2173.

[117] M. Hron, J. Stöckel, L. Kryška, J. Horáček, Proc. IAEA Technical Committee Meeting on Research Using Small Tokamaks, Prague (1996), p. 54.

[118] R. D'Arcy, J. Phys. D: Appl. Phys. **7** (1974) 1391.

[119] E.P. Szuszczewicz, J.C. Holmes, J. Appl. Phys. **46** (1975) 5134.

[120] S. Klagge, Beitr. Plasmaphys. **15** (1975) 309.

[121] J. Kalčík, Czech. J. Phys. **45** (1995) 241.

[122] P. Špatenka, R. Studený, H. Suhr, Meas. Sci. Technol. **3** (1992) 704.

[123] K. Shimizu, K. Yano, H. Oyama, H. Kokai, Proc. ISPC-9, Bari (1989), p. 831.

[124] Y. Kobayashi, T. Ohte, M. Katoh, M. Sugawara, Trans. IEE of Japan **109-A** (1989) 69.

[125] M. Niionomi, K. Yanagihara, ACS Symp. **108** (1979) 87.

[126] C. Winkler, D. Strele, S. Tscholl, R. Schrittwieser, Proc. 12th SAPP, Liptovský Ján (Slovakia), J.D. Skalný, M. Černák (Eds.), p 121 (1999).

[127] E.Y. Wang, N. Hershkowitz, T. Intrator, C. Forest, Rev. Sci. Instr. **57** (1986) 2425.

[128] P. Špatenka, H. Suhr, Plasma Chemistry and Plasma Processing **13** (1993) 555.

[129] P. Špatenka, Z. Beneš, Frontiers in Low Temperature Plasma Diagnostic, Book of Papers, Bad Honnef, Germany (1997), p. 227.

[130] S. Chen, T. Sekiguchi, J. Appl. Phys. **36** (1965) 2363.

[131] T. Okuda, K. Yamamoto, J. Appl. Phys. **31** (1960) 158.

[132] P. Špatenka, H.-J. Endres, J. Krumeich, R.W. Cook, Surface and Coating Technology **116–119** (1999) 1228.

[133] E.A. Mosburg, R.C. Abelson, J.R. Abelson, J. Appl. Phys. **54** (1983) 4916.

[134] R. Johnsen, Rev. Sci. Instr. **57** (1986) 428.

[135] A. Brockhaus, A. Schwabedissen, Ch. Soll, J. Engemann, Proc. Frontiers in Low Temperature Plasma Diagnostic III, Saillon (Switzerland, 1999), p. 89.

[136] N.St.J. Braithwaite, J.P. Booth, G. Cunge, Plasma Sources Sci. Technol. **5** (1996) 677.

[137] U. Flender, K. Wiesemann, Plasma Chem. Plasma Process. **15** (1995) 123.

[138] H. Sabadil, S. Klagge, M. Kammayer, Plasma Chem. Plasma Process. **8** (1988) 425.

[139] E.O. Johnson, L. Malter, Phys. Rev. **80** (1950) 58.

[140] D. Bradley, K.J. Matthews, Phys. Fluids **6** (1963) 1479.

[141] R.H. Kirchhoff, E.W. Peterson, L. Talbot, AIAA Journal **9** (1971) 1686.

[142] J.S. Chang, J.G. Laframboise, Phys. Fluids **19** (1976) 25.

[143] M. Šícha, M. Tichý, V. Hrachová, P. David, T. David, J. Touš, Contrib. Plasma Phys. **34** (1994) 51.

7 Diagnostics of non–equilibrium molecular plasmas using emission and absorption spectroscopy

J. Röpcke*, P.B. Davies[†], M. Käning* and B.P. Lavrov[‡]

* Institut für Niedertemperatur–Plasmaphysik, Friedrich–Ludwig–Jahn–Straße 19,
D–17489 Greifswald, Germany
[†] Department of Chemistry, University of Cambridge, Lensfield Road,
Cambridge CB2 1EW, U.K.
[‡] Faculty of Physics, St.–Petersburg State University, St. Petersburg, 198904, Russia

7.1   Introduction

Low pressure, non-equilibrium molecular plasmas are of increasing interest not only in fundamental research but also in plasma processing and technology. Molecular plasmas are used in a variety of applications such as thin film deposition, semiconductor processing, surface activation and cleaning, and in materials and waste treatment. The investigations of plasma physics and chemistry *in situ* require detailed knowledge of the plasma parameters, which can be obtained by appropriate diagnostic techniques. The need for a better scientific understanding of plasma physics and chemistry has stimulated the improvement of established diagnostic techniques and the introduction of new ones. Methods based on traditional spectroscopy rank among the most important, and it is the goal of this article to focus particularly on those developed over the last decades.

The majority of molecular plasmas is characterised by high chemical reactivity due to the large concentrations of transient or stable chemically active species present. Non-invasive diagnostic methods have been developed for investigating this type of plasma, particularly the methods of optical emission spectroscopy (OES) and absorption spectroscopy (AS); with these this article is concerned. Methods based on OES and AS not only make use of resolved, discrete spectral lines (line positions and line profiles) but also of continuous spectral features. Since light emission is an inevitable consequence of the existence of a plasma it should be considered as a natural (non-invasive) source of information for diagnostic purposes. When illuminated by low intensity light the probing of plasmas by AS can also be considered as non-invasive. Emission and absorption spectroscopy can provide information about atom, molecule and ion densities in excited and ground states, as well as about gas, rotational and vibrational temperatures. High temporal resolution can be achieved, while spatial resolution requires complementary techniques, because OES and AS only provide integrated emission or absorption intensities along the line of sight.

7.2   Instrumental techniques

Electromagnetic radiation can be emitted from low temperature plasmas over a wide spectral frequency range extending from the vacuum ultraviolet (VUV) or even shorter wavelengths, through the ultraviolet and visible regions and out to the infrared (IR). Modern OES spectrometers for detecting the radiation from molecular low temperature plasmas feature high spectral resolution (down to picometers), sensitivity near the single photon detection limit, and time resolution as short as nanoseconds. A survey of the most important emission spectroscopy techniques and the principles of spectral photometry are given in standard text books [1]–[5]. Figure 7.1 shows an example of part of the high resolution spectrum of H_2 emitted from a direct current (dc) capillary-arc discharge recorded with a 1m double monochromator and a thermoelectrically cooled charge coupled device (CCD) matrix detector. Signal recovery with CCD detectors has the advantage of direct manipulation of digital data, simultaneous detection across a specific spectral window, and the high sensitivity and dynamic range (of order 10^5). A wide variety of light sources, dispersive elements, detectors and data aquisition methods can be used for absorption spectroscopy [5]. The classic dispersion experiment for measuring the density of atomic or molecular states in a plasma by AS is based on continuous light sources and a spectrograph with a suitable detector for the spectral range of interest. Fourier Transform Infrared (FTIR) technique with a Michelson interferometer also uses a continuous light source for absorption measurements. The transmitted light intensity depends on the (variable) optical path difference between the mirrors in the two interferometer arms and yields the in-

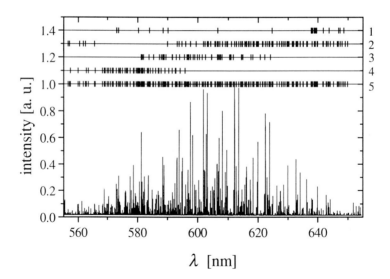

Figure 7.1: Part of the H_2 emission spectrum from a dc capillary-arc discharge, recorded with a 1m double monochromator (U1000, Jobin Yvon). Lines from different electronic transitions are identified as: 1, $e^3\Sigma_u^+ \rightarrow a^3\Sigma_g^+$; 2, $d^3\Pi_u^\pm \rightarrow a^3\Sigma_g^+$; 3, $i^3\Pi_g^- \rightarrow c^3\Pi_u$; 4, $j^3\Delta_g^- \rightarrow c^3\Pi_u$; 5, 130 identified lines of other transitions.

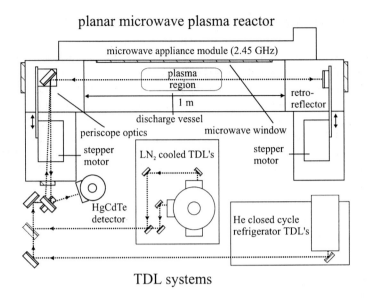

Figure 7.2: Experimental arrangement of a planar microwave plasma reactor and tunable diode laser (TDL) infrared source used for hydrocarbon (CH_4, CH_3OH) plasma diagnostics. The path of the diode laser beam is indicated by dotted lines (with permission from Ref. [7]).

terferogram carrying the spectroscopic information. In contrast to dispersion techniques the FTIR spectrometer records the whole spectrum simultaneously (multiplex or Fellgett advantage). In principle the resolution of FTIR can be as high as 0.002 cm^{-1}, determined by the distance scanned by the movable mirror, but at the expense of time resolution. Fractional absorptions as small as 10^{-4} can be measured.

With the development of tunable, narrow-band light sources such as tunable dye lasers and infrared diode lasers, these have been substituted for continuous light sources in AS experiments. These narrow-band laser sources have the advantage of high spectral intensity, narrow bandwidth, and continuous tunability over the absorption profile. Figure 7.2 shows an example of an experimental arrangement used to investigate plasma chemistry and kinetics in a planar microwave reactor [6] using a tunable diode laser (TDL) spectrometer [7]. The discharge configuration in planar microwave plasmas has the advantage of being well suited for end-on observations because considerable homogeneity can be achieved over relatively long plasma path lengths. The diode laser light beam enters the plasma through a KBr window and after retro-reflection is sent to a mode selection monochromator (not shown) before detection with a mercury cadmium telluride (HgCdTe) detector. Another highly sensitive newer laser technique is cavity ringdown (CRD) absorption spectroscopy. This method is based on the measurement of the intensity decay rate of a laser pulse injected into an optical cavity formed by two most highly reflective mirrors which also enclose the plasma. Absorptions as low as 10^{-9} can be measured with an acceptable signal to noise ratio [8]. Recently, near infrared tunable diode lasers have been used as the light sources for CRD spectroscopy [9].

As well as the implementation of entirely novel techniques like CRD, the performance of the spectrometers used in plasma diagnostics has been greatly enhanced by improved technology over the last decades. The introduction of CCD cameras, holographic gratings, tunable sources and microprocessor instrument control and data collection, are a few examples of the experimental innovations, which have become commercially available within the last decade.

7.3   Emission spectrometry

7.3.1   General considerations

In low-temperature molecular plasmas a variety of processes such as electron collisions, ion-molecule reactions, charge exchange, and chemical processes lead to a large number of atoms, molecules, radicals and ions in excited states. The generation of photons is mainly governed by the de-excitation of these species. Provided the influence of self absorption inside the plasma is relatively small or can be neglected (referred to as optically thin plasma conditions) the emitted photons can be detected outside the plasma. Even if self-absorption does occur, the spectral distribution of the emitted light carries with it a representation of the plasma conditions. It contains information about the different kinds of plasma species, their number density and temperature, as well as about the strengths of internal or external fields.

The intensity of the spontaneous emission from molecules can be expressed as the number of photons emitted by a unit volume per second over all solid angles. The intensity of a spectral line from the $n', \nu', N' \to n'', \nu'', N''$ rovibronic (rotational vibrational electronic) transition may be written as

$$I_{n''\nu''N''}^{n'\nu'N'} = N_{n'\nu'N'} A_{n''\nu''N''}^{n'\nu'N'} \qquad (7.1)$$

where n are the quantum numbers describing an electronic state of the molecule, and ν and N are the vibrational and rotational quantum numbers, respectively. The initial state is denoted by primes and the final state by double primes. $N_{n'\nu'N'}$ is the population density of the initial rovibronic level and A the corresponding transition probability for spontaneous emission.

A plasma diagnostic technique based on emission spectroscopy has the characteristics of an inverse problem. Usually, integral intensities of emission lines in the line of sight are measured within a certain solid angle with a selected spectral resolution. Only if the plasma is homogeneous over the solid angle being investigated, the local value of the intensity can be determined. Otherwise theoretical inversion methods have to be used such as the Abel inversion in the case of cylindrical symmetry, or tomography. The measured, spatially localised line emission intensities allow the calculation of population densities of the electronic or, for molecules, rovibronic levels of the species of interest provided the transition probabilities are known, and the spectrometer functions have been calibrated. The calculation of species densities in the *ground* state from measured line intensities often is also an inverse problem. It requires a theoretical model for the excitation and de-excitation processes, all necessary cross sections, transition probabilities, etc., as well as knowledge about the electron energy distribution function (EEDF). These requirements are often non-trivial to achieve, and in practice an easier approach is to use actinometry, described in the following section, in which the emission intensity is referenced to a known standard.

7.3.2 Actinometry

The principle of actinometry can be formulated succinctly as follows [10], [11]. The concentration of the species of interest n_{spe} can be calculated from the ratio of an emission line of the species I_{spe} and a line of the admixed actinomer I_{act} with known concentration n_{act}

$$n_{spe} = K\, n_{act}\, \frac{I_{spe}}{I_{act}}. \tag{7.2}$$

The factor K contains the spectral and geometrical factors of the detection system and the emission cross sections of the observed transitions. The actinomer frequently is a rare gas atom such as Ar. Equation (7.2) shows that, unless the factor K can be evaluated absolutely, the calculated n_{spe} values are only relative ones. The following main conditions should be fulfilled to ensure the validity of actinometry: (i) excitation of emitting states of the species of interest and of the actinomer only by direct electron impact from their ground state, (ii) de-excitation by spontaneous emission and (iii) similarity of threshold and shape of the electron-impact cross sections.

The applications of actinometry in plasma diagnostics are too numerous to mention all of them so only a few representative examples will be cited. Actinometry was developed to investigate the dissociation of CF_4 in glow discharges and to measure the dependence of the density of atomic fluorine generated on discharge parameters such as pressure p and flow rate [10, 11]. It has also been used in the time resolved mode to analyse the kinetics of O and F atoms in oxygen based plasmas [12, 13]. Actinometry has been used in an etching plasma to determine the atomic chlorine concentrations in Cl_2 and CF_3Cl plasmas and the results compared with infrared absorption measurements [14]. Deviations between the results from both methods were mainly caused by the generation of excited Cl atoms due to dissociation of Cl_2. Besides plasma etching the modification of surface properties is of great importance in plasma technology, and actinometry has been used in studies of plasma polymerisation and for modifying polyethylene surfaces [15]–[17].

Actinometry is an established and widespread diagnostic method in the field of plasma enhanced diamond and diamond-like carbon film deposition. Frequently, different types of hydrogen plasmas with admixtures of hydrocarbons, mainly methane, and also other components such as rare gases or oxygen are used for carbon layer deposition. Actinometry has been particularly useful for determining the degree of hydrogen dissociation [19, 20], and the spatial and temporal distributions of atoms, radicals and molecules in plasma reactors [21]–[24]. The correlation with electron density [25], electron temperatures [26], and film properties [19], [27]–[30], including the effect of changing the precursor gas mixtures [31], has also been studied using actinometry. Recently, actinometry was proposed as a method of absolute hydrogen atom concentration measurement [32]. This requires the exact determination of K in equation (7.2). Absolute actinometry of oxygen atom density has been used by Etamadi et al. [33], to gain insight into the mechanism of silicon oxide film deposition in microwave/radiofrequency discharges. They discussed different formation and removal mechanisms of atomic oxygen, which involved precisely measured excitation cross-sections and calculated electron energy distribution functions.

As well as time resolved studies spatially resolved investigations of the relative concentrations of atomic hydrogen, and of the rotational temperature distributions in planar microwave plasma sources, have been used to improve our knowledge of the scaling laws for plasmas

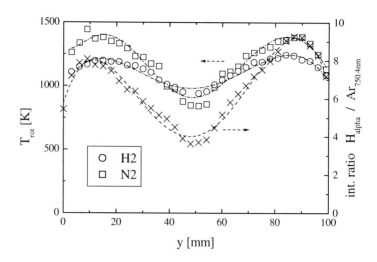

Figure 7.3: H_2 rotational temperature (from the Q(1-1) Fulcher–α band), relative H atom density distribution (H_2 + 18% CH_4 + 9% Ar, p = 100 Pa), and N_2 rotational temperature distribution (ν = 1–4 transition, 2nd pos. system, N_2, p = 100 Pa) 8 mm from and parallel (y-dimension) to the microwave input windows of a microwave double plasma source (with permission from Ref. [18]).

when their lengths are increased or when two or more plasma sources are used in parallel [7], [34]–[38]. Figure 7.3 shows (i) the spatial distribution of the relative density of the H atom using the intensity ratio of H_α:Ar_{750nm} atomic lines, and (ii) the rotational temperature derived from H_2 and N_2 band emission [18]. These measurements show the remarkable temperature and density gradients due to the specific nature of the inhomogeneous excitation in planar microwave reactors.

Within the last two decades actinometry has been developed from a steady state technique in CF_4 plasmas to a flexible dynamic method in a variety of different molecular systems with the potential for spatial measurements. Actinometry is at its most powerful when it is combined with other plasma diagnostic techniques such as mass spectrometry, coherent anti-Stokes Raman scattering (CARS), or laser induced fluorescence (LIF), and also with kinetic modeling.

7.4 Absorption spectroscopy

7.4.1 Background theory

The methods of absorption spectroscopy are of great importance in plasma diagnostics because they provide a means of determining the population densities of species in both ground and excited states. The spectral line positions provide species identification while line profiles, often connected with gas temperature and relative intensities, provide information about population densities. An important advantage of AS over OES methods is that only relative intensities need to be measured to determine absolute concentrations, avoiding the problems of complete instrument calibration inherent in the OES methods. Absorption spectroscopy has been applied right across the spectrum from the VUV to the far infrared (FIR). Continuous lamps (e.g. the Xe-lamp for the VIS and NIR, and the D_2-lamp for the UV) and tunable narrow-band light sources (e.g. tunable dye lasers, diode lasers) can be used as external light sources. Two principle cases need to be distinguished, the measurement of the absorption of light emitted by: (i) the plasma itself (*self-absorption or reabsorption*), or (ii) an external light source. If the former is the case the light from the plasma is measured with and without a retro-reflector or compared with the light emitted directly from an identical plasma. This ensures identical line profiles for the emitted and absorbed light. This method is particularly important for checking the optical thickness of the plasma. Further discussions about self-absorption can be found in Refs. [2, 3].

The change in light intensity dI_ν when a beam passes through a homogeneous plasma layer of thickness dl is given by the net balance of the intensity due to absorption and emission within the layer dl:

$$dI_\nu = (\varepsilon(\nu) - \kappa(\nu)I_\nu)dl \,, \tag{7.3}$$

where $\varepsilon(\nu)$ is the emission coefficient and $\kappa(\nu)$ the absorption coefficient per unit length. The absorption coefficient $\kappa(\nu)$, describing light absorption in an infinitesimally thin layer of a plasma, is given by

$$\kappa(\nu) = \sum_i N_i \sigma_i(\nu) \,, \tag{7.4}$$

where a summation over all absorbing species and states is included. N_i are the species population densities and $\sigma_i(\nu)$ their absorption cross sections at frequency ν. The line profile P_ν is normalised by :

$$\int_{line} P_\nu d\nu = 1 \,. \tag{7.5}$$

Assuming the upper level k is not populated, then the absolute value of $\kappa(\nu)$ results from:

$$\int_{line} \kappa(\nu)d\nu = \frac{h\nu}{c_0} N_i B_{ik} \,. \tag{7.6}$$

The line absorption coefficient then is given by

$$\kappa(\nu) = \frac{h\nu}{c_0} N_i(\alpha, \beta, \gamma, \ldots) B_{ik} P_\nu \,, \tag{7.7}$$

where $N_i(\alpha, \beta, \gamma, \ldots)$ is the population density of the ith level. It depends on different plasma parameters $(\alpha, \beta, \gamma, \ldots)$, and B_{ik} is the Einstein coefficient for the transition between the levels i and k.

In the case where an external light source has a much higher intensity than that of the plasma itself, the absorption of radiation can be described by the Beer-Lambert law:

$$I_\nu(l) = I_\nu(0) \exp(-\kappa(\nu)l) \,. \tag{7.8}$$

$I_\nu(0)$ and $I_\nu(l)$ are the fluxes of the radiation entering and leaving the plasma, and l is the length of the absorbing (homogeneous) plasma column. It should be noted that in plasma spectroscopy the definition of the effective length of the plasma column is not a trivial question. Inhomogeneity can lead to serious errors.

7.4.2   Infrared absorption spectroscopy

The increasing interest in processing plasmas containing hydrocarbons, fluorocarbons or orga-no-silicon compounds has lead to further applications of infrared AS techniques as most of these compounds and their decomposition products are infrared active. FTIR spectroscopy has been used for *in situ* studies of methane plasmas for a number of years, but it is generally insufficiently sensitive for detecting free radicals or ions in processing plasmas. More recently CRD spectroscopy has been used for density measurements of the SiH_2 radical and of nanometer sized dust particles in silane plasmas [39], and for detecting N_2^+ cations in nitrogen discharges [40]. Infrared absorption spectroscopy with tunable diode lasers (TDLs) is increasingly being used in the spectral region between 3 and 20 μm for measuring the concentrations of free radicals, transient molecules, and stable products in their electronic ground states. TDL spectroscopy can also be used to measure neutral gas temperatures [41], and to investigate dissociation processes of molecular low-temperature plasmas [7], [42]–[44]. The main applications of TDL absorption spectroscopy up till now have been for investigating molecules and radicals in fluorocarbon etching plasmas [41, 44, 45] and in plasmas containing hydrocarbons [7], [46]–[50]. A wide variety of low molecular weight free radicals and molecular ions have been detected by TDLAS in purely spectroscopic studies, e.g., Si_2^- [51] and SiH_3^+ [52] in silane plasmas. Most of these spectroscopic results have yet to be applied in plasma diagnostic studies. One of the most successful applications of TDLAS is for studying the methyl radical in a variety of Plasma Enhanced Chemical Vapour Deposition (PECVD) processes used to deposit thin carbon films. The hypothesis that this radical is the most likely precursor for forming such films is now gaining acceptance, hence the great importance of suitable diagnostic methods for its detection.

Most of the measurement techniques for detecting the methyl radical are based on absorption spectroscopy either with 216 nm ultraviolet radiation or in the infrared near 606 cm^{-1}. For example, the ultraviolet absorption of CH_3 at 216 nm was used for number density measurements by Menningen and Childs et al. [53, 54] in different CVD diamond growth environments, hot-filament, dc and microwave plasmas. The infrared TDLAS technique has proven to be highly useful because it can also measure the concentrations of related species provided they are IR active. Wormhoudt demonstrated this flexibility by measuring CH_3 and C_2H_2 in a CH_4-H_2 radiofrequency (rf) plasma using a long path plasma absorption cell [55]. Actually for several reasons, TDLAS is probably the best method for detecting the methyl radical. The ν_2 out-of-plane bending mode is not only intense but has many lines between 600 and 650 cm^{-1}. Thus it is possible to derive rotational and vibrational temperatures from their relative intensities. The (J=K) Q-branch lines of the ν_2 fundamental band near 606 cm^{-1} are particularly useful because several of them lie within 0.5 cm^{-1} of each other. Therefore rotational temperatures in the plasma are easily measured from them. A highly important study concerned to quantifying the concentrations of methyl radicals was the determination of the line strength of the Q(8,8) line of methyl at 608.3 cm^{-1} by Wormhoudt and McCurdy [56].

Systematic TDLAS measurements of several different hydrocarbons, including methyl, in a 20 kHz methane plasma in a parallel plate reactor were reported by Davies and Martineau [42, 57, 58]. Goto and co-workers have published numerous studies of methyl and methanol concentrations in rf and electron cyclotron resonance (ECR) plasmas under different condi-

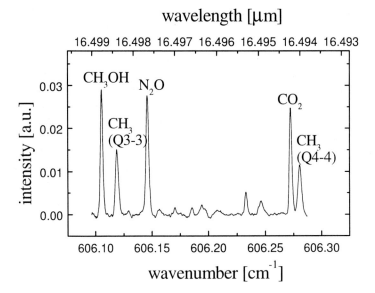

Figure 7.4: The TDL spectrum of some of the molecules in a microwave plasma containing methanol. The line due to N_2O is from a calibration cell placed in the path of the laser beam (with permission from Ref. [7]).

tions, e.g., by investigating the influence of rare gases on the plasma. They have also combined IR absorption with emission spectroscopy, and investigated the effect of water vapour on the methyl radical concentration in argon/methane and argon/methanol rf plasmas using TDLAS [43], [47]–[49], [59, 60]. Kim et al. measured CH_3, C_2H_2 and CH_3OH concentrations in methanol/water rf plasmas by TDLAS and found that methanol was almost completely dissociated even when medium power levels are applied [61].

The most recent applications of TDLAS to plasma diagnostics include studies in which many different species have been monitored under identical plasma conditions [7]. In fact it is sometimes possible to detect the IR spectra of more than one species in a single laser mode using TDLAS, as shown in Fig. 7.4. A key objective of this type of study is to be able to model the chemistry of the plasma, for which it is necessary to detect as many plasma species as possible. Systematic investigations of the plasma chemistry in molecular microwave plasmas containing methane or methanol using TDLAS have been reported.

In addition to the primary species, and the methyl radical, TDLAS was used to detect ten stable molecules in H_2–Ar–O_2 microwave plasmas containing up to 7% of methane or methanol under as well flowing as static conditions (Fig. 7.5). They included the C-2 hydrocarbons, formaldehyde and formic acid, as well as the major products CO, CO_2 and H_2O. The degree of dissociation of the hydrocarbons (CH_4, CH_3OH) varied between 30 and 90%, and the methyl concentration was measured to be in the range of $10^{10} - 10^{12}$ molecules cm^{-3} [7]. Similar TDLAS studies have been carried out on CH_4–H_2–O_2 plasmas in a 20 kHz parallel plate reactor [62]. Chemical modeling has been used to predict successfully the concentra-

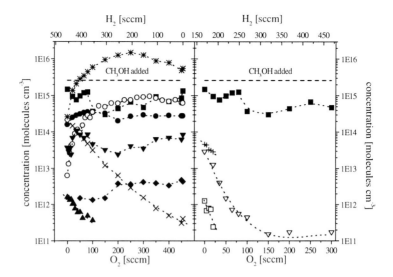

Figure 7.5: Molecular concentrations as a function of oxygen flow rate in a microwave plasma under flowing conditions measured by TDLAS (Φ_{total} = 555 sccm, $CH_3OH-O_2-H_2-Ar$; $\Phi(CH_3OH)$= 40 sccm; $\Phi(Ar)$= 60 sccm; p= 150 Pa). For clarity the C-2 product hydrocarbons are shown separately in the right hand panel. $\times$, CH_4; $\triangledown$, C_2H_2; $\square$, C_2H_4; $+$, C_2H_6; $\bullet$, CO; $\circ$, CO_2; $\blacktriangle$, CH_3; $*$, H_2O; $\blacktriangledown$, CH_2O; $\blacksquare$, CH_3OH; $\blacklozenge$, HCOOH (with permission from Ref. [7]).

tions of molecular species in methane plasmas in the absence of oxygen, and the trends for the major chemical products as oxygen were added. The observation of many different species by diode laser spectroscopy not only shows the versatility of this type of absorption spectroscopy but enabled the progress to be made in identifying major and minor constituents of the plasma under oxidising as well as oxygen-free conditions.

7.5 Results and applications: physical properties of plasmas

Some of the main objectives of plasma diagnostic studies are the determination of internal properties and conditions, and the values of plasma parameters. This section shows how the spectroscopic techniques, in particular OES, have been used to achieve such objectives. Plasmas consist of fields as well as of particles. In direct current (dc) and alternating current (ac) plasmas both the electric and magnetic field strengths and their spatial and temporal evolution are of interest. Generally the fields and particles in the plasma determine the properties of the plasma. Excited and ground state particles (atoms, ions, molecules, etc.) are characterised by their densities and velocity distribution functions, which may not be the same for different states and species in non-equilibrium plasma. Distribution functions may also be introduced for describing populations of various excited states. The concept of temperature is only valid, if a Maxwell-Boltzmann situation applies (local thermodynamic equilibrium). This often is not the case for low-pressure plasma and so the idea of temperature may not be meaningful, or the values should be considered only as effective quantities – characteristical for the average energy. In the literature (especially in application-oriented works) one finds references to many different effective temperatures as are: *gas* (also called *translational* or *kinetic*), *rotational, vibrational*, and even the *excitation* temperature of certain groups of atomic excited levels. In non-equilibrium plasmas the concept of temperature may have a certain physical meaning, but often it is too artificial.

7.5.1 Temperatures and distribution functions

7.5.1.1 Translational temperature

The gas (kinetic) temperature is a very important plasma parameter, since the rate coefficients of chemical reactions usually show an exponential dependence on T. On the other hand, even in a monoatomic gas the plasma density distribution cannot be calculated without knowledge of the temperature distribution. Determination of the gas temperature or attempts to do that by emission and absorption spectroscopy are rather common in plasma diagnostics [18, 19, 53], [63]–[81]. A measurement of the Doppler broadening of spectral lines is a well-known and widely used method of temperature determination in gas discharge plasmas [3, 4]. For a single spectral line the full width half maximum (FWHM) of the Doppler profile is

$$\Delta \lambda_d = \sqrt{\frac{8k\,T\,\ln 2}{mc^2}}\,\lambda_m\,, \tag{7.9}$$

where $\Delta\lambda_d$[nm] is the FWHM of the Gaussian line profile, T[K] the neutral gas temperature, m[kg] the particle mass, and λ_m[nm] the wavelength at the line centre.

One can see that for plasma diagnostics it is better to use lines with longer wavelengths emitted by lighter species. It should be noted that a realisation of the method in practice is not that simple. The main problems are: (i) the determination of the instrumental function of the spectrometer, (ii) taking into account the multiplet and fine structure of the lines, and (iii) extraction of the Doppler broadening contribution from the measured line profile (so-called deconvolution). Moreover, in non-equilibrium molecular plasmas the velocity spectra of excited and ground state species can be different.

Nevertheless, the determination of translational temperatures from Doppler broadening of atomic and molecular hydrogen lines is often used in non-equilibrium plasmas [19, 66, 67, 71, 74, 76]. In the case of H_α and D_α spectral lines, which are of special importance, it is necessary to take into account their fine structure [79]. To obtain dependable results the Doppler broadening is often used in combination with methods based on other physical grounds. A comparison of rotational (recalculated for the ground state [72, 74]), and translational temperatures is most convenient [67, 77, 79, 80]. As a recent example Fig. 7.6 shows the results obtained in a planar microwave discharge [18]. Agreement was found to be within 10–15 % of the rotational temperature, derived from the (0-0), (1-1), (2-2) Fulcher–α bands, with the translational temperatures of hydrogen and deuterium atoms and hydrogen molecules.

Since the first observation of an abnormal Doppler broadening of hydrogen Balmer lines [82], many more observations and attempts to understand the nature of the effect have been

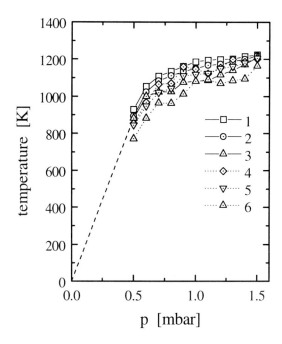

Figure 7.6: Translational and rotational temperatures in a planar microwave discharge (64% H_2 + 16% D_2 + 20% Ar, P = 1.5 kW) and their dependence on the pressure p, 1, 2 and 3 – translational temperatures derived from Doppler broadening of D_α, H_α, and H_2 spectral lines, respectively; 4, 5 and 6 – rotational temperatures of $X^1\Sigma_g^+$, $\nu = 0$ state of H_2 obtained from Q(0-0), Q(1-1) and Q(2-2) branches, respectively, of the Fulcher–α bands (with permission from Ref. [18]).

performed [30, 83]. Detailed studies of glow and asymmetric high-voltage hollow-cathode discharges in mixtures of hydrogen with inert gases (Ne, Ar, Kr) gave evidence of the existence of strongly non-equilibrium velocity distributions of excited hydrogen atoms [83]. A significant part of them (under certain conditions the overwhelming majority) has kinetic energies up to several keV, exceeding the thermal energy by orders of magnitude. It was shown that abnormally fast excited atoms are generated as a result of charge transfer, an effect being especially pronounced in H_2–Ar mixtures. Interaction of fast atoms and ions with the cathode surface can also be substantial in glow hollow-cathode [83] and rf [30] discharges. The kinetic energy distribution functions of excited (3l) and ground state (1s) hydrogen atoms were determined in the range of 0–120 eV [83]. They were found to be different and in a substantially non-equilibrium state. The analysis of H_α line profiles under various plasma conditions made it possible to distinguish contributions of H^+, H_2^+ and H^- ions and to propose a spectroscopic method for studies of the ionic composition [83].

It is obvious that an asymmetric Doppler broadening may be used not only for determination of anisotropic velocity distributions but also for studies of macroscopic flows in plasmas. This can be of importance for special applications like flames, explosions, jet engines, etc.

7.5.1.2 Rotational temperature

When the populations of rotational levels in the excited vibronic states n', ν' are close to a Boltzmann distribution, then the rotational temperature $T_{rot}(n', \nu')$ for this state is given by

$$N_{n'\nu'N'} = c_{n'\nu'} g_{a,s}(2N' + 1) \, \exp\left(-\frac{hc \, E_{n'\nu'N'}}{kT_{rot}(n', \nu')}\right), \qquad (7.10)$$

where $c_{n'\nu'}$ is a normalising constant, $g_{a,s}$ the nuclear spin degeneracy of the $n'\nu'N'$ level ($g_{a,s} = 1$ for heteronuclear diatomics), $E_{n'\nu'N'} [\text{cm}^{-1}]$ is the term value, and h, c, and k are universal constants.

The rotational temperatures have definite physical meaning, if the time of rotational relaxation in a certain vibronic state is much smaller than the mean lifetime of the rotational levels. In this case the rotational distribution of the populations is close to a Boltzmann distribution, with the temperature $T_{rot}(n', \nu')$ equal to the translational temperature T of the colliding particles (if they have Maxwell velocity distribution!). Strictly spoken this situation only can be achieved in an isolated gas under thermodynamic equilibrium.

A peculiarity of rotational energy spectra (in comparison with those of vibrational or electronic transitions), important for plasma spectroscopy, consists in the increase of the level separation with increasing rotational quantum number [72]. Therefore in non-equilibrium plasmas the Boltzmann distribution describes only approximately the populations of a limited number of lower rotational levels even in long-living ground and metastable vibronic states [72, 74, 84]. Non-equilibrium population distributions over rotational levels of excited electronic states of molecules repeatedly have been observed in gaseous discharge plasmas and in flames [63, 74, 76, 77, 85]. Their characteristic feature is a large excess of the population in higher rotational levels. Such a non-Boltzmann behaviour of the populations should be taken into account in the determination of gas temperature from the beginning part of the distributions. A detailed analysis of various mechanisms responsible for this effect can be found in Refs. [72, 74].

In low-pressure plasmas the radiative lifetimes of rotational levels in excited electronic-vibrational states are often much smaller than those of the rotational relaxation in the specified excited vibronic states. If this is the case the rotational population distributions [72]–[75] as well as corresponding rotational temperatures [63, 67] in ground and excited vibronic states can be related in the framework of certain kinetic models. Most simple is the so-called *corona-like* model, based on the assumption of dominance of direct electron impact excitation and spontaneous decay of rovibronic levels [67], [72]–[75]. In this case the measurement of an intensity distribution of the rotational structure of emission bands may be used for the determination of the ground state rotational temperature, which is often close to the gas temperature [63, 67, 73, 74, 76, 77, 79, 84, 86, 87].

From Eqs. 7.1 and 7.10 follows that for the derivation of $T_{rot}(n', \nu')$ or $N_{n'\nu'N'}$ it is necessary to know at least the relative dependence of the transition probability for spontaneous emission on the rotational quantum number. Experimental data are usually unavailable, and researchers have to use theoretical data obtained by the Born-Oppenheimer approximation in which the dependence is described by the so-called Hönl-London factors. That could be the reason for many strange results obtained so far, for example for obtaining different temperatures under the same conditions from emission of various bands and of different

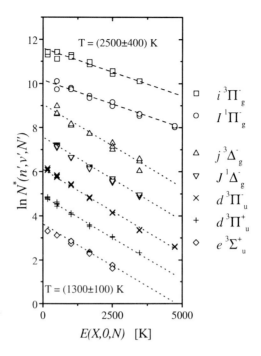

Figure 7.7: Population density distributions in the $i^3\Pi_g^-$, $I^1\Pi_g^-$, $j^3\Delta_g^-$, $J^1\Delta_g^-$, $d^3\Pi_u^-$, $d^3\Pi_u^+$ and $e^3\Sigma_u^+$ electronic states of H_2 measured in a H_2 dc discharge ($I = 200$ mA). For clarity, data corresponding to different vibronic states are arbitrarily shifted in the direction of the ordinate axis (with permission from Ref. [63]).

molecules [74]. Due to rigorous semiempirical studies of rovibronic transition probabilities and radiative lifetimes [85, 88, 89] the situation is more or less satisfactory in the case of H_2 and D_2 molecules. Convenient for application, tables of data for H_2 isotopomer obtained during the last decade and references of the original papers have been collected in Ref. [63]. Recently, these data have been applied to intensities of the $e^3\Sigma_u^+, d^3\Pi_u^- \to a^3\Sigma_g^+$, $i^3\Pi_g^-, j^3\Delta_g^- \to c^3\Pi_u$, and $I^1\Pi_g^-, J^1\Delta_g^- \to C^1\Pi_u$ band systems measured in dc capillary-arc, tube and planar microwave discharges [18, 63]. It was observed that values of the rotational temperature (recalculated for the $X^1\Sigma_g^+$, $\nu = 0$ vibronic ground state) derived from the populations of $e^3\Sigma_u^+, d^3\Pi_u^+, d^3\Pi_u^-, j^3\Delta_g^-$ and $J^1\Delta_g^-$ states coincide within 10–20% and therefore can be used in plasma diagnostics. In contrast, the rotational population density distributions in the $I^1\Pi_g^-$ and $i^3\Pi_g^-$ states show abnormal behaviour corresponding to effective temperatures, which are significantly higher. Figure 7.7 shows the population density distribution for these seven electronic states of H_2 measured in a capillary-arc discharge [63].

In the case of unresolved or partly resolved rotational structure the band shape has to be calculated theoretically and compared with experimental values to get information about the temperature. Many papers describe how nitrogen band emission can be used to determine the temperature in low-pressure discharges [65]. Also at higher pressures, such as in barrier discharges, rotational temperatures have been derived from N_2 band shapes [71]. Stalder and Sharpless [69] compared experimental and calculated spectral emission of C_2 Swan band transitions to estimate the temperature in H_2–CH_4 plasmas used for thin film diamond deposition. In Fig. 7.3 the spatial distribution of the rotational temperature derived from H_2 Fulcher–α emission ((1-1) Q-branch) is compared with that from N_2 band emission ($\nu = 1$–4 transition of the 2nd positive system). Both temperature distributions show maxima at a position roughly below the middle of the microwave windows of the double microwave plasma source [18].

In recent publications [19, 53], [64]–[66], [68] the opportunity was proposed and analysed to use intensities of R-branch lines of (0-0) band in $G^1\Sigma_g^+ \rightarrow B^1\Sigma_u^+$ electronic transition of H_2 for the gas temperature determination of gaseous discharges in pure hydrogen and mixtures H_2–CH_4, H_2–N_2. Hönl-London factors have been used as line strengths. Recently the problem has been analysed in detail on the basis of experimental data about: (i) the intensity distributions in both, R and P, branches of various (0-0), (0-1), (0-2), and (0-3) bands of the $G^1\Sigma_g^+ \rightarrow B^1\Sigma_u^+$ transition, (ii) the intensity distributions in Q-branches of (0-0), (2-2) Fulcher–α bands, and (iii) the Doppler broadening of atomic and molecular lines [90]. It was shown that (i) Hönl-London factors cannot be used in this case because of strong perturbations, and (ii) the agreement between rotational and gas temperatures observed in [19], [53], [64]–[66], [68] is most likely just an occasional coincidence.

The possibility of the determination of the gas temperature from measured rotational intensity distributions is directly connected with the applicability of a certain theoretical model to the plasma under study. The rates of involved elementary processes are, as a rule, not known. Therefore, nowadays it is not possible even to estimate the applicability a *priori*. So in each particular case it is necessary to accompany the intensity measurements with additional methods of temperature determination (Doppler broadening, CARS, thermoprobes, etc.). The other possibility (especially suitable for applications in plasma technology) is the application of internal checks within the method to use simultaneously the emission of several different bands and band systems [63, 73], different isotopomers (like H_2, HD and D_2) [77] or different molecules [18, 76] for a determination of T, which has obviously a specific value for every condition.

7.5.1.3 Vibrational temperature

Population density distributions over vibrational levels have a non-Boltzmann character in gas discharge plasmas [30]. Nevertheless, the vibrational temperature T_{vib} is widely used for very rough estimations of the distributions. Since the measurement of the emission band intensities has been proposed for determination of T_{vib} in N_2 discharges by OES [91], the method was often used in low pressure plasma diagnostics. In particular, it looks prospective for the *in situ* control of plasma processing. The method was generalised by taking into account the resolved rotational structure of H_2 bands [81], what is important for light molecules. Later the method of determination of T_{vib} from relative intensities of Q-branch lines in Fulcher–α bands was seriously corrected and simplified due to the revision of $d^3\Pi_u^- \rightarrow a^3\Sigma_g^+$ transition probabilities and cross sections of the $d^3\Pi_u^- \leftarrow X^1\Sigma_g^+$ electron impact excitation [92].

Recently, the emission of the H_2 radiative dissociation continuum ($a^3\Sigma_g^+ \rightarrow b^3\Sigma_u^+$) has been proposed and used as a source of information for the spectroscopic diagnostics of non-equilibrium plasmas [92, 93]. It was shown that in the 220–400 nm wavelength range the shape of the continuum is sensitive to the vibrational population distribution in the ground $X^1\Sigma_g^+$ state (for T_{vib} = 2000–8000 K). For relatively low input powers the shape of the continuum was observed to be independent of the discharge conditions and close to that calculated by the Franck-Condon approximation (FCA) for electron impact excitation in accordance with previous results [81]. The difference between the FCA calculation and the observed shapes of the continuum for higher input power P has been proposed as a means of determining T_{vib} in plasmas containing hydrogen. This new method has been used in studies of capillary arc [91] and pulsed microwave discharges [87]. In both cases a remarkable difference between T_{vib} and T was observed. This observation is in accordance with previous CARS measurements in other discharges [30], [94]. In pure H_2 plasmas the values of T_{vib} obtained by two independent OES methods are in rather good accordance for T_{vib} = 3000–5000 K [92].

7.5.2   Degree of dissociation

In the past, dissociation processes in low temperature hydrogen-containing plasmas were investigated by several different approaches namely actinometry (see above), absolute atomic line intensities [91, 95, 96], relative atomic and molecular line intensities [97, 98], laser induced fluorescence [99], or vacuum ultraviolet absorption methods [100]. In the case of the approach using atomic emission the electron impact dissociative excitation should be taken into account in molecular plasmas which have a low degree of dissociation [96]. Recently, various methods based on relative intensity measurements have been analysed and compared [98].

   The results obtained for a symmetrically coupled 4 MHz discharge in H_2 are presented in

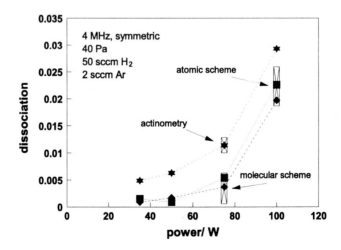

Figure 7.8: Comparison of 3 different emission methods for determining the degree of dissociation of H_2 in a rf parallel plate reactor, based on line intensities of H, H_2 and Ar. Representative uncertainties are shown for each method (with permission from Ref. [98]).

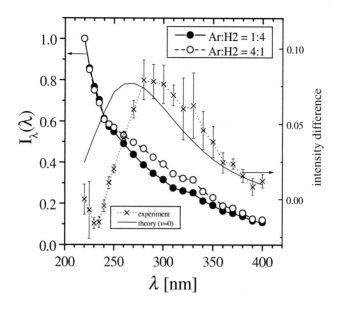

Figure 7.9: Dissociation continua in a planar microwave discharge for 2 different mixtures of H_2 and Ar ($p = 50$ Pa, $P = 1.5$ kW). The difference in the dissociation continua is represented by × symbols, and the theoretical curve is the excitation contribution of the $\nu = 0$ level of the ground state (with permission from Ref. [18]).

Fig. 7.8. One can see an acceptable agreement between different methods. Recently it was shown that absolute values of the H_2 continuum intensities can be used to determine the rate of radiative dissociation via the $b^3\Sigma_u^+$ repulsive state and for estimating the degree of hydrogen dissociation [92, 93]. In the case of a pure hydrogen plasma the method is relatively simple, but in H_2–Ar mixtures it is necessary to take the excitation transfer from metastable states of argon atoms to H_2 molecules into account [101]. Recently, in H_2-Ar microwave plasmas it was observed that the shape of the continuum depends on the ratio of the mixture components [93].

Figure 7.9 shows dissociation continua in a planar microwave discharge for two different mixtures of H_2 and Ar [18, 92]. Absorption measurements of the population of the Ar $3s^2 3p^5 4s$ levels completed by a computer simulation showed that the $Ar^* \rightarrow H_2$ excitation transfer plays a significant role [92].

7.5.3 Electric field, electron temperature, density and distribution function

Several groups have carried out spatially resolved electric field measurements in the electrode regions of dc hydrogen glow and obstructed discharges using the Stark broadening of Balmer lines. For example, electric field values of 250 V/cm were measured 1 cm from the cathode surface in a 80 Pa hydrogen discharge by Barbeau and Jolly [103].

p [Pa]

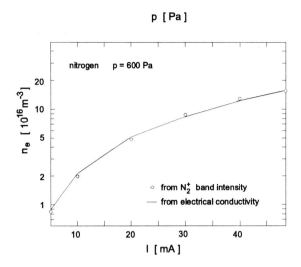

Figure 7.10: Electron density n_e from spectroscopic measurements and from the electrical conductivity of a N_2 glow discharge (with permission from Ref. [102]).

To understand the physical and chemical processes taking place in non-equilibrium plasmas information about electron parameters, their density and energy distribution function (EEDF), are of the highest importance. Langmuir probes are widely used for this purpose, but their range of applicability under different plasma conditions and at higher pressures is frequently limited. Microwave interferometry, a line of sight method, can only deliver data about electron density, while Thomson scattering needs sophisticated laser equipment.

Since the generation of a photon is the result of an earlier excitation event, which in many cases is due to electron impact, this photon can carry the desired information about the electrons. This general relationship between excitation and radiative de-excitation is the underlying concept behind spectroscopic determination of electron parameters. Lavrov et al. [84] analysed the non-equilibrium rotational population distributions of H_2 and derived plasma parameters from the shape of the rotational distribution including the degree of ionisation, the mean electron energy and the gas temperature. Hope et al. [104] monitored the variation of electron temperature in argon and oxygen etching plasmas using Langmuir probes and by the ratios of ArI and ArII emission lines which were greatly different in their excitation energies. Kondo et al. [105] measured the electron temperature in a microwave plasma from the intensity ratio of Fulcher–α bands and the H_α line emission and obtained agreement, within a factor of 2, with the results obtained by a Langmuir probe. Thomas et al. [26] used the ratio of the hydrogen Balmer H_α and H_β lines to measure the electron temperature in low-pressure H_2–Ar microwave discharges, but in order to be able to do this they had to assume local thermodynamic equilibrium.

Already in 1971 the determination of the EEDF by OES was proposed and realised [106]. This method consists of (i) the measurement of the intensities of many spectral lines having different thresholds and energy dependencies of the electron impact cross-sections, and (ii)

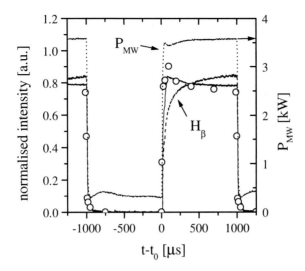

Figure 7.11: Time evolution of H_2 emission: (o) sum of the integrated intensities of Fulcher–α, $\nu'=\nu''=2$, $N'=$ 1–5 lines, (—) dissociation continuum and (– – –) H_β emission with microwave power P_{MW} as a reference [87].

the solution of the reverse in the framework of a kinetic modeling of excitation-deactivation processes. During the last decade one can see more and more activities in this field (e.g., Refs. [102, 107], and references therein). Behringer and Fantz [102] derived electron energy distribution functions in glow discharges from the intensities of Ar and He spectral lines and N_2 molecular bands. Figure 7.10 shows the agreement between the electron density results from spectroscopic measurements and from the electrical conductivity for a N_2 glow discharge. Recently, Bibinov et al. [107] used spectroscopic observations for the determination of the electron energy distribution function in helium and/or nitrogen low-pressure glow discharges and ECR plasmas. Reasonable agreement with Langmuir probe measurements was achieved for electron energies needed for kinetic modeling. Mainly due to further developments of personal computers it can be expected that the method of determination of the EEDF by OES will become more common in plasma spectroscopy.

7.5.4 Time-resolved spectroscopy

Time-resolved emission spectroscopy can yield information about excitation and relaxation processes in ac or pulsed discharges. Spatially and temporally resolved OES of ArI, H_α and CH emission in H_2–CH_4 rf discharges for amorphous carbon deposition was used by Kokubo and co-workers to study the temporal discharge excitation behaviour, the dynamics of electron transport, and the drift of the dominant ions [108, 109]. The influence of negative ions on the discharge structure in SiH_4–CH_4 rf discharges has been discussed, based on H_α, H_2, Si and SiH emission measurements [110]. Hatta et al. modulated a combined ECR-microwave methanol-hydrogen plasma and increased the diamond film growth rate by a factor of two

[111]. The temporal peak intensities of CO, OH, CH, C_2, H_2 and H_α lines were monitored. In the case of a 500 Hz pulsed discharge it was found that a H_2 molecular line (603.2 nm) was excited considerably faster than an atomic H_α line [111]. Recently, different time characteristics of the excitation of atomic and molecular hydrogen levels were also found in pulsed microwave plasmas [86, 87]. Figure 7.11 shows how the sum of the integrated intensities of the Fulcher–α, $v'=v''=2$, $N'=$ 1–5 lines and the hydrogen dissociation continuum precedes the integrated intensity of the H_β line. The specific time evolution of atomic hydrogen lines is most likely caused by the competition between direct and dissociative electron impact excitations and by a delayed increase of the degree of hydrogen dissociation.

7.6   Conclusions

During the past decade several new techniques have been successfully introduced for diagnostic studies of chemically reactive plasmas in which many short-lived and stable species are produced. It has been possible to determine absolute concentrations of ground states using spectroscopy thereby providing a link with chemical modeling of the plasma, the ultimate objective being better to understand the chemical processes occurring in the plasma. The other essential component needed for reaching this objective is to determine physical parameters of the plasma, and this article mentions many of the successful methods used to achieve this. Much further work is required, however, before we can hope to achieve a comprehensive physical and chemical description of non–equilibrium molecular plasmas.

Acknowledgements

The authors are indebted to the Deutscher Akademischer Austauschdienst (DAAD) and the British Council for support of this project as part of the British–German Academic Research Collaboration Program (Project 572). Part of this project was also supported by the Deutsche Forschungsgemeinschaft in Sonderforschungsbereich 198, by the Russian Foundation for Support of Basic Research (grant No 95–03–09394a), and by the Programme of Support of International Collaboration of the Ministry of Higher Education of the Russian Federation.

References

[1] S. Svanberg, Atomic and Molecular Spectroscopy; Basic Aspects and Practical Applications, Springer Series on Atoms and Plasmas **6**, Springer-Verlag, Berlin (1990)

[2] W. Neumann, Spektroskopische Methoden der Plasmaphysik, in R. Rompe, M. Steenbeck, Editors, Ergebnisse der Plasmaphysik und Gaselektronik, Akademie-Verlag, Berlin (1970)

[3] W. Lochte-Holtgreven, Plasma Diagnostics, American Institute of Physics, New York (1995)

[4] H.R. Griem, Plasma Spectroscopy, McGraw-Hill Book Company, New York (1964)

[5] W. Demtröder, Laserspektroskopie, Springer-Verlag, Berlin (1991)

[6] A. Ohl, Large Area Planar Microwave Discharges, in Microwave Discharges: Fundamentals and Applications, C. M. Ferreira, Editor, Plenum, New York (1993)

[7] J. Röpcke, L. Mechold, M. Käning, W.Y. Fan, P.B. Davies, Plasma Chem. Plasma Process. **19**, 395 (1999)

[8] D. Romanini, K. K. Lehmann, J. Chem. Phys. **99**, 6287 (1993)

[9] D. Romanini, A. A Kachanov, F. Stoeckel, Chem. Phys. Lett. **270**, 358 (1997)

[10] J.W. Coburn, M. Chen, J. Appl. Phys. **51**, 3134 (1980)

[11] J.W. Coburn, M. Chen, J. Vac. Sci. Technol. **18**, 353 (1981)

[12] J.P. Booth, O. Joubert, J. Pelletier, J. Appl. Phys. **69**, 618 (1991)

[13] J.P. Booth, N. Sadeghi, J. Appl. Phys. **70**, 611 (1991)

[14] A.D. Richards, B.E. Thompson, K.D. Allen, H.H. Sawin, J. Appl. Phys. **62**, 792 (1987)

[15] S.F. Durrant, R.P. Mota, M.A. Bica de Moraes, Vacuum **47**, 187 (1996)

[16] W.S. Shih, N.C. Morosoff, Plasmas and Polymers **1**, 17 (1996)

[17] P. Favia, M.V. Stendardo, R. d'Agostino, Plasmas and Polymers **1**, 91 (1996)

[18] J. Röpcke, M. Käning, B.P. Lavrov, Journal de Physique IV **8**, 207 (1998)

[19] T. Lang, J. Stiegler, Y. von Kaenel, E. Blank, Diamond & Related Materials **5**, 1171 (1996)

[20] A. Rousseau, A. Granier, G. Gousset, P. Leprince, J. Phys. D: Appl. Phys. **27**, 1412 (1994)

[21] Y. Catherine, A. Pastol, L. Athouel, C. Fourrier, IEEE Trans. Plasma Sci. **18**, 923 (1990)

[22] A. Giquel, K. Hassouni, S. Farhat, Y. Breton, C.D. Scott, M. Lefebre, M. Pealat, Diamond & Related Materials **3**, 581 (1994)

[23] J. Laimer, S. Matsumoto, Int. J. of Refractory Metals & Hard Materials **14**, 179 (1996)

[24] J. Laimer, S. Matsumoto, Plasma Chem. Plasma Process. **14**, 117(1994)

[25] A. Pastol, Y. Catherine, J. Phys. D: Appl. Phys. **23**, 799 (1990)

[26] L. Thomas, J.L. Jauberteau, I. Jauberteau, J. Aubreton, A. Catherinot, Plasma Chem. Plasma Process. **17**, 193 (1997)

[27] K. Ebihara, T. Ikegami, T. Matsumoto, H. Nishimoto, S. Medea, J. Appl. Phys. **66**, 4996 (1989)

[28] Y. Muranaka, H. Yamashita, H. Miyadera, Thin Solid Films **195**, 257 (1991)

[29] E. de la Cal, D. Tafalla, F.L. Tabares, J. Appl. Phys. **73**, 948 (1993)

[30] A.L. Capelli, R.A. Gottscho, T.A. Miller, Plasma Chem. Plasma Process. **5**, 317 (1985)

[31] J. A. Mucha, D. L. Flamm, D. E. Ibbotson, J. Appl. Phys. **65**, 3448 (1989)

[32] G.N. Kurtynina, D.K. Otorbaev, J. Phys. D: Appl. Phys. **30**, 2223 (1997)

[33] R. Etamadi, C. Godet, J. Perrin, Plasma Sources Sci. Technol. A **6**, 323 (1997)

[34] A. Ohl, J. Röpcke, Diamond & Related Materials **I**, 243 (1992)

[35] A. Ohl, J. Röpcke, W. Schleinitz, Diamond & Related Materials **2**, 298 (1993)

[36] J. Röpcke, A. Ohl, M. Schmidt, Journal of Analytical Spectrometry **8**, 803 (1993)

[37] J. Röpcke, A. Ohl, Contrib. Plasma Phys. **34**, 575 (1994)

[38] J. Röpcke, A. Ohl, Contrib. Plasma Phys. **31**, 669 (1991)

[39] A. Campargue, D. Romanini, N. Sadeghi, J. Phys. B: Appl. Phys. **31**, 1168 (1998)

[40] M. Kotterer, J. Conceicao, J. P. Maier, Chem. Phys. Lett. **259**, 233 (1996)

[41] M. Haverlag, E. Stoffels, W.W. Stoffels, G.M.W. Kroesen, F.J. de Hoog, J. Vac. Sci. Technol. A **14**, 380 (1996)

[42] P.B. Davies, P.M. Martineau, Adv. Mater. **4**, 729 (1992)

[43] S. Naito, N. Ito, T. Hattori, T. Goto, Jpn. J. Appl. Phys. **34**, 302 (1995)

[44] M. Haverlag, E. Stoffels, W.W. Stoffels, G.M.W. Kroesen, F.J. De Hoog, J. Vac. Sci. Technol. A **12**, 3102 (1994)

[45] M. Haverlag, E. Stoffels, W.W. Stoffels, G.M.W. Kroesen, F.J. De Hoog, J. Vac. Sci. Technol. A **14**, 384 (1996)

[46] C. Yamada, E. Hirota, J. Chem. Phys. **78**, 669 (1983)

[47] S. Naito, N. Ito, T. Hattori, T. Goto, Jpn. J. Appl. Phys. **33**, 5967 (1994)

[48] S. Naito, M. Ikeda, N. Ito, T. Hattori, T. Goto, Jpn. J. Appl. Phys. **32**, 5721 (1993)

[49] M. Ikeda, N. Ito, M. Hiramatsu, M. Hori, T. Goto, J. Appl. Phys. **82**, 4055 (1997)

[50] G.M.W. Kroesen, J.H.W.G. den Boer, L. Boufendi, F. Vivet, K. Khouli, A. Bouchoule, F.J. de Hoog, J. Vac. Sci. Technol. A **14**, 546 (1996)

[51] Z. Liu, P.B. Davies, J. Chem. Phys. **105**, 3443 (1996)

[52] D.M. Smith, P.B. Davies, J. Chem. Phys. **100**, 6166 (1994)

[53] K.L. Menningen, M.A. Childs, H. Toyoda, Y. Ueda, L.W. Anderson, J.E. Lawler, Diamond & Related Materials **3**, 422 (1994)

[54] M.A. Childs, K.L. Menningen, P. Chevako, N.W. Spellmeyer, L.W. Anderson, J.E. Lawler, Phys. Lett. A **171**, 87 (1992)

[55] J. Wormhoudt, J. Vac. Sci. Technol. A **8**, 1722 (1990)

[56] J. Wormhoudt, K.E. McCurdy, Chem. Phys. Lett. **156**, 47 (1989)

[57] P.B. Davies, P.M. Martineau, Appl. Phys. Lett. **57**, 237 (1990)

[58] P.B. Davies, P.M. Martineau, J. Appl. Phys. **71**, 6125 (1992)

[59] M. Ikeda, K. Aiso, M. Hori, T. Goto, Jpn. J. Appl. Phys. **34**, 3273 (1995)

[60] M. Ikeda, M. Hori, T. Goto, M. Inayoshi, K. Yamada, M. Hiramatsu, M. Nawata, Jpn. J. Appl. Phys. **34**, 2484 (1995)

[61] S. Kim, D.P. Billesbach, R. Dillon, J. Vac. Sci. Technol. A **15**, 2247 (1997)

[62] W.Y. Fan, P.F. Knewstubb, M. Käning, L. Mechold, J. Röpcke, P.B. Davies, J. Phys. Chem. A **103**, 4118 (1999)

[63] S.A. Astashkevich, M. Käning, E. Käning, N.V. Kokina, B.P. Lavrov, A. Ohl, J. Röpcke, J. Quantitative Spectroscopy & Radiative Transfer **56**, 725 (1996)

[64] H.N. Chu, E.A. Den Hartog, A.R. Lefkow, J. Jacobs, L.W. Anderson, M.G. Lagally, J.E. Lawler, Phys. Rev. A **44**, 3796 (1991)

[65] G.M. Jellum, J.E. Daugherty, D.B. Graves, J. Appl. Phys. **69**, 6923 (1991)

[66] A. Giquel, K. Hassouni, Y. Breton, M. Chenevier, J.C. Cubertafon, Diamond & Related Materials **5**, 366 (1996); A. Giquel, M. Chenevier, Y. Breton, M. Petian, J.P. Booth, K. Hassouni, J. Phys. III **6**, 1167 (1996)

[67] B.P. Lavrov, J.K. Otorbaev, Sov. Tech. Phys. Lett. **4**, 574 (1978)

[68] J. Laimer, F. Huber, G. Misslinger, H. Störi, Vacuum **47**, 183 (1996)

[69] K.R. Stalder, R.L. Sharpless, J. Appl. Phys. **68**, 6187 (1990)

[70] L. Tomasini, A. Rousseau, G. Gousset, P. Leprince, J. Phys. D: Appl. Phys. **29**, 1006 (1996)

[71] I.P. Vinogradov, K. Wiesemann, Plasma Sources Sci. Technol. **6**, 307 (1997)

[72] B.P. Lavrov, Opt. Spectrosc. **48**, 375 (1980)

[73] B.P. Lavrov, M.V. Tyutchev, Acta Phys. Hung. **55**, 411 (1984)

[74] B.P. Lavrov, Electronic-rotational spectra of diatomic molecules and diagnostics of non-equilibrium plasma, in: Plasma Chemistry, B. M. Smirnov, Editor, Energoatomizdat, Moscow, 45 (1984)

[75] A.I. Dratchev, B.P. Lavrov, High Temp. **26**, 129 (1988)

[76] N.N. Sobolev, Editor, Electron-excited molecules in non equilibrium plasma, Nova Science Publishers, New York (1989)

[77] B.P. Lavrov, A.A. Solov'ev, M.V. Tyutchev, J. Appl.Spectrosc. **32**, 316 (1980)

[78] K.H. Krysmanski, Beitr. Plasmaphys. **23**, 505 (1983)

[79] B.P. Lavrov, M. Käning, V. L. Ovtchinnikov, J. Röpcke, Frontiers in Low Temperature Plasma Diagnostics II, Bad Honnef, Conf. Proc. (1997), p. 169

[80] J. Röpcke, B.P. Lavrov, A. Ohl, Frontiers in Low Temperature Plasma Diagnostics II, Bad Honnef, Conf. Proc. (1997), p. 201

[81] B.P. Lavrov, V.P. Prosikhin, Opt. Spectrosc. **58**, 317 (1985)

[82] W. Benesch, E. Li Ayers, Opt. Lett. **9**, 338 (1984); Phys. Rev. A **37**, 194 (1988)

[83] B.P. Lavrov, A.S. Melnikov, Opt. Spectrosc. **75**, 676 (1993); **79**, 842 (1995)

[84] B.P. Lavrov, V.N. Ostrovsky, V.I. Ustimov, Sov.Phys.–Tech.Phys. **25**, 1208 (1980)

[85] B.P. Lavrov, M.V. Tyutchev, V.I. Ustimov, Opt. Spectrosc. **64**, 745 (1988)

[86] M. Käning, N. Lang, B.P. Lavrov, V.L. Ovtchinnikov, J. Röpcke, Europhys. Conf. Abstr. **22H**, 396 (1998)

[87] N. Lang, M. Kalatchev, M. Käning, B.P. Lavrov, J. Röpcke, Frontiers in Low Temperature Plasma Diagnostics III, Saillon, Conf. Proc. (1999), p. 253

[88] I. Kovacs, B.P. Lavrov, M.V. Tyutchev, V.I. Ustimov, Acta Phys. Hung. **54**, 161 (1983)

[89] B.P. Lavrov, L.L. Pozdeev, Opt. Spectrosc. **66**, 479 (1989)

[90] S.A. Astashkevich, M.V. Kalachev, B.P. Lavrov, V.L. Ovtchinnikov, Opt. Spectrosc. **87**, 203 (1999)

[91] R. Bleekrode, IEEE J. Quantum Electronics **5**, 57 (1969)

[92] B.P. Lavrov, A.S. Melnikov, M. Käning, J. Röpcke, Phys. Rev. E **59**, 3526 (1999)

[93] M. Käning, B.P. Lavrov, A.S. Melnikov, A. Ohl, J. Röpcke, M.A. Tchaplygin, Frontiers in Low Temperature Plasma Diagnostics II, Bad Honnef, Conf. Proc. (1997), p. 197

[94] M. Pealat, J.P.E. Taran, J.Taillet, M. Bacal, A.M. Bruneteau, J. Appl. Phys. **52**, 2687 (1881)

[95] M.A. Abroyan, Yu.M. Kagan, N.B. Kolokolov, B.P. Lavrov, Proc. XII. ICPIG, Eindhoven (1975), p. 101

[96] B.P. Lavrov, Opt. Spectrosc. **42**, 250 (1977)

[97] A.M. Devjatov, A.V. Kalinin, S.R. Miiovich, Opt. Spectrosc. **71**, 525 (1991)

[98] V. Schulz-von der Gathen, H.F. Döbele, Plasma Chem. Plasma Process. **16**, 461 (1996)

[99] B.L. Preppernau, D.A. Dolson, R.A. Gottscho, T.A. Miller, Plasma Chem. Plasma Process. **9**, 157 (1989)

[100] D. Wagner, B. Dikmen, H.F. Döbele, Rev. Sci. Instrum. **67**, 1800 (1996)

[101] B.P. Lavrov, A.S. Melnikov, Opt. Spectrosc. **85**, 666 (1998)

[102] K. Behringer, U. Fantz, J. Phys. D: Appl. Phys. **27**, 2128 (1994)

[103] C. Barbeau, J. Jolly, Appl. Phys. Lett. **58**, 237 (1991)

[104] D.A.O. Hope, T.I. Cox, V.G.I. Deshmukh, Vacuum **37**, 275 (1987)

[105] K. Kondo, K. Okazaki, H. Oyama, T. Oda, Y. Sakamoto, A. Iiyoshi, Japan. J. Appl. Phys. **27**, 1560 (1988)

[106] A.S. Arbuzov, A.M. Devyatov, S.P. Shushurin, Proc. X. ICPIG, Oxford (1971), p. 387

[107] N.K. Bibinov, D.B. Kokh, N.B. Kolokolov, V.A. Kostenko, D. Meyer, I.P. Vinogradov, K. Wiesemann, Plasma Sources Sci. Technol. **7**, 298 (1998)

[108] T. Kokubo, F. Tochikubo, T. Makabe, Appl. Phys. Lett. **56**, 818 (1990)

[109] T. Kokubo, F. Tochikubo, T. Makabe, J. Phys. D: Appl. Phys. **22**, 1281 (1989)

[110] F. Tochikubo, A. Suzuki, S. Kakuta, Y. Terazono, T. Makabe, J. Appl. Phys. **68**, 5532 (1990)

[111] A. Hatta, K. Kadota, Y. Mori, T. Ito, T. Sasaki, A. Hiraki, S. Okada, Appl. Phys. Lett. **66**, 1602 (1995)

8   Mass spectrometric diagnostics

Martin Schmidt, Rüdiger Foest and Ralf Basner

Institut für Niedertemperatur-Plasmaphysik, Friedrich–Ludwig–Jahn–Straße 19, D–17489 Greifswald, Germany

8.1   Introduction

Plasma mass spectrometry is useful for the diagnostics of heavy particles in plasmas. It completes the optical spectroscopy regarding the qualitative and quantitative investigation of atoms, molecules, radicals and ions. Their nature, concentration and energy can be determined by mass spectrometry. In general a mass spectrometer consists of an ion source for the generation of ions from neutral atoms or molecules, an analyzer system for the separation of the ions according to their mass-to-charge ratio, and the ion detector. The study of the kinetic energy of the ions requires an ion energy analyzer. Mass spectrometry has been connected with the plasma or better gas discharge physics since its beginning. Thomson [1] constructed the first apparatus, the well known parabola spectrograph. A direct current (*dc*) glow discharge served as ion source and the canal ray ions were investigated. Series of truncated parabolic curves indicated the different ions. The blackening of a photographic plate was a measure for the ion intensity, the length of the parabola was connected with the ion energy distribution. Thomson detected the Ne isotopes ^{20}Ne and ^{22}Ne. Today such a device for the measurement of intensity and energy distribution of mass identified ions would be called a plasma monitor. However, several types of discharges like arcs, radiofrequency (*rf*) and laser sparks, as well as *dc* and capacitively and inductively coupled *rf* plasmas are used in ion sources of mass spectrometers especially for the investigation of solids which are transferred into the gaseous state by evaporation, spark erosion, or cathode sputtering.

Mass spectrometers with high resolution power were developed and sometimes these big single and double focusing machines were used for the investigation of the ions in plasmas [2]–[5]. Isotopes and molecular ions of rare gases were detected and molecular gases were studied. A landmark for the introduction of the mass spectrometry as a method of plasma diagnostic was the invention of the quadrupole mass spectrometer by Paul and Steinwedel in 1953 [6]. This small, lightweight and relatively inexpensive device with sufficiently high resolution power is easily coupled to a plasma source and gives valuable information on plasma properties and the reactions in the plasma, and as well enables the control of plasma processes. This new powerful tool for plasma diagnostics is also necessary because complicated reactive gases and gas mixtures are used in plasma processing for thin film deposition, etching, and surface treatment.

Reviews of this diagnostic method were given by several authors, e.g., by Drawin [7], Schmidt and Hinzpeter [8], Märk and Helm [9] and Vasile and Dylla [10] in the past. There

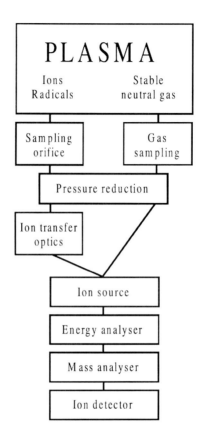

Figure 8.1: Scheme of the measurement of the particle flux (ions, radicals) and neutral gas composition of a plasma.

are several instructive textbooks on mass spectrometry, e.g., Refs. [11, 12, 13].

Two types of mass spectrometric analysis of plasmas are used, the flux analysis and the partial pressure analysis [10]. The flux analysis investigates the particles flying directly without any collisions from the plasma through a small orifice into the analyzer. A study of ions and radicals including their kinetic energy is possible (Fig. 8.1). For the partial pressure analysis of the stable components only a common vacuum tube combined with a sufficient pressure reduction between the plasma volume and the mass spectrometer is required. Collisions with the wall and between the particles do not disturb the results (Fig. 8.1). In the following the structure of a mass spectrometer will be described with its various types of ion sources, analyzing systems and ion detectors. Problems with the coupling of the mass spectrometer to the plasma system will be discussed. Examples of results of mass spectrometric investigations using different techniques are presented. A short survey is given on the application of mass spectrometry for the determination of data of elementary processes in plasmas.

8.2   Instrumentation

8.2.1   Ion source

The investigation of the neutral gas of the plasma by mass spectrometry requires the ionization of the gaseous molecules. The standard ionization method in most mass spectrometers is the electron impact. Photoionization, field ionization in a high electrical field near a tip shaped electrode, or chemical ionization in a charge transfer process, and ion molecule reaction are sometimes advantageous for special applications.

The electron impact ion source contains a heated filament which emits electrons. Tungsten, thoriated tungsten, and rhenium cathodes have been examined [14]. The electrons are accelerated by static electric fields and the resulting electron beam crosses the ionization chamber with the gas probe inside. Additional electrodes are used for the extraction of the ions out of the ionization region and for the formation of an ion beam.

The electron impact ionization is a common method for ion formation. The ionization of atoms and molecules happens by the following processes:

ionization of atoms

$$e^- + A \quad \rightarrow \quad A^+ + 2e^- \,, \tag{8.1}$$

ionization of molecules and molecule ion formation,

$$e^- + AB \quad \rightarrow \quad AB^+ + 2e^- \,, \tag{8.2}$$

dissociative ionization and fragment ion formation,

$$e^- + AB \quad \rightarrow \quad A + B^+ + 2e^- \,, \tag{8.3}$$

and dissociative ionization and ion pair formation,

$$e^- + AB \quad \rightarrow \quad A^+ + B^+ + 3e^- \,. \tag{8.4}$$

Electrons with higher kinetic energies also can produce multiply charged ions like A^{++} and AB^{++}. The ion current i_ν^+ from the electron impact ion source equals

$$i_\nu^+ = i_e^- n_n \sigma_\nu l \,, \tag{8.5}$$

with the electron current i_e^-, the neutral gas density n_n, the partial ionization cross section σ_ν for the formation of the fragment ion ν, and the length l of the ionization region.

The ionization probability is quantified by the value of the ionization cross section. For most molecules ionization starts at a threshold energy between 5 and 15 eV, and the cross section reaches a broad maximum in the region of 70 eV. Therefore commercial instruments are operated with an electron energy of typically 70 eV. The electron impact generates the parent ion of the molecule (Eq. 8.2) or fragment ions (Eqs. 8.3 and 8.4). In this way each molecule has a characteristic mass spectrum. The decomposition pattern contains information on the structure of the original molecule, but the formation of fragment ions complicates the identification of the composition of gas mixtures. The electron impact ionization has the advantage that the energy of the ionizing electrons can be changed easily. The use of lower electron energies near 20–30 eV reduces the decomposition of the molecule (Fig. 8.2). A measurement of the ionization energies of parent ions or appearance energies of fragment ions is possible. The gas pressure in the ionization chamber has to be small enough that single collision conditions are ensured:

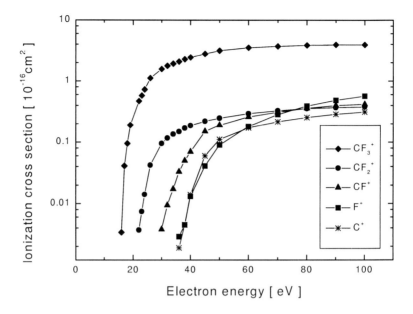

Figure 8.2: Partial ionization cross sections of CF_4 for production of singly charged ions measured with a time-of-flight mass spectrometer.

– A molecule only collides once with an electron such that stepwise processes do not occur.

– Secondary reactions of the ions with other particles are excluded.

This usually leads to gas pressures of 10^{-5}–10^{-3} Pa at electron beam densities of about 10 μA mm^{-2}.

Negative ions can be formed by electron attachment and by pair formation

$$e^- + A \quad \rightarrow \quad A^- , \tag{8.6}$$
$$e^- + AB \quad \rightarrow \quad A^- + B^+ + e^- . \tag{8.7}$$

Electron attachment is most effective for low electron energies whose controlled production is difficult, however. The advantage of ion formation by attachment is the low probability of dissociation of the molecules. The negative ion mass spectrum is then dominated by the molecule ion.

The fragmentation of the parent molecule is suppressed during chemical ionization, too. The chemical ionization of the gas under study is induced by charge transfer processes, ion attachment [15], or ion molecule reactions in a reactant gas which is ionized by electron collisions [16]. Examples for chemical ionization of the target gas M or RNH_2, respectively, are the following processes with Ar and CH_4 as reactant gases, as well as ion attachment of

Li^+:

$$Ar^+ + M \quad \to \quad Ar + M^+ \tag{8.8}$$
$$CH_5^+ + RNH_2 \quad \to \quad RNH_3^+ + CH_4 \tag{8.9}$$
$$Li^+ + M \quad \to \quad LiM^+. \tag{8.10}$$

This ion source is usually operated at higher pressures in the range of 30–100 Pa with sample admixtures in the 10–10^4 ppm region.

Field ionization is also an interesting aspect for plasma mass spectrometry [17]. Ion formation in an electric field of 10^8 V/cm is connected with a small energy transfer to the molecule and results in a preferred formation of molecule ions without additional fragmentation. This is useful in the analysis of gas mixtures.

Photoionization is suitable for exact measurements of the ionization energy because the energy of the ionizing photons is precisely known [18].

8.2.2 Mass analyzer

There are two types of mass analyzers: the static and the dynamic types. In the first type of analyzers static magnetic and electric fields are used for the separation of the ions with different mass. The mass dispersion of the magnetic field for ions with the same energy is based on the Lorentz force which is proportional to the momentum of the ion. The movement of an ion with mass m, charge $z\,e$, where e is the elementary charge, and kinetic energy $e\,U$ in a magnetic field B perpendicular to the ion velocity is determined by the balance of the centrifugal and the Lorentz force. The radius of the circular path equals

$$r = \sqrt{\frac{2mU}{eB^2}}. \tag{8.11}$$

The resolution power $m/\Delta m$, where Δm is the line width, and the sensitivity i_ν^+/n_n can be increased by the combination of a magnetic and an electric sector field. Such double focusing instruments achieve resolution powers of 10^5 and more. The resolution power and the ion transmission is independent of the ion mass. This is an essential advantage of magnetic mass analyzers. Disadvantages are the dimensions and the weight of these spectrometers. Due to the relatively large dimensions and, hence, long ion pathway pressures of about 10^{-5} Pa in the analyzer are needed to avoid ion losses by collisions with the residual gas molecules.

In the past various types of dynamic mass spectrometers were developed [19], e.g., time–of–flight spectrometers (TOF), omegatrons, *rf* and quadrupole spectrometers. The mass separation in dynamic systems is achieved by time varying electro-magnetic fields. The most simple (at least in principle) device is the TOF mass spectrometer. A monoenergetic burst of ions is formed by a pulsed electric field, and the mass separation occurs in a field-free drift tube. The faster moving light ions are registered by the ion detector first, the heavy ions arrive later. An essential advantage of this spectrometer is the large mass range.

The quadrupole mass spectrometer (QMS, Fig. 8.3) is the most widely used analyzer for low resolution applications and is most important in plasma diagnostics [6, 13]. The mass filter of a QMS consists of four parallel conducting cylindrical rods which are positioned in a square with the distance $2r$. Ideally, the cross section of the rods should be hyperbolic, but in

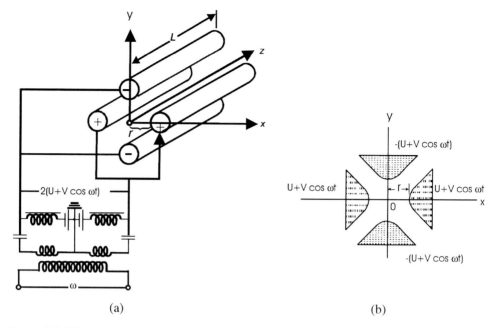

(a)   (b)

Figure 8.3: The quadrupole mass filter showing (a) cylindrical rod assembly and (b) idealized configuration (after Ref. [12]).

practice a circular cross section suffices. Opposite rods are electrically connected with each other. The voltage U between both pairs is given by

$$U = 2(U_{dc} + V_{rf} \cos \omega t),$$ (8.12)

where U_{dc} is a dc voltage, and V_{rf} the amplitude of a rf voltage of frequency $f = \omega/2\pi$.

The ion movement in the axial direction near the centre of this system is described by the Mathieu differential equation [13]. The ion path is characterized by transversal oscillations in the direction of the rods. If the amplitudes are small, ions can pass through the rod system, for larger oscillations the ions strike one electrode and are neutralized. In the Mathieu equation the ion trajectories are determined by the parameters

$$a = \frac{8eU_{dc}}{mr^2\omega^2} \quad \text{and} \quad q = \frac{4eV_{rf}}{mr^2\omega^2}.$$ (8.13)

The region for which stable solutions exist is illustrated in Fig. 8.4. On top of the stable region near $a = 0.23699$ and $q = 0.70600$ for preselected U_{dc}, V_{rf}, r and ω values only ions with a given e/m value can pass the field. By changing U_{dc} and V_{rf} with constant ratio U_{dc}/V_{rf} other ions reach the detector. A mass spectrum is scanned. A constant U_{dc}/V_{rf} ratio determines an operation line. Its slope is responsible for the mass range of the ions which can pass the field, and determines the resolution power. The higher the resolution the lower the sensitivity. The stability diagram illustrates that ions with masses below the stable solution are deflected along the x–direction to the electrodes. Ions with masses above this

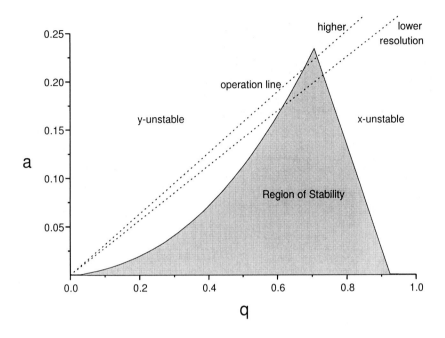

Figure 8.4: Diagram of stability of the ion movement in the quadrupole field.

solution are turned towards the y–direction, and only ions with the suitable mass can pass the mass filter. An entrance energy of the ion beam into the quadrupole field of few eV ensures a sufficient flight time with enough oscillations in the analyzing field. The transmission of the quadrupole field and therefore the sensitivity decreases with increasing ion mass. Additional quadrupole fields at the entrance and exit of the analyzer field improve the sensitivity and resolution power. QMS have several advantageous properties. They are small, light and short (about 30 cm). The small pathway of the ions permits higher pressures ($10^{-3} - 10^{-4}$ Pa) in the analyzer. Commercial devices with a mass range of 1–100, 1–300, even 1–2500 amu are available. The mass scale is linear, the resolution power $m/\Delta m$ increases linearly with the mass number, therefore the peak width is constant over the whole mass range. The line width is somewhat smaller than one mass unit. Therefore the determination of the nominal mass number is possible. By changing the electrical parameters of the *dc* and *rf* field the resolution power and the transmission and in this way the sensitivity can be varied. Sometimes this behaviour is electronically corrected by the manufacturer. It is possible to perform high scanning speeds because just the variation of electrical fields is necessary. These mass spectrometers are widely used as gas analyzers in vacuum-technique and for chemical analysis, e.g., in combination with gas chromatographs. Therefore, these instruments are reliable and available with a good price-performance ratio.

8.2.3 Ion energy analyzer

The knowledge of the kinetic energy of the plasma ions is important in many applications. Therefore modern mass spectrometers for plasma diagnostics are coupled with an energy analyzer to perform the so called energy dispersive mass spectrometry (EDMS). These instruments are known as plasma monitors. The ions of a plasma are not monoenergetic, instead they show typical energy distributions. The measurement of this distribution is accomplished either by retarding field analyzers [20], or by ion optical elements which force the ion beam to off-axis trajectories. In the first case, the energy distribution function of an ion beam moving parallel to the electric field is received by the first derivative of the current voltage characteristic. The second case is realized in commercial instruments where this distribution is directly measured with an electrical sector field [21], or a cylindrical mirror analyzer [22] with an electrical field perpendicular to the beam direction using its energy dispersive properties. Usually the investigation of ion energies up to 1000 eV is possible with a resolution power of 0.5 eV. The energy analyzer is located between ion source and mass analyzer. This also allows the measurement of energies of ions formed in the ion source.

8.2.4 Ion detector

The ion detection with a Faraday cup is insensitive to the nature of the ions, but slow and limited to higher ion currents ($> 10^{-12}$ A). Normally electron multipliers and channeltrons are used. These devices are characterized by high current gain and nanosecond response time. The continuous dynode or channel electron multiplier is now usual in quadrupole mass spectrometers. The advantage of these detectors are the low dark currents and the small dimensions. Moreover, the active surface coating is stable and can be exposed to air. Microchannel plates enable high sensitivity combined with spatial resolution of the registered ions. The reliable counting techniques ensure a high sensitivity independent of mass and chemical nature of the ion [12].

8.3 Coupling of the mass spectrometer with the plasma system

8.3.1 Mechanical coupling

The way of coupling the mass spectrometer with the plasma is strongly influenced by the aim of the study. For the investigation of the neutral gas composition of plasmas by flux measurements the gas, consisting of atoms, molecules, or radicals, flows through a sampling orifice in the wall or one electrode of the plasma reactor into the ion source of the mass spectrometer. The gas pressure should be lower than 10^{-3} Pa. In the ideal case the coupling ensures collision free transfer of the particles from the plasma to the detector. For this purpose the diameter of the sampling orifice should be smaller than the mean free path of the gaseous molecules. A very short sampling orifice minimizes the probability of wall collisions and chemical surface reactions. This orifice serves simultaneously as a pressure stage. In order to avoid distortions of the gas flow composition, a molecular flow through it is required accompanied by a molecular gas flow in the high vacuum pumping system. This condition is limited to pressures of up to 10^3 Pa due to minimal orifice diameters of 10 μm. For measurements at higher pressures differential pumping is necessary. The change of the gas composition by a viscous flow towards the ion source and a molecular flow through the pump has to be taken into account. The partial pressure measurement by mass spectrometry demands a viscous gas flow out of the reactor.

For time resolved measurements a special helium flushed probe system with a short response time (about 10 ms) and a time resolution of about 20 ms is developed [23]. Helium flows through a separately pumped silica capillary and a tube system. By means of a small orifice on the tip of the capillary in the reactor vessel and a second orifice downstream near the ion source the gas sample is transferred. Techniques of GC-MS coupled gas analysis are discussed in Ref. [12]. This method is applied for the investigation of neutral components in atmospheric pressure discharges [24].

The investigation of ions demands flux measurements. Ions are transmitted from the plasma through the small and short sampling orifice into the well evacuated vacuum vessel with the analyzer system. An ion beam is formed by an ion optical transfer system consisting of electrostatic lenses. The ions then pass the energy filter, and/or the mass filter and are measured by the detector. Disturbances of the plasma are possible due to the orifice and also due to external electrical potentials in the orifice region in conjunction with the ion extraction. They can be avoided, if the orifice diameter is small in comparison with the thickness of the sheath d_s which is a small multiple of the Debye-length of the plasma near the sheath edge. For a mean free path much larger than the sheath thickness follows [25]:

$$d_s = 1.1 \sqrt{\frac{\varepsilon_0}{n_e e^2}} \frac{(eU_s)^{3/4}}{(kT_e)^{1/4}} . \qquad (8.14)$$

The sheath voltage U_s is determined by

$$U_s = \frac{kT_e}{2e} ln \left(\frac{2.3m}{m_e} \right) , \qquad (8.15)$$

where ε_0 is the permittivity of free space, k the Boltzmann constant, m_e and m electron and ion mass, respectively, n_e the electron density, and T_e the electron temperature.

For plasma processing conditions orifice diameters of 0.1 mm are smaller than the sheath thickness near the wall (Fig. 8.5). The sheath thickness in front of the cathode or of a powered electrode of an *rf* discharge with essential self-bias is distinctly greater due to the higher sheath voltage. The ions impinging on the surface are collected by the sampling orifice. They are representative for the ions of the plasma bulk, if the sheath is free of collisions. For a collision dominated sheath the energy distribution of the ions is changed dramatically, and the ion population may be influenced by ion molecule reactions. The evaluation of the data must include a discussion of the possible sheath conditions. The properties of the ions should not change during the path through the orifice and the analyzer system. What was said for the neutrals is valid also for the ions: The orifice must be as short as possible to avoid wall losses of charge carriers.

The probability of collisions between the ions and gas molecules outside the plasma, i.e., behind the orifice and in the vacuum system of the analyzer, depends on the orifice diameter and the gas pressure in the reaction vessel. A higher pumping speed with a resulting lower

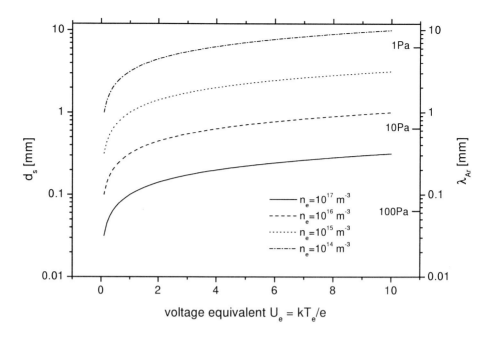

Figure 8.5: Thickness of the plasma sheath d_s versus electron temperature T_e for different electron densities. This is valid only for lower pressures when the ion mean free path is longer than the thickness of the plasma sheath. On the right axes the mean free path λ of Ar atoms for selected pressures is shown.

residual pressure is of lower influence. The neutral gas density decay $n_n(x)$ in the space behind an ideal orifice with the radius R is described by (see Fig. 8.6) [26]–[28]:

$$n_n(x) = \frac{n_0}{2}\left(1 - \frac{x}{\sqrt{R^2 + x^2}}\right) + n_a , \tag{8.16}$$

where x is the distance from the orifice, n_0 the gas density with the mean free path λ in the plasma vessel, and n_a the gas density with the mean free path λ_a according to the residual gas pressure in the analyzer.

The collisions of the beam ions with neutral gas in the space behind the sampling orifice can be estimated as [26]–[29]:

$$\frac{n^+}{n_0^+} = \exp\left(-\frac{x}{\lambda}\left(\frac{1}{2} + \frac{\lambda}{\lambda_a}\right) + \frac{1}{2}\sqrt{\frac{R^2}{\lambda^2} + \frac{x^2}{\lambda^2}} - \frac{R}{\lambda}\right) . \tag{8.17}$$

n^+/n_0^+ is the relative number of ions which travel collision-free from orifice to collector over a distance x. In Fig. 8.7 n^+/n_0^+ is presented in relation to the relative orifice radius R/λ for a residual gas density $n_a/n_0 = 10^{-5}$ and a collector distance of $x = 1000\lambda$. A collision probability near 20% is received for an orifice radius of about 0.5λ. The collision probability in a real orifice with a not negligible length is discussed in Refs. [29, 30]. It is necessary

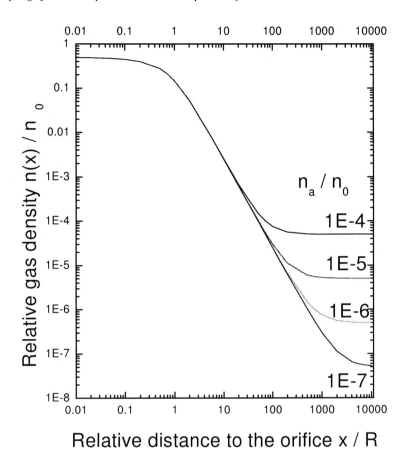

Figure 8.6: Relative gas density $n_n(x)/n_0$ in the analyzer, where n_0 is the gas density in the discharge vessel, versus relative distance x/R to the sampling orifice, where R is the radius of the sampling orifice, for different residual gas densities n_a, according to Eq. 8.16 [28].

to keep in mind that the mean free path is related to a specific process with a characteristic energy-depending cross section. Summarizing this discussion the orifice diameter must be as small as possible, maximum is the mean free path. Within the technological limits the orifice must be as short as feasible. It is common to use a very thin metallic foil and to drill the sampling orifice into this material. Differential pumping is necessary for investigations at higher pressures to reach a sufficiently low residual pressure in the analyzer. Ion losses must be discussed carefully.

Another important viewpoint for the evaluation of the results of ion current measurements is the acceptance angle of the ion collecting system consisting of the orifice, the ion transfer optics, and the analyzer. The maximum acceptance angle is determined by the diameter and the length of the orifice. The energy dependence of this angle is discussed in detail on the

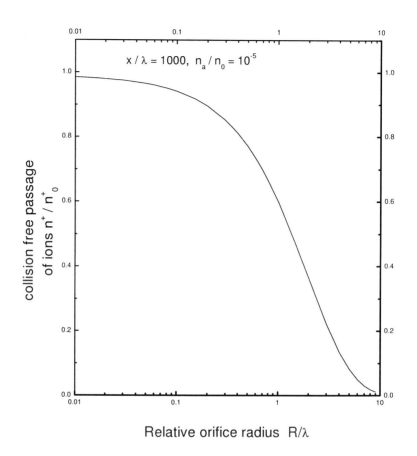

Figure 8.7: Relative number of ions n^+/n_0^+ which reach the collector without collisions versus the relative orifice radius R/λ for a collector distance $x = 1000\lambda$ and a relative residual gas density $n_a/n_0 = 10^{-5}$, according to Eq. 8.17 [28].

basis of SIMION [31] simulations by Hamers et al. [32]. It could be shown that the angle is decreasing with increasing ion energy what is of remarkable influence on the measured ion distribution. In the given example the angle varies from 20 degrees for ions with 0.1 eV down to near 1 degree for 100 eV ions. Measured ion distributions at the grounded electrode of an *rf* planar reactor for collecting conditions with known acceptance angle in dependence of the ion energy are presented in Fig. 8.8a. Fig. 8.8b shows this energy distribution divided by the corresponding solid angle of acceptance. It is obvious that there is an essential influence of ions with low energy which reach the surface with other than perpendicular velocity components. A careful discussion of the angular acceptance of the quadrupole mass spectrometer connected with the collecting orifice and the ion transfer optics given in Refs. [33, 34] shows an acceptance angle near 1 degree for ion energies higher than 100 eV; for an energy of 10 eV

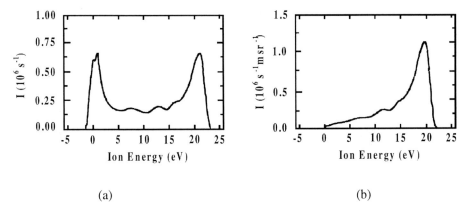

(a)   (b)

Figure 8.8: (a) Energy distribution of Ar^+ ions measured in an Ar *rf* discharge at the grounded electrode. The power input is 30 W at a gas pressure of 5 Pa [32]. (b) The energy distribution of argon ions as measured in Fig. 8.8 (a), divided by the corresponding solid angle of acceptance (after Ref. [32]) .

this angle increases up to 3 degrees.

8.3.2 Electrical coupling

The plasma usually is part of an electric circuit, and the mass spectrometer is a complex electrical device. The electrical coupling of the mass spectrometer to the plasma should disturb neither the plasma nor the transfer of particles from the plasma into the mass spectrometer. Neutral gas analysis of plasma reactors acts without electrical coupling. The connection of the mass spectrometer with a plasma for ion measurements depends on the kind of the spectrometer and on the investigated plasma region. The situation is most simple for a quadrupole mass spectrometer which works with ions of low energy.

Today plasma chemical devices are mostly metallic, and the mass spectrometer is connected with the grounded wall [20, 35, 36], the grounded electrode of an *rf* discharge [37]–[39] (see also figure 19.1 in chapter 19), or the grounded *dc* cathode [40]. A connection of the quadrupole mass analyzer with the powered *rf* electrode with *rf* and *dc* (bias) voltages against ground is solved by floating the whole mass spectrometer on a potential equal to the bias potential [41]. The *dc* potential difference between the powered electrode and mass spectrometer entrance is zero. The influence of the remaining *rf* field on the measured ion energy distribution is calculated to be smaller than 10% for the energetic ions accelerated in the high voltage sheath in front of the powered electrode [41].

The situation is much more difficult, if magnetic mass spectrometers for ion energies in the keV region and with a grounded entrance slit are used. Here it is necessary to operate the discharge circuit on a high potential. The potential of the sampling orifice must be controlled to such a value that the ion energy is optimal for the operation of the mass spectrometer. In all these cases the ion acceleration field in the ion transfer optics may not disturb the electrical

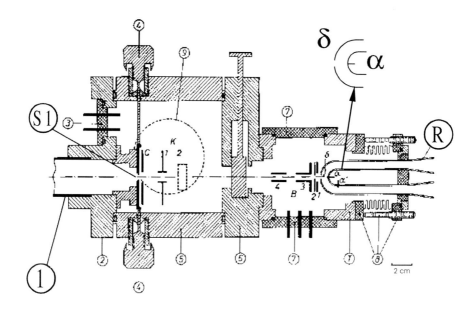

Figure 8.9: Coupling of a static magnetic sector field mass spectrometer 1 with a cylindrical discharge tube R. Sampling orifice in the wall δ, wall probe α for electrical coupling, ion transfer optic B, entrance slit of the mass spectrometer S1, after Ref. [42].

field in the sheath region.

The investigation of ions in the positive column of a *dc* glow discharge may be mentioned as typical for the coupling of a mass spectrometer to a plasma in an insulating vessel. The reference potential for the acceleration of the ions is the wall potential given by the floating potential of the metallic wall probe with the sampling orifice [3] or a wall probe opposite to the orifice in the cylindrical glass tube [42] (Fig. 8.9). The mass spectrometric investigation of ECR plasmas or other plasmas with external magnetic fields demands a magnetic shielding of the mass spectrometer [43].

One should know that the results of the ion measurements sensitively depend on the surface conditions of the orifice and the ion transfer optics. Disturbances of the surface conditions are commonly observed by the investigation of plasma chemical systems for the deposition of thin isolating films. The discussion of the effect of various *clean* surfaces of different orifice materials started recently [44]. The ion movement is very sensitive to small changes of the potentials of the components of the ion optical system especially in the low energy range.

8.4 Neutral gas mass spectrometry

The first step in a mass spectrometric analysis of the neutral gas mixture in a plasma system is the identification of the components. Each substance is characterized by its molecular ion with mass number m/z and its fragmentation pattern, i.e., its mass spectrum with a series of fragment ions of different mass numbers m/z and different intensities [45]. The identification may be complicated in gas mixtures where superposition of peaks is often observed. As already mentioned a reduction of the decomposition of the molecule is possible by using a lower energy of the ionizing electrons, e.g., 20 eV. Also the determination of the ionization energy of the molecule or the appearance energy of the selected ion by changing the energy of the ionizing electron beam in the ion source is helpful. A list of ionization energies of atoms and molecules is given in Ref. [46].

The energy dispersive mass spectrometry of neutral gas mixtures includes an additional possibility for the identification of the parent molecule of a fragment ion [47]. The electron impact ionization is coupled with the formation of ions with kinetic excess energy, if the ionization occurs via a higher excited state [48]. The conservation of momentum leads to a distribution of the kinetic excess energies E_1 and E_2 between the fragments with the masses m_1 and m_2 according to

$$\frac{m_1}{m_2} = \frac{E_2}{E_1}. \tag{8.18}$$

Accordingly, the light fragment receives a higher fraction of kinetic energy and vice versa. An example is presented in Fig. 8.10 [49]. Energy distributions of the ions CH_3^+ (m/z 15) are given together with the distribution of the base peak $Si(CH_3)_3^+$ (m/z 73) of tetramethylsilane (TMS, $Si(CH_3)_4$) produced in the neutral gas flow from an rf discharge in an Ar/TMS mixture in the ion source of a plasma monitor. Without discharge the energy distribution of the m/z 15 ions shows a significant high energy tail [50]. These ions are small fragment ions of the large TMS molecule. With discharge the m/z 15 ions have the same distribution as the base peak m/z 73 of the TMS, at the same time the concentration of the large molecules (TMS) is strongly decreased by plasma chemical processes. The absence of kinetic excess energy indicates that the CH_3^+ ion is now either produced from a CH_3 radical or is a heavy fragment ion of a light molecule such as methane. The formation of CH_3^+ ions by dissociative ionization of other large molecules can be excluded.

The second step in the mass spectrometric neutral gas analysis is the determination of the relative concentrations of the various gaseous components. With known sensitivities this is a classical application of mass spectrometry. A calibration of the mass spectrometer with the gases under study is necessary if these sensitivities are not known. If the gas in question is commercially not available, the sensitivity relative to a reference gas S_x/S_{ref} can be calculated using the partial ionization cross sections $\sigma_{x\nu}$. The sensitivity is given for the corresponding electron impact energy by

$$S_x = \frac{i_{x\nu}^+}{n_x}. \tag{8.19}$$

The relative sensitivity is

$$\frac{S_x}{S_{ref}} = \frac{i_{x\nu}^+ n_{ref}}{i_{ref\mu}^+ n_x} = \frac{\sigma_{x\nu}}{\sigma_{ref\mu}}, \tag{8.20}$$

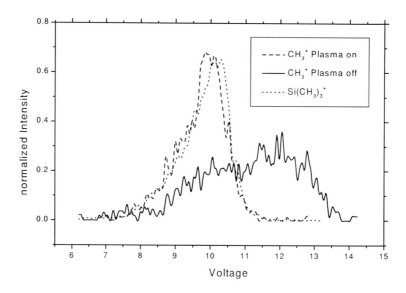

Figure 8.10: Energy distribution of $Si(CH_3)_3^+$ and CH_3^+ ions formed in the ion source of a plasma monitor in the neutral gas flux out of a sampling orifice in the wall of an *rf* planar reactor with an argon–tetramethylsilane mixture with and without plasma. The energy of 10 eV corresponds to the potential of the ion source in relation to the energy analyzer. The dotted curve from $Si(CH_3)_3^+$ serves as a reference for the energy resolution of the analyzer (after Ref. [49]).

where different ions are labeled by μ and ν. It is useful to choose the most intensive peak (base peak) in the spectrum of this substance for the determination of the sensitivity. If the partial cross section is not known it can be estimated. A determination of the total ionization cross section is possible by semi-empirical methods like the additivity rule [51]–[53]. The partial ionization cross section may then be deduced from the relative peak intensity in the mass spectrum using the decomposition pattern for an electron energy of 70 eV. For many substances these decomposition patterns are collected in numerous compilations of mass spectral data [45].

The plasma mass spectrometry beyond the detection of stable neutrals allows the measurement of free radicals in the gas flux from the plasma. Free radicals essentially determine the chemical reactions in reactive plasmas. Measurements of these species are possible by optical emission, absorption, and laser spectroscopy. Some of these methods are discussed in Ref. [54]. One mass spectrometric method for the measurement of radical densities is the threshold ionization mass spectrometry. Examples are the measurements of CH_3, CH_2 and CF_3, CF_2 radicals in CH_4 and CF_4 *rf* discharges, respectively, by Sugai et al. [55, 56], of CH_3 in a CH_4 ECR plasma by Pecher et al. [43], of CH_3 and CH_2, and SiH_3, SiH_2, SiH and Si radicals in the SiH_4-CH_4-H_2 *rf* discharge plasma by Kae-Nune et al. [39] (Fig. 8.11), and of O radicals in an O_2 helicon plasma by Granier et al. [57]. The condition for the application of this method is the difference in the appearance energy of the fragment ion and the ionization energy of the

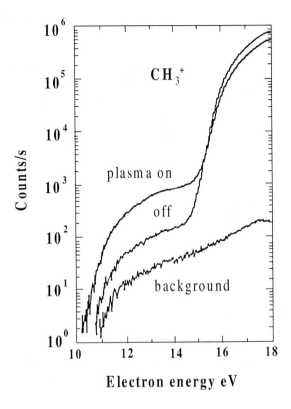

Figure 8.11: Threshold ionization detection of CH_3^+ under plasma-on and plasma-off conditions for a CH_4 discharge at 8 Pa, and 5 W *rf* power, after Ref. [39].

corresponding radical. For example, the appearance energy of CH_3^+ ions by electron impact of CH_4 is 14.3 eV, the ionization energy of CH_3 radical is 9.8 eV. The part of CH_3^+ ions that comes from the radicals can be separated from the fragment ions CH_3^+ of CH_4 molecules by mass spectrometric analysis of the neutral gas flow from the plasma reactor with discharge on and off as a function of the energy of the ionizing electrons (as shown in Fig. 8.11 [39]). The knowledge of the partial ionization cross sections of the stable molecule and of the radical for the formation of the CH_3^+ ion is necessary for a calculation of the concentration of the radicals in relation to the molecule concentration. For this method a small distance between the sampling orifice and the electron impact ion source is necessary to avoid losses of the radical flux by collisions with the residual gas in the vacuum chamber or with walls and electrodes. Error sources like different sticking coefficients on the orifice surface, pyrolysis on the hot filament in the ion source are discussed in [56].

Other methods for the measurement of free radicals by mass spectrometry are developed but they require other ionization techniques or additional installations. One example is the measurement using the paramagnetic properties of the radicals by modulation of the particle

Table 8.1: Radical products and molecules formed by a micro wave discharge in pure CH_4 and their relative intensities, after Ref. [59].

Type	Chemical species
C_nH_{2n-1}	$C_2H_3(100), C_3H_5(68), C_4H_7(36), C_5H_9(28)$ $C_6H_{11}(16),$ $C_7H_{13}(13), C_5H_{15}(12), C_9H_{17}(6), C_{10}H19(4), C_{11}H_{21}(3)$
C_nH_{2n+1}	$C_2H_5(48), C_3H_7(42), C_4H_9(32), C_5H_{11}(29), C_5H_{13}(20),$ $C_7H_{15}(12), C_8H_{17}(9), C_9H_{19}(6), C_{10}H_{21}(3), C_{11}H_{23}(2)$
Others	$C_4H_3(13), C_4H_5(12), C_5H_5(10), C_5H_7(8), C_6H_5(8),$ $C_6H_7(6), C_6H_9(6)$
Molecules	$C_2H_2(10), C_4H_2(11), C_4H_4(7), C_4H_6(8), C_5H_6(11),$ $C_6H_4(9), C_6H_6(12), C_6H_6(12), C_6H_8(5), C_6H_{10}(3)$

beam in an inhomogeneous time alternating magnetic field [58].

An interesting possibility for the measurement of radicals is introduced by Fujii and coworkers [15, 59, 60]. The chemical ionization is based on the attachment of Li^+ ions by the radicals R. The Li^+ ions are produced by heating a glass bead that contains lithium oxide in an alumosilicate matrix. The gas pressure in this ionization source is in the 10 Pa region. The product ions (RLi^+) are analyzed in a quadrupole mass spectrometer. This method is not restricted to radicals, but not applicable to non-polar molecules or molecules with low polarizability, because of their weak Li^+ affinities. A scheme of the experimental device and a summary of the products measured by this method at a micro wave discharge in pure CH_4 are given in figure 8.12 and table 8.1, respectively.

Examples for the application of field ionization with a low decomposition rate are the neutral gas mass spectrometry of *rf* plasmas in styrene, benzene and ethylene as well as in silane [61, 62].

8.5   Ion mass spectrometry

Ion mass spectrometry allows the determination of sorts, concentration and energy distribution of ions after they have crossed the plasma sheath. This knowledge can support the study of ion reactions, and serves to control and optimize technological plasma processes like thin film deposition and plasma etching. Another application is the glow discharge mass spectrometry as a method for analysis of solids [63].

Because the transmission of the mostly used quadrupole mass spectrometers is not independent of the mass number m/z of the ions a quantitative ion mass spectrometry demands a calibration of the transmission in dependence on the mass number. This is possible in princi-

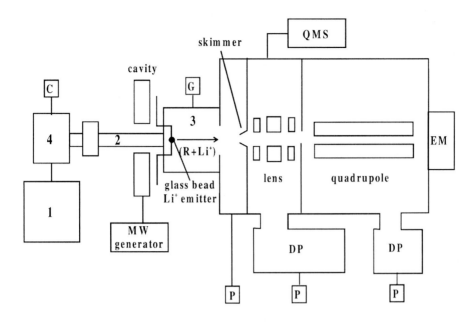

Figure 8.12: A quadrupole mass spectrometer with a Li$^+$ ion attachment reaction chamber (3). A stream of a reaction products from a quartz discharge flow tube (2) is directed into the reaction chamber, after Ref. [59].

ple by the application of a calibrated ion source with a known intensity distribution of a broad ion spectrum. In practice a calibration is performed with different neutral gases like rare gases with consideration of their gas pressure and ionization cross sections. Another possibility is the application of a substance with a sufficient number of fragment ions with known intensity distribution. It is recommended to use fragment ions without remarkable kinetic excess energy to avoid transmission losses. The organic silicon compound tetraethoxysilane Si(OCH$_2$CH$_3$)$_4$ was successfully applied for this purpose [49, 64].

A practicable calibration of the energy analyzer is obtained by evaporation of sodium water glass (Na$_2$O $\bullet$ nSiO$_2$) with a heated evaporator. The energy of the thermally emitted Na$^+$ ions can be controlled by changing the electrical potential of the emitter [49].

The ion flux at the wall or electrode of a plasma contains information on some important plasma properties: the ion population in the plasma, the position of ion formation, the ion loss rate at the wall or electrode, and the properties of the sheath in front of the sampling orifice.

If secondary ion reactions can be excluded for a diffusion determined plasma, the current density j_ν^+ is a function of the ion density n_ν^+ inside the plasma [65]:

$$j_\nu^+ = b_\nu E n_\nu^+ \tag{8.21}$$

$$\frac{n_\nu^+}{n_\mu^+} = \frac{j_\nu^+ b_\mu}{j_\mu^+ b_\nu}, \tag{8.22}$$

where b_ν indicates the ion mobility and E the electrical field strength.

In case that secondary reactions are included the ion concentration in the plasma can be calculated by balance equations. Then the plasma mass spectrometry gives information on the losses of the ions at the wall and electrodes.

The localization of ion formation inside the plasma reactor is possible by measuring the ion energy distribution. One of the first detailed measurements of ion energies at the cathode of a *dc* glow discharge was presented by Davis and Vanderslice [66]. The sampling orifice with a diameter of 0.1 mm was located in the center of the cathode. The experiments were carried out using a double focusing mass spectrometer with 90 degrees sector fields. The investigation of He, Ne, Ar, and H_2 discharges showed the influence of the resonant charge exchange collisions on the ion energy distribution. The intensity of the Ar^+ ions increased monotonously with decreasing energy. The authors therefore concluded that Ar^+ ions are created inside the cathode dark space by charge exchange collisions. The distribution of Ar^{++} ions with their smaller charge exchange collision cross section is characterized by a sharp peak at an energy corresponding to the cathode fall potential. Thus the origin of the Ar^{++} ions is the negative glow. An instructive example for the identification of the position of ion formation by measuring the energy distribution is given by Zeuner et al. [38]. In a capacitively coupled *rf* discharge they observed negative ions with maximum energies of a few hundreds eV at the grounded electrode. The comparison with the self-bias voltage of the driven electrode indicated that these ions are generated in the sheath in front of the driven electrode. They travel through the plasma bulk, the sheath in front of the grounded electrode and reach the sampling orifice.

The sheath voltage is another important parameter. This voltage equals in *dc* discharges the volt equivalent of the maximum energy of the ions. In *rf* discharges the situation is more complicated. In capacitively coupled *rf* discharges ion energy distributions with characteristic structures (Fig. 8.13) are observed. Many features are well understood, and their dependence on ion mass, sheath thickness, collisions inside the sheath, the amplitude and frequency of the *rf* voltage, the *dc* voltage across the sheath is thoroughly studied [67]–[71]. The formation of a saddled distribution with two single peaks right and left of the energy value corresponding to the *dc* sheath potential in a collision-free sheath was discussed in Ref. [67]. The relation between sheath properties and the energy difference ΔW between these peaks

$$\Delta W = \frac{8eU_{rf}}{3\omega d_s} \sqrt{\frac{2eU_{dc}}{m}}, \tag{8.23}$$

derived in Ref. [68], was used to calculate the sheath thickness d_s [41]. Collisions in the sheath lead to complex structures with additional peaks as a result of ion formation at different positions with respect to the wall [41, 72].

The results of the studies of the ion energy distribution have consequences concerning the determination of the ion fluxes to the wall or the electrodes. Correct measurements are only possible by a mass spectrometer with a constant collection efficiency over a broad energy range. Measurements with a plasma monitor with a built-in energy analyzer demand an integration of the ion flux over the whole energy distribution. Conclusions on the ion densities near the wall must consider the different velocities resulting from the specific energies of the various ion types.

In the following some examples of mass spectrometric investigations of ions in different

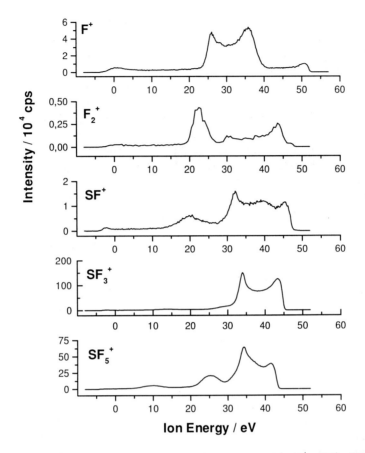

Figure 8.13: Kinetic energy distributions for the plasma ions F^+, F_2^+, SF^+, SF_3^+, and SF_5^+ sampled at the grounded electrode of an *rf* plasma in SF_6 at 13.3 Pa and an applied peak-to-peak *rf* voltage of 200 V, after Ref. [72].

plasmas are presented. Studies of the plasma of the positive column of a *dc* glow discharge show the specific ion formation processes at plasma conditions. Molecule ions of noble gases are formed in collisions of highly excited atoms [42]. Ion molecule reactions are responsible for proton transfer in the formation process of H_3^+ [3] or ArH^+ ions [42]. As shown in Fig. 8.14 the ion population in a silane *rf* discharge is strongly influenced by ion molecule reactions. The increasing contribution of SiH_3^+ ions with enhanced pressure is explained by the reaction [73]

$$SiH_2^+ + SiH_4 \rightarrow SiH_3^+ + SiH_3 . \tag{8.24}$$

SiH_2^+ is the dominant ion in electron impact ionization [74]. In the *rf* discharge $Si_2H_n^+$ ions are observed as products of ion molecule reactions. In the expanding $Ar–H_2–SiH_4$ plasma of

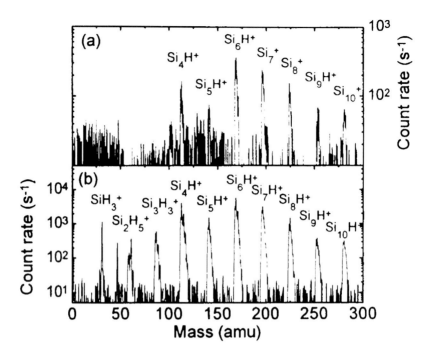

Figure 8.14: Positive ion mass spectra as obtained in the expanding argon-hydrogen-silane plasma for (a) 1 sccs and (b) 10 sccs hydrogen added to the arc. The most dominant ions are indicated, after Ref. [74].

a cascaded arc clusters up to $Si_{10}H_n^+$ were observed as a result of the chain reaction [75]

$$Si_nH_m^+ + SiH_4 \rightarrow Si_{n+1}H_p^+ + qH_2 \ . \tag{8.25}$$

These measurements were performed with a quadrupole mass spectrometer positioned in the reactor chamber with a pressure of 0.2 mbar in front of the plasma source (Fig. 8.15).

The electron impact ionization is expected to be the dominant ion formation process in non-thermal plasmas due to the high electron energy. A comparison between the observed ion distribution in the ion flux from a *rf* plasma in mixtures of Ar with vapors of organic silicon compounds with calculated ion formation rates is shown in Fig. 8.16 [48, 76]. The ion fluxes of the various ions are received by integration over the measured ion energy distributions. The formation rates $Z_{n\nu}$ of the fragment ions ν of the added organic silicon compound n are given by

$$Z_{n\nu} = k_{n\nu}n_e n_n \tag{8.26}$$

with n_e and n_n the electron and neutral gas density, respectively. The rate coefficients $k_{n\nu}$ are calculated with the known electron impact ionization cross sections [50, 64, 77], and a Maxwell distribution of the electrons with an estimated temperature $T_e = eU_e/k$. U_e, U_i,

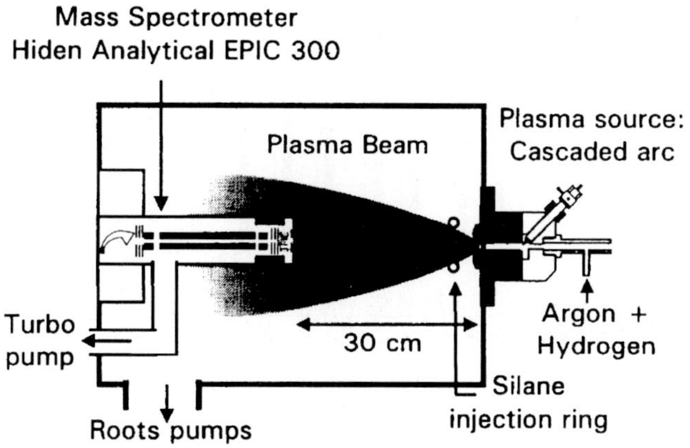

Figure 8.15: The expanding thermal plasma set-up used for fast deposition of hydrogenated amorphous silicon. The substrate holder is replaced by a mass spectrometer with similar geometry, after Ref. [74].

and U are the voltage equivalents of the electron temperature, the ionization energy and the electron energy, respectively:

$$k_{n\nu} = \sqrt{\frac{8e}{\pi m_e}} \int_{U_i}^{\infty} \sigma_{n\nu}(U) \frac{U}{U_e^{3/2}} \exp\left(-\frac{U}{U_e}\right) dU \quad \text{where} \quad eU = \frac{1}{2} m_e v^2 . \quad (8.27)$$

The observed distribution of fragment ions is different from the calculated distribution (Fig. 8.16). The existence of more low mass ions as expected from the calculations indicates the effect of secondary processes in the plasma like additional ionization of neutral products of the primary dissociative ionization, ion molecule reactions, and wall reactions.

The angular distribution of ions striking the surface is important for applications like plasma etching of structures for microelectronic circuits. Janes et al. [33] developed a device for the measurement of the ion flux through the sampling orifice for angles between ± 20 degrees. The quadrupole mass spectrometer coupled with an energy analyzer was rotatable with the fulcrum in the sampling orifice. The angular resolution was about 1 degree. This device was applied to an *rf* plasma in CF_4 for the measurements at the powered electrode [78]. The angular distributions of the different ions are not the same. CF_3^+ ions have an angular width between 3 degrees and 4 degrees in the whole energy range. The angular width of the CF_2^+ and CF^+ ions is near 5 degrees in the high energy range and near 15 degrees for low energies. These differing angular distributions explain observations at trench etching of silicon. The etch rate is maximal for the CF_3^+ ions. The CF_2^+ and CF^+ ions are responsible for the deposition of passivating layers on the trench walls.

The results of the plasma mass spectrometry are often used as basis for discussions of the mechanism of plasma chemical reactions. Helpful for this task is the application of an

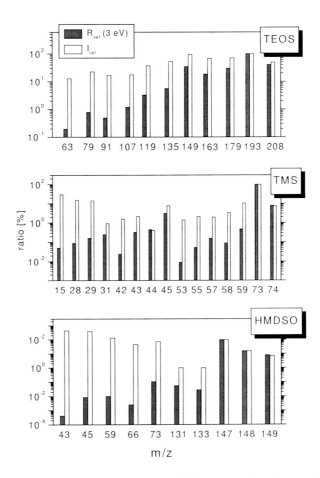

Figure 8.16: Calculated rate coefficients $R_{rel} = k_i/k_{base\ peak}$ for electron impact ionization compared with measured relative ion intensities I_{rel}, for an Ar rf discharge (5 Pa, 45 W) with admixtures of 2% TEOS, TMS, and HMDSO, after Ref. [49].

isotopic labeling technique. The reactions in SiH_4 plasmas were studied for example by the investigation of SiH_4-SiD_4-He and SiH_4-H_2-D_2 plasmas [79].

8.6 Mass spectrometry for the determination of elementary data for plasma physics

The understanding, simulation and theoretical treatment of plasma physical systems demand the knowledge of a lot of elementary data for the calculation of the transport and balance coefficients. Examples are presented for the application of mass spectrometry for the measurement of such data.

The fundamental process of ionization by electron impact is characterized by the ionization cross section. The total cross section describes the total ion production. The partial cross section gives the probability for the formation of a specific kind of ions. Partial cross sections in principle are obtained with mass spectrometers by measuring the ion currents produced in the ion source. The gas density, the current of the electron beam, and the length of the ionizing region must be known [80]. A high resolution power is necessary for the investigation of larger molecules with their variety of fragment ions, complicated by the existence of isotopes. Special efforts are necessary to extract and collect all ions from the ion source, if some of these ions have remarkable kinetic excess energy. The cross section has to be measured as a function of the electron energy. Also electron impact induced dissociation is studied by mass spectrometry. For this an ionization of the neutral radicals is necessary to detect the reaction products. An interesting way to overcome the difficulty of additional dissociative ionization is the mass spectrometric detection of the radicals after they have been transferred into organo-tellurides by chemical surface reactions on a tellurium mirror [81].

The study of ion-molecule reactions demands mass spectrometry for the identification of the ions. One possibility is the observation of the decay of the concentration of the reactant ion and the increase of the product ions in the ion trap of the Fourier transform mass spectrometer [82]. Ion-molecule reactions are studied also in drift tube experiments. The ion movement is measured in a drift tube in dependence on the electrical field and the gas pressure. Mass selected reactant ions are injected into a drift tube. Downstream at the end of the tube the reactant and product ions are measured by mass spectrometry. Such drift tubes are originally developed for the measurement of ion mobilities and diffusion coefficients [83].

8.7 Conclusions

Plasma mass spectrometry is an efficient tool for the investigation of ion population and neutral gas including radical composition in a plasma. The kinetic energy of the ions can be determined. Properties of the sheath in front of the electrodes and walls can be studied. Concerning the investigation of heavy particles plasma mass spectrometry is in competition with optical spectroscopy [84]. Mass spectrometry collects data immediately at the wall, has a high dynamic range, the interpretation of the spectra is mostly easy, and it gives information on the positive and negative ions in the plasma. Data on the neutral components in a plasma reactor are available before the plasma and the process is ignited. Disadvantages are the complicated mechanical and electrical coupling with the plasma reactor, the restricted application concerning the pressure, and the difficulties arising from deposition of disturbing layers in chemical active plasmas. Like all diagnostic methods mass spectrometry gives only limited information, and it is always recommendable to choose additional diagnostic methods to support and complete the data of the investigated plasma. Steady developments in instrumentation concerning performance and cost coupled with the continuous need of accurate data of plasma properties promotes the application of mass spectrometry for plasma diagnostics and process control.

References

[1] J.J. Thomson, Rays of positive electricity and their application to chemical analyses, Longmans, Green and Co. Ltd.: London (1913)

[2] O. Tüxen, Z. Phys. **103**, 463 (1936)

[3] H.D. Beckey, H. Dreeskamp, Z. Naturforschg. **9a**, 735 (1954)

[4] M. Pahl, Z. Naturforschg. **18a**, 1276 (1963)

[5] P.F. Knewstubb, A.W. Tickner, J. Chem. Phys. **38**, 1031 (1963)

[6] W. Paul, H. Steinwedel, Z. Naturforschg. **8a**, 448 (1953)

[7] H.W. Drawin, Mass Spectrometry of Plasmas, in Plasma Diagnostics, W. Lochte-Holtgreven, ed., North-Holland: Amsterdam (1968), p. 777

[8] M. Schmidt, G. Hinzpeter, Beitr. Plasmaphys. **10**, 183 (1970)

[9] T.D. Märk, H. Helm, Acta Phys. Austr. **40**, 158 (1974)

[10] M.J. Vasile, H.F. Dylla, Mass Spectrometry in Plasmas, in Plasma Diagnostics, O. Auciello, D.L. Flamm, eds., Academic Press: Boston, Vol. **1**, 185 (1989)

[11] H. Kienitz (ed.), Massenspektrometrie, Verlag Chemie: Weinheim (1968)

[12] F.A. White, G.M. Wood, Mass Spectrometry, Applications in Science and Engineering, Wiley: New York (1986)

[13] P.H. Dawson (ed.), Quadrupole Mass Spectrometry and Its Applications, Elsevier: New York (1976)

[14] J.B. Hasted, Physics of Atomic Collisions, Butterworths: London (1964)

[15] T. Fujii, M. Ogura, H. Jimba, Anal. Chem. **61**, 1026 (1989)

[16] A.G. Harrison, Chemical Ionization Mass Spectrometry, 2nd edition, CRC: Boca Raton (1992)

[17] H.-D. Beckey, Field Ionization Mass Spectrometry, Akademie-Verlag: Berlin (1971)

[18] J. Berkowitz, Photoabsorption, Photoionization, and Photoelectron Spectroscopy, Academic Press: New York (1979)

[19] E.W. Blauth, Dynamische Massenspektrometer, Vieweg: Braunschweig (1965)

[20] R.J. Seeböck, W.E. Köhler, M. Römheld, Contr. Plasma Phys. **32**, 613 (1992)

[21] M. Zeuner, J. Meichsner, H. Neumann, F. Scholze, F. Bigl, J. Appl. Phys. **80**, 611 (1996)

[22] L. Shi, H.J. Frankena, H. Mulder, Rev. Sci. Instrum. **60**, 332 (1989)

[23] J. Laimer, O. Schnabl, C.G. Schwärzler, H. Störi, J. Vac. Sci. Technol. **A 14**, 2315 (1996)

[24] Hakone VI. International Symposium on High Pressure, Low Temperature Plasma Chemistry, Contributed Papers: Cork (1998)

[25] A. Grill, Cold Plasma in Materials Fabrication From Fundamentals to Applications, IEEE Press: New York (1994)

[26] W. Walcher, Z. Phys. **122**, 62 (1944)

[27] M. Schmidt, Beitr. Plasmaphys. **6**, 147 (1966)

[28] M. Schmidt, R. Foest, R. Basner, J. Physique IV **8**, Pr 7-231 (1998)

[29] T.D. Märk, H. Helm, Acta Phys. Austr. **40**, 158 (1974)

[30] G. Franck, H. Lüdemann, Z. Naturforschg. **27a**, 1278 (1972)

[31] D.A. Dahl, J.E. Delmore, A.D. Appelhans, Rev. Sci. Instrum. **61**, 607 (1990)

[32] E. Hamers, W. van Sark, J. Bezemer, W. Goedheer, W. van der Weg, Int. J. Mass Spectrom. Ion Proc. **173**, 91 (1998)

[33] J. Janes, U. Bänzinger, C. Huth, P. Hoffmann, G. Neumann, H.-C. Scheer, U. Köhler, Rev. Sci. Instrum. **63**, 48 (1992)

[34] J. Janes, C. Huth, J. Vac. Sci. Technol. **A 10**, 3522 (1992)

[35] J.K. Olthoff, R.J. Van Brunt, S.B. Radovanov, J. Appl. Phys. **72**, 4566 (1992)

[36] R. Foest, M. Schmidt, M. Hannemann, R. Basner, Proc. Gaseous Dielectrics VII, Knoxville, L.G. Christophorou, D.R. James, eds., Plenum Press: New York (1994)

[37] R. Foest, J.K. Olthoff, R.J. Van Brunt, E.C. Benck, J.R. Roberts, Phys. Rev. E **54**, 1876 (1996)

[38] M. Zeuner, J. Meichsner, J. A. Rees, J. Appl. Phys. **79**, 9379 (1996)

[39] P. Kae-Nune, Perrin, J. Guillon, J. Jolly, Plasma Sources Sci. Technol. **4**, 250 (1995)

[40] S. Peter, R. Pintaske, G. Hecht, F. Richter, Surface and Coatings Technol. **59**, 97 (1993)

[41] F. Becker, I.W. Rangelow, R. Kassing, J. Appl. Phys. **80**, 56 (1996)

[42] M. Pahl, U. Weimer, Z. Naturforschg. **13a**, 745 (1958)

[43] P. Pecher, W. Jacob, Appl. Phys. Lett. **73**, 31 (1998)

[44] J.K. Olthoff, R.J. Van Brunt, S. B. Radovanov, Appl. Phys. Lett. **67**, 473 (1995)

[45] Eight Peak Index of Mass Spectra, 2nd edition, Mass Spectrometry Centre: Aldermaston (1974); NIST Chemistry WebBook (http://webbook.nist.gov/chemistry)

[46] D.R. Lid (ed.), CRC Handbook of Chemistry and Physics, CRC Press: Boca Raton (1994)

[47] R. Foest, R. Basner, M. Schmidt, Plasma and Polymers **4**, 259 (1999)

[48] L.G. Christophorou (ed.), Electron-Molecule Interactions and Their Applications, Academic Press: Orlando (1984)

[49] R. Foest, Ph.D. Thesis, Universität Greifswald (1998)

[50] R. Basner, R. Foest, M. Schmidt, F. Sigeneger, P. Kurunczi, K. Becker, H. Deutsch, Int. J. Mass Spectrom. Ion Proc. **153**, 65 (1996)

[51] W.L. Fitch, A.D. Sauter, Anal. Chem. **55**, 832 (1983)

[52] H. Deutsch, M. Schmidt, Contrib. Plasma Phys. **25**, 475 (1985)

[53] H. Deutsch, T.D. Märk, V. Tarnovsky, K. Becker, C. Cornelissen, L. Cespiva, V. Bonacic-Koutecky, Int. J. Mass Spectrom. Ion Proc. **137**, 77 (1994)

[54] J. Röpcke, M. Käning, B.P. Lavrov, J. Physique IV **8**, Pr 7-207 (1998)

[55] H. Toyoda, H. Kojima, H. Sugai, Appl. Phys. Lett. **54**, 1507 (1989)

[56] H. Sugai, H. Toyoda, J. Vac. Sci. Technol. **A 10**, 1193 (1992)

[57] A. Granier, F. Nicolazo, C. Vallée, A. Goullet, G. Turban, B. Grolleau, Plasma Sources Sci. Technol. **6**, 147 (1997)

[58] N.I. Butkovskaya, E.S. Vasiliev, I.I. Morozov, V.L. Talrose, Proc. XVth Intern. Conf. on Phenomena in Ionized Gases, Minsk, **1**, 299 (1981)

[59] T. Fujii, Ken-ichi Syouji, J. Appl. Phys. **74**, 3009 (1993)

[60] T. Fujii, H.S. Kim, Chem. Phys. Lett. **268**, 229 (1997)

[61] H. Hayashi, H. Iwai, F. Okuyama, J. Phys. Colloq. **48**, C6-247 (1987)

[62] H. Hayashi, H. Iwai, F. Okuyama, Surf. Sci. **246**, 408 (1991)

[63] G. Holland, A.N. Eaton (eds.), Applications of Plasma Source Mass Spectrometry, Royal Soc. Chem.: Cambridge (1991)

[64] R. Basner, R. Foest, M. Schmidt, K. Becker, Adv. Mass Spectrom. **14** (1998); R. Basner, M. Schmidt, K. Becker, H. Deutsch, Adv. At. Molec. Opt. Phys., M. Inokuti (ed.), Academic Press: San Diego (2000), p. 147

[65] M. Pahl, Z. Naturforschg. **12a**, 632 (1957)

[66] W.D. Davis, T.A. Vanderslice, Phys. Rev. **131**, 219 (1963)

[67] R.T.C. Tsui, Phys. Rev. **168**, 107 (1968)

[68] P. Benoit-Cattin, L.-C. Bernard, J. Appl. Phys. **39**, 5723 (1968)

[69] S. Biehler, Proc. 10th Intern. Symp. Plasma Chem. Bochum **2**, 2.1-52 (1991)

[70] K.-U. Riemann, U. Ehlemann, K. Wiesemann, J. Phys. D: Appl. Phys. **25**, 620 (1992)

[71] H.H. Hwang, M.J. Kushner, Plasma Sources Sci. Technol. **3**, 190 (1994)

[72] R. Foest, J.K. Olthoff, R.J. Van Brunt, E.C. Benck, J.R. Roberts, Phys. Rev. E **54**, 1876 (1996)

[73] G. Turban, Y. Catherine, B. Grolleau, Thin Solid Films **67**, 309 (1980)

[74] R. Basner, M. Schmidt, V. Tarnovsky, K. Becker, H. Deutsch, Intern. J. Mass Spectrom. Ion Proc. **171**, 83 (1997)

[75] W.M.M. Kessels, M.C.M. Van de Sanden, D.C. Schram, Appl. Phys. Lett. **72**, 2397 (1998)

[76] R. Foest, R. Basner, M. Schmidt, P. Kurunczi, K. Becker, Proc. 12th Int. Conf. Gas Discharges & their Applications, Greifswald, **II**, 547 (1997)

[77] R. Basner, R. Foest, M. Schmidt, K. Becker, H. Deutsch, Intern. J. Mass Spectrom. Ion Proc. **176**, 245 (1998)

[78] J. Janes, J. Appl. Phys. **74**, 659 (1993)

[79] G. Turban, Y. Catherine, B. Grolleau, Thin Solid Films **77**, 287 (1981)

[80] T.D. Märk, G. H. Dunn (ed.), Electron Impact Ionization, Springer Verlag: Wien (1985)

[81] S. Motlagh, J.H. Moore, J. Chem. Phys. **109**, 432 (1998)

[82] J. Holtgrave, K. Riehl, D. Abner, P.D. Haaland, Chem. Phys. Lett. **215**, 548 (1993)

[83] E.A. Mason, E.W. McDaniel, Transport Properties of Ions in Gases, John Wiley: New York (1988)

[84] Ch. Hollenstein, Course on Low Temperature Plasma Physics, TU Eindhoven (1998), unpublished

9 Ellipsometric analysis of plasma-treated surfaces

Wolfgang Fukarek

Institute of Ion Beam Physics and Materials Research, Forschungszentrum Rossendorf, P.O. Box 510119, D–01314 Dresden, Germany

9.1 Introduction

Ellipsometry is a more than 100 years old optical technique for determining film thicknesses, dielectric functions, and structural properties of thin films, surfaces, and interfaces. It measures the change in the polarisation state of a light beam after it has interacted with a medium. The polarisation can be modified by reflection, transmission (and refraction), and scattering. Here we only consider reflection ellipsometry, which simply will be called ellipsometry. The name comes from the state of polarisation after interaction, which is generally elliptical.

Two parameters, related to the change in the state of polarisation $\underline{\chi}$, are measured at a given wavelength λ and the angle of incidence Φ. They are expressed as a complex number, $\underline{\rho} = \underline{\chi}_{in}/\underline{\chi}_{out}$, or as the two (real) ellipsometric angles Ψ and Δ:

$$\underline{\rho}(\lambda, \Phi) = \tan \Psi \exp(i\Delta) = \underline{\chi}_{in}/\underline{\chi}_{out} = \underline{R}_p/\underline{R}_s . \tag{9.1}$$

$\underline{R}_p$ and $\underline{R}_s$ are the complex amplitude reflection coefficients parallel (p) and perpendicular (s) to the plane of incidence. This description is not sufficient to characterise complex anisotropic (biaxial or uniaxial with off-normal optical axis), nematic or chiral materials. The analysis of ellipsometric data requires optical multilayer model calculations in combination with regression procedures to obtain a close fit of calculated to measured data.

Ellipsometry can be employed *in real time* (dynamically) and, like all optical techniques, it is compatible with (almost) any environment. Particularly, it can be used for *in situ* measurements under vacuum conditions, in gases at any pressure, and under plasma conditions. Ellipsometry is largely non-destructive, except for photo-chemical reactions that might be induced by high-intensity UV-light beams. Requirements for the applicability of *in situ* ellipsometry are: (i) the sample must be planar and of optical quality (e.g., not a textile), and (ii) optical ports are required in the chamber to provide optical access to the sample. The windows should be strain-free (largely free of birefringence) and protected from becoming either coated with absorbing films, or being modified by etching due to processes occurring in the chamber.

As only a very short introduction can be given here, we refer the interested reader to the excellent books on ellipsometry [1]–[5]. Much more detailed information on the technique and the data analysis, and good examples of applications can be found there.

9.2   Comparison with other techniques

Ellipsometry has a number of advantages over the other *in situ* techniques. In reflectometry the reflected intensity is measured, but in order to extract information about the film significant optical interference or absorption must occur. This precludes its use when very thin layers are to be analysed. The phase parameter Δ in ellipsometry has no counterpart in reflectivity and transmission measurements and is largely responsible for the extraordinary monolayer sensitivity of this technique.

By reflectometry measured data ought to be normalised to source intensity and detector sensitivity; in addition, intensity losses within the optical system, like misalignment or absorption by coated windows, need to be corrected. These sources of measurement error do not influence ellipsometric data as the ellipsometric angles are calculated from intensity ratios. Reflectometry data require model calculations, like in ellipsometry, to extract information on the sample structure.

Reflectance anisotropy (RA) measures surface optical anisotropy and provides information similar to that gained by reflection high-energy electron diffraction (RHEED). It can identify surface chemical bonds and detect layer-by-layer epitaxial growth, without the need for ultra-high vacuum (UHV). However, as a monitoring tool, it is insensitive to the overall layer thickness and dielectric functions, making it complementary to, rather than competitive with ellipsometry.

The widely used crystal film thickness monitor (CM) measures the change in resonance frequency of a crystal oscillator, when a film (mass) is deposited on it. It is a simple and inexpensive method for determining the approximate layer thickness, if the density of the film is known. Due to the remote position of the CM sensor a calibration is required. CMs are very difficult to use under ion bombardment, direct current (DC) or radiofrequency (RF) bias, at elevated temperature, or to monitor plasma etching, because a film, identical to the material to be etched, has to be deposited on the CM sensor before.

Other *in situ* techniques, like electron spectroscopy or scanning probe microscopy can be employed without braking the vacuum but not in a dynamic mode during plasma processing. When particle and energy fluxes towards the sample surface are terminated and temperature is changed, the dynamic equilibrium at the surface under plasma conditions may undergo dramatic changes and end up in a considerably different state. Processes like desorption, relaxation of excited states, defect annihilation, recombination or motion of charged particles, reconfiguration of bonds (aging), etc. may be involved. The above mentioned *in situ* techniques only provide information on the final state, or on processes that take place after the sample has been transferred to the analysis chamber, but no information on the dynamics of plasma processing can be gained.

Besides optical diagnostics, there are only a few other techniques known that can be employed *dynamically in situ* during processing, e.g., ion beam analysis (IBA) [6], or x-ray based techniques (see chapter 10), but here only ellipsometry is addressed to.

9.3 Experimental technique

9.3.1 Instrumentation

The ellipsometric data can be acquired using several different optical configurations.

The scheme of an ellipsometer set-up is shown in Fig. 9.1. It consists of a light source with a polarisation state generator (PSG) to form a light beam of precisely defined state of polarisation. This unit consists at least of a polariser. Single wavelength ellipsometers usually employ a laser, while for spectroscopic ellipsometers Xe short arc lamps in combination with a monochromator are in common use. The detection unit consists of a polarisation state analyser (PSA) and a photodetector. The most common are ellipsometers employing mechanically rotating optical elements, such as rotating analyser, rotating polariser, and rotating compensator (retarder) ellipsometers (RAE, RPE, and RCE, respectively). The RAE is a simple and relatively inexpensive configuration. When configured with a multiple-detector array and spectrograph, the minimum acquisition time for a spectrum is about 20 ms, but typical acquisition times may be much longer, depending on the amount of averaging needed for an acceptable signal-to-noise ratio. The disadvantage of the RAE is that measurements on low-absorbing substrates ($\Delta \approx 0°$) are difficult. This can be improved by adding a compensator, as indicated in Fig. 9.1.

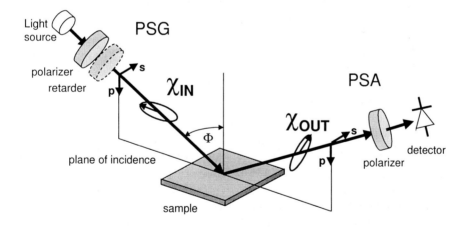

Figure 9.1: Scheme of an ellipsometer set-up consisting of source, polarisation state generator (PSG), sample, and polarisation state analyser (PSA). The change in the state of polarisation due to reflection at the sample is measured.

The spectral range of spectroscopic ellipsometers may cover the near-infrared to the near-ultraviolet (0.7 to 6 eV). Recently, Fourier-transform-infrared-spectrometer- (FTIR) based infrared ellipsometers became available. Powerful computers are necessary to run such systems. The highest resolution and widest spectral ranges can be achieved by means of single detector/scanning monochromator systems. These, however, are relatively slow, and therefore more suited for *ex situ* applications. Better for *real-time* applications is an optical multi-channel analyser- (OMA) based system, which can acquire data at many wavelengths simultaneously. Cost-effective, single-wavelength RAEs are usually sufficient for *real-time* control applications of well-defined processes at moderate rates up to some nm/s, where only thickness and refractive index need to be controlled. With laser-based, single wavelength ellipsometers employing sophisticated signal processing, very low statistical error levels with a standard deviation of $\sigma \Psi \approx 0.001°$ can be obtained [7]. This is about one order of magnitude better than the accuracy of multi-channel spectroscopic ellipsometers.

High-rate plasma processes like etching or cathodic arc deposition that run at rates of 10 to 100 nm/s or higher, require the application of a polarisation modulation ellipsometer (PME) [8]. A PME has no moving parts but uses a photo-elastic element to modulate the state of polarisation at about 50 kHz. Single-wavelength data can be acquired at a two or three orders of magnitude higher rate with a PME than with a RAE, at the price of a more complex and expensive instrument.

A plasma chamber should meet a few requirements to allow *in situ* ellipsometry to be applied. Two strain-free optical windows, as mentioned above, should give access to the sample at an angle of incidence Φ in the range of 65° to 75°, with angles closer to 75° being preferable for semiconductors. Mechanical vibrations from pumps like Roots and cryo pumps and other machinery can be a problem as they may cause changes in the position of the reflected beam at the detector, and hence additional noise in the data. A sample alignment stage is useful, as the angle of incidence is at times difficult to calculate, depending on the properties of the sample. If the sample is being rotated for uniform growth, the rotation should be wobble-free. The size of the chamber is not a critical issue, if a laser is utilised but becomes a critical parameter, if light from a lamp is employed.

Many commercial ellipsometers are available and most of these instruments utilise a microcomputer. Personal computers (PC) are powerful enough for *real-time* spectroscopic data acquisition and analysis. Sophisticated software is available; for certain tasks, the system can be used in a turnkey manner without knowing very much about how the ellipsometer works. However, the tasks which can be done in a turnkey manner represent only a small subset of the possible applications of ellipsometry.

9.3.2   Data analysis

Ellipsometric data, recorded spectroscopically at a number of wavelengths and different angles of incidence, or dynamically at a number of time steps, must be modeled to extract information about the sample. The skilled ellipsometer user is usually able to *read* Ψ, Δ–plots, but this is sometimes ambiguous and lacks of quantification of the results. Nevertheless, the ability to read dynamically recorded Ψ, Δ–plots can yield important information on the process, especially in cases where on-line model calculations are too difficult or time consuming. An example is given in section 9.4.2.3

Measured ellipsometric data are sometimes presented as (complex) pseudo-dielectric function data $\langle \underline{\varepsilon} \rangle = \langle \varepsilon_1 \rangle + i \langle \varepsilon_2 \rangle$, or pseudo-refractive index data $\langle \underline{n} \rangle = \langle n \rangle + i \langle k \rangle$ with $\langle \underline{\varepsilon} \rangle = \langle \underline{n} \rangle^2$. To avoid confusion, a short comment on this kind of data presentation is added here. The calculation of $\langle \underline{\varepsilon} \rangle$ from Ψ, Δ (or ρ) is based on applying a most simple optical model which consists of a flat semi-infinite, homogeneous and isotropic substrate without any surface features like a native oxide or roughness. This simple model has only two unknown parameters (the complex dielectric function $\underline{\varepsilon}$ of the material) and allows inversion of Eq. 9.1:

$$\underline{\rho} = \tan \Psi \exp(i\Delta) = \frac{\underline{R}_p}{\underline{R}_s}$$

$$= \frac{\underline{n}^2 \cos \Phi - \sqrt{\underline{n}^2 - \sin^2 \Phi}}{\underline{n}^2 \cos \Phi + \sqrt{\underline{n}^2 - \sin^2 \Phi}} \times \frac{\cos \Phi + \sqrt{\underline{n}^2 - \sin^2 \Phi}}{\cos \Phi - \sqrt{\underline{n}^2 - \sin^2 \Phi}} \quad (9.2)$$

$$\Rightarrow \underline{\varepsilon} = \underline{n}^2 = \tan^2 \Phi \times \left\{ 1 - \frac{4\underline{\rho}}{(1 + \underline{\rho})^2} \sin^2 \Phi \right\}. \quad (9.3)$$

If the substrate is perfect and without overlayers, as mentioned above, $\underline{R}_p$ and $\underline{R}_s$ simply are the Fresnel reflection coefficients for light polarised parallel (p) or perpendicular (s) to the plane of incidence. Only in this case, the $\langle \underline{\varepsilon} \rangle$-data represent the dielectric properties ($\underline{\varepsilon}$) of the solid. If the sample is only assumed to be a bare substrate, and Eq. 9.3 is used to transform Ψ, Δ–data, then $\langle \underline{\varepsilon} \rangle$ is simply a mathematical transformation of ρ and does not necessarily have direct physical significance, although the data might have some similarity to dielectric function data. Inferring dielectric properties from (raw) pseudo-dielectric data without modeling is ambiguous.

Like all optical techniques for thin film analysis, the evaluation of ellipsometric data generally requires optical multilayer model calculations based on the application of Fresnel's equations for reflection and transmission including coherent multiple reflections (interference). The unknown parameters of the model, which have been selected from the best knowledge of the sample structure or just by experience, are determined by the use of regression algorithms that provide the best fit of calculated with measured data. If the fit is satisfactory, then the best-fit model parameter values are taken as representative of the actual sample. Confidence limits for each parameter can be calculated, as well as the correlation between the parameters.

Analyses of *ex situ* and *in situ real-time* data differ markedly. In an analysis of *ex situ* data for a multilayer sample, not all of the parameters in the model may be fitted, since the variables must be overdetermined (the number of data points must be greater than or equal to the number of model variables). If all dielectric functions are known and only layer thicknesses are to be derived, these may easily be overdetermined by performing spectroscopic measurements. If none of the materials in the multilayer have known dielectric functions, then a series of samples must be analysed, starting with the bare substrate, and adding layer by layer. Even with only one unknown dielectric function in a given structure, there may be a problem. If the film is absorbing, but semitransparent, it has two unknowns at each wavelength ($\varepsilon_1 + i\varepsilon_2$), plus the unknown thickness. This is one more unknown than the number of data points (Ψ and Δ at each wavelength), so the system is underdetermined. Simply taking data at more than one angle of incidence, thus providing more data than unknowns exist, seems to solve the problem. However, the information content of data at different angles is highly correlated, so that little independent information is added. Often the problem can be addressed by using a model for the dielectric function with a small number of wavelength-independent parameters that can be fitted. This technique is known as parameterisation of the dielectric function. Another technique that works well is to fit simultaneously data from two or more samples with different film thicknesses, called multi-sample analysis. The films must have a common dielectric function, which is fitted together with the individual layer thickness. An example is shown in section 9.4.3.2. The situation becomes even more complicated, if graded layers are involved. Gradual variations in composition, structure, density, etc. can result from changes in plasma process parameters during processing, structural evolution during film growth, or ion implantation. Under such circumstances parameterisation of the depth profile and the dielectric function are necessary to reduce the number of unknown parameters.

In situ real-time measurements during deposition or etching are inherently overdetermined, because *snapshots* of the sample are taken as a layer is built up, removed, or modified. For example, ten spectra might be measured during the growth of a single film. This is the same as having data from ten samples, each with an independent thickness, but the same film dielectric function (assuming uniform film composition throughout the growth). As in the case of *ex situ* multi-sample analysis, then the system is over-determined, and both, the dielectric function and the thicknesses, can be extracted. This technique also allows anisotropic films to be analysed, as will be shown in section 9.4.2.2. In all cases, spectroscopic or multiple wavelength measurements are at least preferable and often necessary, since they provide much more independent information than single wavelength data.

From the above discussion it should be clear that ellipsometry is an indirect technique whose results depend on the choice of an appropriate model. Needless to say, inappropriate models are easily constructed and dutifully produce numerical results (which perhaps fit the experimental data well). *In situ real-time* spectroscopic measurements place the highest constraints on the model by producing the greatest amount of independent data to be fitted.

We restrict ourselves to this very short and general discussion of the basics of modeling ellipsometric data. The very important subjects of parameterisation of dielectric functions and depth profiles, effective medium theories, as well as the treatment of complex anisotropic materials by generalised ellipsometry (GE) cannot be addressed here in detail due to the limited space. We refer the reader to Refs. [1]–[5] and references therein.

9.4   Examples

A number of very good collections of examples for the application of ellipsometry has been already published (see Refs. [1]–[5]). In the following we report on applications that cannot be found in other textbooks on ellipsometry. The examples have been selected to demonstrate the unique capabilities of *dynamic in situ* ellipsometry in the analysis of parameter dependencies of plasma processes which are unstable, hardly reproducible, or require a *burn-in* period. It will be shown that ellipsometry can give an insight into details of processes and optical properties that are hidden to the view of other techniques.

9.4.1 *In situ* single wavelength ellipsometry examples

There are at least three good reasons to employ cost-effective, single wavelength *in situ* ellipsometers instead of spectroscopic systems: (i) large chambers, resulting in large distances between source and detector, (ii) the necessity of collecting high precision ellipsometric data to obtain the information of interest, and (iii) single wavelength data are sufficient to gain the information of interest on the sample.

9.4.1.1 Direct current (DC) magnetron sputter deposition of indium-tin-oxide (ITO) films

In reactive magnetron sputtering there occurs the well-known problem of target poisoning and related hysteresis effect. The mechanisms involved are quite well understood. The practical problems related to the hysteresis effect can be solved by the application of optical emission spectroscopy using emission lines from sputtered target material and reactive gas ions in combination with a feedback control of the magnetron power supply [9]. As the preparation of a well-defined target state is a critical issue, the reproducibility of film properties at the same macroscopic deposition parameters turns out to be low. Therefore, information about the dependence of film properties on deposition parameters can hardly be obtained from series of single deposition experiments. The method of performing a parameter scan in one and the same experimental run (to overcome the run-to-run reproducibility problem), and employing simultaneously dynamic *in situ* ellipsometry, perfectly meets the needs. ITO films are conductive and transparent ($k < 0.001$ at $\lambda = 632.8$ nm). This reduces the number of unknown parameters in the optical model. In the case of significant deviations from stoichiometry, the films become strongly absorbing and show metallic properties.

The aim of the *in situ* ellipsometric investigation on ITO deposition was to gain information (i) on the effect of target state (oxidised or metallic) on sputter rate in dependence on discharge power, (ii) on the dependence of the optical properties of ITO films on growth rate and changes in the film forming particle fluxes, and (iii) on the relation between optical and electrical properties.

Fig. 9.2 (a) shows Ψ, Δ–data recorded dynamically during the deposition of an ITO film as the discharge power is increased from 15 to 60 W. The indium/tin target was sputter-cleaned prior to the addition of the reactive gas flow using a pure Ar discharge. After O_2 addition the shutter in front of the substrate was kept closed until optical emission lines and electrical discharge parameters were stabilised. More experimental details are given, e.g., in Ref. [7]. The results of the analysis of the Ψ, Δ–plot are shown in Fig. 9.2 (b). Two distinct regions of the discharge power are identified. In the low-power region (15–30 W) the deposition rate increases almost linearly with discharge power at 0.25 nm/min/W. At high powers (50–60 W) the specific rate is 1 nm/min/W. The difference in the sputter rate is characteristic of the transition from oxidised (poisoned) to metallic target. Interestingly, the refractive index is independent of the discharge power and growth rate in both regions. This means that two different modifications of ITO with distinct optical properties are formed, and their properties do not depend sensitively on changes in the flux of sputtered material and reactive gas. Conductivity measurements showed, that ITO films with a refractive index $n > 2.1$ were conductive without additional thermal treatment. For more details see Ref. [7].

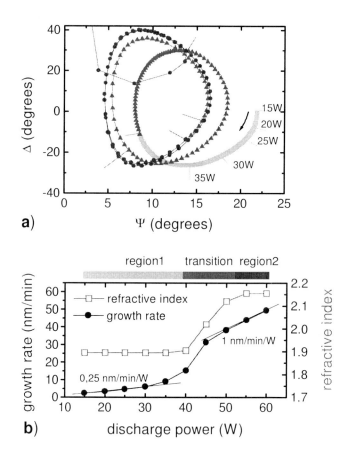

Figure 9.2: (a) Ψ, Δ-plot of a discharge power scan (15 to 60 W) recorded during deposition of ITO, and (b) results of model calculations. Two regions of discharge power are observed where the refractive indices are independent of growth rate and discharge power.

9.4.1.2 Temperature-dependence of a-C:H film growth

The dependence of the growth rate of amorphous hydrogenated carbon (a-C:H) films, deposited from a methane electron cyclotron resonance (ECR) plasma, has been analysed employing the technique of dynamical parameter scanning in combination with *in situ* ellipsometry. Amorphous C:H films are isotropic and non-absorbing ($k < 0.02$), if grown at low ion energies ($E_I < 100$ eV). Therefore, the information from single wavelength ellipsometry data is sufficient. In contrast to the previous example (section 9.4.1.1), in this case the scanning parameter is varied continuously, as a temperature step-scan is difficult to perform. The temperature ramp of heating the substrate holder was recorded simultaneously using a ther-

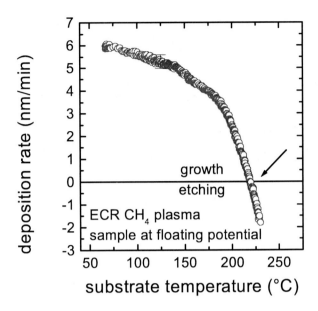

Figure 9.3: Deposition rate of a-C:H from an ECR CH_4 plasma in dependence on substrate temperature, as obtained from a *dynamic in situ* ellipsometry parameter scan. Note the smooth transition from growth to etching.

mocouple mounted directly below the substrate surface. Analysis of the ellipsometric data provides the film thickness as a function of time $d(t)$. The growth rate $\Gamma(t)$ is obtained by numerically calculating the 1st derivative. These data can be correlated to $\Gamma(T)$.

The results in Fig. 9.3 show a smooth curve with a small error bar, which of course depends on the data spacing used in calculating the 1st derivative. Absolute errors are mainly determined by the choice of the optical model and errors involved in the temperature measurements. The most interesting result of this dynamic parameter scan is the transition from deposition to etching at 220 °C without any change in slope. This in turn means, that the temperature dependence of the growth rate in this temperature region is mainly due to a temperature-dependent etching process. At this point, we will not discuss details of the underlying plasma-surface interaction phenomena. However, it should be pointed out, that this experimental finding could not be obtained by a series of single deposition experiments at different substrate temperatures, as above 220 °C no film would grow and etching could not be detected. One would find $\Gamma(T) = 0$ for $T > 220$ °C. The same (wrong) result would stem from a *dynamic in situ* ellipsometry scan under the same experimental conditions but with decreasing substrate temperature.

Reliable data on the temperature dependence of plasma-surface interaction processes recorded with small data spacing enables accurate activation energy calculation and separation of processes that differ only slightly in activation energy.

9.4.1.3   The role of low-energy hydrogen ions in plasma-enhanced chemical vapour deposition (PECVD) of hydrocarbon films

During plasma processing (deposition, sputtering, etching, etc.) the surface of a solid is exposed to ion bombardment and fluxes of photons, electrons, radicals, excited particles, neutrals, etc. The knowledge about the state of the surface layer during plasma processing is still rather incomplete, except for some simple model systems. This is due to the fact, that most analytical techniques can be applied only *quasi in situ*, i.e., after terminating the process but without breaking the vacuum.

The next example shows the application of *in situ* ellipsometry to the precise measurements of the thickness of the modified surface layer of polymer-like a-C:H due to energetic ion bombardment from a plasma. The films have been deposited on silicon substrates at floating potential utilising a CH_4 ECR plasma. The thickness of the modified layer due to energetic ion bombardment from CH_4 and H_2 plasmas has been investigated. Reliable thickness measurements of very thin films (in the range of some nm) by ellipsometry require exact knowledge of the refractive index of the film. Also, when the ion energy is increased during deposition, it is difficult to separate growth under altered conditions from modification of already deposited material. In this application it is demonstrated that these two difficulties can be overcome by applying *dynamic in situ* ellipsometry to deposition/etching sequences.

The procedure is as follows: (i) A polymer-like a-C:H film is grown first to a thickness that allows very accurate measurement of its refractive index. (ii) The optimal film thickness at which the ellipsometric data are most sensitive to changes in the optical properties is calculated. (iii) When the film has been grown to that thickness, the ion energy is set to the value of interest resulting in an abrupt change in direction and spacing of the ellipsometric data. (iv) After deposition, the film is gently removed employing an O_2 plasma etch process with the sample at floating potential. This process has been shown to act only chemically at the surface and does not cause deep-reaching surface modification. The ellipsometric data recorded during etching follow exactly the trace recorded during growth except in that region where the ions caused a modification of previously deposited material. In Fig. 9.4 (a) the *in situ* ellipsometric data recorded during this procedure for a bias voltage of –60 and –90 V are shown. In the Ψ, Δ–traces in Fig. 9.4 (a) a region appears, where etching does not match deposition. This is the Ψ, Δ–region where already deposited material has been modified due to the influence of higher-energy ions. As the optical properties of the a-C:H film before modification are known precisely, the modification depth can be calculated accurately.

A similar procedure can be applied to measure the thickness of the modified layer in a-C:H due to a H_2 plasma [10] or due to ion implantation into semiconductor material [12]. The thickness of the modified layer in dependence on the ion energy is compared to TRIM.SP [13] calculations. TRIM is a Monte Carlo-based computer code which models the ion-solid interaction (see chapter 4). The results of measurements and simulations are compared in Fig. 9.4 (b), showing that the thickness of the modified surface layer corresponds to the projected range of H^+-ions and not to that of C from CH_3^+. Hydrogen ions create dangling bonds by displacement or chemical abstraction of bonded hydrogen. Reconfiguration of the carbon network results in densification, as indicated by the higher refractive index.

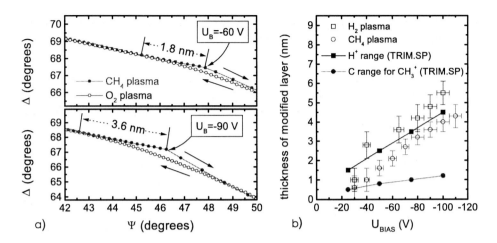

Figure 9.4: (a) Ψ, Δ–data recorded during deposition of a-C:H from a CH_4 ECR plasma with an abrupt increase in ion energy and gentle removal of the film using an O_2-plasma. The thickness of the modified layer is indicated. (b) Comparison of the thickness of the modified layer by CH_4- and H_2-plasma with computer simulations of the projected range of H and C in a-C:H.

9.4.2 *In situ* spectroscopic ellipsometry examples

In this section applications of *in situ* ellipsometry which require the higher content of information from spectroscopic ellipsometry (SE) are presented. The application of SE is essential, (i) if three or more unknown parameters must be calculated simultaneously, (ii) if the dielectric function and not only the refractive index at a single wavelength is of interest, (iii) if the film material is anisotropic, and (iv) if the higher sensitivity at shorter wavelengths is of importance.

9.4.2.1 Surface temperature and oxide thickness during argon sputter-cleaning

Sputter cleaning by means of an argon plasma at increased substrate bias levels is widely employed to clean substrates before deposition. The dynamics and final result of such a sample treatment under specific experimental conditions are usually not exactly known, although the ion-solid interactions involved can be easily simulated. Recipes like *30 min sputter cleaning is essential* are found in the literature without explanation. In this example the dynamics of Ar sputter cleaning, the dependence on the initial oxide thickness at the Si substrate, and the time-dependence of the surface temperature during the treatment are analysed by *in situ* SE. The importance of a shutter in front of the sample and the processes during ignition of the plasma in an ECR source are discussed.

In the experiments, a microwave power of 300 W, Ar pressure of 0.2 Pa, and RF bias voltage of –100 V were used. The base pressure in the chamber was 10^{-4} Pa. In Fig. 9.5 oxide thickness and surface temperature as obtained from *in situ* SE data analysis are shown for three different experimental conditions (cases I–III). In case I, in contrast to the other two

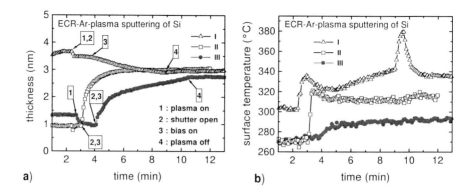

Figure 9.5: (a) Time-dependence of the oxide thickness on Si during ECR-Ar-plasma treatment for different initial states, with and without the use of a shutter, and (b) time-dependence of the surface temperature. Experimental details are described in the text.

experiments, the native oxide at the Si sample surface was not chemically removed before processing. When the plasma was ignited in case I at $t = 2.5$ min the shutter was open and no additional bias was applied. The oxide thickness decreased by about 0.3 nm, which is due to desorption (or etching) of an adsorbed surface contamination layer. In the optical model this layer has not been considered and its thickness can deviate from the calculated value. As can be seen in Fig. 9.5 (a), the Ar plasma treatment of the sample at floating potential for 1 min did not affect the oxide thickness. After increasing the bias to −100 V at $t = 3.5$ min the oxide thickness decreased slowly to a final value of about 3 nm due to sputtering. At $t = 8.5$ min the plasma was extinguished by shutting down the Ar gas flow and switching off the microwave. This procedure had no effect on the oxide thickness, although the plasma parameters changed considerably.

The surface temperature as obtained from *in situ* SE data shows two interesting features to be seen in Fig. 9.5 (b) (case I). This information cannot be obtained easily by other techniques. A temperature spike of about 40°C occurred when the plasma was ignited. This results from processes in the ECR source and is due to the microwave propagating through the chamber and the heating of the substrate until the discharge is ignited and reaches its final density. Under conditions (pressure, matching, wall condition, etc.) where the ignition is delayed, temperature spikes of more than 100°C have been observed. In temperature sensitive plasma processes (growth or etching) this effect can cause severe problems and can be avoided only by implementation of a shutter to protect the sample. Due to the ion flux the substrate temperature increased by about 40°C in case I. Extinction of the discharge by switching off the gas flow leads to dramatic changes in the plasma parameters which in turn cause a temperature increase by about 40°C. In the cases II and III the native oxide was chemically removed some days before the experiments started. Interestingly, the sputter process does not result in a damaged or amorphous surface layer, as one would expect from collision cascade calculations (TRIM), but the build-up of an oxide layer is observed. Amorphous Si and SiO_2 have significantly different optical properties and can be clearly distinguished by SE. Sputtering of oxides in the chamber, water release from walls that are additionally heated by the plasma, as well as the residual gas, result in an oxygen concentration in the chamber, that is sufficient to oxidise the amorphous Si. The final oxide thickness does not depend on the initial state of the Si wafer, as is seen in Fig. 9.5 (a). The time needed to reach equilibrium depends on the plasma density (discharge power and matching) which was higher in case II than in case III as is apparent from the time dependence in Fig. 9.5 (a) as well as from the increase in surface temperature shown in Fig. 9.5 (b).

In case III the sample was contaminated with more than one monolayer of organic solvents. These contaminants are removed within some seconds after ignition of the plasma, although the shutter is closed. This is in agreement with results of plasma cleaning investigations, showing that a very short treatment is sufficient. In the ECR source used the ions are strongly confined due to electron confinement by the magnetic field and ambipolar diffusion. The fact that the cleaning process works also with closed shutter (at 0.2 Pa) indicates that radicals are heavily involved in this process.

The *in situ* SE investigations showed that, under the experimental conditions used, the *best* Si surface (least deviation from a bare c-Si wafer) is obtained with a chemical cleaning process as established in microelectronics fabrication. Eventually, removal of adsorbed organic residuals with the aid of a plasma may improve cleanness. Preparation of a "clean" surface by sputter cleaning requires not only a very low base pressure in the chamber but also a plasma process that does not produce significant concentrations of oxygen by sputtering oxygen containing material in the source or at the substrate holder. Extended plasma treatment (for half an hour or longer) will only help to clean the plasma source and chamber, but may roughen and contaminate the sample.

9.4.2.2 Analysis of unstable plasma processes

ECR discharges often lack long term stability, and the plasma density is hard to reproduce. These discharges are sensitive to changes in the state of the chamber walls, the magnetic field, and matching of the microwave resonator. The reflected microwave power at given forward power is not a sufficient parameter to control the discharge. Without reliable monitoring of the plasma density, reproducible results are difficult to obtain. The following examples demonstrate how this problem can be overcome by the application of *dynamic in situ* SE and how the dependence of the growth process on parameters like gas composition and bias voltage can be efficiently analysed, which would otherwise take weeks or even months of laboratory work using *ex situ* techniques.

The deposition of BN films by PECVD utilising an ECR source with gas mixtures of BF_3, N_2, Ar, and H_2 is a quite complex system with a number of parameters to be varied. The nano-crystalline BN films, if not cubic, are turbostratic (tBN) and optically anisotropic. The orientation of the c-axis of the tBN nano-crystals depends on the ion energy.

The growth rate in dependence on experimental time as obtained from the analysis of the *dynamic in situ* SE data is shown in Fig. 9.6 (a). The flow of H_2 was increased stepwise from 0 to 30 sccm. The shutter was closed during the gas flow change and rematching of the ECR source, causing abrupt changes in the growth rate. Fluctuations in the growth rate appear in Fig. 9.6 (a), which are caused by instabilities of the source (e.g., at 20 and 30 sccm), or due

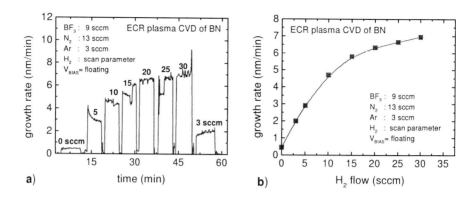

Figure 9.6: a) Growth rate of turbostratic BN from an ECR plasma with stepwise variation of the H_2 gas flow. Rate instabilities and variations are apparent. b) The extracted dependence of growth rate on H_2 flow taking into account instabilities and drift.

to necessary rematching during the process (e.g., at 15 and 25 sccm). The main advantage of the time-resolved information on the growth rate from SE data is that *proper data* can be identified and separated from artefacts due to instability and mismatch. The growth rate is constant without hydrogen but may change over 10 min, if hydrogen is present (e.g., at 5 and 3 sccm) due to long-term changes in the H partial pressure caused by the low pumping speed of the turbo pump for light elements.

The dependence of the growth rate on the H_2 gas flow as extracted from the *proper data* in Fig. 9.6 (a) (taking into account instabilities and drift) is shown in Fig. 9.6 (b). It is a smooth curve with small error bars that can provide information on the underlying mechanisms. Series of single deposition experiments would result in error bars that are much too large to obtain detailed information on the functional dependence. After about 1h of film growth, all the necessary information is stored in the computer of the ellipsometer and the film is no longer needed for analyses. The film can be removed by plasma etching and the next parameter scan can be performed immediately.

Fig. 9.7 shows the growth rate in dependence on H_2 gas flow for low ion energy (sample at floating potential) and additional RF bias of -100 V under (otherwise) identical experimental conditions. Above 10 sccm H_2 flow the growth rate is significantly higher at -100 V bias voltage. The increase in growth rate with increasing H_2 flow and increasing ion energy may have two reasons: (i) The hydrogen influences the plasma parameters, like electron temperature and electron energy distribution, in such a way that the flux of film forming ions (mainly BF_2^+) increases. (ii) As in this growth process a competition between deposition and etching by F radicals exists, hydrogen can remove fluorine efficiently from the near-surface region by HF formation. This scenario explains the increase in BN growth rate with increasing H_2 flow. Argon is added to the gas mixture in order to cause momentum transfer to the growing film via energetic heavy ion bombardment. This is known from PVD processes to be essential for the formation of the cubic BN phase. But Ar reduces the growth rate due to sputtering.

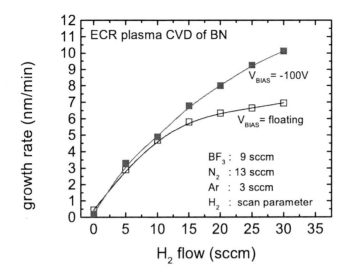

Figure 9.7: Dependence of tBN growth rate on the H$_2$ flow with the sample at floating potential and at -100 V RF bias, respectively.

On the other hand, we have found that a H$_2$/Ar plasma can etch BN films very efficiently, whereas without Ar the etch rate is negligible. In this process the Ar is obviously required to produce defects (to break bonds) enabling the hydrogen to saturate those bonds and form volatile species (ion induced chemical etching).

Again, the technique of a *dynamic in situ* ellipsometry parameter scan was employed to investigate the dependence of the growth rate on the ion energy. In Fig. 9.8 (a) the growth for a step-wise increase of the RF bias voltage is shown. During the first few minutes of growth a significantly increased and exponentially decaying growth rate is observed as in Fig. 9.6 at 5 sccm H$_2$. The decrease in growth rate probably results from a decreasing H$_2$ partial pressure after a spike in the pressure. This in turn results from the characteristics of the gas flow controller used which opens to large flow values before reducing the flow to the preset value. This long-term change in growth rate would cause large errors in the growth rate, if measured on the basis of single deposition experiments. When the bias voltage was set to -260 V a thermal instability in the RF network caused a decrease in bias voltage, resulting in an increased growth rate. Nevertheless, the extracted *proper* data show a perfectly linearly decreasing growth rate with increasing bias voltage (see Fig. 9.8 (b)).

In Fig. 9.9 (a) SE results of the growth rate in dependence on the BF$_3$ flow rate are shown. As in the other experiments (see Figs. 9.6 (a) and 9.8 (a)), a significantly increased growth rate during the first minutes of deposition, which then decreases exponentially, is observed. The extracted *proper* data from Fig. 9.9 (a) show a linear dependence of the growth rate on the BF$_3$ flow with a deviation towards smaller rates up to 2 sccm, as can be seen in Fig. 9.9 (b).

Summarizing the results of the *dynamic in situ* ellipsometry investigations on BN film growth using BF$_3$ containing gas mixtures, the following scenario is proposed: (i) Fluorine

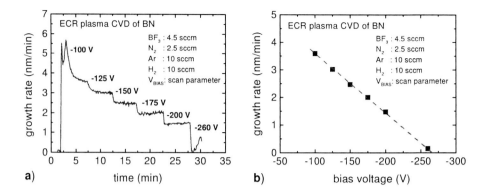

Figure 9.8: (a) tBN growth rate versus time with a step-wise increase in RF bias voltage, as obtained from dynamic *in situ* SE data. Artificially increased growth rates are observed in the first 3 min of the growth process and after increasing the RF bias to −260 V. (b) The growth rate in dependence on RF bias voltage as extracted from the data in Fig. 9.8 (a) shows perfect linear behaviour.

radicals and ions etch effectively the growing BN film and limit the growth rate. (ii) This problem can be solved by adding hydrogen to the gas mixture that will react with the fluorine and form HF. The growth rate increases by more than one order of magnitude. (iii) For the formation of the cubic BN phase a certain momentum transfer per B atom is essential [14].

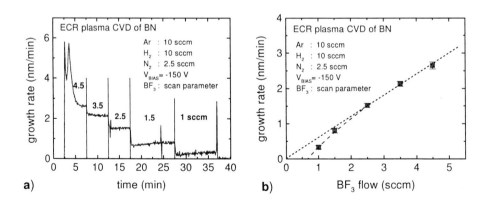

Figure 9.9: (a) tBN growth rate versus time with a step-wise decrease in the BF_3 flow rate, as obtained from *dynamic in situ* SE data. An increased growth rate during the first 2 min of the growth and two instabilities are observed. (b) BN growth rate in dependence on BF_3 flow rate as extracted from the data in a), showing a linear dependence and deviation at low BF_3 flow.

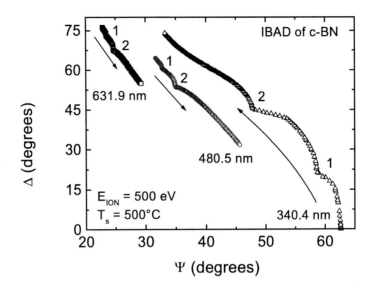

Figure 9.10: SE data (only 3 out of 44 wavelengths) recorded during the growth of cBN by IBAD. The dips in the Ψ, Δ contours (labeled 1 and 2) are due to fluctuations in the boron evaporation rate. See text for details.

In PVD systems this is achieved by inert gas ion bombardment (mainly Ar). However, in PECVD at increased bias voltage the combination of Ar and H also etches BN very efficiently. In PECVD it is difficult (if possible at all) to meet the conditions for the growth of cubic BN as the necessary ion bombardment reduces the net growth rate due to sputtering and ion enhanced etching. As the focus of this paper is on the analytical technique, a more detailed discussion of these results will not be given here.

9.4.2.3   Monitoring of ion beam assisted deposition processes

In ion beam assisted deposition (IBAD), species evaporated either thermally by means of an electron beam evaporator, or sputtered by an ion beam, are deposited on a substrate and simultaneously bombarded by an ion beam of inert gas ions, reactive gas ions, or both. Ion sources usually consist of a plasma source and a system of grids for extraction, acceleration and formation of the ion beam. IBAD is not a plasma process in the classical sense, but firstly, ion beams are a good tool for the investigation of ion surface interaction, because well defined ion fluxes with quite narrow ion energy distributions are easily produced, and secondly, IBAD systems are in common use for deposition.

   The next example demonstrates (i) the usefulness of the application of *dynamic in situ* SE to IBAD to obtain detailed information on the process which would be impossible using other techniques, (ii) the need of simultaneously acquiring data at multiple wavelengths (SE), as sensitivity can be strongly wavelength-dependent, and (iii) the interpretation of ellipsometric data when exact model calculation is rather difficult.

Fig. 9.10 shows SE data recorded during the deposition of cubic BN (cBN) using IBAD. Boron is evaporated onto a Si substrate using an electron beam evaporator which is controlled by a CM. Simultaneously, the deposited boron is bombarded with a beam of Ar^+ and N_2^+ ions (Ar:N_2 gas flow = 2:1) at an energy of 500 eV and a current density of 100 $\mu A/cm^2$ from a Kaufman ion source. This system possesses two sources of instability: (i) Erroneous rate data from the CM, fluctuations in the evaporation rate, and the large time constant of the evaporator due to its heat capacity make electronic control of the growth rate quite difficult. (ii) Fluctuations in the current density of the Kaufman ion source can result from dust particles on the grids or contamination of insulators. These fluctuations can be small ($< 10 \%$) and escape the operator's notice easily. In Fig. 9.10 two dips (labeled 1 and 2) in the Ψ, Δ contours are apparent. Dip No. 1 is of the order of the experimental error at 631.9 nm, whereas at 340.4 nm a clear shoulder appears. This demonstrates the importance of spectroscopic data. The data can be interpreted quite easily whereas exact model calculation turns out to be difficult. When the Ψ, Δ data run into the dip, absorbing boron-rich material is deposited onto the transparent cBN. This is the result of a temporary increase in the boron evaporation rate. After about 45 s in the case of dip 2 the Ψ, Δ data return to the track as if nothing had occurred. The penetration depth of 500 eV N_2^+ and N^+ ions in BN is sufficient to nitride this about 1.5 nm thick boron-rich layer during further growth. But, also if almost stoichiometric BN is formed, the growth of cBN may change to tBN growth, as only tBN could be synthesised by nitrogen implantation into boron or boron rich BN. The growth of tBN can proceed for a certain thickness before it returns to cBN growth. This scenario remains completely hidden to all other analytical techniques. The *in situ* SE data provide an explanation as to why at times (instabilities in the evaporation rate or arcing in the Kaufman ion source) higher tBN fractions are found or intermediate layers of higher tBN content are seen in TEM graphs.

9.4.3   *Ex situ* **spectroscopic ellipsometry examples**

Dynamic in situ SE data contain more information about a sample and its history than *ex situ* SE can provide. The reasons have been discussed in section 9.3.1. Therefore, dynamic SE should be preferable to post-processing analysis. However, there are also chambers and processes that do not allow an *in situ* ellipsometer to be mounted, or the resulting sample structure is too complex to allow adequate analysis with a *dynamic in situ* ellipsometer as in the case of in-plane anisotropic films. In principle it is possible to analyse such samples *in situ*, but this would require much more data to be recorded at a number of different input polarisation states, different angles of incidence, or rotational angles (azimuth) of the sample. The examples in this section demonstrate the capabilities of generalised ellipsometry (GE) and multi-sample analyses.

9.4.3.1   **In-plane anisotropic turbostratic boron nitride films**

The geometry of the set-up used in IBAD causes a different angle of incidence of the evaporated neutrals and ions at the sample surface. Usually the sample is aligned normal to the ion beam. Under such conditions the average orientation of crystals or grains will be between the sample normal (direction of the ion beam) and the direction of the vapour stream. The orientation depends on ion energy, ion current density, and neutral flux. This has been observed in IBAD of metals (e.g., Cr [15]), nitrides, and oxides (e.g., TiO_x [16]) under ion irradiation using energies in the range of 0.5–40 keV.

As the orientation of tBN nano-crystals cannot be properly analysed by XRD or TEM, the information from optical analyses is of major importance. It has been confirmed by polarised infrared reflection spectroscopy (PIRR), that tBN films can show in-plane anisotropy [17]. The tBN basal planes are not only aligned normal to the sample surface, but show a preferred orientation parallel to the plane of incidence of boron vapour and ion beam. This causes biaxial optical anisotropy of the thin film material in the visible region and can be adequately analysed only by generalised ellipsometry (GE) ([1, 18]). GE is similar to Mueller matrix ellipsometry and allows the analysis of complex anisotropic structures. All elements of the transfer matrix for the electromagnetic field components are determined. Practically, this is done by measuring the change in the state of polarisation for a number of different input polarisation states and the use of regression analysis [18].

In Fig. 9.11 (a) the ellipsometric angle $\Psi(R_{pp}/R_{ss})$ that compares to the angle Ψ as measured in conventional ellipsometry [18], is shown for an angle of incidence $\Phi = 30°$ and three azimuth angles around the sample normal. For an isotropic or uniaxial sample with its optical axis (OA) normal to the sample surface, the ellipsometric data would be independent of sample rotation. Model calculations confirm that the tBN film is optically biaxial. Fig. 9.11 (b) shows the refractive indices of the tBN film material which are significantly higher as would be expected from effective medium calculation based on the data reported for the ordinary and extraordinary component of single crystalline hexagonal BN (hBN) (n_o=2.13, n_e=1.65 at 632 nm). The relatively small difference between the components of the refractive index is attributed to the wide spread in the orientation of the nano-crystallites. Additionally, the tBN thin film material is slightly boron rich which results in an absorption tail that extends from the UV into the visible region.

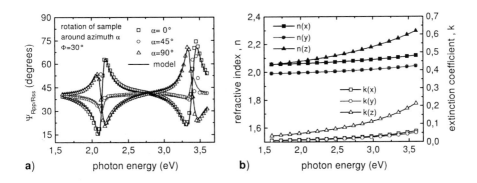

Figure 9.11: (a) A subset of GE data recorded at $\Phi = 30°$ from a biaxial tBN film on Si at three azimuth angles α of sample rotation around its normal. The solid line is the result of the model calculation. (b) The three components of the tBN thin film refractive index, and the extinction coefficient (x: parallel to the evaporator–ion beam plane, z: normal to film surface).

9.4.3.2   Reactive cathodic arc deposition of aluminium oxide films

Cathodic arcs offer high deposition rates and exhibit a high degree of ionisation in the plasma. The utilisation of a magnetic filter prevents the incorporation of macro-particles into the film. The deposition rate for aluminum oxide of the set-up used in this study is in the range from 100 nm/min to 1 μm/min depending on the location of the reactive gas injection as well as on the magnetic field configuration, which in turn determine the lateral uniformity of the film. The deposition rates in cathodic arc systems reach the speed limit of rotating element ellipsometers. This example demonstrates the application of multi-sample analysis of *ex situ* SE data to determine reliable optical data of the film, the film uniformity, and thus to validate the optical model.

Fig. 9.12 shows measured and calculated SE data from 7 positions along the sample diameter. It is evident that this high number of data points imposes strong constraints on the optical model. Note that all the data in Fig. 9.12 have been simultaneously modeled with only 12 independent parameters (7 thickness paramters and 5 parameters of the dielectric function model). The dielectric function of Al_2O_x has been described using a Kramers-Kronig consistent dielectric function with only 5 independent parameters based on the Tauc joint density of states and the Lorentz oscillator [19].

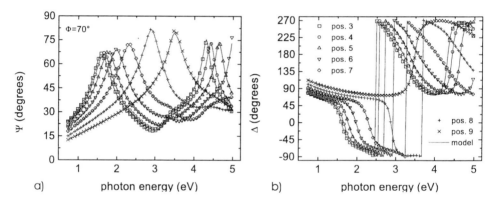

Figure 9.12: Measured SE data (a) Ψ and (b) Δ recorded at 7 positions along the diameter of a Si wafer coated with Al_2O_x using a reactive cathodic arc process. The model calculation is based on multi-sample analysis using identical model parameters except for the film thickness.

The resulting refractive index and extinction coefficient are shown in Fig. 9.13 (b). For comparison, reference data for the ordinary refractive index n_o of synthetic sapphire (α–Al_2O_3) are plotted [20]. Sapphire is uniaxial with a small (less than 2%) anisotropy in the visible region. The refractive index of the film material is about 0.1 lower than that of sapphire. The density of the thin film material is estimated to be $\approx$ 87 %. Weak absorption extending from UV into the visible range ($\approx$ 2.5 eV) is observed. This may be due to the microstructure of the material or may be indicative of a slight oxygen deficiency. In any case, the absorption will result in a steeper increase of the refractive index towards the UV due to Kramers-Kronig (KK) consistency (see Fig. 9.13 (b)). The film thickness data in Fig. 9.13 (a) fit perfectly a Gaussian distribution, although it would be desirable to have more data.

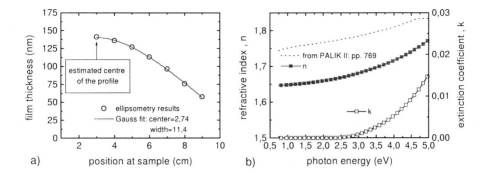

Figure 9.13: (a) Film thickness along the sample diameter as obtained from the Ψ, Δ data in Fig. 9.12, and Gaussian fit to the profile. (b) Refractive index and extinction coefficient based on a Tauc/Lorentz dispersion model, and reference data for Al_2O_3 [20].

9.5   Limitations and remaining issues

The most recent developments in both, hardware and software, make the application of ellipsometry much more convenient, and ellipsometry is at a stage to become a standard technique. On the other hand the availability of sophisticated software may pose a serious problem [21]. The user can easily develop a very complex optical multi-layer model with a large number of free parameters that can be used to fit almost every set of data. It must be shown that the problem is not underdetermined and the model is physically correct, i.e. obeying basic laws like Kramers-Kronig consistency. Additionally, it must be verified that the model, providing a good fit, is the only proper model in the crowd of all possible models. Therefore, any additional information about the sample, e.g., data concerning composition, surface roughness, homogeneity, etc., is helpful and necessary to proof the model. Sophisticated software cannot replace the experienced user because a model that provides a good fit to measured data is not necessarily meaningful.

A serious (unsolved) practical problem is related to measurements on patterned samples. Focusing has been used for both, single wavelength and spectroscopic systems. This has some problems which are related to the wide spread of the angles of incidence around the centre of the beam. This may introduce serious errors around the Brewster angle. There have been attempts to use a broad area light beam and perform ellipsometric data analysis on a selected area of the image (imaging ellipsometry) [22]. However, traditional analysis of data from patterned reflecting surfaces becomes extremely difficult, as the transistor components are at the diffraction limit for visible light. Thus, one is trying to analyse polarised light from irregular and grating-like structures with irregular periods. Another serious problem is related to the calibration of the ellipsometer. The procedures used require isotropic samples ($\underline{R}_{ps}=\underline{R}_{sp}=0$) and work well with semiconductor samples like Si and oxidised Si. Calibration with an *unknown* sample is dangerous as the sample may be depolarising or anisotropic. This can cause severe measurement errors. Recalibration during measurement of complex anisotropic samples is impossible.

In situ SE is a more recent application than *ex situ* SE and entails new problems. Substrates are often rotated for uniform processing. If the axis of rotation is not steady the beam can wobble, resulting in two effects: a change in the angle of incidence with time, and a loss of some fraction of the light at the detector. *In situ* ellipsometry with closed loop feedback control requires not just fast data acquisition, but also fast analysis. This should be a challenge for the software and computer hardware developers. The majority of the vacuum chambers has no optical ports at the necessary angle of incidence. This should be kept in mind when designing chambers in the future. Spectroscopic ellipsometers can be quite expensive, up to half of the price of the chamber, and cost can be a deterrent to more extensive use of ellipsometry.

In spite of the unsolved problems, spectroscopic ellipsometry is an extremely powerful technique for materials analysis of surfaces and thin films for both, *in situ* and *ex situ* environments.

Acknowledgements

The author would like to thank A. von Keudell for providing part of the data in section 9.4.1 and J. Brückner for preparation of the Al_2O_x films. The author is grateful to R.A. Yankov for the careful reading of the manuscript and numerous helpful comments and to W. Möller and E. Richter for kindly supporting this work.

References

[1] R.M.A. Azzam and N.M. Bashara, Ellipsometry and Polarized Light, North-Holland Publishing Co., Amsterdam (1977/1987)

[2] H.G. Tompkins, A user's guide to ellipsometry, Academic Press Inc., Boston (1993)

[3] K. Riedling, Ellipsometry for Industrial Applications, Springer, Berlin (1988)

[4] A.C. Boccara, C. Pickering and J. Rivory, Editors, Proceedings of the First International Conference on Spectroscopic Ellipsometry (Paris, 1993), Thin Solid Films **233/234** (1994)

[5] R.W. Collins, D.E. Aspnes, E.A. Irene, Editors, Proceedings of the Second International Conference on Spectroscopic Ellipsometry (Charleston, 1997) and Thin Solid Films **313/314** (1998)

[6] W. Möller, W. Fukarek, S. Grigull, O. Kruse, S. Parascandola, Nucl. Instr. Meth. Phys. Res. **B 136/138**, 1203 (1998)

[7] W. Fukarek, H. Kersten, J. Vac. Sci. Technol. **A 12**, 523 (1994)

[8] B. Drevillon, J. Perrin, R. Marbot, A. Violet, J.L. Dalby, Rev. Sci. Instrum. **53**, 969 (1982)

[9] S. Schiller, U. Heisig, Chr. Korndörfer, G. Beister, J. Reschke, K. Steinfelder, J. Strümpfel, Surf. Coat. Technol. **33**, 405 (1987)

[10] A. von Keudell, W. Jacob, W. Fukarek, Appl. Phys. Lett. **66**, 1322 (1995)

[11] W. Möller, W. Fukarek, K. Lange, A. von Keudell, W. Jacob, Jpn. J. Appl. Phys. **34**, 2163 (1995)

[12] W. Fukarek, W. Möller, N. Hatzopoulos, D.G. Armour, J.A. van den Berg, Nucl. Instr. Meth. Phys. Res. **B 127/128**, 879 (1997)

[13] W. Eckstein, Computer Simulation of Ion Solid Interactions, Springer Series in Materials Science Vol.10, Springer, Berlin, Heidelberg (1991)

[14] P.B. Mirkarimi, K.F. McCarty, D.L. Medlin, Mat. Sci. Engineering **R21**, 47 (1997)

[15] N. Kuratani, A. Ebe, K. Ogata, J. Vac. Sci. Technol. **A 16**, 2489 (1998)

[16] F. Zhang, Z. Zheng, Y. Chen, Y. Mao, X. Liu, Thin Solid Films **312**, 1 (1998)

[17] W. Fukarek, O. Kruse, A. Kolitsch, W. Möller, Thin Solid Films **308/309**, 38 (1997)

[18] M. Schubert, Phys. Rev. **B53**, 4265 (1996); Thin Solid Films **313/314**, 323 (1998)

[19] G.E. Jellison Jr., F.A. Modine, Appl. Phys. Lett. **69**, 371 (1996); and Erratum, Appl. Phys. Lett. **69**, 2137 (1996)

[20] I.H. Malitson, F.V. Murphy, W.S. Rodeny, J. Opt. Soc. Am. **48**, 72 (1958); F. Gervais, in: E.D. Palik, Editor, Handbook of Optical Constants of Solids II, Academic Press Inc., San Diego (1991)

[21] WVASE software package by J.A. Woollam Co., Inc., Lincoln, Nebraska, USA.

[22] D. Beaglehole, Rev. Sci. Instrum. **59**, 2557 (1988)

10   Characterization of thin solid films

Harm Wulff[†] and Hartmut Steffen[‡]

[†]Institut für Chemie und Biochemie, Ernst-Moritz-Arndt-Universität Greifswald,
Soldmannstrasse 16–17, D-17487 Greifswald, Germany
[‡]Institut für Physik, Ernst-Moritz-Arndt-Universität Greifswald,
Domstraße 10a, D-17487 Greifswald, Germany

10.1   Introduction

The deposition of films by plasma techniques is a well-known and widely used method. Nevertheless the fundamental mechanisms of plasma-wall interaction are not understood in detail, yet. In order to understand the complexities involved in film formation it is necessary to take a close look at the deposition region. This is usually the substrate area that is subjected to plasma radiation and, in particular, to fluxes of energetic and reactive charged and neutral particles in various excited states. Therefore results of detailed plasma diagnostics have to correlate with film properties.

Of particular interest is on the analysis of film properties in the nanometer range. Grazing incidence x-ray diffractometry (GIXD) and grazing incidence x-ray reflectometry (GIXR) have been established as well-suiting tools for investigations of chemical, physical and crystallographic properties of thin films. These x-ray methods have a number of advantages compared to other commonly used techniques. They are non-destructive techniques, therefore a sample can be reused, and can be measured with other techniques, such as x-ray photoelectron spectroscopy (XPS), or atomic force microscopy (AFM). Table 10.1 gives an overview of the possibilities of x-ray thin film analysis. A usable combination of GIXD and GIXR requires that the film thickness does not exceed the upper physical absorption limit for the reflectometry measurements. If this condition is fulfilled, x-ray methods can give a range of interesting information how plasma deposition techniques and plasma parameters influence deposited films. From the investigations described below information about phase and chemical composition, defect structure features as point defects, microstrains, dislocation densities, domain sizes, grain boundaries in TiN films, and about chemical and phase gradients in Ti/TiSi films were obtained. In indium doped tin oxide films (ITO) we have detected phase compositions, structure parameters as atomic positions, bond lengths, bond angles, deposition rates, mass densities, and from *in situ* methods the kinetic process parameters diffusion coefficients, crystal growth rate and activation energies.

Besides these x-ray methods, also other analytical thin film techniques can be used to give information on deposited films. In our plasma-wall interaction studies we have used x-ray photoelectron spectroscopy (XPS) for chemical analysis, and atomic force microscopy (AFM) for surface morphology characterization.

Table 10.1: X-ray analyses of thin solid films.

Film property	X-ray method		Alternatives
phase composition	GIXD	Bragg angle, intensity	TEM
chemical composition (mixed crystals)		Bragg angle	EDX, XPS, RBS
macrostress		Bragg angle	substrate curvature, laser optics
grain size		line profile, line width	TEM, SEM
microstrain		line profile	
preferred orientation		intensity, polfigure	
crystal structure		Rietveld analysis, structure refinement	
thickness	GIXD	intensity	interferometry, ellipsometry, TEM
	GIXR	Kiessig fringes	
density		critical angle of total reflection	ellipsometry
surface roughness interface roughness	GIXR	amplitude of Kiessig fringes	SEM, AFM, ellipsometry
diffusion behavior	*in situ* GIXD, time- and/or temperature-resolved	intensity	SIMS, AES, XPS, combined with sputtering
crystallization rate		intensity	

10.2 X-ray methods for thin film analysis

10.2.1 Grazing incidence x-ray diffractometry (GIXD)

The observation of x-ray diffraction from very thin films is often hampered by weak diffraction intensities due to the smallness of the diffraction volume. Thin polycrystalline films can be studied with advantage in a highly asymmetric Bragg case. In this technique the diffraction volume can be increased by decreasing the angle of incidence. The optical path in grazing incidence x-ray geometry (GIXD) is depicted in Fig. 10.1. The x-ray patterns of a 50 nm titanium layer on a Si(100) wafer measured in normal Bragg-Brentano geometry (BB) and in the asymmetric Bragg case (GIXD) are displayed in Fig. 10.2. The reflection positions are equal in both techniques for the polycrystalline film, but in the GIXD technique the substrate reflections are suppressed, and the film intensities strongly increase.

In the analysis of surface layers, the information depth of x-rays is an important parameter, in particular, if gradients of structure parameters are observed. The contribution of a volume $dV = A dx$ in the depth x_m to the measured integral intensity of the phase m is given by

$$dI = K I_0 B_m A_m A dx,$$ (10.1)

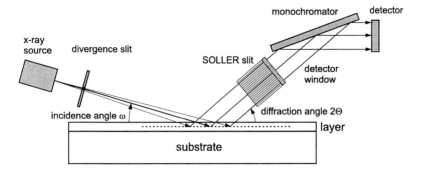

Figure 10.1: Schematic diagram of grazing incidence x-ray diffractometry.

where K is a constant, I_0 the intensity of the incident x-ray, and B_m the theoretical intensity factor. The absorption factor A_m is obtained from

$$A_m = \exp(-\mu x_m z),\tag{10.2}$$

where μ is the linear absorption coefficient,

$$z = \frac{1}{\sin\ \omega} + \frac{1}{\sin(2\Theta - \omega)},\tag{10.3}$$

and ω the angle of incidence of the x-ray beam.

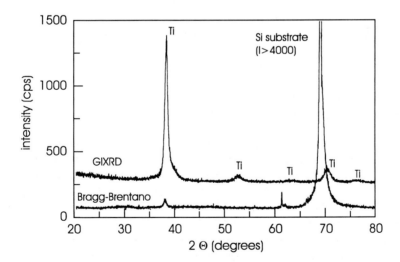

Figure 10.2: The x-ray patterns of a 50 nm titanium layer on a Si(100) wafer measured in normal Bragg-Brentano geometry (BB) and in the asymmetric Bragg case (GIXD).

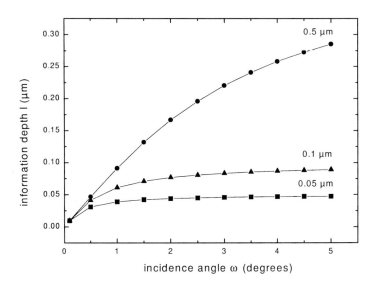

Figure 10.3: Information depth of Cu Kα radiation vs. angle of incidence, calculated for In films of thickness 0.05, 0.1 and 0.5 μm.

The film information depth l is obtained by integration of Eq. 10.2 as

$$l = \int_0^d A_m \, dx = \frac{1}{\mu z} \left(1 - \exp(-\mu z d) \right) \tag{10.4}$$

and strongly depends on the thickness d, the mean absorption coefficients of the film, and the incidence angle of the x-ray beam. The measured integral intensity is proportional to l. Figure 10.3 displays the information depth of Cu Kα radiation at varying incidence angles calculated for 0.05, 0.1 and 0.5μm thick indium films. A further advantage of GIXD compared to the Bragg-Brentano geometry is that the information depth is independent of the Bragg angle 2Θ, if the incidence angle ω is small.

10.2.2 Grazing incidence x-ray reflectometry (GIXR)

The measurement of x-ray reflectivity around the critical angle for total reflection allows the accurate determination of film thickness, mass density, and surface and interface roughness irrespective of the crystalline structure. Grazing incidence x-ray reflectometry is equally well applicable to crystalline, polycrystalline and amorphous materials; it only requires a sufficiently flat sample. The basic principle is shown in Fig. 10.4. In the case of thin films on a substrate constructive interference occurs between the beam reflected at the surface and the beams reflected at the interfaces. Constructive interference results in intensity maxima called *Kiessig fringes*, whose angular spacing is characteristic for the thickness of the layers. The reflected intensity can be calculated by a recursive formula taking into account the Fresnel coefficients for all corresponding interfaces. Using Debye-Waller factor modified Fresnel coefficients to account the roughness in x-ray reflectometry, the reflectivity R of a layer (refractive

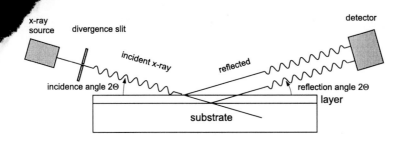

Figure 10.4: Schematic diagram of grazing incidence x-ray reflectometry (GIXR).

index $n_2 = 1 - \delta_2 - i\beta_2$, thickness d_2) on a substrate (refractive index $n_3 = 1 - \delta_3 - i\beta_3$) with the roughness σ_1 (air-layer) and roughness σ_2 (layer-substrate) is

$$R = \left| \frac{r_{12} + r_{23}\gamma_2}{1 + r_{12}r_{23}\gamma_2} \right|, \tag{10.5}$$

where

$$r_{jj+1} = \frac{\vartheta_j - \vartheta_{j+1}}{\vartheta_j + \vartheta_{j+1}} \exp\left(-\frac{8\pi^2 \delta_1^2}{\lambda^2} \vartheta_j \vartheta_{j+1} \right) \tag{10.6}$$

and

$$\gamma_2 = \exp\left(i\frac{4\pi}{\lambda} \vartheta_2 d_2 \right). \tag{10.7}$$

Here ϑ_j is a function of the incidence angle and of the refractive index n_j. On the basis of such an algorithm the modeling of reflectometry measurements yields thickness, surface and substrate roughness, and mass density [1]. Fig. 10.5 shows reflectometry simulations for a 30 nm thick aluminum layer on a silicon substrate and the influence of various surface and interface roughnesses. The value for the critical angle of total reflection Θ_c can be used to determine the mass density ρ of a deposited film [2]

$$\Theta_c = \sqrt{2\delta}, \tag{10.8}$$

with

$$\delta = \sqrt{\frac{N_A r_0 \rho}{2\pi} \frac{\sum_i (Z_i + f_i')}{\sum_i A_i}} \lambda, \tag{10.9}$$

where N_A is the Avogadro constant, r_0 the classical electron radius, $(Z_i + f_i')$ the scattering factor of an atom i, and A_i the atomic weight.

10.3 X-ray photoelectron spectroscopy (XPS)

Surface analysis by x-ray photoelectron spectroscopy (XPS) is accomplished by irradiating a sample with monoenergetic soft x-rays and analyzing the energy of the emitted electrons. Mg Kα (1253.6 eV) or Al Kα (1486.6 eV) x-rays are usually used. The x-ray photons interact

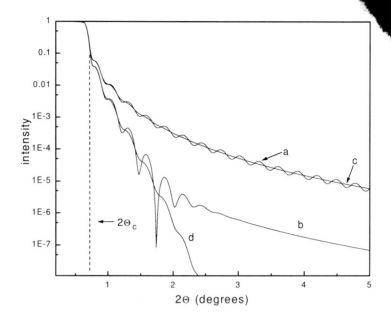

Figure 10.5: Simulation of the reflectivity of a 30 nm Al-layer on Si with varied roughnesses σ_1 (air-Al) and σ_2 (Al-Si); curve a: $\sigma_1 = \sigma_2 = 0$ nm; curve b: $\sigma_1 = 2$ nm, $\sigma_2 = 0$ nm; curve c: $\sigma_1 = 0$ nm, $\sigma_2 = 2$ nm; curve d: $\sigma_1 = \sigma_2 = 2$ nm.

with the atoms in the surface region, causing emission of electrons by the photoelectric effect. The measured kinetic energy E_{kin} of the electrons is given by [3]

$$E_{kin} = h\nu - E_b - \Phi , \tag{10.10}$$

where $h\nu$ is the energy of the photons, E_b is the binding energy of the atomic orbital from which the electron originates, and Φ is the spectrometer work function.

The binding energy may be regarded as the energy difference between the initial and final states after the photoelectron has left the atom. Because there is a variety of possible final states of the ions from each type of atom, there is a corresponding variety of kinetic energies of the emitted electrons.

Because each element has a unique set of binding energies, XPS can be used to identify and determine the concentration of the elements in the surface. The spectrum from a mixture of elements is approximately the sum of the peaks of the individual constituents. Quantitative data can be obtained from the peak heights or peak areas. Variations in the elemental binding energies (chemical shift) arise from differences in the chemical potential and the polarizability of compounds. These chemical shift can be used to identify the chemical state of the materials being analyzed.

Probabilities of electron interaction with matter far exceed those of the photons, so while the path length of the photons is of the order of micrometers, that of the electrons is of the order of several nm. Thus, while ionization occurs at a depth of a few micrometers, only those electrons that originate within several nm below the surface can leave the surface without energy loss. These electrons produce the peaks in the spectra and are the most useful. The electrons that undergo inelastic loss processes before emerging form the background [4, 5, 6].

There are several methods to obtain information on the depth distribution of an element in the sample. Depth profiling can be accomplished using controlled erosion of the surface by ion sputtering. Chemical states are often changed by the sputtering technique, but useful information on elemental distribution can still be obtained. It is possible to change the angle between the plane of the sample surface and the angle of entrance to the analyzer. At 90° with respect to the surface plane, the signal from the bulk is maximized relative to that from the surface layer. At small angles, the signal from the surface becomes greatly enhanced, relative to that from the bulk. The location of an element can thus be deduced by noting how the magnitude of its spectral peaks changes with sample orientation in relation to those of other elements [7]. Furthermore the background can be analyzed to get information on the depth distribution of different elements [8].

10.4 Examples

In the following we present selected examples for the characterization of plasma-deposited solid films by x-ray techniques together with details of the theoretical analysis. The GIXD and GIXR techniques can be performed on normal Bragg-Brentano focusing $\Theta - 2\Theta$ or $\Theta - \Theta$ diffractometers. For GIXD measurements the diffractometers were equipped with special parallel beam attachments. Time and temperature resolved investigations were made on a $\Theta - \Theta$ diffractometer (XRD 3000 Seifert). For *in situ* GIXD measurements the parallel beam attachment was combined with a high-temperature chamber (Bühler HDK 2.4) with surround-ing heater. The GIXR measurements were performed on a diffractometer (Siemens D 5000) equipped with a special reflectometry sample stage or on a $\Theta - \Theta$ diffractometer (XRD 3000 Seifert). The program REFSIM (Siemens) was used for the simulation of the reflectometry measurements.

10.4.1 Phase analysis of plasma-deposited TiN$_x$ films

Plasma wall interactions were studied in the complex system of titanium and nitrogen [9]. The investigated TiN$_x$ films were deposited in a hollow cathode arc evaporation device (HCAED) described in Ref. [10]. Si $< 100 >$ wafers were the substrate material. The nitrogen flux was varied from 0 to 90 sccm at a constant argon flow of 60 sccm and a discharge power of 5.5 kW. Plasma diagnostic results show an increasing energy flux to the substrate surface due to ion bombardment for increasing N$_2$ gas flow. The most frequently used and the most important problem in the thin film analysis is to prove the existence of a desired phase after film deposition or to find out which phase or phases were formed.

Phases and chemical composition of the films were investigated by GIXD and XPS. The structures of α–Ti and δ–TiN are closely related. Titanium possesses a hexagonal close packed lattice (hcp) with a = 2.95 Å and c = 4.866 Å. Titanium nitride possesses a face centred cubic lattice (fcc) with a = 4.242 Å. Nitrogen can be incorporated on the octahedral sites of both the hcp lattice and the fcc sublattice and thus form TiN$_x$. In stoichiometric TiN$_x$ it is assumed

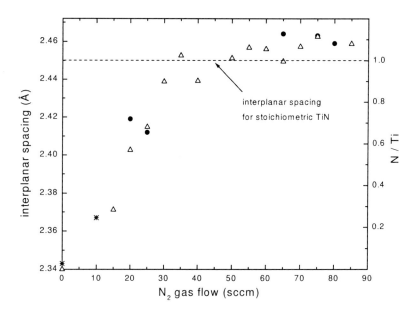

Figure 10.6: Film composition determined by XPS ($\triangle$) and interplanar spacings of (002) α–Ti ($*$) and (111) δ–TiN$_x$ planes ($\bullet$) versus N$_2$ gas flow.

that all the octahedral sites are filled with nitrogen [11]. Titanium nitride is an intercalation compound whose stoichiometry in TiN$_x$ can vary between $0 < x < 1$. The GIXD results clearly show the transition from hcp Ti to fcc TiN$_x$ at an N$_2$ gas flow below 20 sccm. The interplanar spacings of the (002) planes in the hcp α–Ti and the (111) planes in the fcc δ–TiN change in the same manner as the N/Ti ratio which is shown in Fig. 10.6.

For N$_2$ gas flows up to about 25 sccm the films show a strong preferred orientation with the close packed lattice planes parallel to the substrate surface. For higher nitrogen gas flows the XRD patterns become similar to the patterns of polycrystalline bulk TiN$_x$. No preferred orientation was observed. The intensity ratio of all reflections remains constant independent of the incidence angle ω. XPS measurements show an increasing N/Ti ratio in the films from 10 to 40 sccm nitrogen flow (Fig. 10.6). At higher gas flow saturation is obtained. That means all the octahedral sites are occupied by nitrogen. The stoichiometric number x becomes one. At low nitrogen flow up to 25 sccm the stoichiometric number of TiN$_x$ is $x < 1$. The deposition mechanism can be described by a simple consideration. The dominant deposition of single titanium layers with subsequent incorporation of reactive atomic nitrogen (or nitrogen ions) on the octahedral sites of the hcp or fcc lattice is possible. Atomic nitrogen with a radius of 0.62 Å fulfills the conditions of the cannon-ball rule for titanium and with a few restrictions also for TiN$_x$. The titanium atoms deposited in a low nitrogen atmosphere continue to settle on the sample surface as they did in a nitrogen free atmosphere. The same titanium lattice is built during the deposition of titanium and titanium nitride. The nitrogen only changes the layer stacking sequence. It is clear that this geometric consideration is only a crude model to explain the deposition mechanism. But this model is sustained by the high sticking coefficient of titanium.

Higher nitrogen amounts prevent the formation of coherent titanium layers and produce polycrystalline TiN. Crystallographic facets and orientations are preserved on islands and interfaces between initially disoriented, coalesced particles. The crystallite positions in the films separated by grain boundaries are statistically distributed.

10.4.2 Characterization of defect structures by x-ray investigations

In the investigation described in section 10.4.1 both the particle composition of the plasma and the energy flux to the growing film were changed by increasing the N_2 gas flow. For obtaining further information on the effect of the energy flux due to ion energy the substrate potential was varied from $U_{sub} = 0$ V to -90 V at constant N_2 gas flow. Only the additional energy flux to the growing films due to the increased ion energy can be responsible for the changes in the film properties observed in the x-ray line profiles and positions. Lattice defects in thin films can influence x-ray data. Defect structures are partially or wholly manifested by diffraction intensity, line shape, or the change in line shape with respect to the diffraction angle 2Θ. Imperfections of the first type, such as point defects, displacement disorders or substitution disorders, shift the line position; imperfections of the second type such as domain sizes or dislocations act on the diffraction line shape.

10.4.2.1 Imperfections of the first type

Displacement disorders are manifest in the volume of the unit cell. Equation (10.11) gives the connection between the relative change in volume, $\Delta V/V$, and the concentration of interstitial sites c_i [12]:

$$\frac{\Delta V}{V} = \frac{9c_i(1 - \nu)(a' - a)}{a(1 + \nu)(1 + 4\mu\chi/3)}, \tag{10.11}$$

where ν is the Poisson ratio, μ is the shear modulus, and χ is the compressibility. For interstitial atoms $a' \cong R$, the atomic radius, and $a = R_\nabla$, the radius of the atoms on the tetrahedral interstices in the cubic close-packed TiN structure is used.

The volume change can be determined experimentally by measurement of lattice constants a_0 in the cubic crystal system. The lattice constants of broadened profiles were defined by the center of gravity. A volume shrinking can also lead to a stress change as the films are attached onto the substrate. The stress change is connected with the relative volume change $\Delta V/V$ according to

$$\Delta\delta = -\frac{E}{3(1 - \nu)}\frac{\Delta V}{V}, \tag{10.12}$$

where $\Delta\delta$ is the stress change, $\Delta V/V$ the relative volume change, and $E/(1 - \nu)$ the biaxial stress modulus.

10.4.2.2 Imperfections of the second type

The shape analysis of diffraction peaks essentially comprises three problems:

(i) extraction of the pure physical line profile from the experimental profile,

(ii) unraveling of the contributions of various types of lattice imperfections to the physical line profile, and

(iii) quantitative estimation of substructure parameters. The experimentally observable diffraction line profile of an x-ray reflection $h(u)$ is the convolution of a physical line profile $f(x)$ caused by lattice disorder of the second type and an instrumental line profile $g(u)$.

The influence of the apparatus function $g(u)$ is separated from the experimental profile $h(u)$. The method employed was the Stokes-Fourier series [13]. The profile $g(u)$ was determined with a standard material that contains no defects, strain or particle size broadening.

The Fourier coefficients $F(L)$ of the physical line profile $f(x)$ contain information about the particle size due to interfaces (P), the mean strains due to internal stresses, which are constant within a crystallite or a subgrain (S), and restricted randomly distributed dislocations (D)

$$F(L) = A_P(L) \cdot A_S(L) \cdot A_D(L). \tag{10.13}$$

Within the framework of the kinematical theory, the Warren-Averbach method [14, 15, 16] and the Krivoglaz-Wilkens method [17, 18, 19] can provide a quantitative description of the imperfect crystalline structure of the films:

$$F(L) = \exp(-L/T) \cdot \exp(-KL^2 < \varepsilon^2 >) \cdot \exp(-BL^2 \ln(L_0/L)) \tag{10.14}$$

and

$$- \ln F(L) = \frac{L}{T} + L^2[K < \varepsilon^2 > + B \ln(L_0/L)]. \tag{10.15}$$

L is the domain of definition of the experimental line profile, $T = T(hkl)$ the effective particle size, $B = B(hkl)$ a factor proportional to the mean total dislocation density ρ_ν, L_0 a length proportional to the core radius r_0 of the strain field of a dislocation, and $< \varepsilon^2 > = < \varepsilon^2(hkl) >$ the mean square microstrain due to internal stress, while K is a reflection constant. In the traditional Warren-Averbach analysis the linear part of the graph (Warren-Averbach plot)

$$- \frac{\ln F(L)}{L} = \frac{1}{T} + KL < \varepsilon^2(L) > \tag{10.16}$$

allows the determination of the domain size from the intercept and the mean square strain $< \varepsilon^2 >$ from the slope.

With the results of Krivoglaz and Wilkens the dislocation densities can be estimated by using the so-called Krivoglaz-Wilkens plot

$$- \frac{\ln F(L)}{L^2} = \frac{1}{TL} + (K < \varepsilon^2 > + B \ln L_0) - B \ln L. \tag{10.17}$$

From the slope of the linear branch at higher L the quantity $B(hkl)$ and the dislocation density ρ_ν can be calculated.

Fig. 10.7 shows the Warren-Averbach plots of four TiN$_x$ films, deposited at 0, −30, −50, and −80 V. There are appreciable differences in the microstructure between the 0 voltage sample and the films deposited at negative U_{sub}. From 0 V to −30 V domain sizes and microstrains increase considerably, attaining a maximum at −30 V and diminish again at −50 V

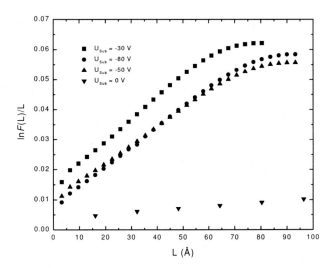

Figure 10.7: Warren-Averbach plot of TiN$_x$ films deposited at constant N$_2$ flow and for different substrate voltages U_{sub}.

and −80 V. With increasing negative substrate voltages XPS measurements show an increased amount of incorporated nitrogen in the films although the supply of nitrogen in the discharge remains constant. Table 10.2 contains the calculated microstructure parameters: domain sizes, microstrains, dislocation density, interstices concentration. The calculation of the interstices concentration from Eq. 10.11 is based on the difference between the measured lattice constant a_o and the lattice constant which represents the nitrogen content in the films measured by XPS. From both the results of nitrogen flow experiments (see section 10.4.1) and the negative substrate voltage experiments it can be concluded that the ion concentration as well as the ion energy determine the film properties. Three effects of ion energy influence film growth and film properties. The formation of activated atoms by interaction of neutrals and charged

Table 10.2: Physical parameters of TiN$_x$ deposited at different substrate voltages.

	U_{sub} (V)			
	0	−30	−50	−80
N/Ti	0.57	0.62	0.68	0.85
d_{111} (Å)	2.419	2.443	2.447	2.448
T (Å)	253	67	99	130
$\sqrt{<\varepsilon^2>} \times 10^3$	4.45	14.9	13.6	14.2
ρ_ν (10^{11} cm^{-2})	2.7	36	28	24
C_i (%)	reference	1.9	1.67	0.39

particles determines the nitrogen content x on the octahedral sites in TiN$_x$. Direct ion impacts could be responsible for the lattice imperfections. At a bias of -30 V the defects have a maximum. Domain size is smallest, and microstrain and dislocation density have a maximum. Increasing ion energy leads to an enhanced mobility of surface species and/or causes an annealing of lattice defects. This energy causes surface atomic displacement but is not enough to cause subsurface damage.

10.4.3 Calculation of depth profiles in plasma-deposited Ti/TiSi films

Extensive analysis of particle and energy flux to the substrate during titanium deposition in a hollow cathode arc evaporation device (HCAED) showed that at low discharge power Ar ions are the dominating species, whereas with increasing discharge power or deposition rate Ti ions become important [20, 21]. At high discharge power most of the ions are Ti ions. A strong increase of the Ti ion fraction can be observed especially up to a discharge power of about 4 kW. The energy flux density of the Ti ions (Ti$^+$, Ti^{2+}) divided by the deposition rate (energy per layer volume) depending on the discharge power seems to be decisive for the film properties, because this value increases by one order up to about 4 kW and remains constant for higher discharge power.

In order to obtain detailed information on the growth mode of Ti deposited on Si $< 100 >$ wafers at discharge powers higher and lower than 4 kW, GIXD techniques in combination with XPS were used. Fig. 10.8 shows GIXD patterns of Ti-films at an incident angle $\omega = 0.5°$. The main reflections of α–Ti of the sample deposited at high discharge power can be seen in Fig. 10.8a. By contrast, the diffraction patterns of the thin film deposited at low discharge power show strongly broadened reflections (Fig. 10.8b). The film thickness is 70 nm in both samples. If very low incidence angles are chosen, α–Ti reflections become observable. The separation of the overlapping peaks is possible using pseudo Voigt-functions. Fit procedures with only three α–Ti peaks, which are sufficient for sample 1, give unsatisfactory results for sample 2. The quality of the fit for sample 2 improves with an additional peak at $2\Theta = 37.60°$ (Fig. 10.8b). This peak is the most intensive of TiSi [22]. It is obvious that the effective intensities of the titanium peaks and the titanium silicide peak change in an opposite manner, from the silicon substrate surface to the titanium film surface. That means there are gradients in chemical composition of the deposited film.

If gradients of chemical, physical or structural properties exist between the film layers on the substrate and the outermost layers, the x-ray parameters measured represent only absorption weighted effective parameters, different from the true values in the film. If the thickness d of the film is known, the true values in the film can be calculated from the effective x-ray data. In general, for absorption weighted means y_{eff} of x-ray parameters, the expression

$$y_{eff}(T_p) = \frac{\int\limits_0^d y(x)\exp(-\frac{x}{T_p})dx}{\int\limits_0^d \exp(-\frac{x}{T_p})dx} \tag{10.18}$$

is valid. The aim of a layer analysis is the calculation of true values $y(x)$ from the effective parameters y_{eff}. The $y_{eff}(T_p)$–values are measured by varying the incidence angle ω in the

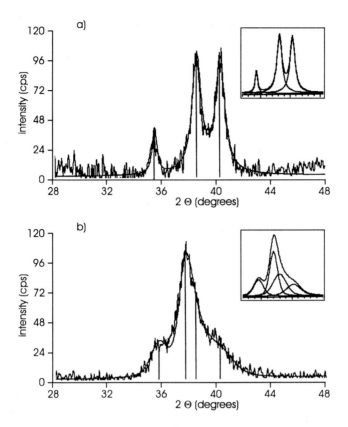

Figure 10.8: GIXD pattern of Ti-films deposited at (a) high (sample 1) and (b) low discharge power (sample 2) at an incident angle $\omega = 0.5°$. Inserts show the deconvolution of the spectra into individual peaks.

GIXD. Here $T_p = (\mu z)^{-1}$ is the penetration depth. Common solutions of Eq. 10.18 are possible, if the $y(x)$ are approximated by Fourier series or by polynominals as used in our investigations [23]:

$$y(x) = \sum_m a_m x^m .$$ (10.19)

The effective parameters for a T_p-variation are then

$$I_{eff}(T_p) = \frac{\sum_m a_m T_p^m \exp(-d/T_p) Q}{\exp(-d/T_p) - 1} - \frac{\sum_m a_m T_p^m m!}{\exp(-d/T_p) - 1}$$ (10.20)

with

$$Q = \left(\frac{d}{T_p}\right)^m + m\left(\frac{d}{T_p}\right)^{m-1} + m(m-1)\left(\frac{d}{T_p}\right)^{m-2} + \ldots + m! .$$ (10.21)

Measurements of effective values, as integral intensities or centers of gravity at different penetration depths give the desired a_m of the polynomial by solving linear equation systems.

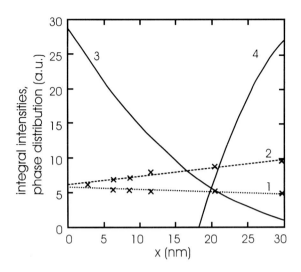

Figure 10.9: Effective integral intensities of the TiSi (210) peak (curve 1) and the α–Ti (100) peak (curve 2), and the calculated distribution of the crystalline phases TiSi (curve 3) and Ti (curve 4).

From the effective integral intensities of the TiSi (210) peak and the α–Ti peak (100) the true distribution of the phases is calculated. In general, the integral intensities are directly proportional to the x-ray irradiated sample volume. If the density is known, the mass can be calculated. By means of the theoretical structure factor (see equation 10.1 in section 10.2.1) calculated for the TiSi (210) peak and α–Ti (100) the corrected intensity ratios are obtained. In Fig. 10.9 the integral intensities y_{eff} of both reflections after correction with the theoretical structure factor are shown (curves 1 and 2). These corrected effective integral intensities are the input values for solving equation (10.20). A fourth degree polynomial was used. The film information depth T_p was estimated for a mixture of Ti and TiSi. Curves 3 and 4 (Fig. 10.9) show the true distribution of the TiSi and Ti phases. Diffraction measurements can only give information about crystalline materials. TiSi is the unique crystalline phase on the substrate surface. Its portion decreases to a layer thickness of about 30 nm and then seems to stay nearly constant up to the film surface. The crystalline α–Ti is observable at first above a layer thickness of 18 nm and then strongly increases. The true particle sizes can be calculated in the same manner from integral breadth.

These results show that the film properties, as chemical composition and particle size, are strongly influenced by the deposition conditions. A strong relation between the energy of the ions striking the substrate and the film composition was detected. The whole energy of the ions does not seem to be decisive for the layer composition, only the Ti/Ar ratio is of importance. At high discharge powers (> 4 kW) most of the ions are titanium ions. Only α–Ti is observed. At low discharge powers (< 4 kW) the Ar ions are the dominating species. A discharge power of 3.4 kW produces films of TiSi and α–Ti. XPS investigations confirm the GIXD results. Detailed XPS spectra of the Si–2p peak and their time evolution during

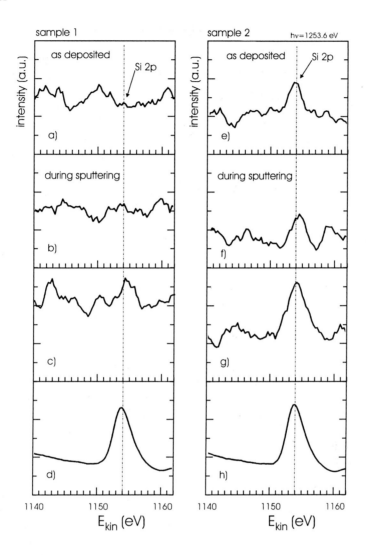

Figure 10.10: XPS of Ti on Si deposited at high discharge power (a)–(d) and low discharge power (e)–(h). Spectra (a) and (e) as deposited; (b)–(d) and (f)–(h) time evolution during sputtering.

sputtering are shown in Fig. 10.10. The XPS measurements indicate that Si can found in the bulk volume of Ti films deposited at low discharge power (sample 2). In contrast to sample 2, a silicon peak was found in sample 1 only after sputtering almost all of the titanium.

10.4.4   Structural studies of thin ITO films

Macroscopic film properties are strongly influenced by the film microstructure, and the microstructure itself depends on deposition conditions. Very often only correlation between deposition conditions and macroscopic properties are described. For instance, in ITO films the tin doping concentration is linked to the electrical conductivity, but there are only a few investigations about the tin insertion mechanism into the In_2O_3 bixbyite structure.

Pure In_2O_3 crystallizes in the bixbyite structure type (space group I a 3) with two non-equivalent In (In1, In2) sites. In this type only 3/4 of the tetrahedral sites are filled with oxygen. Each In1 is coordinated sixfold in a regular oxygen octahedron, each In2 is surrounded by six oxygen atoms at three different distances [24]. The ratio of In1:In2 is 1:3.

The model developed for tin insertion into In_2O_3 by Franck and Köstlin results from an analysis of the carrier concentration and additional measurements of the lattice constants dependent on the doping concentration [25]. Parent et al. used extended x-ray absorption fine structure (EXAFS) investigations and found that the In atomic environment was modified by the Sn doping [26].

The application of the Rietveld refinement to GIXD patterns provides information about the crystallographic structure of thin films. Lattice constants, atomic positions, and bond lengths and bond angles can be calculated. The Rietveld refinement is a well-known method used to calculate structural parameters of polycrystalline bulk materials from diffraction patterns in Bragg-Brentano geometry.

During a Rietveld refinement, atomic positions, temperature and occupancy factors, scale factors, unit cell parameters and background coefficients, along with profile parameters describing peak widths and shape, are varied in a least-squares procedure until the calculated powder pattern, based on the structure model, best matches the observed pattern:

$$\sum_i \frac{(y_i - y_{ci})^2}{y_i} \to min,$$
(10.22)

where y_i is the measured, y_{ci} the calculated value. The y_{ci}–values are calculated according to

$$y_{ci} = S \sum_i M_i PL_i |F_i|^2 \Phi(\Theta_i) p_i + y_{bi}$$
(10.23)

with the scale factor S, the multiplicity M_i, the polarization-Lorentz factor PL_i, the structure factor F_i, the profile function $\Phi(\Theta)$, the texture function p_i, and the background correction y_{bi}. The application of the Rietveld method to GIXD patterns requires consideration of all the peculiarities of this technique causing significant deviations from the intensities recorded in Bragg-Brentano geometry:

(i) a peak shift $\Delta\Theta$ at small incident angles due to refraction and reflection [27],

$$2\Delta\Theta = \omega - \sqrt{\omega^2 - \Theta_c^2},$$
(10.24)

where Θ_c is the critical angle of total reflection of the upper layer,

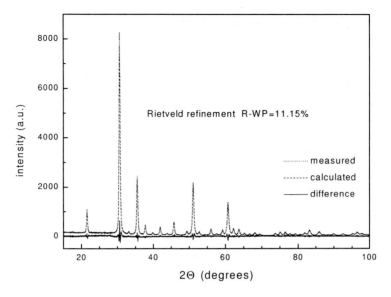

Figure 10.11: Rietveld refinement pattern of tin-doped indium oxide.

(ii) a change of the x-ray absorption behavior due to the asymmetric beam geometry

$$\frac{I}{I_0} = \frac{A}{\mu} \left(1 + \frac{\sin\omega}{\sin 2\Theta}\right)^{-1}, \tag{10.25}$$

where I and I_0 are the intensities of diffracted and incident x-rays, A is the irradiated sample area, and μ the linear absorption coefficient,

(iii) the influence of the finite film thickness [28], and

(iv) a peak shift due to macroscopic stress.

If one takes the points (i)–(iv) into account, the Rietveld results can be matched [29]. In_2O_3 powders doped with different SnO_2 concentrations deposited onto glass substrates were investigated. The Rietveld program WYRIET 3.0 was used for the refinement [30]. Cubic ITO films without SnO_x phases were deposited (Fig. 10.11). The In:Sn ratios were obtained from XPS measurements. All deposited films were polycrystalline. They did not show any preferred orientation. The crystallographic parameters of the films changed with the tin concentration. With increasing doping level the x-ray peak positions shifted and the line profiles became broader. Lattice constants and bond lengths in the doped indium oxide lattice were calculated. The Rietveld refinement results are shown in Fig. 10.12 and 10.13. Small tin amounts (up to 5% Sn) cause a decreasing lattice constant as well as a decrease of the In1–In2 bond length. The In1–O bond length increases and the distortion of the non-regular In2–O octahedron becomes stronger. This distortion causes a decrease of the octahedron volume and is responsible for the decrease in the lattice constant and the In1–In2 distance. Doping concentrations between 5 and 10 % Sn effect an increase in lattice constants and metal-metal

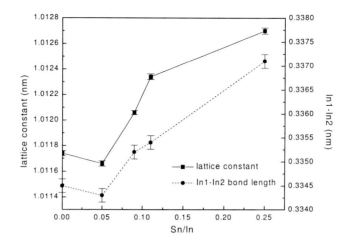

Figure 10.12: Lattice constants and In1–In2 bond length versus tin concentration.

bond lengths. The In1–O distance decreases, and the In2–O octahedron distortion is reduced. At larger doping concentrations a further increase in the lattice constant and the In1–In2 bond length was observed. The In1–O distance increases. The oxygen octahedrons surrounding the In2 atoms become more and more distorted. From these In2–O bond lengths it was calculated that the volumes of the distorted octahedrons were larger. On the basis of the atomic and defect structure results two mechanisms for the tin insertion into the indium oxide lattice can be discussed. The radius of the Sn^{4+} ions (0.71 Å) is smaller than that of the In^{3+}

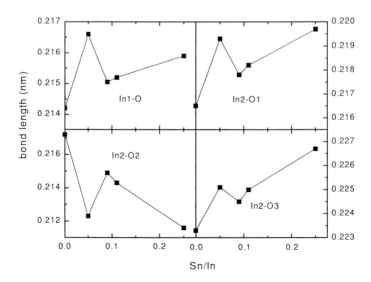

Figure 10.13: In–O bond lengths versus tin concentration.

ions (0.80 Å) [31]. Therefore a Sn insertion on regular In positions should decrease the lattice constant and bond lengths. However, decreasing lattice constants were observed only for small tin concentrations up to 5%. This decrease can only be explained, if the tin ions are assumed predominantly to occupy In2 positions. The stoichiometric balance can be expressed as $O_i^{..} \leftrightarrow 1/2O_2 + 2e^{.}$, and the conductivity increases appreciably due to the free electrons. At larger doping concentrations (5–10 % Sn) the lattice constant increases rapidly. Then the tin ions occupy In1 and In2 sites. The stoichiometric balance can be maintained only, if additional oxygen is inserted into the lattice. These oxygen atoms can fill the empty tetrahedral sites in the bixbyite structure. They tend mainly to gather near the Sn ions on In2 sites and can form strongly bound O–Sn2–O units. This occupation mechanism does not produce additional free electrons. The increase in conductivity is only small. Larger doping concentrations cause a further oxygen reception. Apart from the O–Sn2–O units O–Sn1–O arrangements can also occur. This effects a further distortion of the bixbyite structure. The microstrain increases, and the domain size decreases. The resistivity behavior does not change significantly.

10.4.5 Investigation of plasma-deposited ITO films

Although ITO deposition by magnetron sputtering is a very mature technique, our basic understanding of this complex process is incomplete. Emphasis is placed on the relation between microstructure and plasma parameters. With GIXD and GIXR supported by AFM we have investigated the effect of several oxygen flows and negative substrate voltages on film formation processes of ITO layers deposited by a DC magnetron. The experimental set-up is described in Ref. [32]. The ITO films were deposited on Si(100) wafers. A metallic In/Sn (90/10) target was used. The flow of the reactive gas oxygen was varied between 0 and 2 sccm. Bias voltages between 0 and −100 V were used. For the GIXR investigations the deposition time was kept constant (30 s). For the GIXD investigations the deposition time was varied to reach constant layer thickness of about 50 nm. Morphological investigations were carried out with atomic force microscopy (AFM) in contact mode.

10.4.5.1 Influence of oxygen flow during film deposition

During the deposition, the oxygen flow influences chemical and phase composition, film density, growth rate, and coating morphology. Deposition without oxygen forms crystalline metallic In/Sn films (Fig. 10.14). No preferred orientation was observed. The intensity ratios are similar to the pattern of polycrystalline bulk In [33]. With increasing oxygen flow only small amounts of crystalline metallic In/Sn can be detected in the layers. The mean observable domain sizes reduce from 26 to 7 nm. Oxygen flows > 0.5 sccm prevent the formation of a crystalline phase. The dependence of film density and growth rate (as determined by GIXR) on oxygen flow in the deposition chamber are presented in Fig. 10.15. The deposition rate decreases with increasing oxygen partial pressure. Without oxygen the growth rate is 0.6 nm/s; it decreases to 0.2 nm/s for an oxygen flow of 2 sccm. Simultaneously the layer density increases from about 4.5 g/cm³ to 7.2 g/cm³. Oxygen flows larger than 1 sccm lead to film densities similar to the x-ray density of In and In_2O_3 (7.28 g/cm³ and 7.12 g/cm³, respectively). The GIXR results show a high amount of voids in the metallic films. Increasing oxygen flows yield more compact layers. From the GIXR measurements additionally a drop

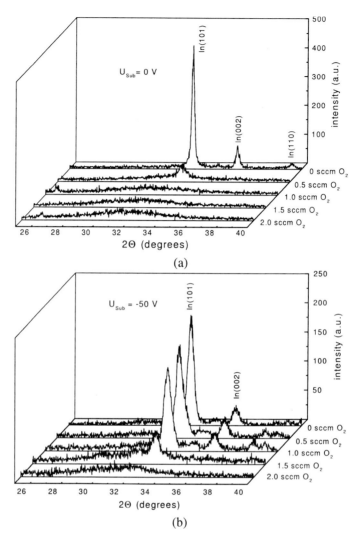

Figure 10.14: Diffraction pattern of films deposited at different oxygen flows and for two substrate voltages. (a) $U_{sub} = 0$ V; (b) $U_{sub} = -50$ V.

of the substrate roughness follows (Fig. 10.16). These results are taken from 5–15 nm thick films. The assumption of voids in the more metallic films is supported by AFM investigations. AFM micrographs of these films demonstrate clearly that grains become smaller with an increasing oxygen flow (Fig. 10.17). The metallic film shows large grain sizes forming a rough film surface. Deposition with higher oxygen partial pressure causes a smooth surface where grain sizes are not clearly observable. XPS measurements confirm the existence of oxygen besides indium and tin. These experimental results show (i) that the coatings become x-ray-amorphous with increasing O_2 flow, (ii) that these amorphous layers seem to contain

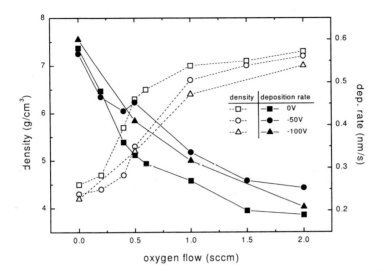

Figure 10.15: Density (open symbols) and deposition rate (closed symbols) of samples de-posite it substrate voltages U_{sub} = 0 V ($\Box$, $\blacksquare$), –50V (o, •) and –100V ($\triangle$, $\blacktriangle$) vs. oxygen flow (Lines only to guide the eye.).

the ITO phase, and (iii) that the nucleation rate of ITO as well as the crystallite growth rate of the metallic In/Sn in an opposite manner is dominated by the oxygen flow.

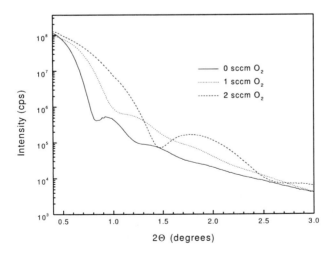

Figure 10.16: GIXR measurements of films deposited at U_{sub} = 0 V and different oxygen flows. 0 sccm O_2 flow: thickness 15.1 nm, roughness 1.52 nm; 1 sccm O_2 flow: thickness 11.5 nm, roughness 0.98 nm; 2 sccm O_2 flow: thickness 7.1 nm, roughness 0.75 nm.

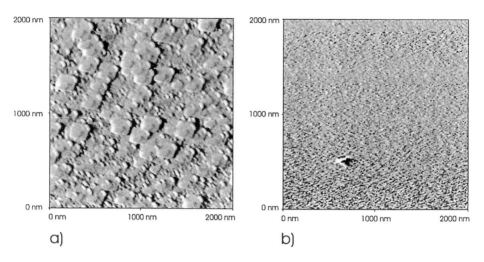

Figure 10.17: Atomic force micrographs (AFM) of samples deposited (a) without oxygen, and (b) at an oxygen flow of 1.5 sccm; $U_{sub} = 0$ V, deposition time 30 s.

10.4.5.2 Influence of the negative substrate voltage

In the experiments described above both the particle composition of the plasma and the energy flux to the growing film were changed by increasing the O_2 flow. To obtain further information on the effect of energy flux due to ion energy, the substrate potential U_{sub} was varied from 0 V to -100 V at the same oxygen flows. The additional energy flux to the growing films (due to the increased ion energy) can be responsible for the changes observed in diffraction patterns and in reflectometry measurements and thus in the film properties. Fig. 10.14 shows the diffraction pattern of ITO films deposited at 0 V and -50 V. There are notable differences in the microstructure between the 0 V samples and the films deposited at -50 V. An increased amount of metallic In/Sn was detected in films deposited with -50 V substrate voltage although the supply of oxygen in the discharge remains the same as in the 0 voltage experiments. Film density and growth rate show that the curvature is similar to the curve which was measured without negative substrate voltage. However, the measured densities have smaller values, particularly in the middle field (0.5–1.5 sccm O_2–flow), and the films grown at 0.5 sccm to 2 sccm oxygen flows exhibit growth rates higher for -50 V and -100 V substrate voltages.

10.4.5.3 Post-deposition annealing

During post-deposition heat treatment the formation of crystalline indium tin oxide in the films can be observed. Fig. 10.18 shows the phase transformation of a metallic In/Sn film into crystalline indium tin oxide. After an annealing time of 1 h at 200 °C no In/Sn is detectable with GIXD. The new peaks can be identified as reflections from the In_2O_3 phase [34]. The growth rate of the crystalline ITO depends strongly on the deposition conditions. Information on the crystallization process itself can be obtained from the increase of the In_2O_3 (222) peak intensity with the annealing time (Fig. 10.18). Two mechanisms determine the formation of

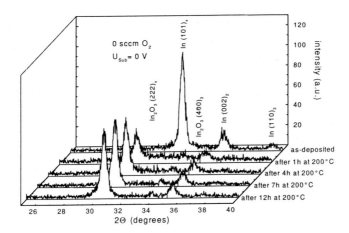

Figure 10.18: Phase transformation of metallic In/Sn into crystalline ITO at 200 °C.

crystalline indium oxide. A fast crystallization and a diffusion limited reaction were observed.

10.4.6 *In situ* studies of diffusion and crystal growth in plasma-deposited thin ITO films

In the previous section 10.4.5 we reported on the formation and growth process of plasma-deposited ITO films. This sections deals with *in situ* grazing incidence x-ray diffractometry to study kinetic processes as diffusion of oxygen into metallic In/Sn films, and the amorphous to crystalline transformation of ITO films [35, 36]. Diffusion information is required to estimate complete homogenization time in thin ITO films at a given temperature, and of course is also of fundamental interest. The crystallization of amorphous ITO is of basic interest since it occurs rapidly at very low temperatures. X-ray diffraction is a well-suited method to study diffusion in thin films. Because of the limited penetration depth of x-ray beams one can only study zones that are near the free surface. Murakami et al. have published an excellent review of diffusion studies using x-ray diffraction involving thin films [37]. While a large number of systems have been investigated via diffraction techniques, no *in situ* study has been reported until now. Solid state crystallization studies are a promising field for high-temperature diffractometry. The amorphous to crystalline transformation kinetics in thin ITO films by monitoring the x-ray integral intensity has not been described before. An analytical model was developed for the investigation of the diffusion process. An evaluation of the activation energies for diffusion and crystallization was made.

10.4.6.1 Determination of kinetic parameters

The time dependence of the x-ray integral peak intensity of a particular reflection during isothermal annealing was analyzed to determine the diffusion coefficients as well as the activation energy of the processes. Since a single crystalline phase is formed, measurements and data analysis of the x-ray intensities are relatively straightforward. We take into account a homogeneous layer with the thickness d which consists of a material n (e.g., In). A phase transformation to phase m (e.g., In_2O_3) takes place through diffusion and crystallization. This process is limited by diffusion, i.e., diffusion is the slower process. The contribution of a volume $dV = A dx$ in a depth x_m to the measured integral intensity of phase m is given by

$$dI_m(t) = B_m K_0 I_0 A_m \frac{C}{C_0} A dx ,$$

(10.26)

where B_m is the theoretical intensity factor, K_0 an apparatus constant, I_0 the intensity of the incident x-ray beam, and A_m given by Eq. 10.2. The ratio C/C_0 is the fraction of phase m in the volume dV. Integration of Eq. 10.26 yields the time-dependence of the x-ray integral intensity of a particular peak of the phase m

$$I_m(t) = B_m K_0 I_0 A \int_0^d \frac{C(x,t)}{C_0} A_m dx$$

(10.27)

The constants in Eq. 10.27 can be calculated from the maximum intensity $I_{max} = I_m(t \to \infty)$. To solve Eq. 10.27 the concentration profile of phase m at any time has to be known. As mentioned before, the observed phase transformation is limited by diffusion. Here we assume a one-dimensional diffusion perpendicular to the film surface. This is justified because the domain sizes are large compared to the small layer thickness. The boundary conditions are: constant concentration C_0 at the layer surface, and no diffusion into the substrate. From Fick's second law we get

$$\frac{C(x,t)}{C_0} = 1 - \frac{4}{\pi} \sum_{n=1}^{\infty} \frac{(-1)^{n-1}}{2n-1} \cos\left(\frac{\pi}{2}(2n-1)\frac{d-x}{d}\right) \exp\left(-\frac{\pi^2}{4}(2n-1)^2 \frac{Dt}{d^2}\right)$$

(10.28)

with D the diffusion constant. For a given film thickness the only adjustable parameter in the peak intensity versus time plot (Eq. 10.27) is the D value. By comparison of the measured with the calculated peak intensities, the D value was adjusted until the best agreement between the two curves was obtained. The kinetics of the crystallization process was described by the Johnson-Mehl-Avrami equation [38]:

$$y_m(t) = 1 - \exp\left(-(k't)^n\right) ,$$

(10.29)

where n is the reaction order, and k' is the rate constant. The Johnson-Mehl-Avrami equation 10.29 describes the time-dependence of the amorphous to crystalline transformation in terms of the crystalline fraction y_m. A value of n between 1 and 2 indicates one-dimensional growth with rapid nucleation, while two-dimensional growth generally has a growth mode parameter between 2 and 3 [38]. The growth mode parameter n can be obtained from the slope of a plot of $\ln(-\ln(1-y_m))$ versus $\ln(t)$, or from a fit procedure employing Eq. 10.29. The activation energies for both the diffusion and the crystallization behavior were determined in the usual way from the Arrhenius plot in which the logarithm of the diffusion coefficient, or for the crystallization process the logarithm of the rate constant k' from Eq. 10.29, is plotted against the inverse of the temperature in Kelvin.

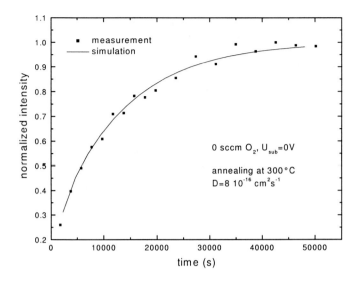

Figure 10.19: The measured normalized x-ray integral intensities (■) and the calculated intensity values (dashed line) of a film deposited at 0 sccm O_2 and 0 V bias voltage versus time.

10.4.6.2 Diffusion

Two sets of annealing experiments were carried out. At first, films deposited with oxygen flows of 0, 0.5, 1.0, 1.5, and 2 sccm and bias voltages of 0, –50V, and –100V were annealed at a temperature of 250 °C. The time for the whole diffusion process was about 10 hours. The diffusion constants were determined as described above. Fig. 10.19 shows the measured normalized x-ray integral intensities and the calculated intensity values of a film deposited at 0 sccm O_2 and 0 V bias voltage versus time. Table 10.3 displays the oxygen flows, the bias voltages, the diffusion constants D, and the factor f that represent the amorphous ITO fraction

Table 10.3: Diffusion constant D and fraction f of amorphous ITO in the as-deposited films at $T = 250\,°C$, calculated from Eq. 10.27.

O$_2$–flow	U_{bias} (V)					
	0		−50		−100	
	D (cm^2/s)	f	D (cm^2/s)	f	D (cm^2/s)	f
0	6.0×10^{-16}	0	1.3×10^{-15}	0	9.4×10^{-16}	0
0.5	5.5×10^{-16}	0	1.7×10^{-15}	0	9.1×10^{-16}	0
1.0	6.0×10^{-16}	0.75	1.4×10^{-15}	0.3	9.6×10^{-16}	0.5
1.5		1	1.5×10^{-15}	0.5	9.6×10^{-16}	0.65
2.0		1	1.6×10^{-15}	0.7	9.6×10^{-16}	0.8

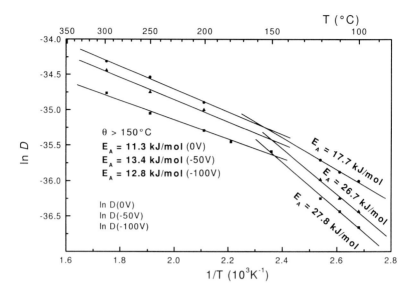

Figure 10.20: Arrhenius plot for the diffusion process. For each curve, two different activation energies were found for $T < 150\,°C$ and $T > 150\,°C$.

in the investigated films. It is obvious that D depends on the bias voltage but neither on the oxygen flow nor the amorphous ITO fraction. The diffusion constants increase considerably from 0 to –50 V reaching a maximum at –50 V and diminishing again at –100 V. In a second set of experiments the activation energy of the diffusion process was determined in films deposited without oxygen. Films deposited with substrate voltages of 0 V, –50 V and –100 V were annealed in the temperature range from 100 °C to 300 °C. Plotting $\ln D$ versus the inverse temperature in Kelvin yields the Arrhenius plot as shown in Fig. 10.20 from which the activation energies were estimated.

Obviously the activation energies E_A depend on the temperature, and for temperatures $T < 150\,°C$ also on the bias voltage. The activation energy E_A for temperatures $T < 150\,°C$ suggests that different defect structures caused by the bias voltages (plasma parameters) lead to distinguishable diffusion mechanisms [39]. For temperatures $T > 150\,°C$ nearly the same activation energies were found, independent of the bias voltage. The melting point of indium is 156 °C. We assume that the layer is partially molten during annealing at higher temperatures. The activation energy for diffusion in liquids is expected to be smaller than in solids.

10.4.6.3 Crystallization

The crystallization process was investigated in complete amorphous ITO films deposited at an oxygen flow of 2 sccm and without negative substrate voltage (see table 10.3). The time for a total amorphous to crystalline transformation is about 20 min. This is 1/30 of the time necessary for the whole diffusion process. To determine the activation energy crystallization experiments were made at three different temperatures 220 °C, 230 °C and 240 °C. In Fig.

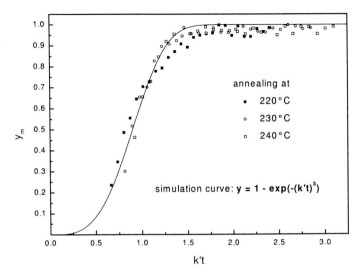

Figure 10.21: Johnson-Mehl-Avrami power law for the reaction order $n = 3$ (solid line) in comparison with experimental values recorded at 220 °C (■), 230 °C (∘) and 240 °C (□).

10.21 the Johnson-Mehl-Avrami power law for the reaction order $n = 3$ (solid line) and the experimental values recorded at 220 °C, 230 °C and 240 °C are shown. Reaction orders between 2.4 and 3.0 were found representing a two-dimensional growth. The activation energy for the crystallization process was estimated from an Arrhenius plot of $\ln k'$ versus T^{-1}. The

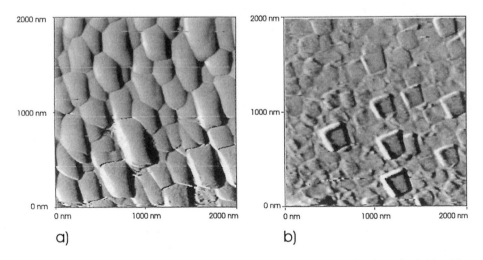

Figure 10.22: Atomic force migrographs (AFM) of annealed samples deposited (a) without oxygen and (b) at an oxygen flow of 2 sccm.

obtained activation energy is E_A = 74 kJ/mol. In all annealing experiments the finally formed films consist of crystalline ITO in the bixbyite structure type. However, corresponding AFM images (Fig. 10.22) show differences in surface morphology. Large grain sizes with rounded edges are observed, if the ITO growth is diffusion limited (Fig. 10.22a, for 0 sccm O_2–flow during deposition), whereas cubic grains were formed in the fast crystallization process for films deposited with 2 sccm O_2–flow (Fig. 10.22b). The surface is very smooth. GIXR measurements show that before and after annealing film thickness and film density do not change. That means the grain distributions in the thin film remain preserved also at temperatures higher than 150 °C. Small ITO layers surround the metallic In drops and prevent the formation of large liquid regions. The AFM micrograph (Fig. 10.22a) confirms this assumption.

10.5   Conclusions

Grazing incidence x-ray diffractometry (GIXD) and grazing incidence x-ray reflectometry (GIXR) in combination with x-ray photoelectron spectroscopy (XPS) and atomic force microscopy (AFM) were used to study non-destructively the microstructure of thin films as well as the influence of plasma parameters on the microstructure and film formation processes. This chapter was aimed to demonstrate the effectiveness of x-ray methods to this purposes, using traditional XRD equipments like parallel beam (GIXD) and Bragg-Brentano (GIXR) geometry. Six types of analyzing methods have been discussed, which are of interest to thin film investigations. Phase analysis, defect structure analysis, analysis of microstructural gradients, and crystal structure investigations with the Rietveld method are useful for microstructure characterization, whereas film formation analysis, and *in situ* studies of kinetic processes mainly are concerned with the growth mechanisms.

Acknowledgements

The authors like to thank Dr. C. Eggs, Dr. J. Klimke, Ms. Dipl.-Chem. A. Quade, Ms. Dipl.-Chem. M. Quaas, and Mr. T.M. Tun (M.Sc.) for their help on parts of the presented examples.

References

[1] L.G. Parrat, Phys.Rev. 95 (1954) 359.

[2] B. Lengeler, Microchim. Acta 1 (1987) 455.

[3] G.C. Smith, Surface Analysis by Electron Spectroscopy, Plenum Press: New York (1994).

[4] D. Briggs, M.P. Seah, Practical Surface Analysis, Vol. 1, Auger and X-ray Photoelectron Spectroscopy, John Wiley & Sons: New York (1990).

[5] J.F. Moulder, W.F. Stickle, P.E. Sobol, K.D. Bomben, Handbook of X-ray Photoelectron Spectroskopy, Perkin-Elmer Corporation (1992).

[6] T.L. Barr, Modern ESCA, CRC Press (1993).

[7] M.F. Ebel, H. Ebel, F. Olcaytug, Advances in X-Ray Analysis, 35 (1992) 857.

[8] S. Tougaard, H.S. Hansen, Surf. Sci. 236 (1990) 271.

[9] H. Wulff, C. Eggs, J. Vac. Sci. Technol. A 15 (1997) 2938.

[10] A. Lunk, Vacuum 41 (1990) 165.

[11] P. Ehrlich, Z. Anorg. Chem. 259 (1949) 1.

[12] P. Paufler, G.E.R. Schulze, Physikalische Grundlagen Mechanischer Festkörpereigen-schaften I, Akademie Verlag: Berlin (1978).

[13] R. Stokes, Proc. Phys. Soc. 61, 382 (1948).

[14] B.E. Warren, Prog. Metal Phys. 9, 147 (1969).

[15] P. Klimanek, Freiberger Forschungshefte B 265, 76 (1988).

[16] P. Klimanek, Mat.Sci. Forum 79-82, 73 (1991).

[17] M.A. Krivoglaz, Theory of X-Ray and Neutron Diffraction by Imperfect Crystals, Kiev (1983).

[18] M. Wilkens, phys.stat.sol.(a) 2, 359 (1970).

[19] M. Wilkens, Cryst. Res. Techn. 11, 1159 (1976).

[20] H. Steffen, H. Kersten, H. Wulff, J. Vac. Sci. Technol. A 12 (1994) 2780.

[21] H. Steffen, H. Wulff, Thin Solid Films 263 (1995) 18.

[22] ICDD PDF-2 17-424.

[23] H. Wulff, J. Klimke H. Steffen, C. Eggs, Thin Solid Films 261 (1995) 25.

[24] M. Marezio, Acta Cryst. 20 (1966) 723.

[25] G. Franck, H. Köstlin, Appl. Phys. A 27 (1982) 197.

[26] P. Parent, H. Dexpert, G. Tourillon, J.M. Grimal, J. Electrochem. Soc. 139(1992) 276.

[27] T. Takayama, Y. Matsumoto, Adv. X-ray Anal. 33 (1990) 91.

[28] W. Pitschke, Mater. Sci. Forum 228-231 (1996) 178.

[29] M. Quaas, C. Eggs, H. Wulff, Thin Solid Films, 332 (1998) 277.

[30] J. Schneider, Acta Cryst. A 43 (1987) 295.

[31] R.D. Shannon, Acta Cryst. A 32 (1976) 751.

[32] H. Wulff, M. Quaas, H. Steffen, Thin Solid Films 355-356 (1999) 395.

[33] ICDD PDF-2 5-642.

[34] ICDD PDF-2 4-416.

[35] H. Wulff, M. Quaas, H. Steffen, R. Hippler, Thin Solid Films 377-378 (2000) 418.

[36] M. Quaas, H. Steffen, R. Hippler, H. Wulff, Surface Science 454-456 (2000) 790.

[37] M. Murakami, A. Segmüller, K.-N. Tu, In Analytical Techniques for Thin Films, Eds. K.-N. Tu, R. Rosenberg, Treatise Mater. Sci. Technol. 27, Academic Press: New York (1988), p. 201.

[38] Y. Shigesato, D.C. Paine, Thin Solid Films 238 (1994) 44.

[39] M. Ohring, Material Science of Thin Films, Academic Press: New York (1992), p.361.

11 Plasma sources

Martin Schmidt and Hans Conrads

Institut für Niedertemperatur-Plasmaphysik, Friedrich–Ludwig–Jahn–Straße 19,
D–17489 Greifswald, Germany

11.1 Introduction

The low temperature plasmas treated in this chapter first emerged as a small number of positive ions and negative electrons in a sea of neutral gas. These carriers of positive and negative charges require the input of a certain amount of power to be generated and sustained [1, 2]. The kind of energy can be different. Very common in technical applications is the use of electric or electromagnetic fields for sustaining an electrical discharge. The ignition of the discharge results from fields acting on some free electrons generated by cosmic rays or radioactivity. These electrons then pick up kinetic energy in the electric field between collisions with neutrals. The energy gain must be at least as large or even larger than the ionization energy of the neutral gas particles. Those charge carriers, which do not recombine, participate in further ionization processes. An avalanching creation of charge carriers starts and proceeds until loss processes, e.g., three-body recombination, radiative recombination, wall recombination, put the system into equilibrium. The discharge is sustained, and a plasma is created. Due to the small energy loss of the light electrons in elastic collisions with the heavy atoms and/or molecules, and the effective energy transfer in elastic ion-neutral collisions, the electron gas becomes hot in the cold neutral gas environment of the non-thermal plasma.

Another way to create a plasma is to have chemical compounds with pronounced exothermic reaction characteristics acting on each other as in hot flames. In this case the tail of the kinetic energy distribution of some of the atoms or molecules is large enough to trigger ionization, and a thermal plasma will be generated. Heating of a gas volume up to temperatures sufficient for plasma generation also occurs by adiabatic compression. For a short time dense non-ideal plasmas are formed [3]. The dense plasmas generated by wire explosions are a result of the thermal energy density originating from the Ohmic heating of the wire [4]. External heating of the walls of the discharge vessel enables by contact ionization the formation of plasmas (Q machines) in alkali vapors with their low ionization energy (e.g., 3.9 eV for Cs) [2]. Such plasmas are important for fundamental research.

The ignition and sustaining of a plasma is possible also by the injection of beams of particles as well as of bulk plasma into neutral gas or even into other plasmas. Neutral beams penetrating electric and magnetic fields without interaction are common tools for generating and sustaining a plasma and shaping its parameters in space and time. Here the energy of the beam has to match the required absorption length. Further plasmas can be created by adiabatic compression by rapidly increasing magnetic fields. This is an effective way to generate

Table 11.1: Applications of plasmas.

Surface Modification	Etching	Structuring (microelectronics, micromechanics) Cleaning (assembly lines)	
	Functionalization	Hydrophilization, Hydrophobization Graftability, Adhesability, Printability	
	Interstitial modification	Diffusion (bonding) Implantation (hardening)	
	Deposition	Change of properties	Mechanical (tribology) Chemical (corrosion protection) Optical (antireflecting and decorative coatings)
		Structuring	Microelectronics Thin film transistors Photovoltaics
		Architecturing	Crystallography (lateral diamonds) Morphology (scaffolds for cells)
Volume-related Transformation	Energy conversion	Electrical energy → electromagnetic radiation	Luminiscence lamps High pressure lamps Excimer lamps Gas lasers
		Nuclear energy → electrical energy	Fusion of DT
	Plasma chemistry	Transformation into specific compounds Clean-up of gases	Precursor production Excimer production Odors, Flue gases Diesel exhaust
Carrier functions	Electrical current Heat		Circuit breakers Spark gap switches Welding/Cutting arcs Plasma spray Thermoelectric drivers
Particle sources	Electrons, Ions, Neutrals, Radicals		

a plasma with a high ion (T_i) and a low electron (T_e) temperature. Ratios of $T_i/T_e = 10$ have been achieved [5].

Very different plasmas can be generated with a broad spectrum of properties with the size and detailed shape of the source determined by its respective applications. Some relevant examples for industrial involvement in plasma processing is the production of microelectronics and the clean-up of exhausts of cars and flue gases in power stations. Table 11.1 summarizes various applications of plasmas. Industrial applications require a high standard for the lateral homogeneity of the plasma, even for sources of linear dimensions of meters. Also, the temporal requirements of the plasma sources have interesting features. A wide range of time constants have to be considered in designing sources. Since most of the plasma induced processes rely on chemical reactions, which on purpose should not be balanced by their counter-reaction, a precise timing in starting and stopping the wanted reaction path is a prerequisite to create a non-equilibrium situation. Feeding power to the source and stopping it has to match the time constant of particular reactions. These time constants can vary from a tenth of a microsecond to fractions of seconds. Further, coupling of the electromagnetic power to a source is limited by the fact that licensed power supplies are available on the market for only a limited range of frequencies, e.g., 13.560 MHz, 27.120 MHz, 2.450 GHz. Finally, power amplification by squeezing more energy in shorter pulses is an effective measure to increase the instant power applied to the plasma without rising the operating costs significantly.

Many parameters for the lay-out of a plasma source as presented above are mingled due to the history of development of a particular source. Some of the sources are quite mature in terms of reliability and performance, and their design is inherited from generation to generation, although not all their design features are required for the application in question.

Plasma properties important for applications are summarized in the next paragraph. Essential plasma properties are briefly discussed for the specific application. After that a review is given of plasma production by electric fields and beams of electrons or photons together with practical examples. This review can be a selection only and deals with non-thermal low-pressure plasmas mostly.

11.2 Properties of non-thermal plasmas

Some plasma properties important for plasma sources will be dealt with in the following. A detailed review of plasma properties is given in chapter 1. For plasma sustaining the power absorption in the electric field is essential. In an electrostatic field E the power absorption P_{abs} per volume unit V reads [6]

$$\frac{P_{abs}}{V} = \frac{e^2 n_e}{m_e \nu} E^2 , \tag{11.1}$$

where e, n_e, and m_e are the charge, density, and mass of an electron, respectively, and ν is the electron-neutral collision frequency.

The power absorption P_{abs} in a high-frequency field $E(t) = E_0 \cos \omega t$, where ω is the angular frequency of the electromagnetic field and E_0 its amplitude, is given by [1, 6, 7]

$$\frac{P_{abs}}{V} = \frac{1}{2} \frac{e^2 n_e}{m_e \nu} \frac{\nu^2}{\nu^2 + \omega^2} E_0^2 . \tag{11.2}$$

In a magnetic field B the Lorentz force causes a circular motion of electrons e and ions i with the corresponding cyclotron frequency ω_c [1, 6]

$$\omega_c = \frac{e B}{m} , \tag{11.3}$$

where m denotes the mass of an electron (m_e) or ion (m_i). The radius r_c of this motion is given by

$$r_c = \frac{mv_\perp}{eB},$$ (11.4)

where $v_\perp$ is the velocity component perpendicular to B. In the presence of a magnetic field B perpendicular to the electric field, the power absorption changes to

$$\frac{P_{abs}}{V} = \frac{1}{4} \frac{e^2 n_e}{m_e \nu} \left(\frac{\nu^2}{\nu^2 + (\omega - \omega_{c,e})^2} + \frac{\nu^2}{\nu^2 + (\omega + \omega_{c,e})^2} \right) E_0^2,$$ (11.5)

where $\omega_{c,e}$ denotes the electron cyclotron frequency, Eq. 11.3 with $m = m_e$.

According to the dielectric properties of a plasma, electromagnetic waves with frequencies below the electron plasma frequency $\omega_{p,e}$ [8]

$$\omega_{p,e} = \left(\frac{e_0^2 n_e}{\varepsilon_0 m_e} \right)^{\frac{1}{2}}$$ (11.6)

will be reflected. Here ε_0 denotes the permittivity of vacuum. Thus, the electron density corresponding to the electron plasma frequency is called cut-off density. However, the skin effect enables to some extent the penetration of the wave into the plasma. The power absorption is limited to the skin sheath of thickness δ_s. For $\nu \gg \omega$ the skin depth δ_s is given by [9]

$$\delta_s = \sqrt{2} c \left(\frac{\varepsilon_0 m_e \nu}{e^2 n_e \omega} \right)^{\frac{1}{2}},$$ (11.7)

where c refers to the velocity of light.

In a non-thermal microwave (2.45 GHz) plasma at 10^4 Pa with $n_e = 10^{11}$ cm^{-3} and $\nu = 3 \times 10^{11}$ Hz the above relation yields a skin depth of 0.09 m. For a radiofrequency (13.56 MHz) plasma at 100 Pa with $n_e = 10^8$ cm^{-3} and $\nu = 3 \times 10^9$ Hz a skin depth of 4 m is obtained.

Typical for non-thermal plasmas is the high temperature or mean energy of the electrons in contrast to the low temperature of the heavy particles of ions and neutrals. Therefore the thermal stress of any substrate is moderate. The high electron energies and therefore their high velocities lead to the formation of a space charge sheath between the plasma and the substrate. The plasma is positively charged with respect to a floating target to ensure that the net electric current is zero. The sheath voltage, given by

$$U_s = \frac{kT_e}{2e} \ln \left(\frac{2.3 m_i}{m_e} \right),$$ (11.8)

is responsible for the kinetic energy of the ions impinging on the target surface [1].

The electron energy decays rapidly out of the excitation region. Therefore the remote afterglow plasma is characterized by low sheath voltage and small ion energy. The charge carrier density is low caused by the absence of ionization processes and counter-effective wall and/or volume losses of charge carriers by recombination processes. In contrast, the density of free radicals may be high. Afterglow plasmas can be generated in time with a pulsed power regimes or in space with a sufficient distance from the active plasma zone.

The excitation, dissociation and ionization rate of non-thermal plasmas depends on the electron energy distribution function. Its control by the various parameters such as gas species, gas density, and electrical field strength is pointed out in chapter 2.

The application of a magnetic field has several effects on a plasma. The power absorption is enhanced by increasing the electron-neutral collision frequency due to the longer circular, helical, or cycloidal pathway inside the low-pressure plasma. This is valid also for ionizing collisions that increase the ionization rate of the electrons. The magnetic field opens additional absorption channels, as shown in Eq. 11.5, with the electron cyclotron resonance that is used in the electron cyclotron resonance (ECR) plasma source. Another example is the excitation of helicon waves in a magnetized plasma [10]. Further, the magnetic field hinders the diffusion of the charged particles, especially of the electrons, moving perpendicular to the magnetic field lines. This confinement decreases the charge carrier losses and enables the creation of higher plasma densities.

The various applications make use of different specific properties of the plasma state. An effective etch process needs an inflow of reactive species, i.e. radicals or ions out of the plasma to the target. A sufficiently high aspect ratio can be achieved only with ions impinging perpendicularly to the target surface (see chapter 18). Therefore, etch plasmas operate at low pressure and the ion energy must be controllable. Plasmas for surface treatment of polymers have to contain a high concentration of free radicals. In contrast the ion energy and the photon energy must be low to avoid damage of the polymer structure (see chapter 19). High radical concentrations are useful also for surface cleaning processes, therefore plasmas at higher pressure are favorable. Cleaning by physical sputtering needs satisfactory ion energies and is most effective at low pressures (see chapter 4). In this way redeposition of the sputtered material can be avoided. Thin film deposition processes by sputtering also require low operating pressures. Plasma assisted chemical vapor deposition processes can work from low up to atmospheric pressure (see chapter 13). In etching and deposition processes the partial pressure of the reactive components must be low enough to avoid volume processes that lead to the formation of dusty particles which may diminish the quality of the product. Chemical processes like flue gas cleaning or acetylene synthesis are only effective at higher gas pressures to achieve a sufficiently high matter throughput. A review of the generation of atmospheric pressure non-equilibrium plasmas is given in Ref. [11].

Plasmas for radiation and especially for consumer lighting (see chapters 16 and 17) should ensure a selective excitation of the desired energy levels at minimum power consumption for sustaining the discharge, e.g., by tailoring the electrode falls.

11.3 Plasma generation by electric fields

Technical plasmas are generated mostly by the electrical breakdown of a neutral gas in the presence of an external electric field. Discharges are classified as direct current (dc) discharges, alternating current (ac) discharges, or pulsed discharges on the basis of the temporal behavior of the sustaining electric field. The spatial and temporal characteristics of a plasma depend to a large degree on the particular application for which the plasma will be used.

11.3.1 Direct current (dc) discharges

Direct current (dc) discharges generally burn in closed discharge vessels between interior electrodes (Fig. 11.1). Various types of discharges, and therefore different plasmas can be obtained depending on the applied voltage and the discharge current (Fig. 11.2) [1, 9, 12]. The transition from the Townsend discharge with a low discharge current to a sub-normal glow discharge and to a normal glow discharge is marked by a decrease in voltage and an increase in current. The normal glow discharge is characterized by a constant current density at the cathode, the discharge covers only partially the cathode surface. The value of the normal cathode fall depends on the nature of the gas and the cathode material. Typical values are presented in table 11.2. An abnormal glow discharge develops as the current is increased even further with increasing voltage. The cathode is covered completely by the discharge. Finally, at higher currents, the discharge undergoes an irreversible transition into an arc (glow-to-arc transition) with low sustaining voltage caused by the heating of the cathode up to thermionic emission. The typical arc plasma is in or near thermal equilibrium with hot neutrals and ions also. A resistor in series with the discharge prevents the transition into an arc. Alternatively, the discharge can be interrupted for a short period of time before the glow to arc transition occurs.

The brightest part of the discharge is the negative glow, which is separated from the cathode by the cathode dark space (*Crookes* or *Hittorf dark space*). The cathode dark space is a region of the discharge where the electrical potential drops drastically (cathode fall). The negative glow is separated from the cathode dark space by a well-defined boundary and it is followed by a diffuse region in direction towards the bright positive column and the anode. The negative glow, where the electric field is close to zero, and the positive column are sepa-

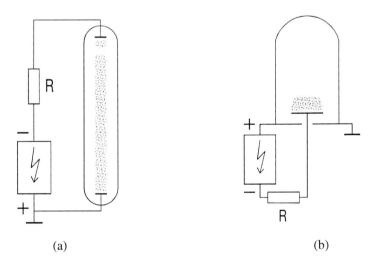

(a)   (b)

Figure 11.1: Direct current (dc) discharge (a) in a glass tube with plane electrodes, high voltage power supply and resistance R, and (b) in a metallic reactor.

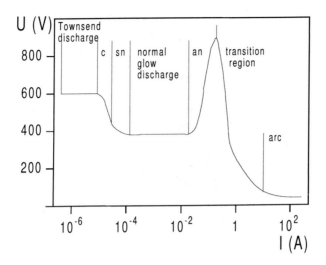

Figure 11.2: Voltage–current $(U - I)$ characteristics of a glow discharge (sn subnormal, n normal, an abnormal glow, c corona, and arc discharge).

rated by the *Faraday dark space*. The homogenous or striated (standing or moving striations) positive column stretches all the way to the anode, which may be covered by a characteristic anode glow.

The microscopic processes in such a discharge can be described as follows: a positive ion from the negative glow is accelerated by the electric field in the cathode fall and directed towards the cathode surface. The collision of the energetic ion with the surface produces secondary electrons (γ process, see chapter 4), which are subsequently accelerated in the cathode fall to comparatively high energies. These energetic electrons transfer most of their energy to heavy particles (atoms, molecules) in inelastic collisions (excitation, dissociation, and ionization), which occur primarily in the cathode fall and in the negative glow region. The positive ions formed by electron impact of neutral atoms and/or molecules in the gas phase (α process) are accelerated in the cathode fall and produce again secondary electrons on the cathode surface. The cathode regions of the discharge play a crucial role in sustaining the glow discharge. The positive column is formed only in the presence of a long, narrow discharge gap with charge carrier losses to the wall. In the homogeneous positive column, a

Table 11.2: Normal cathode fall (in Volt) for selected cathode materials and gases [13].

	air	Ar	He	Hg	Ne	H_2	N_2	O_2
Al	229	100	140	245	120	170	180	311
Fe	269	165	150	298	150	250	215	290
Pt	277	131	165	340	152	276	216	364

constant longitudinal electrical field is maintained. The electrons gain energy in this field and form an electron energy distribution with an appreciable number of energetic electrons for the formation of a sufficiently large number of ions and electrons to balance the charge carrier losses to the wall.

Etching of the cathode material and the deposition of a thin film on a separate substrate by cathode sputtering is an important process with technological applications, that occurs in the cathode region of the dc discharge [1, 6]. This process is carried out at low pressure ($10^{-1} - 10$ Pa) in order to avoid the re-deposition of the sputtered material on the cathode. If additional magnetic fields are present (magnetron), the efficiency of this process can be increased significantly. A magnetic field perpendicular to the electric field increases the path-length of the electrons and ensures a sufficiently high ionization rate [6]. The confinement of the secondary electrons by the magnetic field to a region near the cathode results in a high plasma density ($n_e > 10^{11}$ cm^{-3} [14]) and an increased discharge current at relatively low discharge voltages.

For a cathode shaped as a hollow cylinder with a diameter of nearly the length of the negative glow, the negative glow is observed inside the cathode cylinder. This results in a significant increase in the discharge current. The high efficiency of the so-called hollow cathode is due to (i) *pendulum electrons*, which are trapped in the negative glow between the retarding cathode sheaths, and (ii) additional electron emission caused by the impact of photons and metastables on the cathode surface [2]. Groups of electrons with low energies ($\sim$0.5 eV) and with higher energies ($\sim$5 eV up to more than 100 eV) are observed. This effect works also with sufficiently small dimensions in micro hollow cathodes up to atmospheric pressure [15, 16].

The cathode temperature increases with increasing discharge current. If this temperature is sufficiently high for thermionic electron emission, the glow discharge changes to an arc discharge. Similar conditions are observed in discharges that use cathode materials with low work functions, such as oxide cathodes, or discharges with externally heated cathodes. Oxide cathodes are used, e.g., in luminescence lamps (see chapter 16). The hollow cathode and the arc mechanism are mixed in the hollow cathode arc with its high electron density [17]. The non-thermal plasma of the gliding arc is promising for plasma chemical applications [18].

Examples for applications of dc discharges are sputtering for surface cleaning and thin film deposition, glow discharge mass spectrometry of element analysis of inorganic solids [19]. One example of the application of dc glow discharges will be discussed more in detail.

The *Bucket-Source* is a plasma source with heated cathodes [20]. A dense plasma is generated by using an array of low-voltage arcs with heated cathodes in conjunction with additional external magnetic fields. Such a source has been jointly developed by JET, ASDEX and TEXTOR as an ion donor for an ion accelerator (100 A, 60 kV) in fusion research. The *Bucket-Source* is a large plasma source with a cross-sectional area of 0.2×0.4 m^2 and consists of a *bucket* made of copper, 24 insulated heated tungsten filaments, and a *checker-board* of permanent magnets on the outside. The gas pressure is about 1 Pa. A voltage is applied between the filaments and the bucket forcing a high dc arc current to flow between the cathode filament and the bucket which acts as the anode. The ionization is sustained by the high current in conjunction with magnetic cusp-fields in the vicinity of the bucket walls produced by the checker-board magnets; this increases the path length of the electrons before they impinge on the wall. The result is a plasma that is characterized by a spatially very flat ion distribution

ıd densities reach values that are typical
m³. Up to now this kind of source has
ırces have the potential to be used in the
oundance of low-energy ions.

ıischarges

11.3 Plasma ger ırge, pulsed dc discharges are also used in plasma-
rces have the following advantages:

ıp ontrol by a variable duty cycle of active plasma regime and
ossible;

– while variations in the neutral gas composition between the plasma boundary and the
plasma center (due to plasma chemical reactions) may cause, e.g., inhomogeneous thin
film deposition in a continuous dc plasma, pulsed operation in conjunction with rapid gas
exchange between pulses can prevent or minimize such effects.

A special type of a high-power pulsed plasma is the plasma focus [21]. The device consists of
two coaxial electrodes separated by a hat-shaped insulator at one end, while the other end is
open. The space between the electrodes is filled with the feed gas. The discharge is powered
by a capacitor bank. In the early stage of the formation of the plasma focus, a plasma sheath
is formed which is driven by an azimuthal magnetic field. This field increases as the current is
forced to flow through an increasingly smaller annular plasma structure. The plasma is pushed
to the open end of the line by the magnetic field of the discharge current. The final state is
called the plasma focus. There is a one-to-one correspondence between the maximum current
that can be achieved and the power density in the final plasma state. Ion and electron beams
are generated by the plasma focus which exceed power flux densities in the order of TW/cm².
The nuclei of light ions fuse and a rich spectrum of electromagnetic radiation is emitted.
Applications of the plasma focus range from pulsed neutron sources for on-line analysis of
volatile components in coal to radiation sources for lithography and microscopy in the range
of soft x-rays.

11.3.3   Radiofrequency (rf) discharges

Radiofrequency (rf) discharges usually operate in the frequency range $f = \omega/2\pi \cong 1 - 100$
MHz. The corresponding wavelengths ($\lambda = 3 - 300$ m) are large compared to the dimensions
of the plasma reactor. The rf plasma reactors mostly work with a frequency of 13.56 MHz.
For microwaves the most common wavelength is 12.24 cm corresponding to a frequency of
2.45 GHz. This wavelength is roughly comparable to the dimensions of a typical microwave
reactor.

For lower frequencies, the ions accelerated in the field move towards the electrodes and
produce secondary electrons, similar to what happens in a dc discharge, here with alternating
electrodes. As the frequency increases, the ions and subsequently also the electrons can no
longer reach the electrode surface during the acceleration phase of the exciting external field.

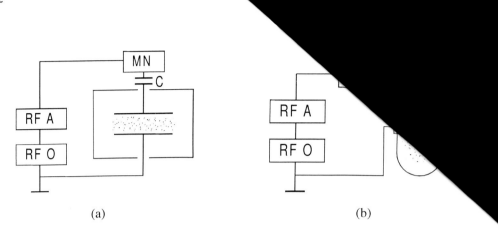

Figure 11.3: Scheme of a capacitively coupled rf discharge with (a) inner electrodes and (b) outer electrodes (electrodeless), with rf oscillator (RFO), amplifier (RFA), matching network (MN), and blocking capacitor C.

The power coupling in rf discharges can be accomplished in different ways, as

– capacitively coupled discharges, E-discharges,

– inductively coupled discharges, H-discharges [22].

A special type of rf discharges at higher pressures, the dielectric barrier discharge (DBD) and the corona discharge are extensively treated in chapter 13. Another type of a rf discharge operating at higher pressure is the plasma torch [23].

11.3.3.1 Capacitively coupled radiofrequency discharges

A reactor vessel with capacitively coupled radiofrequency discharge [9, 22, 24, 25, 26, 27] may have interior circular disc-shaped parallel electrodes which are separated by a distance of a few centimeters. They may be in contact with the discharge (Fig. 11.3a) or insulated from it by a dielectric barrier (Fig. 11.3b). In the case of insulating chamber walls, outer electrodes, i.e. electrodes on the outside of the vessel, are sometimes used. Such discharge is denoted also as electrodeless. Gas pressures are typically in the range of $1 - 10^3$ Pa. A conventional rf-system for sustaining a discharge consists of a generator, usually combined with an impedance matching network, and the reactor with the electrodes. The generator type has to be licensed in terms of the frequency band for commercial use. A matching network is necessary to match the impedance of the generator to that of the discharge. In this case, the power transfer from the generator to the discharge is at peak efficiency and the reflected rf power is minimized (Fig. 11.3) [25]. The electrodes in rf discharge are covered by sheath regions, which is similar to the cathode dark space in a dc glow discharge. The space between the electrodes is filled with the bulk plasma. For moderate pressures, capacitively coupled rf discharges exist in two forms, the α and the γ mode [26, 27]. The α mode is characterized by lower currents and a positive voltage-current $(U - I)$ characteristics, whereas the γ mode corresponds to higher

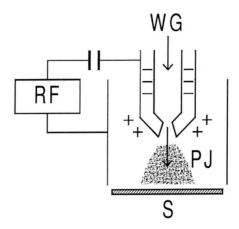

Figure 11.4: Scheme of a plasma jet (PJ) reactor with rf power supply (RF), nozzle with working gas (WG) inflow, and substrate (S), after Ref. [28].

currents and a partially negative $U - I$ characteristics. The denotations α and γ mode refer to the first Townsend ionization coefficient α for the avalanching of charge carriers in the volume, and to the γ coefficient for the release of secondary electrons from a target surface by incident positive ions, respectively. The sheath regions in front of the electrodes are quite different in the two modes. In the α sheath, in contrast to the γ sheath, electrical conductivity and charge carrier concentration are very small. The α discharge shows a weak luminous region in the center of the gap between the electrodes with maximum glow intensity near the electrodes. In the γ mode, the emission is generally much more intense. The plasma at the gap center is separated from the bright *negative glow* by the *Faraday dark space*. Rf-discharges at intermediate pressures are used, e.g., in CO_2-lasers.

A characteristic feature of this type of plasma is the *self-bias* [1, 6, 9]. The *self-bias* is a negative dc potential that develops between the plasma and the powered electrode as a consequence of (i) the use of a coupling capacitor between the rf generator and the powered electrode (the electrode is floating) and (ii) the use of appropriately shaped areas of the (smaller) powered electrode and the (larger) grounded electrode. This feature can be understood roughly assuming that the currents from the plasma to both electrodes must be equal. The higher current density at the small electrode demands a higher voltage between plasma and electrode. In other plasma devices, the application of an additional rf bias to a capacitively coupled sample holder produces a negative self-bias. The net current to the holder vanishes only for a negative bias potential due to the greater mobility of the electrons compared to that of the ions. In a capacitively coupled rf discharge, the electron density is in the range of $n_e = 10^9 - 10^{10}$ cm^{-3}, and densities up to 10^{11} cm^{-3} are possible at higher frequencies [10]. The ion energy near the powered electrode can reach energies of a few hundred electron-volts due to the self-bias. Such discharges are successfully applied to thin film deposition and plasma etching as well as to the sputtering of insulating materials.

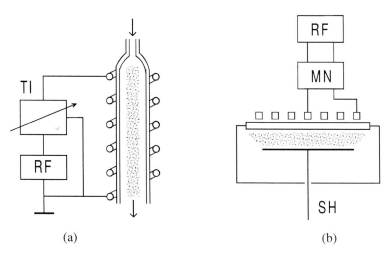

(a)   (b)

Figure 11.5: Scheme of inductively coupled plasma reactors (a) with a helical coupler and (b) with a spiral coupler, with rf generator (RF), tuning impedance (TI), matching network (MN) and substrate holder (SH), after Ref. [29].

A hollow cathode effect can also be observed for rf discharges and applied for the design of plasma reactors. Such a scheme has been used in a supersonic rf plasma jet, which was successfully tested in thin film deposition and etching experiments (Fig. 11.4) [28]. The application of capacitively coupled rf discharges for etching and polymer treatment is discussed in detail in chapters 18 and 19, respectively.

11.3.3.2 Inductively coupled radiofrequency discharges

An inductively coupled radiofrequency plasma (ICP) [9, 22, 23, 29] is excited by an electric field generated by a transformer from a rf current in a conductor. The changing magnetic field of this conductor induces an electric field in which the plasma electrons are accelerated. Schemes of two inductively coupled plasma reactors are shown in Fig. 11.5. The coil is formed as a helix (Fig. 11.5a) or as a spiral (Fig. 11.5b). The current-carrying coil or wire can be either outside or inside the plasma volume. It is also possible to use a ring-shaped plasma volume as a single-turn secondary winding of a transformer. The effect of the electric field of the wire can be shielded resulting in a suppression of the capacitive coupling.

Inductively coupled plasmas can achieve high electron densities ($n_e = 10^{12}$ cm^{-3}) at low ion energies. Several applications are reported such as thin film deposition, plasma etching, and ion sources in mass spectrometric analysis [19].

Inductively coupled plasmas can also operate in static magnetic fields. The helicon discharge as a special type [10, 30, 31] is usually generated in a cylindrical vacuum vessel in a longitudinal homogeneous magnetic field of 0.005–0.03 T [10] or higher [32]. The electromagnetic energy is transferred to the plasma source with frequencies between 1–50 MHz, usually with 13.56 MHz for processing plasmas [9]. Helicon waves are generated in the

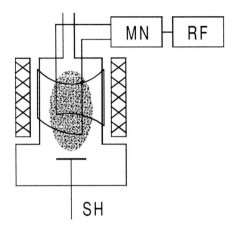

Figure 11.6: Scheme of a helicon reactor, after Ref. [31].

plasma column by specially-shaped antennas. The damping of this wave can be explained by collisional theory alone [32], but collision-less (Landau) damping of helicon waves has also been discussed [9]. This type of discharge achieves electron densities up to $10^{12} - 10^{13}$ cm^{-3} in the 0.1 Pa pressure range. A schematic diagram of a reactor [31], and several antenna constructions [33] are presented in Figs. 11.6 and 11.7, respectively.

The Neutral Loop Discharge (NLD) source, which was developed for plasma processing of large wafers for the production of microelectronics [34], is an interesting variant of the inductively coupled plasma source technology for low processing pressures (~ 0.1 Pa). A

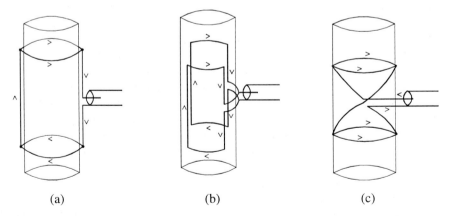

| (a) | (b) | (c) |

Figure 11.7: Schemes of helicon antennas of (a) Nagoya, (b) Boswell, and (c) Shoji type, after Ref. [33].

neutral magnetic loop ($B = 0$) is produced inside the reactor by a set of coils. The plasma is generated by applying a rf electric field along the loop that is induced by a one-turn rf antenna. It was reported [34], that the etch rate of a NLD-source can be three times higher than that of a conventional ICP source under otherwise identical conditions in the low-pressure regime

11.3.4   Microwave discharges

Plasma generation using microwaves is widely employed in many applications [7, 35, 36, 37, 38, 39]. A characteristic feature of microwaves is the wavelength, which is comparable to the dimensions of the plasma apparatus (2.45 GHz, $\lambda = 12.24$ cm), and the short period of the exciting microwave field. The power absorption (Eq. 11.2) depends on the electron-neutral collision frequency, i.e. on the gas pressure and the nature of the gas. The absorption efficiency in a 2.45 GHz discharge is high for He in the region between 10^3 and 10^4 Pa, whereas the maximum efficiency for Ar is reached for 200 Pa. However, microwave discharges can be operated at higher pressures as well, even at atmospheric pressure. The microwave frequency of 2.45 GHz corresponds to a cut-off density of the electrons of about 10^{11} cm^{-3} (Eq. 11.6). Waves of this frequency can penetrate into plasmas with higher densities only up to the thickness of the skin sheath (Eq. 11.7), which under these conditions equals few centimeters. The microwave power absorption inside the skin sheath transfers energy into the plasma. A microwave plasma reactor consists in principle of a microwave power supply, a circulator, the applicator, and the plasma load. The transmission lines are rectangular waveguides or, at lower powers, coaxial cables. The applicator should optimize the energy transfer into the plasma, and minimize the power reflection. The circulator protects the power supply against reflected power.

Marec distinguished three types of microwave reactors: discharges are produced in closed structures, or in open structures, or in resonance structures with a magnetic field [36]. In closed structures, the plasma chamber is surrounded by metallic walls. Resonant cavities of high quality with their high electric field allow an easy ignition of discharges, even at higher pressures. Examples for discharges in open structures are microwave torches, slow wave structures, and surfatrons. The ECR (electron cyclotron resonance) plasma is a typical example of a microwave plasma in magnetic fields. Methods for the coupling of waveguides to discharge

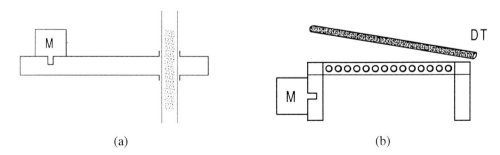

(a)   (b)

Figure 11.8: Coupling of a microwave discharge with (a) a rectangular waveguide and (b) a slow wave structure, with magnetron (M) and discharge tube (DT), after Ref. [37].

Figure 11.9: Elements of a surface wave plasma source. The surface wave propagates on the plasma surface in the dielectric discharge tube, with matching network (MN) and field applicator (FA), after Ref. [35].

tubes are presented in Fig. 11.8 [37]. Fig. 11.8(a) presents a closed configuration. The discharge tube is located at the point of maximum electric field. The slow wave structure is an open-type configuration which is shown in Fig. 11.8(b). This principle can also be used for the excitation of rf plasmas. The excitation of surface waves is another way of generating plasmas by microwaves [35]. The essential elements of a surface-wave plasma source are shown in Fig. 11.9. The surface wave propagates along the boundary between the plasma column and the dielectric vessel. The wave energy is absorbed by the plasma. This type of plasma excitation is realized in the Surfatron. The slot antenna (SLAN) plasma source (Fig. 11.10) transfers the microwave energy from a ring cavity through equidistantly positioned resonant coupling slots into the plasma chamber which is made of quartz [7]. As a plasma of high conductivity, this configuration can be treated as a coaxial waveguide with standing waves. For lower conductivity (caused by lower power), traveling waves mode can be observed. Large volume plasmas with electron concentrations up to 10^{12} cm^{-3} in broad pressure ranges can be created based on this principle.

A sketch of the microwave applicator for the generation of large-area planar microwave plasmas is shown in Fig. 11.11 [38, 39]. It consists of two waveguides. The first one is connected to the microwave power source and closed by a matched load. The second one is connected to matched loads on both ends. Traveling waves only exist in this interface

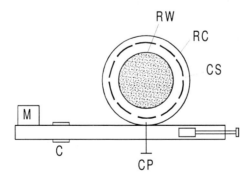

Figure 11.10: Applicator of the slot antenna (SLAN) plasma source. A homogeneous power distribution is achieved by the resonant ring cavity (RC). Microwave power is transmitted by coupling slots (CS), with M magnetron, C circulator, CP coupling probe, RW reactor wall, and the plasma inside the reactor (after Ref. [7]).

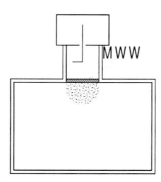

Figure 11.11: Cross section of the planar microwave plasma source with two rectangular waveguides, coupling hook and a quartz microwave window (MWW) to the vacuum vessel where the plasma is excited (after Ref. [38]).

waveguide. Microwave energy is coupled from one waveguide to the other using discrete, adjustable coupling elements distributed along the waveguides. The plasma which is separated from the second waveguide by a quartz window acts as the fourth wall for this waveguide. An adjustment of the coupling elements opens possibilities to create a homogeneous planar plasma. Plasmas with a length of 1.2 m have been produced. The generation of a planar plasma of 30×30 cm^2 was possible by using an array of such waveguide arrangements. Substrate surfaces are modified, e.g., by etching or by thin film deposition, either in the active plasma zone or in the remote plasma outside of the active zone.

The generation of microwave discharges in the low-pressure regime with a low collision frequency and thus low power absorption is difficult. The ignition of the microwave discharge can be aided by a magnetic field B where the electrons rotate with the electron cyclotron frequency $\omega_{c,e}$ in a magnetic field (Eq. 11.3) [1, 6, 10]. If the cyclotron frequency equals the microwave frequency, the power absorption reaches a maximum (Eq. 11.5). For a magnetic field of 0.0875 T, the electron cyclotron frequency becomes 2.45 GHz, and the rotational movement of the electrons is in resonance with the microwaves of 2.45 GHz. The mean free path of the electrons between collisions should be larger than the radius r_c (Eq.11.4). Fig. 11.12 shows the construction principle of an ECR plasma source. The plasma is excited in the upper part of the apparatus and the magnetic field is generated by the magnetic coils. The remote plasma generated in this device can be used for thin film deposition applications. ECR plasma sources work in a pressure range of 10^{-3} Pa to 1 Pa [6]. Typical values of electron temperatures and ion energies at the substrate are near 5 eV and between 10–25 eV, respectively [1]. In such plasmas, collisions between the atoms, molecules and ions are reduced and the generation of macroscopic particles in the plasma volume (dusty plasmas) is prevented.

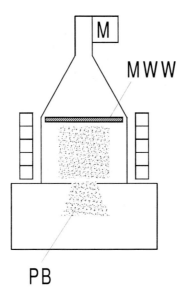

Figure 11.12: Scheme of an ECR plasma source with magnetron (M), microwave window (MWW), magnet coils, and plasma beam (PB).

11.4   Plasma generation by beams

A beam-produced plasma discharge [40] is sustained, e.g., by the interaction of an electron beam with a gaseous medium. Collective effects produce turbulent plasma oscillations with high amplitudes. The heating of the plasma electrons in this turbulent field is sufficient to sustain the beam-produced discharge plasma. The energy transfer is very effective as up to 70% of the beam energy can be transferred to the plasma. It is possible to create plasmas with high degrees of ionization in low-pressure environments. The plasma properties may be controlled by the electron beam current, the acceleration voltage, the gas pressure, and by the shape of the beam. Electron-beam generated plasmas are being used for large-area material processing [41]. A spatially flat plasma with a processing area of about 1 m^2 and a thickness of about 1 cm is sustained by an electron beam confined by a magnetic field of 0.01–0.02 T. For process gas pressures between 1 and 100 Pa, plasmas with electron densities up to 10^{12} cm^{-3} and electron temperatures of about 1 eV can be obtained.

The interaction of laser beams of sufficient energy with matter is connected with the formation of plasmas. It is used, e.g., for fusion research, for cutting of metal, and for chemical analysis of evaporated solid state materials by optical spectroscopy or mass spectrometry. The discussion of the particularities of such plasmas will go beyond the scope of this chapter.

11.5   Conclusions

The generation and maintenance of a plasma is one of the main challenges in plasma technology. Plasma parameters such as densities, temperatures, potentials, chemical composition, flows, pulse shaping, the position relative to the target, additional bias potentials, etc. have to be designed specifically for each given application. The choice of the plasma source and its particular design depend on the specific requirements of each application. Each of the various plasma sources discussed above has its own peculiarities, advantages and disadvantages. The choice of the proper source for a specific task requires the study of the characteristics of the various plasmas. Table 11.3 presents essential parameters and applications of various plasmas used in plasma sources. The dc discharge has the advantage that the microscopic processes are rather well-known and understood and that such plasmas can be diagnosed in great detail. Interior electrodes are required, and the possibility of reactions with reactive and corrosive gases must be considered in certain applications. The ion energy at the cathode is usually comparatively high. Power sources are well-developed and widely available. By contrast, rf discharges can operate with insulated or external electrodes i.e. electrode-less. Therefore, reactive processes with metal electrodes can be avoided. The lifetime of devices with electrode-less rf discharges is long. Rf sources can be operated over a wide range of pressures. Low gas pressure is possible in discharges using magnetic fields (helicon discharges). The ion energy can be controlled over a wide range. The microscopic processes in rf discharges are rather complex. Diagnostics are well-developed, but sometimes difficult to use due to interference by the rf sustaining voltage. Microwave discharges also operate without electrodes. Plasmas of high density can be generated in a pressure range from 10 Pa up to atmospheric pressure. The plasma excitation at very low pressures (less than 1 Pa) is effective in microwave-excited and magnetic-field-supported ECR discharges. The ion energy in the microwave plasma is generally low and can be controlled by additional dc fields or a rf bias voltage. Suitable power supplies are readily available. Plasma generation in the low pressure range (< 100 Pa) by electron beams involves more or less complicated beam sources. The shape of the plasma may be controlled by the shape of the exciting beam. The dielectric barrier discharge operates at or near atmospheric pressure. The plasma that is produced is highly non-thermal. Various applications in the area of surface treatment, cleaning, and modification are feasible, because no vacuum environment is required.

Plasma sources are usually operated with static electric or alternating electromagnetic fields. An additional static magnetic field performs two tasks. Firstly, the plasma confinement is enhanced by limiting the diffusion of charged particles perpendicular to the magnetic field. Secondly, the power absorption is enhanced by increasing the electron-neutral collision rate due to the longer trajectories of the electrons in the plasma at low pressures. The magnetic fields also open new absorption channels, caused by, e.g., ECR or helicon waves. The installation of static magnetic fields may be cheap, if permanent magnets are used.

These sources cover a wide range of plasma parameters and linear dimensions, but most of the plasma sources are not scalable over a large range of parameters, i.e. there are no simple, known formulas according to which power densities, homogeneity of the plasma parameters or linear dimensions, etc. can be predicted for sources not yet built and diagnosed. Furthermore, the state of maturity of the various plasma sources discussed in this paper is quite different from source to source. The development and optimization of plasma sources

Table 11.3: Plasmas used in plasma sources, typical parameters and applications.

Plasma type	Pressure (Pa)	n_e (cm^{-3})	T_e (eV)	Bias	Application
dc glow					
cathode region	$10^{-1} - 10^4$	$10^5 - 10^9$	10^2	yes	Sputtering, Deposition, Implantation, Surface analysis, Ion sources
negative glow	$10^{-1} - 10^4$	10^{12}	0.5	no	Chemistry, Radiation
positive column	$1 - 10^4$	10^{10}	$1 - 10$	no	Radiation
hollow cathode	$1 - 10^5$	$10^{12} - 10^{15}$	0.5	no	Radiation, Chemistry
magnetron	10^{-1}	$10^{10} - 10^{11}$	2–5	yes	Sputtering, Deposition
Arc, hot cathode					
ext. heating low voltage	10^2	10^{12}	1	no	Radiation
int. heating	10^5	10^{13}	0.5	no	Radiation, Welding
Focus	10^3	5×10^{18}	10^3		Radiation
RF, capacitive					
low pressure	$10^{-1} - 10$	10^{11}	$1 - 10$	yes	Processing, Sputtering, Ion sources
moderate pressure	$10 - 10^3$	10^{11}	$1 - 10$	no	Processing, Deposition
hollow cathode	10^2	10^{12}	0.5	no	Processing, Radiation
magnetron	10^{-1}	10^{11}	2–5	yes	Sputtering
RF, inductive	$10^{-1} - 10^3$	10^{12}	1	no	Processing, Etching
helicon	$10^{-2} - 1$	10^{13}	1	no	Processing, Etching
Microwave					
closed structure	10^5	10^{12}	3	no	Chemistry
SLAN	10^5	10^{11}	5	no	Processing
open structure					
Surfatron	10^5	10^{12}	5	no	Processing
Planar	10^4	10^{11}	2	no	Processing
ECR	10^{-1}	10^{13}	5	no	Processing
Electron beam	$1 - 10^2$	10^{12}	1	no	Processing
DBD, dielectric barrier discharge	10^5	10^{14}	5	no	Chemistry (Ozone), Processing, Radiation

include a profound knowledge of the plasma parameters; this demands the application of the various methods of plasma diagnostics, as probe diagnostics, optical emission and absorption spectroscopy, laser diagnostics, and mass spectrometry. Many challenges remain for further scientific and engineering work in order to meet the demands of the various diverse plasma technological applications.

Acknowledgements

Helpful discussions with K. Becker, Stevens Institute of Technology, Hoboken, NJ, and our colleagues R. Foest and L. Mechold are gratefully acknowledged.

References

[1] A. Grill, *Cold Plasma in Material Fabrication, From Fundamentals to Applications*, New York: IEEE Press (1994)

[2] A. Rutscher, H. Deutsch, Eds., *Wissensspeicher Plasmatechnik*, Leipzig: Fachbuchverlag (1983)

[3] H. Hess, Contrib. Plasma Phys. 26, 209 (1986)

[4] H. Hess, A. Kloss, A. Rakhel, H. Schneidenbach, Int. J. Thermophys. 20, 1279 (1999)

[5] R.L. Morse, Phys. Fluids 16, 545 (1973)

[6] S.M. Rossnagel, J.J. Cuomo, W.D. Westwood, Eds., *Handbook of Plasma Processing Technology*, Park Ridge: Noyes Publication (1990)

[7] D. Korzec, F. Werner, R. Winter, J. Engemann, Plasma Sources Sci. Techn. 5, 216 (1996)

[8] F. Chen, *Introduction to Plasma Physics and Controlled Fusion, Vol. 1 Plasma Physics*, New York: Plenum Press (1984)

[9] M. Lieberman, A. Lichtenberg, *Principles of Plasma Discharges and Materials Processing*, New York: Wiley (1994)

[10] O. Popov, Ed., *High Density Plasma Sources*, Park Ridge: Noyes Publ. (1995)

[11] E.E. Kunhardt, IEEE Trans. Plasma Sci. 28, 189 (2000)

[12] G. Francis, *The Glow Discharge at Low Pressure*, in: S. Flügge, Ed., *Handbuch der Physik Band XXII*, Berlin: Springer (1956)

[13] S.C. Brown, *Basic Data of Plasma Physics: The Fundamental Data on Electrical Discharges in Gases*, New York: Am. Inst. Phys. (1994)

[14] E. Passoth, P. Kudrna, C. Csambal, J.F. Behnke, M. Tichý, V. Helbig, J. Phys. D (Appl. Phys.) 30, 1763 (1997)

[15] G. Schaefer, K.H. Schoenbach, *Basic Mechanisms Contributing to the Hollow Cathode Effect*, in *Physics and Applications of Pseudosparks*, M. Gundersen, G. Schaefer, Eds., New York: Plenum (1990), p. 55

[16] K.H. Schoenbach, A. El-Habachi, W. Shi, M. Ciocca, Plasma Sources Sci. Techn. 6, 468 (1997)

[17] G. Rohrbach, A. Lunk, Surf. & Coat. Techn. 123, 231 (2000)

[18] O. Mutaf-Yardimci, A.V. Saveliev, A.A. Fridman, A.L. Kennedy, J. Appl. Phys. 87, 1632 (2000)

[19] G. Holland, A.N. Eaton, Eds., *Applications of plasma source mass spectrometry*, Cambridge: Roy. Soc. Chem. (1991)

[20] H. Conrads, H. Euringer, U. Schwarz, in K.H. Spatschek, Ed., *Contributions to High Temperature Plasma Physics*, Berlin: Akademie Verlag (1994)

[21] A. Bernard, P. Cloth, H. Conrads, A. Coudeville, G. Gourlan, A. Jolas, C. Maisonnier, J.P. Rager, Nucl. Instr. Meth. 145, 191 (1977)

[22] G.G. Lister, J. Phys. D: Appl. Phys. 25, 1649 (1992)

[23] G. Lins, Contributions Plasma Phys. 40, 147 (2000)

[24] J.R. Hollahan, A.T. Bell, *Techniques and Applications of Plasma Chemistry*, New York: Wiley (1974)

[25] M. Hirose, *Plasma-deposited films: Kinetics of formation, composition, and microstructure*, in J. Mort, F. Jansen, Eds., *Plasma deposited films*, Boca Raton: CRC Press (1986)

[26] Yu.P. Raizer, *Gas Discharge Physics*, Berlin: Springer (1991)

[27] Yu.P. Raizer, M.N. Shneider, N.A. Yatsenko, *Radio-Frequency Capacitive Discharges*, Boca Raton: CRC Press (1995)

[28] L. Bardos, Proc. XXIth Intern. Conf. Phen. Ion. Gases 3, 98 (1993)

[29] J. Hopwood, Plasma Sources Sci. Techn. 1, 109 (1992)

[30] R.W. Boswell, Proc. XXIth Intern. Conf. Phen. Ion. Gases 3, 118 (1993)

[31] F. Nicolazo, A. Goullet, A. Granier, C. Valée, G. Turban, B. Grolleau, Surf. Coat. Techn. 98, 1578 (1998)

[32] F.F. Chen, I.D. Sudit, M. Light, Plasma Sources Sci. Techn. 5, 173 (1996)

[33] F.F. Chen, J. Vac. Sci. Techn. A 10, 1389 (1992)

[34] H. Tsuboi, M. Itoh, M. Tanabe, T. Hayashi, T. Uchida, T., Jpn. J. Appl. Phys. 34, 2476 (1995)

[35] C.M. Ferreira, M. Moisan, Eds., *Microwave Discharges, Fundamentals and Applications*, NATO ASI Series B: Physics, Vol. 302, New York: Plenum Press (1993)

[36] J. Marec, P. Leprince, in C.M. Ferreira, M. Moisan, Eds., *Microwave Discharges, Fundamentals and Applications*, NATO ASI Series B: Physics, Vol. 302, New York: Plenum Press (1993), p. 45

[37] M. Konuma, *Film Deposition by Plasma Techniques*, Berlin: Springer (1992)

[38] A. Ohl, in C.M. Ferreira, M. Moisan, Eds., *Microwave Discharges: Fundamentals and Applications*, NATO ASI Series B: Physics, Vol. 302, New York: Plenum Press (1993), p. 205

[39] A. Ohl, J. Phys. IV (France) 8, Pr 7-83 (1998)

[40] M. Schmidt, T. Föste, P. Michel, V.V. Sapkin, Beitr. Plasma Phys. 22, 436 (1982)

[41] R.A. Meger, R.F. Fernsler, M. Lampe, D. Leonhardt, W.M. Manheimer, D.P. Murphy, R.E. Pechacek, Bull. Am. Phys. Soc. 44, 38 (1999)

12 Reactive non-thermal plasmas — chemical quasi-equilibria, similarity principles and macroscopic kinetics

Hans-Erich Wagner

Institut für Physik, Ernst-Moritz-Arndt-Universität Greifswald, Domstraße 10a
D-17487 Greifswald, Germany

12.1 Introduction

Atomic and molecular physics of ionized gases quickly escalates to extreme complexity, if plasma chemical processes are included. Especially this is the case under non-thermal plasma conditions. Examples of these are the different types of gas discharges (glow discharge, microwave discharge, corona discharge, dielectric barrier discharge, ...) operating in reactive gas mixtures. In spite of the considerable progress in microphysical modeling of plasma chemical reactions during the last decade (e.g., two dimensional models) their comprehensive theoretical description is a not mastered problem. The main problems consist in the high complexity of the processes (discharge physics, bulk plasma chemistry, reactive plasma-wall interactions, ...) as well as in the missing knowledge of various basic data (cross sections of electronic collision processes, rate coefficients, ...) of elementary processes. Some general treatments based on statistical or thermodynamical models have promoted insight in the extremely complex reaction mechanisms of non-equilibrium plasma chemical processes (e.g., Refs. [1, 2]). But with the exception of heterogeneous reactions in chemical transport phenomena of low-pressure plasmas [3], any practically utilizable results could not be obtained.

To bypass the enormous difficulties of microscopic kinetics in plasma chemistry over and over again a more coarse method was put to the test: macroscopic kinetics. This method connects the chemical conversions in non-thermal plasma chemical reactors immediately to the operation parameters of the reactor (e.g., the power input, flow rate, ...) by means of macroscopically determined rate coefficients in the relevant reaction equations. So the method allows a very compact description of different plasma chemical reactors. For a long period the macroscopic kinetics was a pure phenomenological method, far away from the foundation by first principles [4, 5, 6]. But it could be shown that of greatest importance for this concept are chemical quasi-equilibrium states, which represent a characteristic phenomenon of the overall behaviour of gas discharge reactors. The existence of quasi-equilibria reflects the basic role of gross reactions even for non-thermal plasma chemical conversions, and inaugurate a more general approach to the method of macroscopic kinetics [4, 7, 8, 9]. Recently the occurrence of quasi-equilibrium states in non-thermal plasma chemical reactors has attained increasing attention [10, 11].

Moreover, on this occasion it became clear that there exists a close connection regarding the formulation of the principles of plasma chemical similarity. Up till now these principles were only considered in special cases. Two examples are mentioned: For high-power arc discharges (thermal plasmatrons and torches) detailed investigations on scaling and similarity were performed, but for the most part without satisfactory physical interpretation [12]. For various low-pressure non-thermal plasmas and ion beams the etching rates have been summarized successfully using a similarity presentation [13]. On the other hand the similarity principles of non-thermal plasmas, e.g., of the low-pressure positive column, are mostly restricted to pure electronic similarity [14, 15]. Plasma chemical similarity (as well as structural similarity) remained marginal problems [16]. Dimensionless parameters proved to be useful in the twentieth century development of several branches of physics. Concerning this only a few approaches exist in plasma chemistry. By modeling the non-thermal positive column plasma (considering all relevant equations) the conditions of chemical similarity of volume reactions have been developed within the frame of microscopic kinetics. The substance of similarity principles is contained in parameters which govern the comparison of different model reactors [17, 18].

The main task of the paper is to give a survey on the fields characterized above: the chemical quasi-equilibrium states, the progress in foundation of parameters of plasma chemical similarity, and the method of macroscopic kinetics, respectively. The significance and capability of chemical quasi-equilibrium states as well as the efficiency of plasma chemical similarity parameters and of the method of macroscopic kinetics are demonstrated by various experimental results.

12.2 Chemical quasi-equilibria

12.2.1 The concept

Contrary to the usual chemical equilibrium which, in principle, exists in an unlimited homogeneous medium, the quasi-equilibrium states are related to the special operation conditions of plasma chemical reactors. In this way they combine the general meaning of the term equilibrium with the particularities of the non-equilibrium plasma chemistry. To understand the formation and the character of chemical quasi-equilibria, it is indispensable to take into consideration the double structure of every plasma chemical reactor and the restriction to only stable reaction products (e.g., a quasi-equilibrium does say nothing about the concentrations of free radicals). The double structure of the reactor means the combination of an active zone with a passive one in flow systems, separated in space, or an active phase combined with a passive one in closed systems, separated in time (see Fig. 12.1). For certain models (e.g., plug-flow model, back-mixing model) both cases can be transformed one into the other, requiring only one mathematical apparatus [6, 7, 19].

Figure 12.1 summarizes the scheme of a plasma chemical flow reactor and of a closed reactor, respectively. In the active zone (phase) the incoming mixture of stable compounds $A_1....A_k$ is activated by hot discharge electrons, producing among others an abundance of unstable intermediate products (radicals, excited species, ions, ...). During its flow through the passive zone (during the passive phase) the mixture is deactivated by afterglow processes

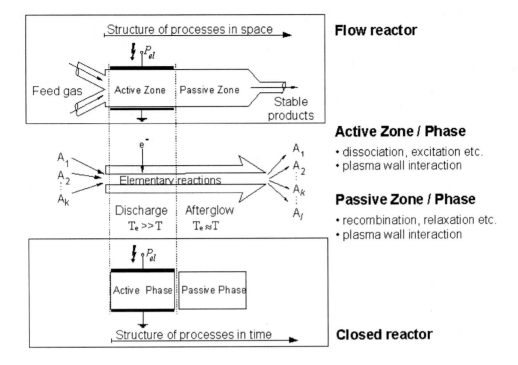

Figure 12.1: Scheme of a plasma chemical flow reactor and of a closed reactor, respectively. P_{el} is the electrical power input.

resulting at the end in stable products again, including the input species and some new compounds $A_{k+1}...A_l$. In every case the statements of a quasi-equilibrium are related to the output of the passive zone (the end of the passive phase). Unstable products (e.g., free radicals) are not included. In practice this corresponds to the flowing afterglow, or late afterglow, of pulsed discharges.

The occurrence of quasi-equilibrium states is noticeable because of the following fact: Independent of the input mixture ratio in the limiting cases, $P \to 0$ and $P \to \infty$, which includes the limits $n_e \to 0$ and $n_e \to \infty$, respectively (P is the electrical power input, and n_e is the electron concentration in the active zone), the set of chemical rate equations results in two final compositions [4, 7, 8, 20]:

- CEEC – Chemical Equilibrium of Electronic Catalysis. Generally, the occurrence of a finite degree of conversion in a closed plasma chemical system at the limit of vanishing ionization is a very typical case, representing a kind of electronic catalysis: In principle, already one hot electron (permanently regenerated) is able to convert a reactant mixture in this equilibrium state of altered composition.

- CECD – Chemical Equilibrium of Complete Decomposition. Large concentrations of hot electrons (high power input) result in the nearly complete decomposition of the reactants in the active zone (phase). In the afterglow, recombination processes etc. yield a final composition of the reactant mixture according to neutral gas reactions starting from the elementary constituents. Obviously this behavior must be a general characteristic equilibrium state of every non-thermal plasma chemical reactor, too.

We note in passing that chemical quasi-equilibria, as illustrated here, are formed in an other type of closed reactors, too. These reactors are characterized by a stationary active zone, separated in space and surrounded by a stationary afterglow.

A necessary condition for the existence of chemical quasi-equilibrium states consists in the existence of two different time scales for the chemical conversions: on a short time scale the unstable components – generated by electron collision processes in the active zone (phase) – recombine in the passive zone (phase) into an altered mixture of stable compounds. The transition into the thermodynamic equilibrium follows for longer time periods caused by thermally initiated conversions of the stable products. Therefore the quasi-equilibria may be interpreted as kinetic frozen states [20]. The formation of the quasi-equilibria from both sides of reversible gross reactions in dependence on power input and flow conditions is pointed out in section 12.4.3.

12.2.2   Chemical quasi-equilibria and the kinetic background

After this introducing description now a definition of the term quasi-equilibrium is given. The reaction system is characterized by a reversible gross reaction of the type $A + B + e \rightleftharpoons C + e$ (e symbolizes the hot electron) and some plasma parameters like the kinetic temperature of electrons T_e and the gas temperature T. A plasma chemical quasi-equilibrium is understood to be

- a stationary state of the composition of stable reaction products,

- attainable from both sides of a reversible gross reaction, as mentioned above,

- independent of reactor parameters, as power input (degree of ionization), geometry and type of the plasma zone as well as of the reactor dead zone (i.e., the part of a closed reactor which is never filled with a plasma in the active phase),

- registrable by a modified mass action law with a constant different from the usual one, only depending on the gas temperature (the most important point). It is determined by a multitude of elementary processes of the reactor and is kinetic by nature.

In the case of the Chemical Equilibrium of Thermal Reactions (CETR) for the conversion

$$A + B \overset{k_+}{\underset{k_-}{\rightleftharpoons}} C$$

the mass action law reads:

$$\frac{n_A \, n_B}{n_C} = \frac{k_-}{k_+} = K(T) \,, \tag{12.1}$$

where n_i is the particle number density of component i.

Including electronic processes

$$A + B + e \rightleftharpoons C + e$$

we get for the closed system in the limit of vanishing electron concentration $n_e \to 0$ and $\tau \to \infty$, where τ is the duration of the active phase [4, 7, 8],

$$\text{CEEC} \qquad \frac{n_A^0 \, n_B^0}{n_C^0} = \frac{k_-^0(n_e, T_e, T)}{k_+^0(n_e, T_e, T)} \stackrel{T_e \geq T_{e,min}}{\Longrightarrow} K^0(T). \qquad (12.2)$$

The ratio of forward and backward rate coefficients depends, first of all, on the concentration n_e and temperature T_e of the electrons, and on the gas temperature T. But in plasma chemistry very often conditions prevail where the electronic parameters approximately disappear. This is the case if the rate coefficients are proportional to n_e and become nearly independent of T_e at values $T_e \geq T_{e,min}$, i.e., $k_\pm^0(n_e, T_e, T) \approx n_e \times f(n, T)$ (n is the total particle number density). The minimum value $T_{e,min} \approx 1$ eV will be surpassed in many plasma chemical systems.

The chemical equilibrium of complete decomposition ($n_e \to \infty$) is governed by afterglow processes mostly including neutral species. Then the gas temperature again controls the conversion:

$$\text{CECD} \qquad \frac{n_A^\infty \, n_B^\infty}{n_C^\infty} = K^\infty(T). \qquad (12.3)$$

To demonstrate the different states of equilibria now a simple reaction system is analyzed, describing the basic mechanism of the ozone synthesis [7], according to the gross reaction

$$3O_2 + e \leftrightarrow 2O_3 + e.$$

Table 12.1: Basic mechanism of ozone synthesis. The constants k_i and γ_i are given in cm^3 s^{-1} and cm^6 s^{-1}, respectively, ε is the recombination probability, $\bar{c}$ the thermal velocity, and r the reactor radius.

	Elementary processes			Constant
1	$O_2 + e$	$\longrightarrow O(^3P) + O(^1D) + e$		$k_1(T_e)$
		$\longrightarrow O(^3P) + O(^3P) + e$		
2	$O_2 + O + M$	$\underset{k_{th}}{\overset{\gamma_2}{\rightleftharpoons}} O_3 + M$		$\gamma_2 = 4.2 \times 10^{-35} \exp(1050/T)$
3	$O_3 + e$	$\longrightarrow O_2 + O + e$		$k_3(T_e)$
4	$O + O + M$	$\longrightarrow O_2 + M$		$\gamma_4 = 2.7 \times 10^{-32} \, T^{-0.41}$
5	$O_3 + O$	$\longrightarrow 2 O_2$		$k_5 = 2.0 \; 10^{-11} \exp(-2400/T)$
6	$O + \text{wall}$	$\longrightarrow 0.5 \, O_2$		$\alpha = \varepsilon \, \bar{c}/2r$

The analytical solution of the system of particle balance equations (given in table 12.1) for a closed reactor results for $n_e \to 0$ and $\tau \to \infty$ (CEEC) in non-vanishing concentrations of ozone:

$$n_{O_2}^0 / n_{O_3}^0 = \frac{k_5 + k_3 \, (n_e/n_O)^0}{\gamma_2 \, n} = K^0(T), \tag{12.4}$$

with the *finite* limiting value of the concentration ratio of electrons n_e and atomic oxygen n_O

$$(n_e/n_O)^0 = \lim_{n_e \to 0} (n_e/n_O) = \frac{1}{k_3} \times \frac{\alpha + \gamma_2 \, n \, n_{O_2}^0 + k_5 \, n_{O_3}^0}{2 \, n_{O_2}^0 \, k_1/k_3 + n_{O_3}^0}. \tag{12.5}$$

For $n_e \to \infty$ (CECD) only a numerical solution is possible. At different gas temperatures the relevant concentrations of O_2 and O_3 are described by:

$$n_{O_2}^\infty / n_{O_3}^\infty = K^\infty(T) \; .$$

Figure 12.2 shows the course of this dependence. Obviously, in the region of higher temperatures an exponential dependence prevails (Arrhenius plot). Together with the CECD in figure 12.2 also the CEEC is shown, and for the used rate coefficients both quasi-equilibrium states coincide:

$$K^\infty(T) = K^0(T) = K_e(T) \approx \frac{k_5}{\gamma_2 n} . \tag{12.6}$$

Of considerable interest is the situation where the chemical compositions of the two quasi-equilibria coincide. Numerous experimental results seem to confirm the suggestion that this is

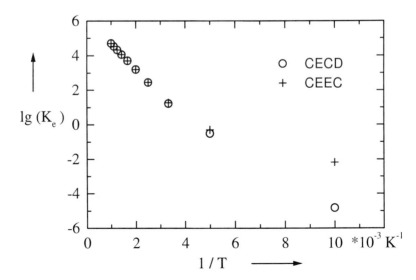

Figure 12.2: The constant $K_e(T)$ for the quasi-equilibrium states CEEC and CECD in the system O_2/O_3 (closed reactor; $n_0 = 3 \times 10^{17}$ cm^{-3}, $3k_1 = k_3 = 9 \times 10^{-10}$ cm^3 s^{-1}, $\alpha = 3$ s^{-1}; other constants as in table 12.1).

a more general case: Non-thermal plasma chemical conversion tends to reach CEEC = CECD. Only then a starting point is given to develop the method of macroscopic kinetics, using the so-called specific energy as the central reactor parameter [5, 21] (see section 12.4).

The kinetic background of the chemical quasi-equilibrium states of the ozone synthesis in low pressure non-thermal plasmas, considering all relevant particle species and elementary reactions, was studied in detail in [22]. Generally, for CEEC = CECD the following conditions must be fulfilled:

- The gain and loss processes for stable final products are determined at vanishing power input ($P \rightarrow 0$) only by the collisions of neutral particles. The most important result of electron collisions is the generation of unstable intermediate particles (atoms, radicals, excited species) which influence the gain and loss of the final products, e.g.,

$$
\begin{aligned}
O_2 + e &\rightarrow O + O + e \\
O + O_2 + M &\rightarrow O_3 + M \\
O_3 + O &\rightarrow 2O_2 .
\end{aligned}
$$

A disturbing process for the equalization is $O_3 + e \rightarrow O_2 + O + e$.

- Quadratic processes of unstable intermediate products are of minor importance, e.g.,

$$
\begin{aligned}
O + O + M &\rightarrow O_2 + M \\
O^- + O &\rightarrow O_2 + e \\
2O_2(a^1\Delta_g) + O_2 &\rightarrow 2O_3 .
\end{aligned}
$$

- The afterglow which transfers the active plasma from complete decomposition to the final state of CECD must be governed by the same collisions of neutral particles as discussed above, too. This means that for both CEEC and CECD a very *similar chemical climate* exists. This includes constant gas temperature and constant wall conditions (in the case of heterogeneous reactions).

An additional interesting point of the quasi-equilibrium states is the following one: although these states are defined as limiting cases ($P \rightarrow 0$ or $P \rightarrow \infty$) in practice they can be observed already at finite values of P. This is of considerable importance with regard to their experimental verification.

12.2.3 Experimental verification

In the following we list some experimental results on the occurrence and the qualities of chemical quasi-equilibria in non-thermal plasma reactors. The experiments were performed in the direct current *(dc)* positive column, in the radiofrequency *(rf)* discharge, and in the so-called Tesla spark discharge, respectively.

Experimental results on the ozone production in a flow reactor (active zone: *dc* positive column within a cylindrical glass tube) according to the gross reaction $3O_2 + e \rightleftharpoons 2O_3 + e$ are shown in Figs. 12.3 and 12.4 [22]. The ozone concentration, detected in the afterglow tube (passive zone) by spectroscopical absorption technique, has a nearly current independent

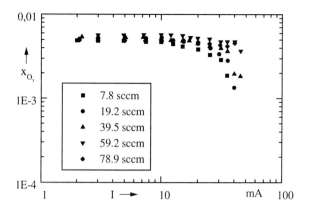

Figure 12.3: Dependency of the relative ozone concentration x_{O_3} on the discharge current I in the passive zone ($p = 1000$ Pa) for different flow rates.

value (Fig. 12.3). At a discharge current above 30 mA strongly decreasing ozone concentration values have been detected. This is caused by the increasing gas temperature in the active plasma, which sensitively affects the endothermic ozone production reactions (e.g., table 12.1). The current-independent plateau of the relative ozone concentration represents the case of identical chemical quasi-equilibria, CEEC = CECD, which have been proven by these measurements. The plateau represents a large variation of the ionization degree $x_e = n_e/n$ of about two orders of magnitude in the discharge. In contrast to this situation the relative concentration, detected in the active zone, decreases with increasing discharge current (Fig. 12.4). At larger discharge current (i.e., larger power input) ozone is produced mainly in the

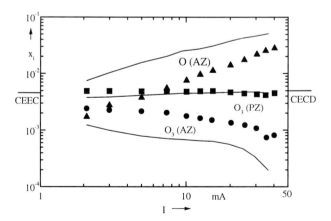

Figure 12.4: Comparison of the current dependencies of the relative ozone concentration in the active (AZ) and passive zone (PZ) as well as of the O atom concentration in the active zone ($p = 1000$ Pa; gas flow rate $F = 79$ sccm; symbols: experiments; lines: kinetic model).

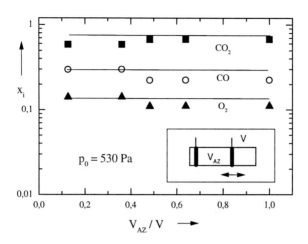

Figure 12.5: Stationary composition of the stable products ($\blacksquare$, CO_2; $\circ$, CO; $\blacktriangle$, O_2) of CO_2 decomposition in a *rf* discharge in dependence on the relative volume V_{AZ}/V of the active zone (glass tube reactor, $P = const.$, $f = 27.12\,\text{MHz}$).

afterglow.

As discussed earlier, the plateau of quasi-equilibria is independent of the reactor dead zone as well as of the plasma type. These results are illustrated for the CO_2 decomposition according to the reversible gross reaction $2\,CO_2 + e \rightleftharpoons 2\,CO + O_2 + e$ in a glass tube reactor under closed conditions [9, 23]. The concentration of the reaction products was measured by a gaschromatographic technique. Figure 12.5 shows this conversion in a *rf* discharge at nearly

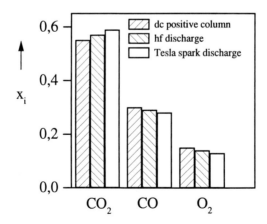

Figure 12.6: Stationary composition of stable reaction products for the CO_2 decomposition in different discharges types (glow discharge: $P = 10\,\text{W}$; *rf* discharge: $P = 10\,\text{W}$, $f = 27.12$ MHz; Tesla spark discharge: $P < 0.1\,\text{W}$; other conditions as in figure 12.5).

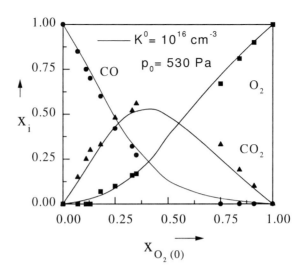

Figure 12.7: The relative concentrations x_i $(i = O_2, CO, CO_2)$ at the plateau of the chemical quasi-equilibrium CEEC in dependence on the relative input concentration $x_{O_2 (0)}$ of O_2 in the O_2/CO mixture (dc positive column, Al electrodes, $I = 30$ mA; symbols: experiments; lines: calculations with $K^0 = const$, according to Eq. 12.2).

constant power input. An extreme variation of the relative portion of the plasma zone does not influence the final composition of stable products. Different types of non-thermal plasmas do not affect markedly the level of quasi-equilibria at constant wall conditions, as is illustrated in figure 12.6.

In Fig. 12.7 the change of the level of the CEEC in dependence on the input ratio of a CO/O_2 mixture (active phase: dc positive column, within a cylindrical glass tube) is shown. The composition of the stable reaction products follows satisfactorily the simple *electronically modified* mass action formula given above with $K^0 = 10^{16}$ cm^{-3} $= const$. Surprisingly, by means of an identical value of K^0 the pressure dependence of the CEEC can be described, too [9]. This fact underlines especially the capability of the discussed formula.

12.3   Plasma chemical similarity

12.3.1   Similarity principles in the chemistry of non-thermal plasmas

A very general starting point of the similarity analysis is the settlement and preparation of all the equations relevant to the problem in question by transformation to dimensionless forms. The central equation for reactors in process engineering is the generalized equation of transport (heat, matter, momentum [16]). In our special case the most important versions of this equation are the particle balance equations of the different chemical species [15, 17]. For particles of kind i we have:

$$\frac{\partial n_i}{\partial t} + \mathrm{div}\,(n_i \vec{v}) - \mathrm{div}\,(D_i\,\mathrm{grad}(n_i)) - S_i = 0\,. \tag{12.7}$$

The thermal energy balance of the neutral gas requires

$$\frac{\partial\,(\rho c_p T)}{\partial t} + \mathrm{div}\,(\vec{v}\rho c_p T) - \mathrm{div}\,(\lambda_W\,\mathrm{grad}\,(T)) - H = 0. \tag{12.8}$$

The particle balance equation of electrons is given by

$$\frac{\partial n_e}{\partial t} + \mathrm{div}\,(n_e \vec{v}_e) - S_e = 0\,, \tag{12.9}$$

where

$$n_e \vec{v}_e = -\mu_e n_e \vec{E} - \mathrm{grad}\,(n_e D_e) \qquad \text{and} \qquad \mu_e = \frac{e_0}{m_e} \tau_e\,. \tag{12.10}$$

The energy balance of electrons reads [24]:

$$\frac{3}{2}\frac{\partial}{\partial t}\,(n_e U_e) + \frac{3}{2}\,\mathrm{div}\,\left(n_e U_e \vec{v}_e^*\right) + n_e \vec{v}_e \vec{E} + n_e H_e = 0\,. \tag{12.11}$$

Here n_i and n_e are the particle number density of heavy particles (i) and of electrons (e), respectively, $\vec{v}$ is the gas flow velocity, $\vec{v}_e$ and $\vec{v}_e^*$ are the drift velocity of the electrons and the corresponding velocity for the energy flux, respectively, D_i and D_e diffusion coefficients of heavy particles and electrons, respectively, T and ρ the gas temperature and density, respectively, c_p is the specific heat, λ_W the heat conductivity, U_e the mean energy of electrons, $\vec{E}$ the electrical field strength, μ_e and τ_e mobility and mean life time of electrons, respectively, S_i and S_e effective source terms for the production of heavy particles and electrons, respectively, H is the heat production, and H_e the effective energy loss of electrons by collisions.

From the detailed discussion of these equations (in a suitable dimensionless form) concerning the chemical point of view and **restricted to volume reactions** the following general principles of plasma chemical similarity for homologous points of two systems of the *active* plasma zone result [17]:

- Similar systems agree in the spatial and temporal distributions of all their components, expressed by identical values of

$$\frac{n_i\left(\frac{\vec{r}}{L_0}, \frac{t}{\tau_0}\right)}{n_0} \qquad \text{with} \qquad n_0 = \sum_i n_i\left(\frac{\vec{r_0}}{L_0}, \frac{t_0}{\tau_0}\right), \tag{12.12}$$

where τ_0 is the residence time of the gas mixture in the active zone, L_0 is a characteristic length (e.g., of the reactor). The subscript $'0'$ indicates reference parameters.

- The similarity is described by the correspondence in:

$$\frac{n_i}{n_0}\,;\ \frac{T}{T_0}\,;\ \frac{n_e}{n_{e0}}\,;\ \frac{U_e}{U_{e0}}\,;\qquad i = 1, 2, \dots\ . \tag{12.13}$$

The conditions for this correspondence are the following:

(i) From outside in chemical similar plasmas two reduced physical fields must be in correspondence

$$\frac{\vec{v}\left(\frac{\vec{r}}{L_0}, \frac{t}{\tau_0}\right)}{v_0} \quad \text{and} \quad \frac{\vec{E}\left(\frac{t}{\tau_0}\right)}{E_0}. \tag{12.14}$$

In this the geometrical similarity is manifested by itself.

(ii) The dimensionless similarity parameters of the problem are:

$$\frac{\tau_0 D_{i_0}}{L_0^2} \quad ; \quad \frac{\lambda_{W_0} T_0}{L_0 p_0 v_0} = P_e \text{ (Peclet-Number)};$$

$$\frac{\lambda_{e_0} E_0}{U_{e_0}} \quad ; \quad \frac{\mu_{ion}}{\mu_{e_0}} ; \tag{12.15}$$

$$\frac{\lambda_{e_0}}{L_0} \quad ; \quad \frac{\tau_{e_0}}{\tau_0}.$$

Both the last ones are a consequence of introducing kinetic reference parameters for the electron gas (subscript $'0'$).

(iii) For plasma chemical similarity the correspondence of all the reduced source terms

$$\frac{\tau_0 S_i}{n_0}, \quad \frac{\tau_0 H}{p_0}, \quad \frac{\tau_{e_0} S_e}{n_{e_0}} \quad \text{and} \quad \frac{\tau_{e_0} H_e}{U_{e_0}} \tag{12.16}$$

is imperative. The widely general principle of plasma chemical similarity reflects the underlying electronical similarity of the plasma. In our case this is the so-called B-invariant similarity [14], which is based on fundamental invariant properties of the Boltzmann-equation: Similar non-thermal plasmas show identical electron energy distribution functions. Summarizing, in this way we are coming to the well-known similarity conditions: Similarity exists for corresponding values of

$$\frac{E_0}{n_0}, \quad L_0 n_0, \quad \tau_0 n_0, \quad \frac{n_{e_0}}{n_0} \quad \text{and} \quad T_0. \tag{12.17}$$

In transferring these principles to plasma chemical reactors some simplifications are necessary and possible. A far-reaching simplification is the restriction to stationarity and spatial homogeneity (e.g., by averaging over space and time). In this way we get from the energy balance of electrons (Eq. 12.11) a very practicable expression to estimate the degree of ionization, which is fundamental for the reactor operation:

$$\frac{n_e}{n} = \frac{2}{3} \frac{\tau_e n}{e_0 \delta U_e} \frac{P/V}{n^2}. \tag{12.18}$$

The mean energy loss of electrons per collision δ for elastic collisions only is $\delta_{el} = 2m_e/M \approx 10^{-4}...10^{-5}$. The energy loss δ_{inel} during inelastic collisions is one to two orders of magnitude larger than δ_{el}.

12.3.2 Application to the flow reactor

Applying the general principles of plasma chemical similarity to non-thermal flow reactors, it is indispensable to subdivide the reactor in two different zones and their joint action (compare Fig. 12.1), i.e., an active zone (AZ, plasma region) and a passive zone (PZ, afterglow). The

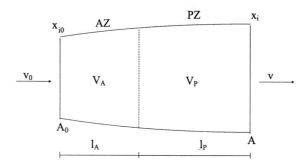

Figure 12.8: Schematic model for non-thermal reactors under flow conditions.

most important question in reactor similarity is to ask for conditions that result in the same spectrum of the reduced output values $x_i = n_i/n$ of the stable reaction products at corresponding input. The detailed distributions inside the combined reactor are of minor interest in this case. Thus the difficulties mentioned can be counteracted by restriction to the so-called back-mixing reactor model which assumes an one-dimensional, steady-state flow (see Fig. 12.8). With a sufficiently long PZ ($l_P \to \infty$) only chemical stable species leave the reactor. Within the scope of the one-dimensional back-mixing model the balance equations of the stable products of kind i become [18]:

$$\frac{V_A}{v_0 A_0} \cdot \frac{\widehat{S}_i}{n} = \frac{vA}{v_0 A_0} x_i - x_{i_0} \quad ; \quad \frac{vA}{v_0 A_0} = \frac{\sum m_j \cdot x_{j_0}}{\sum m_j \cdot x_j} , \tag{12.19}$$

where

$$V_A \widehat{S}_i = \int\!\!\int_{V_A}\!\!\int S_{i_A} \, dV + \lim_{l_P \to \infty} \int\!\!\int_{V_P}\!\!\int S_{i_P} \, dV .$$

V_A and V_P are the respective volumes of the reactor zones, v_0 and v the corresponding flow velocities, A_0 and A the respective cross sections, x_{i_0} and x_i relative concentrations, $x_i = n_i/n$, $n = \sum n_j = const.$, while x_{i_0} are input values. The term $\widehat{S}_i$ stands for an averaged effective (regarding particle gain and loss) and summary (regarding the combination of AZ and PZ) source rate. The local rates S_{i_A} and S_{i_P} contain the well-known expressions for gain and loss by collisions, e.g., $k_{ij} n_i n_j$.

Considering real plasma chemical conversions the microphysical analysis of $\widehat{S}_i$ generally includes an overwhelming number of different species and reactions. The specialization of

Eq. 12.19 with constant reactor cross-section results in

$$\frac{\tau_0 \widehat{S}_i}{n} = \frac{\sum m_j \, x_{j_0}}{\sum m_j \, x_j} \, x_{i_P} - x_{i_0} \quad \text{where} \quad \tau_0 = \tau_A = \frac{V_A}{v_0 A}, \tag{12.20}$$

$$\tau_0 \widehat{S}_i = \tau_0 S_{i_A} + \lim_{l_P \to \infty} (\tau_P S_{i_P}). \tag{12.21}$$

According to equation 12.20 the condition for similarity is now:

- Two reactors are plasma chemical similar (i.e., equal input values x_{i_0} result in equal output values x_{i_P} of the passive zone), if the reduced summary effective source terms $\tau_0 \widehat{S}_i / n$ correspond in both cases.

Within the microscopic picture the detailed analysis of the different source terms is extremely expensive. The *analysis of various kinetic models* showed that the output values are given by five determining variables $T, T_e, n, \tau_0, x_e = n_e/n$, which however appear in two combinations only: $k_i(T_e) n \tau_0 x_e$ and $k_j(T) n \tau_0$. With corresponding temperatures T and T_e, which is a general premise for similarity, we retain the combinations $n \tau_0 x_e$ and $n \tau_0$. Most surprisingly the first one, $n \tau_0 x_e = n_e \tau_0$, alone was sufficient to represent with high precision the output values. This is a direct consequence of the combined action of the active and passive reactor zones. Generally, similarity rules for an active zone can be drastically changed by the tandem connection of a passive zone (e.g., with regard to the role of forbidden processes and so on). Here from the first the unique characteristic value for plasma chemical similarity is $\tau_0 n_e$, and $\tau_0 \widehat{S}_i / n = f(\tau_0 n_e)$. Because the electron concentration n_e is an inherent plasma parameter of the active zone and not given immediately, it would be very desirable to process a connection to the operating parameters of the reactor. This is possible via Eq. 12.18. Introducing the pressure $p = nkT$, we obtain

$$\tau_0 n_e = \epsilon \frac{\tau_0 P}{p V_A} \quad ; \quad \epsilon = \frac{2}{3} \frac{\tau_e n}{\delta} \frac{kT}{e_0 U_e}. \tag{12.22}$$

Obviously, the central similarity quantity $\tau_0 n_e$ is mainly determined by a new basic dimensionless reactor parameter

$$R = \tau_0 P / p V_A = \tau_0 P / (V_A \, nkT) = (W/N)/(kT), \tag{12.23}$$

where W is the energy input, and N the number of particles. Regarding the physical meaning of this parameter, we note that

- R represents the energy invested per particle of the gas mixture during the flow through the active reactor zone in relation to the thermal energy kT, i.e. it is proportional to the well-known *specific energy* [5, 21].

The factor of proportionality ϵ in Eq. 12.22 is a true constant for constant gas and electron temperature, which is a premise for similarity. Of course to guarantee corresponding values of T and U_e, extra conditions must be fulfilled, as already mentioned. But under some restrictions on the operation of non-thermal plasma chemical reactor variations of T and U_e remain small, and $\epsilon \approx const$ is a tolerable approximation. The most important of these restrictions are sufficiently high gas pressure and sufficiently low Joule heating of the gas. The latter implies nearly constant reduced field stength E/n and U_e (or T_e), respectively. At small $R \to 0$ the reduced summary effective source terms are proportional to R (i.e., $\tau_0 \widehat{S}_i/n \sim R$, for generated species with $\tau_0 n_e \sim R$). Large $R \to \infty$ results in $\tau_0 \widehat{S}_i/n \to const$, and then the quasi-equilibrium states mentioned in section 12.2 are reached and can be interpreted as the outcome of an electronically modified mass action law, which describes a complex chemical situation by a reversible gross reaction (compare section 12.4).

12.3.3 Comparison with experimental results

To prove in a systematic manner the applicability of the dimensionless reactor parameter, the separate variations of all the four quantities which are of influence should be analyzed with regard to changes of the output values $x_i = n_i/n$ at equal input. According to the analysis given, only such plasmas should be included, which show nearly equal gas and electron temperatures. The energy distribution function of electrons, neglecting Coulomb interaction, is mainly determined by the reduced electric field strength E/n. Therefore we have to look on such plasma conditions which do not differ too much in E/n. Exemplarily we give the similarity representation of our measurements, using geometrical similar positive columns as the active zone of the flow reactor [17]. Figure 12.9 outlines the experimental arrangement. Very clearly the junction of E/n with a constant value at higher pressures is shown in figure 12.10 for discharges in CO_2. It is a rather general experience, that the value of E/n in electrical discharges controlled by bulk processes depends only slightly on current and pressure. In chemical active plasmas with many channels for generation and loss of charge carriers the approximation $E/n \approx const$ should be useful over rather extended regions of the operation

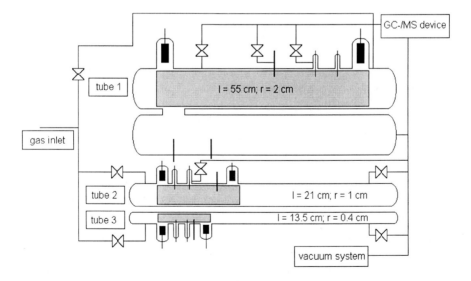

Figure 12.9: Schematic diagram of the positive column flow reactor with geometrical similar discharge tubes.

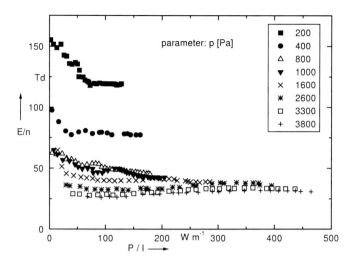

Figure 12.10: Reduced field strength E/n versus P/l in the positive column of CO_2 discharges ($r = 0.4$ cm).

parameters. The analysis of the chemical conversions was realized by the gaschromatographic technique (detection of CO_2, CO and O_2). Figures 12.11 and 12.12 summarize some results for the decomposition of CO_2. There is a clear compression of different experimental dates to one curve by using the similarity quantity $\tau_0 n_e$ or the reactor parameter R.

Experimental results on the chemical conversion of CH_4 in a microwave discharge ($f =$

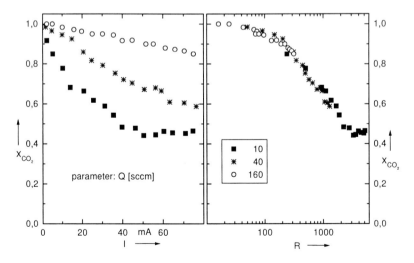

Figure 12.11: Dependence of the relative CO_2 concentration x_{CO_2} on the discharge current I and the reactor parameter R (tube 2 of figure 12.9, $p = 1600$ Pa).

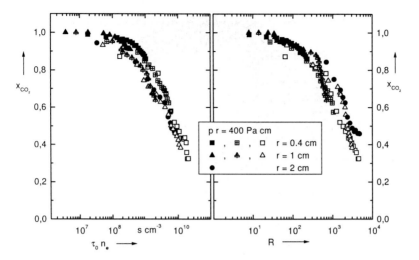

Figure 12.12: Dependence of the relative CO_2 concentration x_{CO_2} on $\tau_0 n_e$ and R at different values of τ_0 (tubes 1, 2, 3 of figure 12.9).

$2330 - 2350$ MHz) in flowing $Ar + 1\% CH_4$ mixtures, presented in Ref. [25], have been analyzed, too. The experiments were performed for different powers, pressures and residence times. The volume of the active reactor zone remained constant ($V_A = 12 \text{ cm}^3$). Figure 12.13 shows the decomposition of CH_4 and the simultaneous generation of H_2 with increasing investment of power as well as the obvious fusion of these different curves by introducing R.

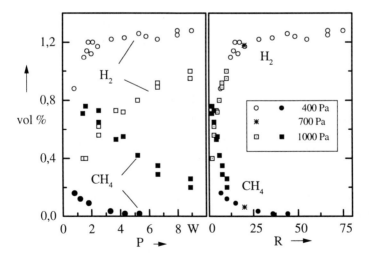

Figure 12.13: Measured concentrations (in volume %) of CH_4 and H_2 in dependence on the power P and on the reactor parameter R.

12.4 The method of generalized macroscopic kinetics

12.4.1 History and concept

The origin of macroscopic kinetics goes back to some papers of Warburg and his school [26], and to Becker [21], who the efficiency of chemical reactions in gas discharges brought in connection to important discharge parameters. Some of these purely empirical considerations include one fundamental point of view: the chemical changes of reactants during the flow through the discharge zone are determined by the power P invested in the plasma volume V_A and multiplied by the residence time τ_0 of the gas in the reaction zone (the Becker formula [21]). The development of such considerations to an independent method which involved basic ideas of chemical kinetics were performed by Eremin and co-workers (e.g., Ref. [5]). The denotion *macroscopic kinetics* first was used in the monograph of Drost [27]. A detailed analysis of Eremin's macroscopic kinetics showed its restriction to selected processes (e.g., small changes of gas composition ↔ small degrees of conversion, no consideration of back reactions, ...) [4, 7]. Therefore it seemed desirable to look for a more general base of macroscopic kinetics. This concept was named *method of generalized macroscopic kinetics* (e.g., Refs. [4, 7, 28]). Parallel efforts to generalize macroscopic kinetics have been attempted in Ref. [29]. Recently the macroscopic description of etching processes was used to predict their uniformity, too [30].

The starting point is a suitable schematized model of the non-thermal plasma chemical reactor. Within this concept both zones of a real reactor (AZ and PZ, compare Figs. 12.1 and 12.8) are summarized to a black box with input and output flows of only stable chemical components and with an effective source term for each of these species, representing a mean value (see section 12.3.2). The reactor model adequate to this physical situation is the back-mixing model, discussed above already.

12.4.2 The particle balance equations

Within the concept of macroscopic kinetics the volume reactions and reactive plasma-wall interactions (deposition, etching) are described – without exception – by reversible gross reactions, taking into consideration all stable components (see table 12.3 below). Transforming Eqs. 12.20 and 12.21, the particle balance equation for a stable component i within the back-mixing flow reactor model reads as follows:

$$\widehat{S}_i = \frac{n_i}{\tau} - \frac{n_{i_0}}{\tau_0}; \qquad \tau_0 = \frac{l_A}{v_0}; \qquad n = \sum n_j = const. \tag{12.24}$$

$$\widehat{S}_i = S_{i_A} + \lim_{l_P \to \infty} \left(\frac{l_P}{l_A} S_{i_P}\right) = G_i - L_i \quad \text{(Gain – Loss)}.$$

The balance equations include the conservation of the atomic components of the particles. Regarding chemical changes in the volume and reactive plasma wall reactions the ratio τ_0/τ is fixed by the equation of the mass flow

$$\frac{\tau_0}{\tau} = \frac{v}{v_0} = \frac{I}{I_0} \cdot \frac{\sum m_j n_{j_0}}{\sum m_j \cdot n_j}; \qquad I_0 = A v_0 \sum m_j n_{j_0}, \tag{12.25}$$

where I_0 and I are input and output mass flow in [g/s], respectively.

The main question regards the structure of the effective source term $\widehat{S}_i = G_i - L_i$. This structure must be correlated in a well-defined manner with the reversible gross reaction of the process j. In contrast to the microscopic kinetics the effective source terms $\widehat{S}_i$ are estimated by macroscopic rate coefficients κ_i which have to be determined by selected experiments [4, 7], i.e.,

$$\widehat{S}_i = \widehat{S}_i\left(\kappa^j_{G\,i}, \kappa^j_{L\,i}, n_1, ..., n_i, ..., n_l\right), \tag{12.26}$$

where the subscript i denotes the stable components, and the particular superscript j gross reactions (i.e., volume reactions, deposition processes and etching processes). The concept is restricted to *first order* reactions with regard to the concerning particles. In the case of the simple gross reaction $A + B + e \rightleftharpoons C + e$ the gain and loss terms, e.g., of the component A, have the form $G_A = \kappa_{G\,A}\,n_C$ and $L_A = \kappa_{G\,A}\,n_A n_B$, respectively.

Of special interest is the dependence $\kappa_i(n_e)$, or $\kappa_i(P/V_A)$. The proportionality $n_e \sim P/V_A$ is typical for non-thermal plasma sources, as is obvious from Eq. 12.18. The following ansatz shows a high flexibility:

$$\kappa_i(P/V_A) = k'_i \cdot (P/V_A) / \left[1 + a \cdot (P/V_A)\right] + k''_i \cdot (P/V_A), \tag{12.27}$$

where k'_i, k''_i, and a are empirical constants. This ansatz reflects the well-defined chemical quasi-equilibrium states CEEC and CECD of the reactor operation, which are characterized by the constants K^0 and K^∞. The ansatz especially describes both possible cases, CEEC = CECD (i.e. $K^0 = K^\infty$) as well as CEEC $\neq$ CECD (i.e. $K^0 \neq K^\infty$) (compare section 12.2).

The deposition and etching processes result in an alteration of the mass flow according to

$$\begin{aligned} I &= I_0 - \Delta I_D + \Delta I_E \tag{12.28} \\ \text{with} \quad \Delta I_D &= I^+_D - I^-_D \\ \text{and} \quad \Delta I_E &= I^+_E - I^-_E, \end{aligned}$$

where ΔI_D and ΔI_E are the rate of deposition (D) and the etching rate (E), respectively. The superscript $'+'$ either denotes deposition on the surface or the etching reaction, while the superscript $'-'$ relates to the respective back reaction. The rates describe volume averaged values. In the case of etching processes an additional information is given by the placement and dimension of the sample. Of course, the rates ΔI_D and ΔI_E of the reactive plasma-wall processes depend in well defined manner on the competent rate coefficients, too,

$$\Delta I_{D,E} = \Delta I_{D,E}\left(m_i, \kappa^j_{G\,i}, \kappa^j_{L\,i}, n_1, ..., n_i, ..., n_l\right). \tag{12.29}$$

Demonstration examples of the presented concept are discussed in the next section.

12.4.3 Demonstration examples

To demonstrate the method we discuss the following simple reversible gross reaction:

$$A + B + e \rightleftharpoons AB + e$$

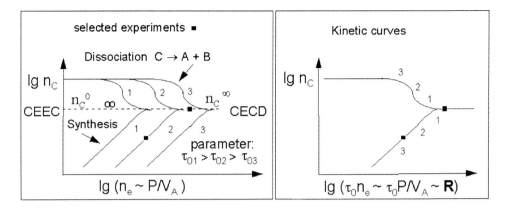

Figure 12.14: The principle set of characteristics n_C versus $n_e \sim P/V_A$, and the kinetic curves n_C versus $\tau_0 n_e$ for the reaction $A + B + e \rightleftharpoons C + e$ for the special situation CEEC = CECD. Full squares represent experimental points needed for the determination of macroscopic constants.

The nonlinear set of particle balance equations 12.24 then reads

$$\frac{n_A}{\tau} - \frac{n_{A_0}}{\tau_0} = \kappa_{GA}\, n_C - \kappa_{LA}\, n_A n_B$$
$$\frac{n_B}{\tau} - \frac{n_{B_0}}{\tau_0} = \kappa_{GB}\, n_C - \kappa_{LB}\, n_A n_B \qquad (12.30)$$
$$\frac{n_C}{\tau} - \frac{n_{C_0}}{\tau_0} = \kappa_{GC}\, n_A n_B - \kappa_{LC}\, n_C\,,$$

where $C \equiv AB$. Because of the conservation of atomic components follows $\kappa_{GA} = \kappa_{GB} = \kappa_{LC}$ and $\kappa_{LA} = \kappa_{LB} = \kappa_{GC}$. Therefore both directions of the reversible gross reaction are determined by only two independent macroscopic rate coefficients, e.g., κ_{LC} and κ_{GC}. The restriction of the ansatz 12.27 to the simple case with $k_i' = 0$ yields

$$\kappa_i = k_i'' n_e \sim k_i\,(P/V_A)\,. \qquad (12.31)$$

Then for the limiting cases of reactor operation, $n_e \to 0$ (at $\tau \to \infty$) and $n_e \to \infty$, from the system of particle balance equations 12.30 identical constants $K^0 = K^\infty = K_e$ (see Eqs. 12.2, 12.3, and 12.6) are obtained, which characterize identical quasi-equilibrium states, CEEC = CECD,

$$K^0 = \frac{n_A^0 n_B^0}{n_C^0} = \frac{k_{LC}}{k_{GC}} \qquad \text{and} \qquad K^\infty = \frac{n_A^\infty n_B^\infty}{n_C^\infty} = \frac{k_{LC}}{k_{GC}}\,. \qquad (12.32)$$

Therefore the knowledge of K_e and k_{GC} alone enables the complete macroscopic description of the discussed reversible gross reaction. In this special case the conversion is estimated by the similarity parameters $\tau_0 n_e$ or $\tau_0\,(P/V_A) \sim R$. All curves coincide into two kinetic curves for the direct as well as for the back reaction. This situation is schematically illustrated

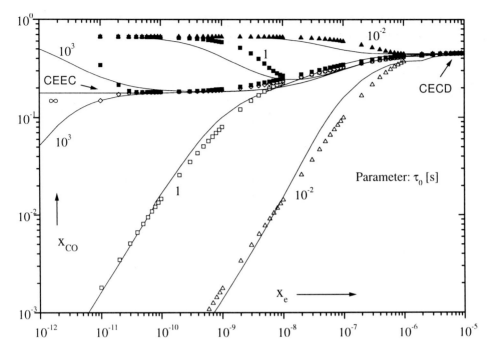

Figure 12.15: Dependencies x_{CO} versus ionization degree x_e, calculated by the generalized macroscopic kinetics (lines) and microscopic kinetics (symbols) methods; conditions: CEEC $\neq$ CECD; $p = 20$ mbar.

in figure 12.14. The necessary constants K_e and k_{GC} can be estimated by (only) two selected experiments (marked by squares).

The flexibility and usefulness of the ansatz Eq. 12.27 is illustrated for different quasi-equilibrium states, CEEC $\neq$ CECD. This typical situation is shown in figure 12.15 for the gross reaction

$$2\,CO_2 + e \rightleftharpoons 2\,CO + O_2 + e.$$

The results of a microscopic model and those on base of the generalized macroscopic kinetics are compared. Both models describe the plasma chemical conversion in a flow reactor with the non-thermal plasma of the positive column (AZ). The microscopic plug-flow model considers the most important elementary processes, including the species C, O, O_2, CO, and CO_2 (10 reaction channels, see Ref. [9]). Further input values are the reduced electrical field strength E/n, the ionization degree x_e, and the residence time τ_0. Within this frame the plasma chemical process was studied in the AZ ($x_e > 0$), followed by the relaxation into the stable products in the PZ ($x_e = 0$). Note that the macroscopic kinetics requires 5 constants only:

$$K^0 = 0.0255\,;\qquad K^\infty = 0.3115\,;$$
$$n^2 k'_{L\,CO} = 2 \times 10^9 s^{-1}\,;\quad n^2 k''_{L\,CO} = 8 \times 10^8 s^{-1}\,;\quad a\,n = 10^7\,.$$

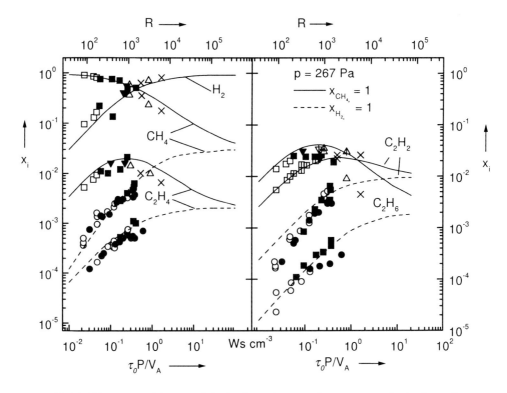

Figure 12.16: The composition x_i of the hydrocarbons at the reactor outlet in dependence on the specific energy and the reactor parameter; symbols: experimental results for various τ_0 (comp. table 12.2); lines: macroscopic kinetics.

The results of the microscopic and the macroscopic model agree astonishing well. The different levels of the quasi-equilibrium states CEEC $\neq$ CECD can be seen clearly. Of course, in this case the reactor parameter R is not a similarity parameter.

12.4.4 Macroscopic modeling of experimental results

Investigations on the formation as well as the decomposition of hydrocarbons in non-thermal plasmas are of significant practical relevance. For the characterization of these complex processes, including etching and deposition reactions, the test of the concept of macroscopic kinetics is of high interest.

The discharge operated in graphite hollow cathodes under flow conditions, starting from methane or hydrogen. The operation parameters are summarized in table 12.2. The composition of the stable reaction products was detected by the gaschromatographic technique. The cathode (as well parts before and behind it) could be removed for the estimation of the etching rate and polymer deposition rate. Beside the input components H_2 and CH_4, the major stable reaction products were the light hydrocarbons C_2H_2, C_2H_4, and C_2H_6. The

Table 12.2: Operation parameters.

graphite hollow cathode	diameter $d = 8$ mm, length $l = 30$ mm
process gas and pressure	hydrogen, methane, $p = 270$ Pa
power density	$P/V_A = 5 \ldots 100$ W cm^{-3}
residence time	$\tau_0 = 2 \ldots 100$ ms

heavier hydrocarbons $C_3H_4, .., C_3H_8$ could be detected in traces only (< 5 mass %). As can be seen in Figs. 12.16 and 12.17 the process is surprisingly well controlled by the parameter $\tau_0 \, (P/V_A) \sim R$.

The macroscopic kinetics starts from 13 parallel reversible gross reactions, which are summarized in table 12.3. According to the concept, presented in section 12.4.2, the particle balance equations must be formulated for every stable component under consideration of all gross reactions $j = V_1 \ldots V_5$ (volume reactions), $j = E_1 \ldots E_4$ (etching processes) and $j = D_1 \ldots D_4$ (deposition processes). With the specified ansatz Eq. 12.27, $\kappa_i = k_i^j \, P/V_A$ from the system of particle balance equations 12.24, and under consideration of the conservation of atomic components a set of macroscopic rate coefficients and quasi-equilibrium

Table 12.3: Considered gross reactions for the plasma conditions given in table 12.2. $\bar{x} = H/C_{\text{layer}}$, $\tilde{x}_{C H4,\max} = 4$, $\tilde{x}_{C2H2,\max} = 1$, $\tilde{x}_{C2H4,\max} = 2$, and $\tilde{x}_{C2H6,\max} = 3$.

V_1	$2\,CH_4 + e$	$\rightleftharpoons$	$C_2H_2 + 3\,H_2 + e$
V_2	$2\,CH_4 + e$	$\rightleftharpoons$	$C_2H_4 + 2\,H_2 + e$
V_3	$2\,CH_4 + e$	$\rightleftharpoons$	$C_2H_6 + H_2 + e$
V_4	$C_2H_4 + e$	$\rightleftharpoons$	$C_2H_2 + H_2 + e$
V_5	$C_2H_6 + e$	$\rightleftharpoons$	$C_2H_4 + H_2 + e$
E_1	$C_{wall} + 2\,H_2 + e$	$\rightleftharpoons$	$CH_4 + e$
E_2	$C_{wall} + 0.5\,H_2 + e$	$\rightleftharpoons$	$0.5\,C_2H_2 + e$
E_3	$C_{wall} + H_2 + e$	$\rightleftharpoons$	$0.5\,C_2H_4 + e$
E_4	$C_{wall} + 1.5\,H_2 + e$	$\rightleftharpoons$	$0.5\,C_2H_6 + e$
D_1	$CH_4 + e$	$\rightarrow$	$(2 - 0.5\tilde{x}_{CH4})\,H_2 + CH_{\tilde{x}CH4} + e$
		$\leftarrow$	$(2 - 0.5\bar{x})\,H_2 + CH_{\bar{x}} + e$
D_2	$0.5\,C_2H_2 + e$	$\rightarrow$	$(0.5 - 0.5\tilde{x}_{C2H2})\,H_2 + CH_{\tilde{x}C2H2} + e$
		$\leftarrow$	$(0.5 - 0.5\bar{x})\,H_2 + CH_{\bar{x}} + e$
D_3	$0.5\,C_2H_4 + e$	$\rightarrow$	$(1 - 0.5\tilde{x}_{C2H4})\,H_2 + CH_{\tilde{x}C2H4} + e$
		$\leftarrow$	$(1 - 0.5\bar{x})\,H_2 + CH_{\bar{x}} + e$
D_4	$0.5\,C_2H_6 + e$	$\rightarrow$	$(1.5 - 0.5\tilde{x}_{C2H6})\,H_2 + CH_{\tilde{x}C2H6} + e$
		$\leftarrow$	$(1.5 - 0.5\bar{x})\,H_2 + CH_{\bar{x}} + e$

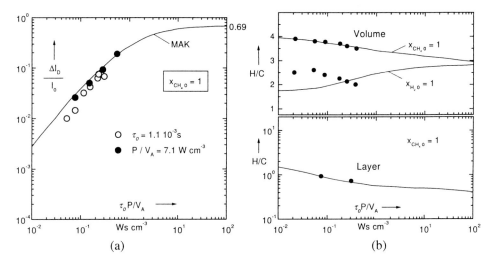

Figure 12.17: (a) The rate of polymer deposition referred to the methane input I_0, (b) the averaged ratio H/C of all the hydrocarbons in the volume (top) and of the polymer layer (bottom), in dependence on the specific energy (symbols: experimental results; lines: macroscopic kinetics).

constants K_i follows. The detailed analysis showed, that for the description of the complex process 13 k_i and 4 K_i are necessary [28]. These constants could be estimated by selected experiments, analogous to the procedure demonstrated in section 12.4.3. The calculated dependencies are shown in Figs. 12.16 and 12.17. As can be seen, the macroscopic kinetics represents in sufficient manner the experimental results. In particular, the model allows to predict the deposition rate (see figure 12.17 a), the etching rate of graphite (not shown here), and the averaged ratio $\bar{x} = H/C$ of the polymer layer (see figure 12.17 b).

12.5   Summary

Under non-thermal conditions the analysis of plasma chemical conversions demonstrated the occurrence of selected states within the corresponding reactors, which could be interpreted as chemical quasi-equilibrium states. The critical condition is the occurrence of hot electrons ($T_e \geq 1$ eV) in the active zone of the reactor. Various experiments and kinetic models support this interpretation. Chemical quasi-equilibria are simple in the verification. For instance, a few measurements with a closed reactive system are sufficient to describe this system as well as the corresponding flow reactor, too. The significance and capability of the chemical quasi-equilibria are founded on the following fact: These states represent the basis and the suppositions for the method of macroscopic kinetics. In the case of identical equilibrium states, CEEC $\equiv$ CEED, the reactor parameter is the decisive parameter. For CEEC $\neq$ CECD a generalized macroscopic kinetics is applicable.

This result may be generalized and is linked with the application of similarity principles for the reactor operation. After a general formulation of such principles, starting from the balance equations of particles and energy, the dimensionless reactor parameter is formulated, characterizing the composition of the effluent gas. The applicability of this parameter is demonstrated by experimental results.

The method of macroscopic kinetics is applicable to non-thermal plasma chemical reactors of different types (glow discharge, radiofrequency discharge, corona discharge, etc.). It allows the compact description of volume reactions as well as of reactive plasma-wall processes (deposition, etching).

References

[1] S. Veprek, J. Chem. Phys. **57**, 952 (1972)

[2] J. Preiss, Collect. Czechosl. Chem. Commun. **38**, 635 (1973)

[3] S. Veprek, Pure and Appl. Chem. **54**, 1197 (1982)

[4] A. Rutscher, H.-E. Wagner, Proc. ICPIG-18, Inv. Papers, Swansea (1987), p. 172

[5] E. N. Eremin, Elementy Gazovoi Elektrochimii, Moskva (1968)

[6] H.-E. Wagner, A. Huczko, Proc. ISPC-10, Bochum (1991), p. 3.2-18

[7] A. Rutscher, H.-E. Wagner, Contr. Plasma Physics **25**, 315 (1985)

[8] A. Rutscher, H.-E. Wagner, Plasma Sources Sci. Technol. **2**, 279 (1993)

[9] A. Sonnenfeld, H. Strobel, H.-E. Wagner, J. Non-Equilib. Thermodyn. **23**, 105 (1998)

[10] T. Lang, J. Laimer, L. Störi, H. Störi, Proc. ISPC-11, vol. **4**, 1338 (1993)

[11] A. Cenian, A. Chernukho, V. Borodin, Contr. Plasma Physics **35**, 273 (1995)

[12] A. Marotta, ISPC-11, vol. **1**, 240 (1993)

[13] N. Hershkowitz, R. Breun, J. Ding, K. Kimse, C. Lai, J. Meyer, T. Quick, J. Taylor, R.C. Woods, J.Z. Wu, A. Wendt, C. Yang, Proc. ISPC-12, Minneapolis, vol. **1**, 533 (1995)

[14] S. Pfau, A. Rutscher, K. Wojaczek, Beitr. Plasmaphysik **9**, 333 (1969)

[15] Y.P. Raizer, Gas Discharge Physics, Springer Verlag (1991)

[16] S. Kattanek, R. Gröger, C. Bode, Ähnlichkeitstheorie, VEB Deutscher Verlag für Grundstoffindustrie, Leipzig (1967)

[17] F. Miethke, A. Rutscher, H.-E. Wagner, Acta Physica Universitatis Comenianae (Bratislava) **XLI**, 3 (2000)

[18] H. Jacobs, F. Miethke, A. Rutscher, H.-E. Wagner, Proc. ISPC-12, Minneapolis (1995), p. 469

[19] A. Rutscher, H. Deutsch, Editors, Wissensspeicher Plasmatechnik, VEB Fachbuchverlag, Leipzig (1983)

[20] F. Miethke, H.-E. Wagner, A. Rutscher, S. Gundermann, J. Phys. Chem. A **103**, 2024 (1999)

[21] H. Becker, Wissenschaftliche Veröffentlichungen SIEMENS-Konzern **7**, 76 (1920)

[22] H. Jacobs, F. Miethke, A. Rutscher, H.-E. Wagner, Contr. Plasma Physics **36**, 471 (1996)

[23] W. Lucke, F. Miethke, A. Rutscher, H.-E. Wagner, Proc. ESCAMPIG-11, St. Petersburg (1992), p. 438; W. Lucke, F. Miethke, S. Pfau, A. Rutscher, H.-E. Wagner, Proc. ISPC-11, Loughborough (1993), p. 1356

[24] T. Ruzika, A. Rutscher, S. Pfau, Ann. Physik **24**, 124 (1970)

[25] Yu. A. Gerasimov, T.A. Gracheva, Yu.A. Lebedev, Chim. Vys. Energii **17**, 270 (1983)

[26] E. Warburg, Jahrbuch der Radioaktivität **6**, 181 (1909)

[27] H. Drost, Plasmachemie, Akademie-Verlag, Berlin (1978)

[28] F. Miethke, A. Rutscher, H.-E. Wagner, Proc. ISPC-13, Peking, vol. 2, 1584 (1997)

[29] H. Schlemm, Wissenschaftliche Schriftenreihe der TU Karl-Marx-Stadt **2** (1987)

[30] D. L. Flamm, J. P. Verboncoeur, Proc. ISPC-13 (Supplements), Peking (1997), p. 2037

13 High-pressure plasmas: dielectric-barrier and corona discharges – properties and technical applications

Ulrich Kogelschatz* and Jürgen Salge†

* ABB Corporate Research Ltd., Segelhof 1, CH-5405 Baden, Switzerland
† Institut für Hochspannungstechnik und Elektrische Energieanlagen, TU Braunschweig, Braunschweig, Germany

13.1 Introduction

Dielectric-barrier discharges (DBDs) and corona discharges represent self-sustained non-equilibrium electrical gas discharges that can be operated at pressures of the order of 1 bar. Dielectric-barrier discharges are characterised by insulating layers on one or both electrodes or on dielectric structures inside the discharge gap. Corona discharges depend on inhomogeneous electric fields at least in some parts of the electrode configuration, thus restricting the primary ionisation processes to a small fraction of the interelectrode region. At atmospheric pressure the physical processes in both types of discharges are similar and resemble those in transient high-pressure glow discharges. Both discharges can be reliably operated up to high power levels in the megawatt range, a property which is of importance for industrial applications. Sometimes a certain mixture between dielectric-barrier discharges and corona discharges exists and, to complete the confusion, dielectric-barrier discharges are often called *corona discharges*.

The first dielectric-barrier discharge, sometimes also referred to as silent discharge or barrier discharge, was proposed by W. Siemens in 1857 [1]. Scientific descriptions of corona discharges can be traced back to E. Warburg [2] and J. S. Townsend [3], while practical engineering formulae for corona onset were formulated by F.W. Peek Jr. [4], and J.B. Whitehead and W.S. Brown [5] in the early 20th century.

13.2 Dielectric-barrier discharges

Usually DBDs at atmospheric pressure consist of many tiny parallel current filaments referred to as *microdischarges*. Under special conditions also homogeneous discharges can be obtained. Most of the applications are operated up to now with filamentary discharges; but homogeneous discharges are coming up and open new perspectives.

13.2.1 Filamentary discharges

13.2.1.1 Electrode configurations and discharge evolution

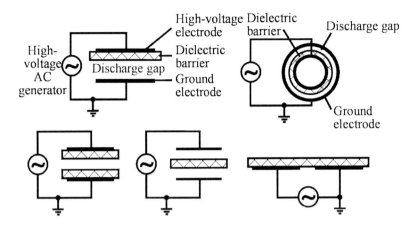

Figure 13.1: Common planar and cylindrical dielectric-barrier discharge configurations.

Typical electrode configurations for DBDs are given in Fig. 13.1. Common materials for dielectric barriers are glass, quartz, ceramics or enamels. Also plastic films, silicon rubber and teflon plates can be used. In the filamentary mode the discharge starts at atmospheric pressure with breakdown processes at many points. The breakdown and the development of a microdischarge has been studied by several authors, e.g., Refs. [6, 7]; the 4 typical stages are shown schematically in Fig. 13.2. If the electric field in the gas gap is sufficiently high to initiate avalanches, the breakdown starts with the Townsend phase. Next a streamer occurs and a conducting channel – the filament – is formed. Now charges are transferred through the channel and accumulate at the dielectric surface. The voltage across the filament is compensated and the discharge dies out (phase No. 4). The complete discharge development takes only about some 10^{-9} s.

Fig. 13.3 shows a characteristic Lichtenberg figure recorded directly on a photographic film together with the experimental arrangement [8]. The emulsion of the film was facing the discharge gap, and a glass plate acted as a dielectric barrier. After applying a single steep rectangular voltage pulse of 15 kV the film was taken out in a dark room and developed. Nearly the whole surface area is involved. Besides numerous footprints of microdischarges gliding discharges fed from the filaments cover the surface. Intensity and extension of footprints and surface discharges differ remarkably. Obviously different conditions exist for each microdischarge. One major reason seems to be that in spite of the steep voltage pulse not all the discharges are initiated at the same time. Those ignited first touch the dielectric surface first, and gliding discharges can spread undisturbed over a large area. The scope of later arriving discharges is limited. Such a situation is shown schematically in Fig. 13.4.

In consequence also microdischarge properties (i.e. current conducting time, channel di-

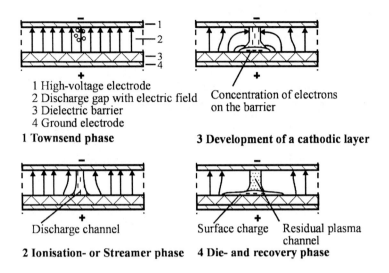

1 High-voltage electrode
2 Discharge gap with electric field
3 Dielectric barrier
4 Ground electrode

Concentration of electrons
on the barrier

1 Townsend phase

3 Development of a cathodic layer

Discharge channel

Surface charge   Residual plasma
channel

2 Ionisation- or Streamer phase   **4 Die- and recovery phase**

Figure 13.2: Microdischarge development.

mensions, channel heating, radiation intensity) are different. The further development of new discharges is determined mainly by the charges deposited on the dielectric surface by preceding microdischarges together with the applied voltage. Four essential situations are illustrated schematically in Fig. 13.5, if a sinusoidal voltage is applied.

At t_1 an electric field is established by the applied voltage. At t_2 it is assumed that microdischarges are established simultaneously, which short-circuit the gas gap; surface dis-

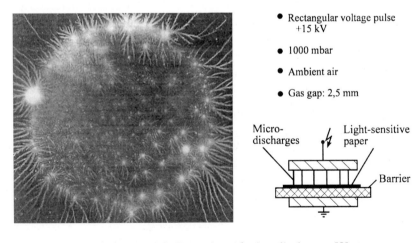

- Rectangular voltage pulse +15 kV
- 1000 mbar
- Ambient air
- Gas gap: 2,5 mm

Micro-
discharges

Light-sensitive
paper

Barrier

Figure 13.3: Footprints of microdischarges [8].

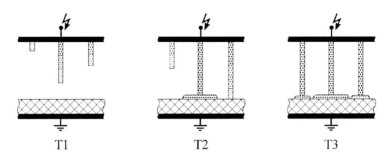

Figure 13.4: Microdischarge development at different discharge onset (schematically).

charges described above cover the whole dielectric barrier. The microdischarges extinguish, and the resulting voltage across the gas gap is close to zero. The whole process lasts only some ns. In the gas gap remain channels of the extinguished microdischarges; their conductivity decreases as the recovery continues. These channels represent privileged locations

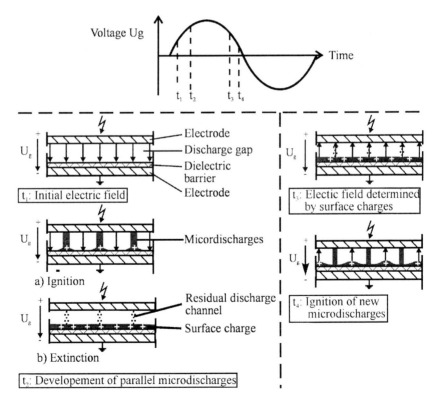

Figure 13.5: Development of filamentary dielectric-barrier discharges.

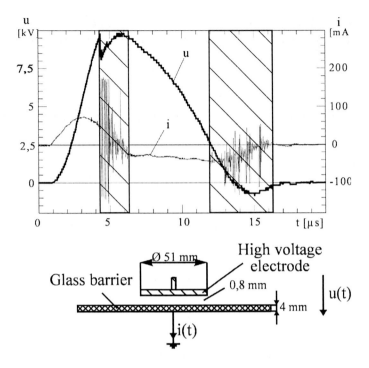

Figure 13.6: Voltage and current wave form of a barrier discharge.

for the ignition of new microdischarges, if the applied voltage is reversed. At t_3 the voltage across the gas gap is dominated by the memory charges deposited at the surface, and at t_4 new microdischarges of opposite polarity occur. These discharges prefer the residual channels of previous microdischarges, if their channels have not recovered sufficiently in the meantime.

In reality the connection between applied voltage, discharge establishment, memory charges on the dielectric barrier, and reignition is more complex. As discussed above the microdischarges develop with differences in time. Fig.13.6 elucidates the situation for a single voltage pulse [7]. The current through the barrier system indicates two discharge periods. The second period starts before the applied voltage has changed its polarity.

13.2.1.2 Microdischarge properties

Having in mind the complex discharge evolution processes it is not easy to define properties valid in general. It can be stated that at atmospheric pressure microdischarges, which have a duration of a few nanoseconds, reach current densities of the order of 100 A/cm^2 in thin cylindrical filaments of about 100 μm radius. In these discharge channels and in the surface discharges belonging to them non-equilibrium plasma conditions are obtained that can be characterised as transient high-pressure glow discharges. These conditions are ideal for the generation of ions and excited atomic and molecular species that can be utilised for volume and surface reactions and for the efficient generation of wavelength selective radiation.

13.2.1.3   Ionisation, dissociation and ensuing plasma chemistry

To initiate the discharge the electrical field is raised above a critical value at which electrical breakdown occurs, and initial electrons gain sufficient energy in the electric field to sustain ionisation processes. At atmospheric pressure breakdown normally immediately leads to spark or arc formation by heating up the breakdown channel. In the two discharge types discussed in this chapter this thermalisation of the plasma is prevented by allowing critical conditions only for an extremely short time (DBDs, pulsed coronas) or only in restricted areas (direct current corona discharges). Under such conditions electron multiplication can be controlled without leading to complete breakdown of the gas, and non-equilibrium plasma conditions can be established in a reliable way. For the applications discussed here energetic electrons are utilised to excite, dissociate and ionise species of the background gas in order to initiate plasma chemical reactions or provide ions for charging solid particles or droplets.

An example for the controlled action of an excited precursor species is the formation of Xe_2^* excimer complexes that provide strong vacuum ultraviolet radiation in excimer lamps and in plasma displays for the excitation of red, green or blue phosphors.

$$e^- + Xe \quad \rightarrow \quad e^- + Xe^* \,(^3P_{1,2}) \tag{13.1}$$
$$Xe + Xe^* + M \quad \rightarrow \quad Xe_2^* \,(^1\Sigma_u^+,\, ^3\Sigma_u^+) + M \tag{13.2}$$
$$Xe_2^* \quad \rightarrow \quad Xe + Xe + \text{VUV radiation (172 nm)}. \tag{13.3}$$

In these reactions M is a third collision partner which can be another xenon atom or that of a buffer gas like, e.g., neon or helium. An example for the utilisation of ionic reactions is the formation of $XeCl^*$ excimer complexes, a process which is used in commercial excimer lamps.

$$e^- + Xe \rightarrow e^- + Xe^+ + e^- \quad \text{and} \quad e^- + Cl_2 \rightarrow Cl^- + Cl \tag{13.4}$$
$$Xe^+ + Cl^- \,(+M) \quad \rightarrow \quad XeCl^* \,(+M) \tag{13.5}$$
$$XeCl^* \quad \rightarrow \quad Xe + Cl + \text{UV radiation (308 nm)}. \tag{13.6}$$

Dissociation by electron collisions is by far the most important process leading to plasma chemical synthesis like ozone formation or to pollution control by decomposition of toxic compounds. The result of the dissociation reactions can be ground state or excited atoms or highly reactive molecular fragments. As an example the dissociation of O_2 molecules by electron impact is discussed. Starting from ground state O_2 molecules two reaction paths towards dissociation are available: via excitation of the A $^3\Sigma_u^+$ level with an energy threshold of about 6 eV and via excitation of the B $^1\Sigma_u^-$ level starting at 8.4 eV.

$$e^- + O_2 \rightarrow e^- + O_2(A\,^3\Sigma_u^+) \rightarrow e^- + O(^3P) + O(^3P) \tag{13.7}$$
$$e^- + O_2 \rightarrow e^- + O_2(B\,^1\Sigma_u^-) \rightarrow e^- + O(^1D) + O(^3P). \tag{13.8}$$

Thus, in principle, electrons of an average energy of about 6–9 eV are required. In oxygen this corresponds to reduced fields E/n of about 150–350 Td (1 Townsend [Td] corresponds to 10^{-17} Vcm^2). Since, according to Paschen, E/n at breakdown is only a function of the product particle density times gap spacing (nd) this parameter can be influenced by adjusting the pressure or the width of the discharge gap.

Discharges in air or flue gases are more complicated. Excitation and dissociation of nitrogen molecules lead to a number of additional reaction paths involving nitrogen atoms and the excited molecular states N$_2$ (A $^3\Sigma_u^+$) and N$_2$ (B $^3\Pi_g$). They result in the formation of different nitrogen oxides, N$_2$O, NO, NO$_2$, NO$_3$, N$_2$O$_5$, and in the formation of additional oxygen atoms that lead to enhanced ozone formation.

In the presence of humidity, reactions of water molecules lead to the generation of OH radicals, a very powerful oxidising species. OH radicals are formed by direct electron impact dissociation of H$_2$O and by fast reactions of electronically excited oxygen atoms and nitrogen molecules. OH radicals are of major importance in flue gas treatment and in tropospheric chemistry.

$$e^- + H_2O \quad \rightarrow \quad H^- + OH \tag{13.9}$$

$$O(^1D) + H_2O \quad \rightarrow \quad 2\,OH \tag{13.10}$$

$$N_2(A\ ^3\Sigma_u^+) + H_2O \quad \rightarrow \quad N_2 + OH + H\,. \tag{13.11}$$

In most atmospheric-pressure non-equilibrium discharges the desired chemical changes are due to free radical reactions rather than to the direct action of electrons or ions. Free radical concentrations can be several orders of magnitude higher than typical charged particle concentrations. In general, excited atoms or molecules are much more reactive than their respective ground states. Since these reactions are not subjected to classical thermodynamic equilibrium limitations plasma chemistry in non-equilibrium discharges has found a number of interesting novel applications [9]–[14].

13.2.1.4   Discharge control

There exist a variety of methods to control barrier discharge. Besides voltage wave form, voltage amplitude, frequency, and capacitive energy stores connected in parallel to the electrodes, the electrode configuration, the design of the dielectric barriers, the kind of gas, the gas flow in the gap determine the microdischarge distribution and their intensity. In Fig. 13.7 different examples are illustrated.

For all applications it is important to obtain a statistically uniform microdischarge distribution in the gas gap and on the surfaces, respectively. The recovery behaviour of the extinguished discharge channels has a dominating influence on this distribution. This can be controlled efficiently either by a gas flow through the discharge gap and/or by time intervals without discharges, which can be achieved by using repetitive voltage pulse trains (see Fig. 13.7) [15].

13.2.1.5   Numerical modeling

Different aspects of dielectric-barrier discharge modeling have been addressed. For a determination of the different time constants involved it often is convenient to disregard spatial gradients in the beginning and treat a homogeneous plasma as a first step. To simulate the action of a short current pulse either a short high-voltage pulse is applied or a monoenergetic electron concentration is assumed as an initial condition. In both cases it is necessary to derive the rate coefficients for electron collisions in the gas mixture under consideration by solving the Boltzmann equation. This requires reliable sets of electron collision cross sections for all major components of the gas mixture. Normally the local field approximation is used, which assumes that the electron energy distribution is in equilibrium with the electric field and that

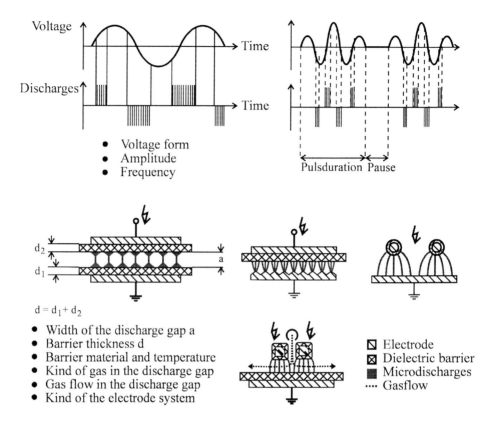

Figure 13.7: Possibilities to control barrier discharges (schematically).

all rate coefficients can be tabulated as a sole function of the mean electron energy or the reduced field E/n. Since, at atmospheric pressure, an electron needs only a few ps to reach equilibrium, in many cases it is justified to work with stationary solutions of the Boltzmann equation [9], [12]–[14].

Excitation and dissociation by electron collision are extremely fast processes followed by free radical reactions that occur at an intermediate time scale, typically 1 to 100 μs, at atmospheric pressure. Most free radical reactions are completed before any substantial displacement of the involved species by diffusion or convection can take place. These processes take longer and occur at ms time scales.

A first approach to simulate the action of successive discharge pulses in complicated gas mixtures is sometimes to disregard the electron kinetics by respectively injecting certain concentrations of free radicals and then compute the following chemical reactions. This approach is taken, if reliable electron collision cross sections are missing for some components of the gas mixture under consideration.

Two-dimensional modeling of microdischarge formation yielded additional information on the properties of individual current filaments [12]–[14], [16, 17]. This is closely related to computing electrical breakdown in atmospheric-pressure gases. When an overvoltage is applied to the discharge gap, an electron avalanche starting at the cathode soon reaches a critical stage where the local *eigenfield* caused by space charge accumulation at the avalanche head leads to a situation, where extremely fast streamer propagation towards both electrodes is initiated. Streamer propagation is caused by ionisation waves traveling at a speed much higher than the electron drift velocity. Such computations come to the conclusion that extremely high electric fields occur at the streamer head, that a slender conductive channel (diameter of the order 100 μm) is formed with maximum electron densities in the range of 10^{14} to 10^{15} cm^{-3}. Charge accumulation at the dielectric surface leads to a collapse of the electric field at this location and thus chokes the microdischarge typically after a few ns.

The transported charge and the energy dissipated in a microdischarge can be influenced by several parameters: gas properties and density, thickness and permittivity of the dielectric barrier. Depending on the application microdischarge properties can be tailored to optimise the ensuing plasma chemical reactions.

13.2.2   Homogeneous discharges

It has also been demonstrated that dielectrically controlled homogeneous or uniform discharges at atmospheric pressure can be established under special conditions [18]–[21]. In principle these discharges allow a completely uniform treatment of gas volumes and surfaces. The Okazaki group members have been pioneers in this field since 1987 [21]. If reliable control can be provided and if an energy transfer into the discharge is obtained comparable to that of filamentary discharges, this type of discharges seems to be of particular interest in view of industrial applications.

Fig. 13.8 shows schematically an electrode configuration developed by Kogoma and Okazaki for the generation of ozone. It consists of two metal foils covered with an metal mesh and ceramics plates. This arrangement allows to generate glow discharges at atmospheric pressure in helium, air, argon and oxygen even when using a 50 Hz power source. A further

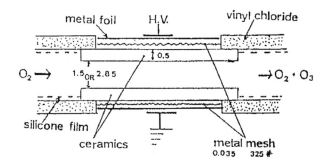

Figure 13.8: Electrode configuration of Kogoma and Okazaki [18].

improvement is obtained by using polyethylene terephthalate plates instead of ceramics; in this case glow discharges have been observed also in nitrogen [21]. At Toulouse dielectrically controlled glow discharges at atmospheric pressure have been obtained in helium and other gases between two plane electrodes, 4 cm in diameter, covered by an alumina layer 0.6 mm thick [20]. The gas gap could be extended up to 2 cm. The discharge evolution is characterised by a single current peak per half cycle of the applied voltage. Amplitude and duration of those peaks depend on a number of different parameters, i.e., electrode configuration, barrier material, dimension of gas gap and dielectric barriers, kind of gas, and the frequency used.

Fig. 13.9 shows a typical current pulse together with the voltage wave form obtained from a uniform discharge in nitrogen at atmospheric pressure generated in an electrode configuration similar to that of Fig. 13.8 [22]. The electrodes consist of copper foils covered with wire mesh and Mylar foils 350 μm thick. A sinusoidal voltage of 50 Hz was applied. The large current peak of nearly 7 A and the short pulse duration of 15 ns seems to be caused mainly by the dielectric material and by the electric field formed by the wire mesh. The barrier is an electret able to accumulate charges on its surface. Supported by the applied voltage charge carriers are trapped more or less uniformly on the surface. If the electric field changes its polarity and a certain threshold is exceeded, the charge carriers are expelled spontaneously from the surface and the development of filaments is prevented. Up to now it is difficult to control homogeneous glow discharges at atmospheric pressure sufficiently. For instance changes of the electrode configuration or small variations of the amplitude of the applied voltage cause a turnover into a filamentary discharge mode. For industrial applications this could be a severe draw back compared to filamentary discharges.

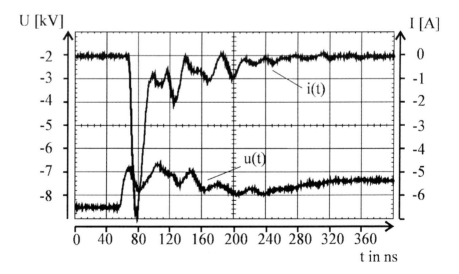

Figure 13.9: Current pulse and voltage waveform of a uniform discharge in nitrogen (gas gap: 2 mm).

13.2.3 Applications

13.2.3.1 Surface treatment and modification, coating

The treatment of surfaces by dielectrically controlled barrier discharges is a well-established method to improve surface properties like wettability or adhesion [23]. Usually the surface to be treated is moved continuously and exposed to filamentary discharges in air at atmospheric pressure. Fig. 13.10 shows schematically the arrangement for the treatment of conductive

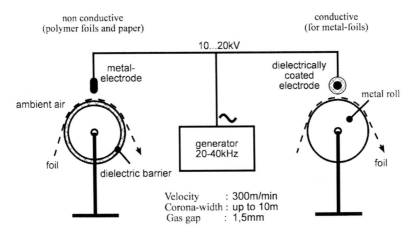

Figure 13.10: Arrangement for the treatment of foils with barrier discharges.

and non-conductive foils. The foils are moved at speeds up to a few hundred m/min and exposed to the discharges established between the electrodes. Assemblies of dielectrically coated electrodes or metallic knife edges are used. Foils up to 10 m width are treated which requires amounts of electric power in the range of 100 kW at frequencies in the range of 20–40 kHz. Beside foils also textiles and non-woven materials, which are porous and partially permeable for discharges, have been treated successfully [24].

In comparison to common treatment in air significant further improvements can be achieved by using reactive gases or gas mixtures instead of air [15]. This treatment can be performed in both closed- and open-air systems.

The basic design of an open-air system which allows a treatment of moving foils is given in Fig. 13.11. The storage coils are arranged in open air. The foil is pulled across a metal roll coated with a dielectric barrier. A box which forms small slits with the coated metal roll and the foil contains a number of electrodes and exhaust systems. The atmosphere in the box and, in particular, in the discharge areas between the electrodes and the foil is controlled by the gases supplied through the electrodes. The gas mixture is exhausted from the box via a gas absorbing and conditioning unit to the open air.

A typical example for the efficiency of this method is the improvement of the surface tension and its stability of biaxial oriented polypropylene (BOPP) foils [15]. Fig. 13.12 shows that a conventional treatment in air is insufficient. If the foil is first exposed to discharges

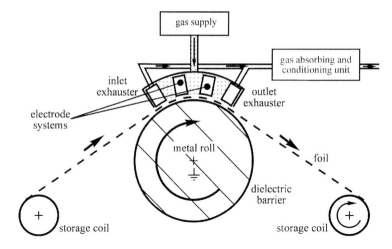

Figure 13.11: Basic design of a system for foil treatment (Softal).

in argon and then to discharges in acetylene, the surface tension increases to 72 mN/m and remains constant with time.

Barrier discharges at atmospheric pressure are also suited to deposit thin films on different substrates. Coatings up to some μm thickness have been obtained using reactive precursors [7, 8, 25]. Compared to surface modification of foils the exposure time has to be increased drastically.

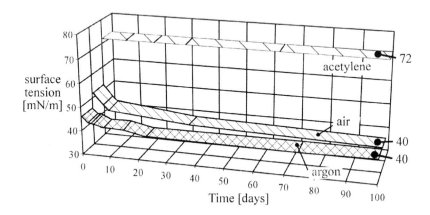

Figure 13.12: Surface tension of BOPP foils vs time for different treatment procedures [15].

13.2.3.2 Ozone generation

Industrial generation of ozone in DBDs has a long tradition. Shortly after 1900 the first larger industrial ozone generating systems for the treatment of drinking water were installed in Nice and Paris, France, and in St. Petersburg, Russia. Today, many drinking water plants throughout the world use ozone generators utilising DBDs to convert the diatomic oxygen molecule O_2 to the chemically much more active triatomic ozone molecule O_3. Larger installations can have a considerable power consumption, on the average more than 1 MW for the top 25 major installations.

The main application of ozone has historically been linked to the development of water purification processes. Ozone is a strong oxidising agent, stronger than all other industrial chemicals with the exception of fluorine. It is a practically colourless gas with a characteristic odour. Ozone finds application as a potent germicide and viricide as well as a strong bleaching agent. In many applications ozone is increasingly used to replace other oxidants like, e.g., chlorine, that present more environmental problems. Since strong oxidising agents are chemically active species the storage, handling and transportation of such chemicals may represent substantial hazards. Another problem is the question of residues and side reactions. In both respects ozone represents the superior choice. Due to its inherent instability ozone is neither stored nor shipped. It is always generated on the site at a rate controlled by the process.

Most technical ozone generators use cylindrical glass tubes of about 1.5–2 mm wall thickness acting as dielectric barriers. These tubes have an internal metal coating serving as the high voltage electrode. The glass tubes are mounted in stainless steel tubes of slightly larger diameter to form annular discharge gaps of 1–2 mm radial width and 1–2 m length. Since large electrode areas are required for mass ozone production several hundred of these discharge tubes are mounted in a steel tank in a configuration similar to that of a cross flow heat exchanger (Fig. 13.13). An important property of DBDs is that many elements can be electrically linked in parallel and can be fed from a common power supply. Unlike other gas

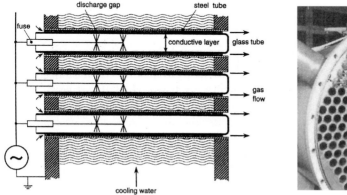

Figure 13.13: Schematic diagram of discharge tubes [29] and photograph of a partly assembled medium size ozone generator.

discharges DBDs do not need external elements to distribute the current evenly to the discharge tubes. Cooling water is passing between the steel tubes, which are welded to two end flanges to form a sealed compartment. Efficient heat removal is essential for good ozonizer performance. Traditionally technical ozone generators were fed from the mains just using step-up transformers to raise the voltage to about 20 kV. Recent developments for high-power ozone generators make use of thyristor-controlled frequency converters operating between 500 and 5000 Hz and impressing square-wave currents.

This change of technology brought several advantages. The power density of these ozone generators could be increased by an order of magnitude resulting in much more compact ozone installations. The peak voltage was reduced to such an extent that problems with dielectric failure of the glass tubes could be practically eliminated. In addition, the implementation of modern power electronics resulted in less bulky high-voltage transformers and compensation networks and provided a simple way to link the ozone generators to superimposed process control.

The first step towards ozone formation in gas discharges is the dissociation of O_2 molecules by electron impact. Detailed calculations about the fraction of the energy going into different electronic reaction paths in oxygen show that about 80% of the electron energy can be utilised for the dissociation process. Ozone is then formed in a three-body reaction involving O and O_2.

$$O + O_2 + M \rightarrow O_3^* + M \rightarrow O_3 + M . \tag{13.12}$$

M is a third collision partner, e.g., O_2, O_3, O, or, in the case of air, also N_2, while O_3^* is a transient excited state in which the ozone molecule is initially formed after the recombination of an O atom with an O_2 molecule. The time scale for this reaction to occur at atmospheric pressure is a few microseconds. Unfortunately, even during this short time, other competing side reactions come into play [9, 28, 29]:

$$O + O + M \rightarrow O_2 + M \tag{13.13}$$
$$O + O_3^* + M \rightarrow 2O_2 + M \tag{13.14}$$
$$O + O_3 + M \rightarrow 2O_2 + M . \tag{13.15}$$

These undesired side reactions pose an upper limit on the desirable atom concentration, i.e., the degree of dissociation, in the microdischarges, and finally on the maximum attainable ozone concentration, too. The drastic onset of these side reactions at relative atom concentrations above 10^{-4} has been demonstrated by calculations based on a fairly extensive reaction scheme of about 70 reactions in an oxygen plasma [9]. From this point of view it would be ideal to work with rather *weak* microdischarges resulting in a relative atom concentration of 10^{-4}.

Unfortunately, there are other considerations that have an influence on this choice. The plasma in the discharge filaments consists in addition to electrons of other charge carriers, in oxygen mainly of the ions O^+, O_2^+, O^-, O_2^-, and O_3^-. Since ionic reactions do not noticeably contribute to ozone formation, energy losses due to ions must be minimised. Calculations and experiments show that a considerable fraction of the discharge energy (up to 50%) can be dissipated by ionic species in weak microdischarges. The situation improves when at higher electron and ion densities the recombination time becomes shorter than the time an ion requires to drift through the gap. In the limiting case of high electron densities the ions are more or less stationary and the energy practically is totally dissipated by electrons.

Thus for optimising the microdischarge properties for the ozone formation process it essentially means to find a compromise between excessive energy losses due to ions in weak microdischarges and the avoiding of undesired chemical side reactions, if the microdischarges become too strong. In oxygen a reasonable compromise can be found at a relative atom concentration of about 2×10^{-3} in a microdischarge channel.

Different operating parameters are optimal when ozone is generated from air, which is the case in many small ozone generators. In this case nitrogen cannot be regarded as a passive carrier gas [26]–[30]. On the contrary, excitation and dissociation of nitrogen molecules result in additional oxygen atoms that can help in ozone formation.

$$N + O_2 \quad \rightarrow \quad NO + O \tag{13.16}$$
$$N + NO \quad \rightarrow \quad N_2 + O \tag{13.17}$$
$$N_2(A) + O_2 \quad \rightarrow \quad N_2O + O \tag{13.18}$$
$$N_2(A, B) + O_2 \quad \rightarrow \quad N_2 + 2O. \tag{13.19}$$

As a matter of fact, about half of the ozone produced in air-fed ozone generators originates from such indirect processes [26, 29].

13.2.3.3 High-power CO_2 lasers

Dielectric-barrier discharges have also found applications in high-power CO_2 lasers. After initial laboratory experiments with pulsed CO_2 lasers by Ishchenko et al. [31], and by Christensen [32] in 1978/79 an industrial high-power laser was developed by Yagi and Tabata [33] at Mitsubishi Electrical Corporation. This SD CO_2 laser (SD for silent discharge) soon became the most successful commercial laser for material processing on the Japanese market. The water cooled plane metal electrodes are covered by glass or alumina dielectrics and are separated by 20–50 mm. A high-velocity cross flow passes the discharge gap at 50–80 m/s for heat removal and discharge stabilisation. Typical gas mixtures are $CO_2/N_2/He$ (1/8/4) at a total pressure of 5–20 kPa. Being operated at 170 kHz there is not enough time for the ions to decay or to be swept out between succeeding half-waves. As a consequence, the discharge behaves very much like a resistive load (ion trapping discharge). Nearly diffraction limited infrared radiation at the wavelength $\lambda = 10.6 \ \mu m$ is obtained with output powers up to 5 kW. The efficiency exceeds 10%. High-speed welding and cutting of metal plates and other materials is the main application of this SD CO_2 laser [34].

13.2.3.4 Excimer lamps

When a dielectric-barrier discharge is operated in rare gases, halogens, or rare gas/halogen mixtures the plasma conditions in a microdischarge channel are similar to those in pulsed excimer lasers. Each microdischarge can act as an intense source of ultraviolet (UV) or vacuum ultraviolet (VUV) radiation. Excimer formation requires three-body collisions and is therefore favoured by high pressure. Efficient excitation and ionisation of precursor species depends on sufficiently high electron energies. Both conditions can simultaneously be met

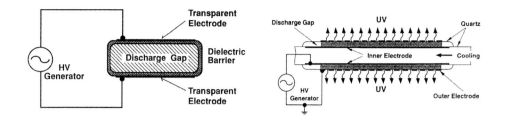

Figure 13.14: Sealed planar and cylindrical excimer lamp configurations.

in non-equilibrium discharges. DBDs conveniently combine these two requirements. Based on earlier work of Tanaka [35], and Volkova et al. [36] simple and efficient excimer lamps were developed in recent years [10, 14, 37, 38]. With different excimer forming gas mixtures a large number of powerful VUV and UV sources with fairly narrow emission bands of a few nm spectral width were tested. When these gas mixtures are sealed in quartz vessels the walls can act as dielectric barriers. Using external transparent or perforated electrodes the discharge can be operated without having metal electrodes in contact with the plasma. This is a considerable advantage for extended operation. Electrodeless lamps can reach lifetimes in exceeding more than 80000 hours.

Fig.13.14 shows a flat planar configuration and a water-cooled cylindrical high power version. These lamps have gap spacings of a few mm, require operating voltages of a few kV, and use frequencies of a few hundred kHz. Reliable switch-mode power supplies can be matched to the characteristics of the discharge to reach plug efficiencies of 90% as far as the energy deposited in the plasma is concerned. In a practical device typically 5–15% of the discharge power can be converted to UV or VUV radiation. Using pulsed discharges measured VUV efficiencies up to 60%, and theoretical efficiencies close to 80% have been reported [39, 40].

Two typical examples are the formation of Xe_2^* or $XeCl^*$ excimer complexes. One of these is formed essentially from neutral excited atoms, the other mainly via recombination of ions as described in section 13.2.1.3. Other excimer complexes obtained are the rare gas dimers Ar_2^* radiating at 126 nm, Kr_2^* at 146 nm, the halogen dimers Cl_2^* at 259 nm, Br_2^* at 289 nm, I_2^* at 342 nm, and the rare gas/halogen exciplexes $ArCl^*$ at 175 nm, $KrBr^*$ at 207 nm, $KrCl^*$ at 222 nm, XeI^* at 253 nm, $XeBr^*$ at 283 nm, and $XeCl^*$ at 308 nm [37, 41]. Commercial DBD excimer lamps are available for different VUV and UV wavelengths or, in connection with phosphors, as mercury-free fluorescent lamps.

These novel UV and VUV sources have found a number of interesting industrial applications: UV curing of printing inks, low-temperature material deposition, surface modification, pollution, and odour control [14],[42]–[48].

13.2.3.5 Plasma display panels

Dielectric-barrier discharges are also used in the new generation of large area flat television screens (see also chapter 15). In these alternating current (ac) plasma display panels (PDPs) the VUV radiation of a xenon plasma in miniature discharge cells is converted by phosphor coatings on the cell walls to red, green or blue. Cell depth and cell separation are of the order of 100 μm. Adjacent cells are separated by dielectric ribs (see Fig. 13.15). Ne/Xe or He/Xe gas mixtures at a pressure of about 40–70 kPa are used. This relatively high pressure is required to obtain the first and second excimer continua of the Xe_2^* at 147 nm and 172 nm, respectively [14, 37]. Only by making use of excimers sufficient radiation can be generated in such small volumes. Either matrix electrodes on opposite sides of the gas gap between two glass plates or coplanar electrodes imbedded in one glass plate are used. In the latter case a surface discharge between adjacent electrodes can be initiated by a trigger electrode on the opposite plate. In both cases individual cells can be addressed by two sets of perpendicular thin electrode strips deposited on the glass plates. In alternating current (ac) PDPs the electrodes are coated by dielectrics and thin protective layers of magnesium oxide. Extremely low sputtering rates of the MgO layers assure long lifetimes, and their high coefficient for the emission of secondary electrons (see also chapter 4) helps to lower the operating voltage. PDPs are operated with 200 V integrated driver circuits. Typical sustaining frequencies for ac displays are in the range 50–100 kHz. Current pulse duration is about 20 ns, depending on the rise time of the square-wave driving voltage, the gas mixture, and the electrode configuration. The intensity of a cell is adjusted by using duty cycle modulation. 256 grey levels and 16 million colours can be obtained, resulting in extremely bright and colourful displays.

Mass production of flat wall-hanging television screens using 40 inch PDPs started at several Japanese manufacturing sites in 1996/97. Relative inexpensive thick-film processes are used. Larger displays up to 60 inches diagonal are conceivable. The display is only 6 mm

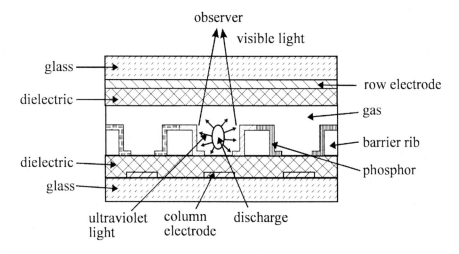

Figure 13.15: Plasma display panel (PDP) configuration.

thick, the complete package presently still about 5–10 cm. It is expected that this technology will eventually replace bulky cathode ray tubes in many television and large-area monitor applications.

13.2.3.6 Pollution control

Applications of DBDs to pollution control in general and to the destruction of poisonous compounds are gaining more and more attention. After initial work on military toxic wastes by Clothiaux et al. in 1984 [49], and Fraser et al. in 1985 [50], an increasing number of investigations have been devoted to the treatment of nitrogen oxides and sulphur oxides in flue gases, and to the decomposition of volatile organic compounds (VOCs) [14], as there are hydrocarbons, chlorocarbons, and chlorofluorocarbons (CFCs). Contamination of exhaust air with gaseous hydrocarbons or organic solvent vapours occurs in many industrial processes, e.g., in chemical processing, in print and paint shops, in semiconductor processing as well as in soil remediation and water treatment. Recent reviews of the subject were published by Penetrante et al. [51, 52] and by Rosocha [53].

Many hazardous organic molecules are readily attacked by free radicals, electrons or UV photons. DBDs are used to provide reactive species such as $N_2^*(A^3\Sigma_u^+)$, $N_2^*(B^3\Pi_g)$, $O_2^*(a^1\Delta_g)$, $O(^1D)$, $O(^3P)$, H, OH, and N. These species initially formed by electron collisions in the microdischarge filaments subsequently initiate a number of reaction paths generating additional O, OH or HO_2 radicals for decomposing pollutants. The aim is to form non-hazardous or less hazardous substances such as O_2, O_3, CO, CO_2, H_2O, simple acids or, for example, by addition of ammonia, solid salt particles.

Dielectric-barrier discharges have advantages over conventional techniques when pollutant concentrations are low, typically in the 10 to 1000 ppm range. The main reason is that for dilute pollutant concentrations the energy required for raising the temperature of the complete carrier gas stream to incineration temperatures or only to temperatures where catalytic destruction can be initiated becomes prohibitive. In DBDs energetic electrons are generated in the microdischarges without noticeably raising the enthalpy of the background gas flow. DBDs may also have advantages over conventional techniques when different pollutants have to be treated simultaneously. Additional effects caused by UV radiation or inserted dielectric pellets, possibly with catalytic coatings, are also under investigation.

13.2.3.7 Greenhouse gas mitigation

Recent research activities in the field of dielectric-barrier discharges also address the problem of global warming and threatening, possibly irreversible, climate changes. At the United Nations Conference on Climate Change held in Kyoto, Japan, in December 1997, most of the industrial nations agreed on legally binding commitments to reduce greenhouse gas emissions to the atmosphere. CO_2, originating from the combustion of fossil fuels, is considered to be the major man-made greenhouse gas. The UN Intergovernmental Panel on Climate Change (IPCC) recommends a 50% reduction of global CO_2 emissions within the next fifty years. Therefore CO_2 disposal and CO_2 utilisation has become a major issue. One proposal is to recycle CO_2 as an energy carrier, perhaps combined with hydrogen in the form of a liquid fuel, or as a feed stock in the chemical industry. New catalysts have been developed for the methanol synthesis from CO_2 and H_2. Methanol could be used as a fuel in a conventional car engine directly or, in connection with a small reformer, to generate hydrogen for a fuel cell driving an electric motor. Methanol, a liquid fuel, can be handled very much like gasoline and would avoid the difficulties connected with storing, transporting and distributing hydrogen [54].

Hydrogen is expensive and can be obtained from the electrolysis of water or by steam reforming of methane. Its generation is often linked to additional CO_2 emissions, if the current is generated by burning fossil fuels. Dielectric-barrier discharges may provide an alternative road using hydrogen donors rather than hydrogen itself. Different groups investigate the hydrogenation of CO_2 and the partial oxidation of CH_4 in DBDs. Methanol formation has been observed in CH_4/oxygen and CH_4/air mixtures [55]–[58]. High conversion rates were obtained in CO_2/CH_4 mixtures resulting in syngas, a mixture of H_2 and CO, which is a valuable feed stock for many chemical processes including methanol synthesis [59].

13.3 Corona discharges

13.3.1 Direct current (dc) discharges

13.3.1.1 Electrode configurations, properties and discharge evolution

Corona discharges are self-sustained gas discharges typically operated in air at atmospheric pressure between metal electrodes. At least one of the electrodes is of a geometric form which causes locally high electric fields. Common configurations are pointed electrodes facing a plane, thin wires in a cylinder or running parallel to a plane, or also knife edge shaped electrodes. When the voltage is raised in such a configuration current starts to flow at corona onset and increases until the potential for spark breakdown is reached. This intermediate range of corona activity is referred to as a partial breakdown of the gap. The corona discharge is characterised by a faint visual glow in the high field region occasionally accompanied by luminous streamers propagating towards the other electrode. The visible glow can cover the whole electrode area (positive corona) or concentrate in certain spots, so-called corona tufts (negative corona). Even with applied direct current (dc) voltages of either polarity the appearance is often that of a burst corona exhibiting very regular current pulses (Trichel pulses). The physical mechanism of these current pulses is a regular build-up and removal of space charge that modulates the intensity of ionisation processes in the high field region [60, 61].

13.3.1.2 Current voltage relations and power consumption

For a given electrode configuration the minimum electric field E_o for corona onset can be approximated by Peek's formula:

$$E_o = c_1 \delta + c_2 (\delta/r_0)^{1/2} , \qquad (13.20)$$

where c_1 and c_2 are experimentally derived constants which slightly depend on polarity, δ is the relative density of air, and r_0 is the radius of curvature of the corona electrode. According to Townsend the current-voltage characteristics can be approximated by

$$I \; = \; c_3 \, (U - U_o) \, U \qquad\qquad (U > U_o) \qquad\qquad\qquad (13.21)$$
$$\;\; = \; 0 \qquad\qquad\qquad\qquad (U \le U_o) \,,$$

where c_3 is a constant depending on the electrode geometry, and U_o is the voltage at corona onset. The dissipated power is obtained by multiplying Eq. 13.21 with the voltage.

13.3.1.3 Charging and transport of particles and droplets

Modern computational tools can be used to calculate the electric fields, the ion densities and the current density distributions also in complex electrode configurations. This task is facilitated by the fact that free electrons rapidly attach to electronegative molecules in air or flue gas mixtures. As a consequence, in many applications the inter-electrode volume can be modeled as a passive unipolar ion drift region which connects the active corona region to the passive corona electrode. At atmospheric pressure the ionisation region where charge carriers are generated is restricted to minute volumes close to the active corona electrode. For other gases Peek's formula can be generalised and based on the ionisation and attachment coefficients of the gas mixture. The active region can be considered as an abundant source of charge carriers that is switched on when the critical breakdown field is reached and which, due to space charge effects, remains close to this field strength. The amount of charge carriers extracted from that zone is determined by the passive ion drift region which has a limited current carrying capability. The ions interact with the gas flow and generate the ionic wind (corona wind) causing strong secondary flows. Droplets or solid particles transported through such corona regions will be charged according to the well-established mechanisms of diffusion charging (dominating for very small particles) and field charging (dominating for particles of diameters $> 1 \mu m$).

13.3.2   Pulsed corona discharges

High concentrations of free radicals can be produced, if, by using fast-rising high-voltage (HV) pulses, positive streamers are generated. When the voltage is high enough the streamers will bridge the whole gap between the electrodes. Positive streamers propagate from the anode, a thin wire or pointed electrode, towards the cathode at high speed ($> 2 \times 10^5$ m/s) and have an optical radius of about 150–200 μm. Up to 600 streamers per meter of high-voltage wire have been observed with steep HV pulses [62]. At the primary streamer head a thin (about 10 μm wide) layer of high-energy electrons with average energies of 10–20 eV efficiently ionises and excites background molecules during their passage through the gas. The neutral gas temperature in the streamer channel stays close to the ambient gas temperature. Thermalisation and breakdown can be avoided, if the applied high-voltage pulse is short enough ($\ll 1 \mu s$). Best energy efficiency for radical formation is obtained when the pulse length roughly equals the streamer transit time. This requires advanced high-voltage switching technologies. Two-dimensional computer simulations of streamer propagation and of the chemical processes initiated by the streamers have been performed by several groups. Positive streamer coronas have been investigated with the aim to induce plasma chemical changes in atmospheric-pressure gases. The obtained results are similar to those obtained by other non-equilibrium processes like dielectric-barrier discharges or electron beam injection.

13.3.3   Applications

Negative coronas with gap spacings of a fraction of a meter are used in large-scale industrial electrostatic precipitators (ESPs) to charge and separate dust particles from gas flows. Similar configurations are used to charge paint droplets or solid particles for spray coating. Positive coronas with electrode spacings in the mm range are used in most copying machines to charge toner particles. Other corona applications using special gas mixtures include voltage stabilisers and Geiger-Müller counters for detecting ionising radiation. Since corona onset leads to current flow and power dissipation, which eventually can even cause flashover, it usually is an unwanted phenomenon in high-voltage equipment. On high-voltage transmission lines typical power losses due to corona discharges can amount to several kW/km. Therefore, in many high-voltage applications care is taken to avoid points or sharp edges or to use auxiliary anti-corona rings and bundled conductors in order to prevent corona onset.

13.3.3.1   Electrostatic precipitators

One of the major technical devices making large-scale use of corona discharges is the electrostatic precipitator [63, 64]. It is used in coal fired power plants to separate the fly ash from the flue gas, in the cement industry and metallurgy to prevent excessive dust emissions to the environment, or in catalytic processes to recover expensive catalysts. In ESPs the dust laden gas flow is channeled through many parallel ducts, in which specially shaped HV corona electrodes are mounted. Typical duct spacings are 30 to 40 cm, and the applied negative high-voltage can reach 50 to 100 kV. In a power plant the ESP may have to clean a flue gas stream

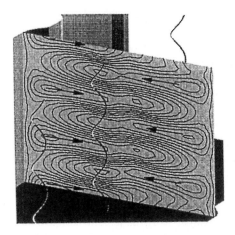

Figure 13.16: Perspective view of ion induced cross flow [65].

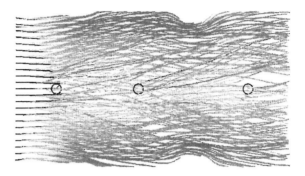

Figure 13.17: Top view of charged particle trajectories [65].

of a few million m^3 per hour and collect tens of tons of fly ash per hour with a cleaning efficiency approaching 99.9%. Such a precipitator may have a height and length of about 15 m
and a width of about 40 m. The corona electrodes and the grounded duct walls are shaped
in such a way that particle charging by the ions in the corona and, at the same time, electric
fields for particle removal are optimised. The charged particles are subjected to electric forces
depending on the electric field in the corona regions, in addition to the hydrodynamic drag
forces in the conveying flow. Fig. 13.16 shows examples of induced secondary flows due
to the ionic wind in an ESP configuration with helical corona electrodes between grounded
collection plates [65]. Computed cross flow velocity components are plotted in a plane perpendicular to the inflow velocity at a position between the second and third electrode. At
higher voltages the cross flow velocities caused by the corona can become comparable to the
main flow velocity ($\cong$ 1 m/s). Fig. 13.17 shows trajectories of charged particles viewed from
above.

Much smaller precipitators normally using positive coronas are used for indoor applications in ventilation systems to remove and collect particles from recirculating gas flows.

13.3.3.2 Pollution control

Positive streamer coronas have been investigated for the reduction of the NO and SO$_2$ content
of flue gases [66, 67]. The idea is to oxidise NO and SO$_2$ with the aid of O, O$_3$, OH and
HO$_2$ to form acids which can be separated or, by addition of ammonia, converted to solid salt
particles which can be collected in ESPs or bag filters. The products, ammonium sulfate and
ammonium sulfate nitrate, can be sold in certain markets as an agricultural fertiliser and soil
conditioner.

Kinetic simulations in air and flue gas mixtures show that the presence of oxygen atoms
can also counteract oxidation processes by reduction of higher oxides to lower oxides.

$$NO_2 + O \quad \rightarrow \quad NO + O_2 \tag{13.22}$$
$$SO_3 + O + N_2 \quad \rightarrow \quad SO_2 + O_2 + N_2 \,. \tag{13.23}$$

Certain gas phase additives like ammonia improve the situation by a fast removal of the higher
oxides. Also, the simultaneous removal of SO$_2$ and NO can be of advantage.

The removal of nitrogen and sulfur oxides from flue gas mixtures by pulsed corona discharges has been extensively studied in the laboratory and in pilot experiments. It turns out that, in addition to gas phase free radical mechanisms, also the role of heterogeneous and thermal reactions is of importance. In a 100 m^3/h slip stream of a combustion flue gas from ENEL's thermoelectric power plant at Porto Marghera 99% SO_2 reduction and 60% NO reduction was achieved with a corona energy deposition of 12–15 Wh/m^3 when ammonia was injected in the flue gas [68, 69]. This corresponds to a power requirement of 4–5% of the generated electricity. Initial concentrations of SO_2 and NO were fairly low: 270 ppm and 250 ppm, respectively. Recent medium-scale investigations starting from similar concentrations reach comparable reduction results at a much lower energy consumption of only 4 Wh/m^3 [70].

Also depollution of industrial off-gases containing toxic volatile organic compounds (VOCs) by positive streamer coronas is considered. In most cases when dilute concentrations of pollutants are involved it is not economic to heat the whole gas stream to combustion temperatures or even to temperatures where a catalytic destruction can be initiated (200–500°C). In these cases non-thermal plasmas may provide the best solution as they are generating electrons, ions and radicals that can decompose toxic molecules. Pulsed streamer coronas have, for example, been investigated for the destruction of trichloroethylene (TCE, $ClHC=CCl_2$), methylene chloride (CH_2Cl_2), carbon tetrachloride (CCl_4), methanol (MeOH, CH_3OH), toluene (methylbenzene, $C_6H_5CH_3$), and xylene (dimethylbenzene, $C_6H_4(CH_3)_2$) [71].

Most experiments with pulsed streamer coronas have been performed in point-plane, wire-cylinder or wire-plane configurations. In many applications the classical ESP configuration with straight wires at the centre plane of a duct is used. The HV pulses can be formed by using triggered or rotating spark gaps or thyratrons, switching devices which are limited in lifetime and repetition rate.

References

[1] W. Siemens, Poggendorffs Ann. Phys. Chem. **102**, 66–122 (1857)

[2] E. Warburg, Wiedemanns Ann. Phys. Chem. **66**, 652–659 (1898)

[3] J.S. Townsend, Phil. Mag. **28**, 83-90 (1914)

[4] F.W. Peek Jr., Trans. A.I.E.E. **30**, 1889-1965 (1911)

[5] J.B. Whitehead, W.S. Brown, Trans. A.I.E.E. **36**, 169-198 (1917)

[6] V. Gibalov, G. Pietsch, Proc. Int. Conf. of Gas Discharges and their Appl., Swansea, GB, p. 552–555 (1992)

[7] R. Schwarz, thesis, TU Braunschweig, Shaker Verlag, ISBN 3-8265-1344-4 (1996)

[8] U. Reitz, thesis TU Braunschweig, Ber. FZ Jülich, Jül-2613, ISBN 0366-0885 (1992)

[9] B. Eliasson, M. Hirth, U. Kogelschatz, J. Phys. D: Appl. Phys. **20**, 1421–1437 (1987)

[10] B. Eliasson, U. Kogelschatz, Appl. Phys. B **46**, 299–303 (1988)

[11] B. Eliasson, U. Kogelschatz, in Nonequilibrium Processes in Partially Ionized Gases, M. Capitelli, J.N. Bardsley, Editors, Plenum Press, New York, 401–410 (1990)

[12] B. Eliasson, U. Kogelschatz, IEEE Trans. Plasma Sci. 19, 309–322 (1991)

[13] B. Eliasson, W. Egli, U. Kogelschatz, Pure & Appl. Chem. 66, 1275–1286 (1994)

[14] U. Kogelschatz, B. Eliasson, W. Egli, 23rd Int. Conf. on Phenomena in Ionized Gases (ICPIG XXIII), Toulouse, France, Invited Papers, M.C. Bordage, A. Gleizes, Editors, C4-47 to C4-66 (1997)

[15] S. Meiners, J.G.H. Salge, E. Prinz, F. Förster, Surface and Coatings Technoloy **98**, 1121–1127 (1998)

[16] D. Braun, V. Gibalov, G. Pietsch, Plasma Sources Sci. Technol. **1**, 166–172 (1992)

[17] A. C. Gentile, M. J. Kushner, J. Appl. Phys. **79**, 3877–3885 (1996)

[18] M. Kogoma, S. Okazaki, J. Phys. D: Appl. Phys. **27**, 1985–1987 (1994)

[19] V. Schorpp, thesis, University Karlsruhe (1991)

[20] F. Massines, A. Rabehi, P. Decomps, R.B. Gadri, P. Ségur, C. Mayoux, J. Appl. Phys., **83**, 2950–2957 (1998)

[21] S. Okazaki, M. Kogoma, M. Uehara, Y. Kimura, J. Phys. D: Appl. Phys. **26**, 889–892 (1993)

[22] J. Tepper, M. Lindmayer, J. Salge, Proc. Hakone VI, Int. Symp. on High Pressure, Low Temperature Plasma Chemistry, Cork, Ireland, p. 123–127 (1998)

[23] J.C. v.d. Heide, H.L. Wilson, Modern Plastics **5**, 199–208 (1961)

[24] J.G.H. Salge, Proc. Aachener Textile Conference, DWI-Report, ISBN 0942–301X, (1998)

[25] J. Salge, Surface & Coatings Technology **80**, 1–7 (1996)

[26] B. Eliasson, U. Kogelschatz, P. Baessler, J. Phys. B: At. Mol. Phys. **17**, L797–L801 (1984)

[27] D. Braun, U. Küchler, G. Pietsch, J. Phys. D: Appl. Phys. **24**, 564–572 (1991)

[28] U. Kogelschatz, in Process Technologies for Water Treatment, S. Stucki, Editor, Plenum Press, New York, p. 87–120 (1988)

[29] U. Kogelschatz, B. Eliasson, in Handbook of Electrostatic Processes, J.S. Chang, A.J. Kelly, J.M. Crowley, Editors, Marcel Dekker, New York, p. 581–605 (1995)

[30] V.G. Samoilovich, V.I. Gibalov, K.V. Kozlov, Physical Chemistry of the Barrier Discharge (Russ.), Moscow State University (1989), English translation: J.P.F. Conrads, F. Leipold, Editors, DVS-Verlag GmbH, Düsseldorf (1997)

[31] V.N. Ishchenko, V.N. Lisitsyn, A.R. Sorokin, Sov. J. Quant. Electron. 8, 453–457 (1978)

[32] C.P. Christensen, Appl. Phys. Lett. 34, 211–213 (1979)

[33] S. Yagi, N. Tabata, Proc. IEEE/OSA Conf. on Lasers and Opto-Electronics, Washington DC (1981), p. 22

[34] K. Yasui, M. Kuzumoto, S. Ogawa, M. Tanaka, S. Yagi, IEEE J. Quant. Electron. **25**, 836–840 (1989)

[35] Y. Tanaka, J. Opt. Soc. Am. **45**, 710–713 (1955)

[36] G.A. Volkova, N.N. Kirillova, E.N. Pavlovskaya, A.V. Yakovleva, J. Appl. Spectrosc. **41**, 1194–1197 (1984)

[37] B. Gellert, U. Kogelschatz, Appl. Phys. B **52**, 14–21 (1991)

[38] M. Neiger, Proc. 6th International Symposium on the Science & Technology of Light Sources (LS-6), Budapest, L. Bartha, F.J. Kedves, Editors, p. 75–82 (1992)

[39] F. Vollkommer, L. Hitzschke, Phys. Blätter **53**, 887–889 (1997)

[40] F. Vollkommer, L. Hitzschke, Proc. 8th International Symposium on the Science & Technology of Light Sources (LS-8), Greifswald, G. Babucke, Editor, p. 51–60 (1998)

[41] J.-Y. Zhang, I. W. Boyd, J. Appl. Phys. **84**, 1174–1178 (1998)

[42] U. Kogelschatz, Pure & Appl. Chem., **62**, 1667–1674 (1990)

[43] U. Kogelschatz, B. Eliasson, H. Esrom, Materials & Design **12**, 251–258 (1991)

[44] U. Kogelschatz, Appl. Surf. Sci. **54**, 410–423 (1992)

[45] H. Esrom, U. Kogelschatz, Thin Solid Films **218**, 231–246 (1992)

[46] H. Esrom, J.-Y. Zhang, U. Kogelschatz, in Polymer Surfaces and Interfaces: Charaterization, Modification and Application, K.L.Mittal, K.-W. Lee, Editors, VSP International Science Publishers, The Netherlands, p. 27–35 (1997)

[47] J.-Y. Zhang, PhD Thesis, TU Karlsruhe (1993)

[48] J.-Y. Zhang, H. Esrom, G. Emig, U. Kogelschatz, in Polymer Surface Modification: Relevance to Adhesion, K.L. Mittal, Editor, VSP International Science Publishers, Utrecht, The Netherlands, p. 153–185 (1996)

[49] E.J. Clothiaux, J.A. Koropchak, R.R. Moore, Plasma Chem. Plasma Proc. **4**, 15–20 (1984)

[50] M.E. Fraser, D.A. Fee, R. Sheinson, Plasma Chem. Plasma Proc. **5**, 163–173 (1985)

[51] B.M. Penetrante, S.E. Schultheis, Editors, Non-Thermal Plasma Techniques for Pollution Control, NATO ASI Series, Vol. G 34: Ecological Sciences, Part A: Overview, Fundamentals and Supporting Technologies, Part B: Electron Beam and Electrical Discharge Processing, Springer, Berlin (1993)

[52] B.M. Penetrante, J. N. Bardsley, M. C. Hsiao, Jpn. J. Appl. Phys. **36**, 5007–5017 (1997)

[53] L.A. Rosocha, in Plasma Science and the Environment, W. Manheimer, L.E. Sugiyama, T.H. Stix, Editors, Am. Institute of Physics, Woodbury, New York, p. 261–298 (1997)

[54] B. Eliasson, in Carbon Dioxide Chemistry: Environmental Issues, J. Paul, C.-M. Pradier Editors, The Royal Society of Chemistry, Cambridge, p. 5–15 (1994)

[55] K. Okazaki, T. Nozaki, Y. Uemitsu, K. Hijikata, Proc. 12th Int. Symp. on Plasma Chemistry (ISPC–12), Minneapolis, p. 581–586 (1995)

[56] U. Kogelschatz, B. Eliasson, Phys. Blätter **52**, 360–362 (1996)

[57] B. Eliasson, U. Kogelschatz, B. Xue, L.-M. Zhou, Ind. Eng. Chem. **37**, 3350–3357 (1998)

[58] L.-M. Zhou, B. Xue, U. Kogelschatz, B. Eliasson, Plasma Chem. Plasma Process. **18**, 375–393 (1998)

[59] L.-M. Zhou, B. Xue, U. Kogelschatz, B. Eliasson, Energy & Fuels **12**, 1191–1199 (1998)

[60] R. Morrow, Phys. Rev. A **32**, 1799–1809 (1985)

[61] A.P. Napartovich, Yu.S. Akichev, A.A. Deryugin, I.V. Kochetov, N.I. Trushkin, J. Phys. D: Appl. Phys. **30**, 2726–2736 (1997)

[62] Y.L.M. Creyghton, PhD Thesis, Technical University Eindhoven (1994)

[63] H.J. White, Industrial Electrostatic Precipitation, Addison-Wesley, Reading (1963)

[64] K.R. Parker, Editor, Applied Electrostatic Precipitation, Blackie, London (1997)

[65] W. Egli, U. Kogelschatz, E.A. Gerteisen, R. Gruber, J. Electrostat. **40 & 41**, 425–430 (1997)

[66] A. Mizuno, J.S. Clements, R.H. Davis, IEEE Trans. Ind. Appl. **22**, 58–64 (1986)

[67] I. Gallimberti, Pure & Appl. Chem. **60**, 663–674 (1988)

[68] G.D. Dinelli, L. Civitano, M. Rea, IEEE Trans. Ind. Appl. **26**, 535–541 (1990)

[69] L. Civitano, E. Sani, in Plasma Technology: Fundamentals and Applications, M. Capitelli, C. Gorse, Editors, Plenum, New York, p. 153–166 (1992)

[70] E.M. van Veldhuizen, L. M. Zhou, W. R. Rutgers, Plasma Chem. & Plasma Proc. **18**, 91–111 (1998)

[71] U. Kogelschatz, Proc. 8th Int. Conf. on Switching Arc Phenomena/Int. Symp. on Electrical Technologies for Environmental Protection (SAP & ETEP '97), Z. Tarocinski, Editor, Lodz, Poland, p. 299–303 (1997)

14 Transient plasma-assisted diesel exhaust remediation

M. Gundersen, V. Puchkarev, A. Kharlov, G. Roth, J. Yampolsky and D. Erwin

University of Southern California, Department of Electrical Engineering-Electrophysics,
Los Angeles, CA 90089-0271, U.S.A.

14.1 Introduction

Non-thermal plasma processing is a promising alternative to exhaust aftertreatment of NO_x and particulate emissions from diesel engines. The most frequently used methods for producing the plasma are the dielectric barrier (silent) discharge and the pulsed corona discharge. It has been reported [1] that under the same experimental conditions, a pulsed corona discharge is more efficient (in terms of energy cost per removed NO_x molecule) than a silent discharge. In this chapter we report a study on the pulsed corona method with an energy cost of $\sim$10–20 eV/molecule, and report laser-induced fluorescence (LIF) studies of the temporal behavior of NO depletion.

We present experimental data on NO/NO_x removal from diesel engine exhaust using a low-energy-cost pulsed corona method. We describe and discuss the effect of various parameters on the energy cost of NO/NO_x removal, temporal dynamics of NO/NO_2 destruction obtained by LIF, by-product analysis, and pulsed power issues. There are several issues that affect the practical application of pulsed plasma devices including:

- energy cost,

- emission of by-products,

- pulsed power implementation, and

- reactor design.

Reported energy costs vary considerably – for example, in terms of energy cost per treated molecule, from 3 to 500 eV/molecule. To be competitive for remediation of diesel engine emission, the energy cost should be <10–20 eV per NO_x molecule for concentrations of $\sim$1000 ppm, which would correspond to an overall power consumption <5% of the total engine power. By-products of plasma-induced chemical reactions should not include harmful components, and the power system and plasma reactor must be robust, reliable and inexpensive.

14.2 Experiment

Two types of experiments have been performed in this program:

- Treatment of actual diesel exhaust, and

- LIF of NO/NO$_x$ produced in a pulsed plasma.

14.2.1 Diesel exhaust treatment

The experimental apparatus was comprised of a discharge chamber, a pulsed power modulator, and a gas manifold with controller gauges and emission analyzer. A pulse generator supplied high-voltage pulses up to 40 kV, with a pulse duration of 50–100 ns and a rise time of 20 ns. The repetition rate was up to 1 kHz. A typical corona reactor included a cylindrical chamber of length 0.4 m, and an inner, thread-wire-type high-voltage electrode. Varying diameters of inner and outer electrodes, reactor lengths, annular dielectric inserts to prevent arcing, flow rates, and pulse repetition rates were used to vary the plasma volume, and the pulsed and mean energy deposition into the gas. These parameters were varied to determine optimum conditions for cost-effective NO/NO$_x$ removal. By monitoring the voltage and current signals related to the discharge the energy deposition into the gas, E_p, was computed. The energy cost, ε, was calculated using:

$$\varepsilon = \frac{250 E_p f}{F \; \Delta \text{NO}_x} \; \text{eV/molecule},$$

where f is the frequency in Hz, F the flow rate in l/s, and ΔNO_x the NO$_x$ removal in ppm. Three sets of experiments using different diesel engines were conducted:

1. Initial experiments used a 60-kW engine (Volkswagen Rabbit [Golf]) operated at idle speed and NO emission concentrations of 100–130 ppm. Here the exhaust temperature was 40–500 C. We monitored the outlet gas composition using an electrochemical NO analyzer (Bacharach, Nonoxor II). No other species concentrations were measured.

2. In subsequent tests a 300-kW engine operated under load with high NO emission (600–1000 ppm) was used. Exhaust temperature was varied from 85 to 1800 °C. A Horiba analyzer was used for the emission monitoring. This system allowed detection of the following components: NO, NO$_2$, HC, CO, CO$_2$, and SO$_2$.

3. Recent experiments were performed with a 10-kW diesel motor generator and variable electrical load. Emission was monitored using a portable electrochemical LANCOM flue gas analyzer. Typical emission constituents and their concentrations are: CO, 400–800 ppm; CO$_2$, 2–4%; NO, 120–400 ppm; NO$_2$, 30–60 ppm; C$_x$H$_x$, 120–600 ppm; O$_2$, 14–18%. Here the higher values in the concentrations of exhaust constituents refer to a loaded engine. Depending on the reactor dimension the flow rate was varied between 60 and 1500 standard liters per minute, with the flow velocity in the reactor between 1 and 8 m/s. In some experiments we also used synthetic gases .

14.2.2   Laser-induced fluorescence (LIF) of NO/NO$_x$

LIF experiments studied the temporal and spatial profiles of NO/NO$_2$ concentration in a corona discharge [2]. The laser beam was focused to a 1 × 6-mm sheet by a cylindrical lens, and passed through the gas volume at variable delays with respect to the discharge excitation pulse. The fluorescence signal was monitored with a CCD camera. The NO and NO$_2$ excitations were done at 226 and 440 nm, respectively. In the LIF experiments two discharge configurations have been used:

1. A needle-plate electrode assembly was used for the study of NO depletion in a single streamer corona. 25–75 ppm NO was seeded either in dry air or in pure nitrogen. The gas flows across the discharge channel at a rate of ~2 cm^3/s. Details of the experiment and LIF images have been published in Ref. [2].

2. In the second set-up, a coaxial cylinder, 1 cm in diameter and 6 mm long with an annual glass insert, was used. This geometry enabled us to study a wide range of space velocity from 0.3 to 30 m/s, corresponding to a gas flow rate of 1–100 sl/min. High-voltage pulses were applied to a central 0.5 mm wire electrode with a repetition rate up to 3000 Hz. Simulated gases were used in all LIF experiments.

14.3   Experimental results

A discussion of the effect of various parameters on energy cost of NO$_x$ removal is presented in detail elsewhere [3]. Here we summarize important former data and discuss new ones. The following issues will be discussed:

- pulsed power and non-thermal plasma formation,
- time/space resolved NO/NO$_x$ depletion, and
- plasma chemistry.

14.3.1   Pulsed power and plasma formation

We found that short pulses (<50 ns) are more advantageous for efficient energy usage than longer pulses. This is plausible because short pulses allow the more efficient utilization of higher electric fields and thus 'hotter' electrons, which are much more favorable for ion and radical formation. The energy cost using positive corona (plus at the central electrode) is 1.5–2 times that for negative corona, although NO$_x$ removal is almost the same. It is well known that a positive corona carries higher current at a given pulse voltage than a negative. In Ref. [3] we measured energy cost versus current density and found that a lower energy cost is realized rather at low than at high current density. This behavior is remarkable, because the number of electrons (and radicals) is proportional to the current density. However, the corona discharge uniformity is much greater for low current density. Due to a strong nonlinear dependence of field emission on the local electric field the number of starting streamers is an extremely sensitive function of electric field and cathode microgeometry, which are difficult to control. It is likely that at higher current density the current is carried by a finite number of filaments,

Table 14.1: Energy parameters and NO_x/NO reduction for various experimental conditions. E_p is the specific input energy per pulse per cm^3, E^* is the total specific energy, and ε is the energy cost for NO_x and NO. ΔNO_x and ΔNO denote NO_x and NO removal, respectively, in absolute values (ppm). The last row is the ratio of NO_x removal to NO_2 production.

Flow, l/s	4	4	2.4	2.4
f, Hz	1000	2000	1000	300
E_p, J/l per pulse	0.03	0.018	0.088	0.15
E^*, J/l	2.5	3	12	6.3
$\varepsilon(NO_x)$, eV/mol.	7.2	8	23	21
$\varepsilon(NO)$, eV/mol.	3.9	4.6	16	10
ΔNO_x, ppm	84	97	113	70
ΔNO, ppm	152	167	180	152
$\Delta NO_x/\Delta NO_2$	1.24	1.38	1.63	0.86

rather than being evenly distributed over the cathode surface. Thus the energetic electrons are generated only in the streamers whose volumes are small parts of the total gas volume; in the extreme limit, all current can flow through a single channel. This can explain the effect of pulse polarity on energy cost: the streamer velocity is higher for a positive hot electrode, which results in strong current filamentation. The most important figure of merit is the energy density in the gas (in J/l). Table 1 lists selected experimental data obtained by varying flow rate and energy deposition in the gas.

From table 1 it follows that the energy-effective regime is achieved at $E^* \sim$ 2–3 J/l. At this energy consumption, only 20–25% NO_x was removed. The energy input is basically related to the pulse current, and the current shows rapid growth with voltage. However, NO/NO_x removal increases more slowly, resulting in increasing energy cost. This is, we believe, due to the non-uniform current distribution over the reactor volume.

To check the effect of current non-uniformity we measured the energy cost in eV/molecule for different plasma reactor volumes while keeping the same energy density (in J/l) in the gas. The plasma volume was varied from 300 to 30 cm^3 at 2.5–3 J/l. It was found that the energy cost for NO removal changed very little: $\varepsilon = 5.5$–7.6 eV/molecule, while for NO_x removal there was a slight increase from $\sim$12 to 25 eV/molecule.

One should note that the energy cost and the total NO/NO_x removal depends on the initial concentration of NO, hydrocarbons (HC), the temperature, and particulate matter. The efficiency increases with NO concentration. Hydrocarbons are favorable for NO/NO_2 oxidation but considerably increase the pathways for various by-products. The NO_x removal and the energy cost depend on particulate matter concentration, which is varying in different engines. For example, we found that a blocking mechanical filter upstream of the reactor resulted in lower NO_x removal and increased the energy cost by a factor of 1.5–3.

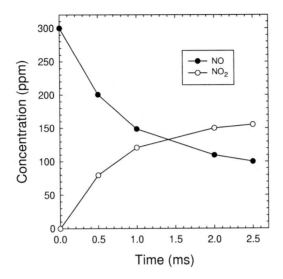

Figure 14.1: Temporal concentrations of NO (•) and NO$_2$ (○) plotted as a function of the time following the pulse. The data indicate that the pulse initiates the process of conversion. The flow rate is 100 l/min and the energy density is 15 J/l.

14.3.2 Time- and space-resolved NO/NO$_x$ depletion

Knowledge of the temporal behavior and the spatial distributions of chemical species in the plasma are important for engineering and optimization of plasma-assisted reactors. In our work LIF was used to obtain temporally and spatially resolved data. Here we summarize the LIF results.

- NO depletion occurs in close proximity to and along the streamer channel, i.e., radical production takes place in the region of high local electric fields.

- In pure nitrogen the energy cost for NO destruction for these LIF experiments is about 150–200 eV/molecule at the optimum current density, which in the streamer channel is about 0.5–1 kA/cm^2. This design is far from optimum, but the observed energy cost corresponds to the observations of others for NO destruction in pure N$_2$ [4]. We observe parenthetically that the measured low energy cost in real diesel exhaust gives evidence that the major reactions go via other plasma-chemical species than are achievable in a pure nitrogen background.

- NO destruction in air occurs within a few milliseconds following the pulsed excitation (see figure 14.1).

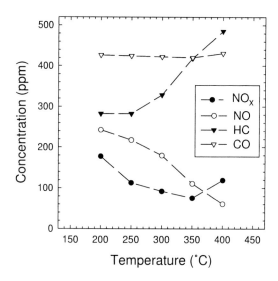

Figure 14.2: Emission components as a function of temperature. Inlet (baseline) concentrations: NO/NO$_x$, 500/510 ppm; HC (propene), 500 ppm; H$_2$O, 7%; CO, 200 ppm. Flow rate: 40 l/min; energy density, 25 J/l. The reactor volume was 300 cm^3. No particulates were present.

14.3.3   Plasma chemistry

Plasma aftertreatment of diesel exhaust principally relies on the removal of nitrogen oxides (NO$_x$) and other hazardous components by highly reactive radicals produced in corona discharges by energetic electrons, such as N, O, OH, and O$_3$. During the high-voltage short pulse, energetic electrons and radicals are generated and subsequently initiate chemical reactions. For this, the pulse parameters (voltage amplitude and pulse width) must be optimized in accordance with the reactor design, and it is observed that they depend in subtle ways on small variations in the design. Short-pulse excitation is followed by a post-pulse period, which lasts from tens of microseconds to hundreds of ms or longer [5].

During this time various remediation and oxidation reactions occur, which are as yet not well characterized. There are two key pathways for NO removal:

1. Reduction: NO + N → N$_2$ + O, and

2. Oxidation: NO + O/O$_3$ → A and NO/NO$_2$ + OH/HC → A, where A refers to HC or other species.

The first reaction is most desirable, and according to Ref. [5] has the highest rate constant. However, recent experiments [1] indicate that only 5 to 20% remediation of NO occurs via this reduction. In the pulsed corona discharge NO is partly converted into NO$_2$ and other species. To study the effect of hydrocarbons and temperature on the performance of a plasma reactor, in Ref. [6] simulated gas was used.

Figure 14.2 shows a plot of exhaust emission concentration versus temperature to determine changes in reaction kinetics at different temperatures. There is very little effect on NO_x reduction. From 60% to 80% of NO is converted into NO_2. There are considerable changes in the level of CO and hydrocarbons; by contrast, without plasma, there are no changes.

14.3.4 Plasma-assisted catalyst

Many studies now suggest that the conversion NO to NO_2 is an important intermediate step in the reduction of NO_x to N_2 [7], and the most efficient way to do this is using a plasma reactor located upstream of a catalytic reactor. In the first step the plasma oxidizes NO to NO_2 in the presence of HC:

$$\text{Plasma} + NO + HC + O_2 \rightarrow NO_2 + \text{HC-products.}$$

In the second stage the catalyst reduces NO_2 to N_2 by selective reduction using hydrocarbons:

$$\text{Catalyst} + NO_2 + HC \rightarrow N_2 + CO_2 + H_2O$$

Our recent experiments indicate that up to 70% NO can be reduced from exhaust using this approach with an energy cost of 10–15 eV/molecule.

14.4 Summary

At present (August 2000) the transient plasma, or *short* pulse approach to plasma remediation represents a distinction in terms of technology from a plasma reactor employing a barrier discharge. The short pulse approach does the same what a barrier discharge does, in effect, with more flexible, although more sophisticated, technology. The most cost-effective approach for applications will be determined by the degree to which these can be optimized. However, the transient plasma work described here, based specifically on the use of short pulses, is an excellent base for research on the subject, and this chapter has described several research areas within the study of transient plasma remediation.

Indeed, the transient plasma (which is not the plasma that has been modeled to date, and for which there is no model at this time) might be more carefully characterized as a *true* non-thermal plasma – this a plasma that is not likely to be modeled based on one or more temperatures. The quasi-steady-state plasma that emerges rapidly may be appropriate for modeling based on an electron temperature of 1 or a few eV, with colder ions and neutrals, but it would be more accurate to convey to researchers who are not specifically plasma modelers that these plasmas are modeled as bi-thermal, rather than non-thermal, which implies no use of temperature. Further, the problem of the true non-thermal plasma needs much attention, has received little, and must be recognized as a clear and distinct research area.

The varied research areas in plasma remediation encourage more interdisciplinary studies – toward developing an understanding of the underlying physics, chemistry and engineering. For example, better understanding is needed of the initial, transient plasma, the subsequent plasma chemistry that occurs, more and better diagnostics including development of simple, *in situ* diagnostics methods for NO_x and particulates, development of improved pulse generators along with plasma reactors, and integration into systems with catalysts and feedback controls. This is indeed quite a range of research activity – the plasma chemistry alone is bewildering, and engineering issues such as pulse generator development may require state-of-the-art advances in solid state switches. The authors of this chapter believe that this is thus a very exciting and promising range of related yet different research, development and engineering problems, with beneficial pay-off for society, as well as satisfying intellectual challenges.

References

[1] T. Hammer, S.Broer, T. Kishimoto, *Pulsed Excitation of Silent Discharge for Diesel Exhaust Treatment*, presented at the 4th Int. Conf. on Advanced Oxidation Technologies for Water and Air Remediation (AOTs-4), Orlando, Florida, USA (1997).

[2] G. Roth and M. Gundersen, *LIF Images of NO Distribution After Needle-Plane Pulsed Negative Corona Discharge*, IEEE Trans. Plasma Sci. 27, 28 (1999).

[3] V. Puchkarev, G. Roth and M. Gundersen, *Plasma Processing of Diesel Exhaust by Pulsed Corona Discharge*, SAE 982516 (1998).

[4] M. A. Tas, R. van Hardeveld, E. M. van Veldhuizen, *Reaction of NO in a Positive Streamer Corona Plasma*, Plasma Chem. Plasma Proc. 17, 371 (1997).

[5] A. C. Gentile and M. J. Kushner, *Reaction Chemistry and Optimization of Plasma Remediation of N_xO_x from Gas Stream*, J. Appl. Phys. 78, 2074 (1995).

[6] R. Slone, M. Ramavajjala, V. Palekar, V. Puchkarev, *Pulsed Corona Plasma Technology for the Removal of NO_x from Diesel Exhaust*, SAE 982431 (1998).

[7] See, for example, articles in the SAE Technical Paper Series *Plasma Exhaust Aftertreatment* (SP-1395), Intl. Fall Fuels and Lubricants Meeting, San Francisco (1998), obtained from the SAE, Warrendale PA, 15096-0001.

15 Plasma display panel

J.K. Lee[†] and J.P. Verboncoeur[‡]

[†] Department of Electronic and Electrical Engineering, Pohang University of Science and Technology, Pohang 790-784, South Korea
[‡] Department of Electrical Engineering and Computer Sciences, University of California, Berkeley, CA 94720-1770, U.S.A.

15.1 Introduction and overview

The color plasma display panel (PDP) is one of the leading candidates for high definition television (HDTV) display. The unique advantage of the PDP panel is in slim, high-resolution, large flat-panel display devices, especially in a television (TV) set market. There is an increasing demand in HDTV for a large (60-80 inch) flat-panel display. Thin-film-transistor liquid crystal displays (TFT-LCD) and conventional cathode-ray tubes (CRT) have limitations in producing large and slim devices over 30-inches. Projection TV has its weakness in resolution. Recently, remarkable progresses in PDP technology made possible the small-scale pilot-plant production of PDP sets in large quantities exceeding 100,000 sets worldwide, and soon expected to observe an exponential rise to 2.6 million sets in 2003, and 5.4 million in 2005. The present status (and the near-term goal) of the 40 inch PDP technology is luminous efficiency of 2 lm/W (5 lm/W), power consumption of 300 W (200 W), brightness of 550 cd/m^2 (700 cd/m^2), and a manufacturing cost of around US$ 5000 (US$ 3000).

The alternating current (AC) type, shown in Fig. 15.1, is currently the dominant type of PDP studied worldwide. The top (the side facing the viewer) has two parallel transparent indium-tin-oxide (ITO) electrodes for sustaining and scanning in addition to two non-transparent Cr/Cu/Cr bus electrodes. The top side is coated with protective materials such as MgO, which has a high coefficient value of secondary electron emission. The bottom side (the rear side to the viewer) has an address electrode perpendicular to the sustaining and scanning electrodes. These two sides are covered with dielectric materials such as PbO and separated by a barrier rib of 100-150 μm.

The barrier ribs and dielectric on the bottom are coated alternately with red, green, and blue phosphors. The phosphor covers only the dielectric at the bottom (or the rear) address electrode, not the dielectric at the top (or the front) sustaining and scanning electrodes where the surface discharge occurs. This surface discharge type PDP thus minimizes the flux of charged particles to the phosphors, important for reducing the degradation of the device that results from damage and sputtering of the phosphors. The PDP cell is filled with a mixture of He, Ne, and Xe gases (or Ar and Kr occasionally), at a pressure of about one half atmosphere.

(a)

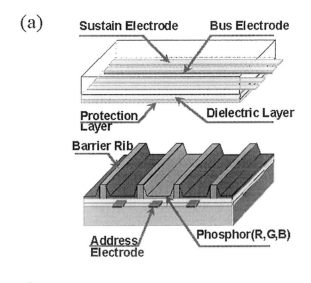

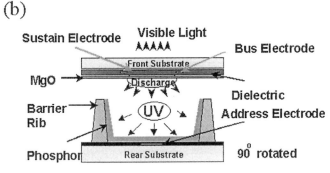

Figure 15.1: AC-PDP cell structures. (a) matrix type, (b) surface discharge type.

15.2   History and background

The alternating current plasma display panel (AC-PDP) was invented at the University of Illinois in 1964 [1], followed by the direct current plasma display panel (DC-PDP) at Philips in 1968. The AC type dominated the PDP research [2]–[5], except for the brief periods of interest in the DC type in the late 1960s and early 1980s. Since 1995 with the 42 inch color AC-PDP panel by Fujitsu, rapid progress has been made to make mass production feasible. Late in 1999, several industrial laboratories including LG Electronics generated impressive products, such as 60 inch PDP sets, and a series of remarkable scientific advances in slimness (78 mm), power consumption (200 W for a 40 inch set), efficiency (up to 2.4 lm/W), and luminance (550 cd/m^2).

15.3 Alternating current plasma display panel (AC-PDP)

15.3.1 The plasma discharge driven by a high voltage

15.3.1.1 Paschen's law for breakdown

In a PDP cell, a gas mixture such as Ne and Xe is ionized by a high voltage in the range of 150-250 volts, depending upon the gas pressure and the electrode gap. The resultant charged particles (i.e., the plasma) produce a sufficient number of excited Xe atoms that radiate vacuum ultraviolet (VUV). This VUV radiation arrives at phosphor at the address electrode side and then is converted to visible photons. The brightness (or luminance) and the luminous efficiency ultimately depend upon these photons. The self-sustaining plasma discharge is governed by Paschen's law of breakdown. For a simple one-dimensional diode with the cathode located at $z = 0$ and the anode at $z = d$, the fluid equation of electron density n_e is

$$\frac{dn_e}{dt} + \frac{d\Gamma_e}{dz} = n_e \nu_{iz} = \alpha \Gamma_e, \tag{15.1}$$

where $\Gamma_e = n_e v_e$ is the electron flux, and v_e is the electron velocity.

The temporal growth rate is denoted by ν_{iz} and the spatial one by the Townsend coefficient α. For a self-sustained plasma discharge the ionization has to be large enough to increase the electron flux exponentially in space and time,

$$\Gamma_e(z) = \Gamma_e(0)e^{\alpha z}. \tag{15.2}$$

For the ion flux $\Gamma_i(0)$ impinging on the cathode, there are $\gamma_{se}\Gamma_i(0)$ secondary electrons emitted:

$$\gamma_{se} = \frac{\Gamma_e(0)}{\Gamma_i(0)}. \tag{15.3}$$

The breakdown criterion is met when the plasma system has sufficient ionization to satisfy

$$\gamma_{se}[\exp(\alpha d) - 1] - 1 = 0. \tag{15.4}$$

To derive this condition, consider a finite but small number of seed electron current density J_{seed} at the cathode ($z = 0$). $J_{seed}/e \approx 10\text{–}100$ cm^{-2} sec^{-1} is supplied by cosmic rays in a laboratory. For a PDP, J_{seed} could be much larger and is given by the previous pulse discharge. The current densities J and the fluxes Γ differ by the charge of the respective carrier,

$$J_{total}(0) = J_e(0) + J_i(0) = J_{seed} + \gamma_{se}J_i(0) + J_i(0), \tag{15.5}$$

$$J_{total}(d) = J_e(d) + J_i(d) = J_e(d) = e^{\alpha d}J_e(0) = e^{\alpha d}\{J_{seed} + \gamma_{se}J_i(0)\}. \tag{15.6}$$

With J_i (0) from Eq. 15.5, and

$$J_{total}(d) = J_{total}(0) \tag{15.7}$$

we obtain the multiplication factor

$$m = \frac{\Gamma_{total}(d)}{\Gamma_{seed}} = \frac{e^{\alpha d}}{1 - \gamma_{se}(e^{\alpha d} - 1)} .$$

(15.8)

To achieve $m \gg 1$, the denominator must vanish, i.e., the condition of Eq. 15.4 is met. Then the electron flux at the location $z = d$ has sufficient spatial (and temporal) growth compared with that at $z = 0$, which is triggered by a small seed current density J_{seed}. This is called a self-sustained discharge or a breakdown (of an insulating neutral gas converted to a conducting plasma).

The exponential growth rate α has been measured to fit the analytic form

$$\frac{\alpha}{p} = A \exp\left(-\frac{Bp}{E}\right) = A \exp\left(-\frac{Bpd}{V_b}\right)$$

(15.9)

as shown in Fig. 15.2. The constants A and B are 26 and 350 for Xe, 4 and 100 for Ne, 3 and 34 for He, and 12 and 180 for Ar [6, 7].

With these values placed into Eqs. 15.4 and 15.9, the breakdown voltage V_b *vs.* pd of the Paschen theory can be calculated for given values of pd and γ_{se}.

$$V_b = \frac{Bpd}{\ln Apd - \ln[\ln(1 + 1/\gamma_{se})]} .$$

(15.10)

These curves for a few typical gases are shown in Fig. 15.3 for $\gamma_{se} = 0.2$, except for Xe, which has $\gamma_{se} = 0.02$. The theoretical predictions of Paschen's law compared well with calculations of a two-dimensional fluid PDP simulation [8]–[11], which solves the continuity equations for charged particles, Poisson's equation, and an equation for dielectric charging.

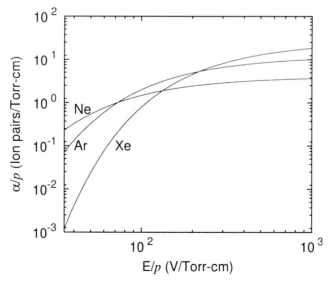

Figure 15.2: α/p–coefficient *vs.* E/p for Ne, Ar, and Xe.

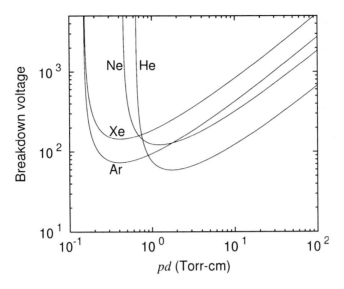

Figure 15.3: Paschen breakdown curves.

15.3.1.2   Collisional mean free paths

In a PDP cell, there are many types of collisions among charged and neutral particles. The mean free path λ is a function of the collisional cross section σ and the neutral density n. Namely,

$$\lambda = \frac{1}{n\sigma} = \alpha^{-1} . \qquad (15.11)$$

These quantities are related to the ionization frequency

$$\nu_{iz} = nK = nv\sigma . \qquad (15.12)$$

The cross section data for Ne and Xe are given in Refs. [8, 9] and those for Ar in Refs. [6, 7]. These values are needed to estimate the mean free paths. For $p = 300$ Torr, electron energy $E_e = 1$ eV, and ion energy $E_i = 30$ eV, the elastic scattering mean free paths for electrons and ions, λ_e and λ_i, are 1.5 and 1 μm, respectively, for He, 5 and 1 μm, respectively, for Ne, and 5 and 0.33 μm, respectively, for Ar. The mean free paths for electrons and ions are short compared with the PDP system size (about 100 μm) and the sheath size (about 10 μm), thus making the low energy particles abundant. These values of σ vs. E_e are used for a kinetic analysis. For a fluid analysis with electron energy evolution [10], the averaged values K vs. E_e over the Maxwell distribution are used. For the popular fluid analysis [8, 11] with the local field approximation, the rate coefficients from the zero-dimensional Boltzmann equation are tabulated as a function of E/p and inserted into the continuity equation. The mobility is also tabulated as a function of E/p, and the diffusion coefficient D is obtained by the Einstein relation.

15.3.2   One-dimensional AC-PDP model

A one-dimensional model of an AC-PDP can provide significant insight into the operating properties while avoiding the complications of geometry important in a real cell. The one-dimensional model is most similar to the opposed discharge type AC-PDP. In this section, the discussion treats the case where the cell is initially in the uncharged state; this can be considered to be the write pulse. The discussion is expanded to include the sustain pulse in section 15.3.4.

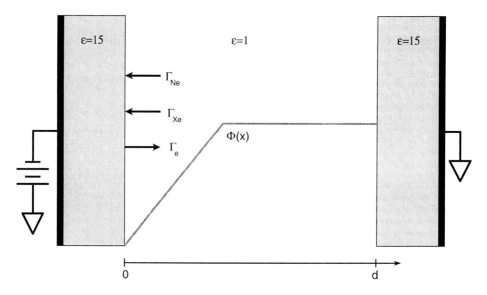

Figure 15.4: A one-dimensional model of an AC-PDP.

A simplified one-dimensional model of an AC-PDP cell is shown in Fig. 15.4. The cell is comprised of opposing electrodes separated by a gap of 150 μm. Each electrode is coated by a 25 μm thick dielectric layer, with a relative permittivity of 15. The resulting vacuum gap is 100 μm wide. The right electrode is grounded, while the left electrode is driven by an ideal voltage source ($V_0 = 200$ Volts here). A typical electric potential is shown schematically in figure 15.4. The dielectric layers are coated with a layer of magnesium oxide (MgO) of negligible thickness. The MgO layer is a good secondary emitter, with secondary coefficients given by $\Gamma_{Ne} = 0.5$ and $\Gamma_{Xe} = 0.05$. The ion and secondary electron fluxes at the left surface are shown in the figure. In this model, the phosphor is taken to lie beneath the MgO layer, and VUV absorption in the MgO layer is neglected.

The vacuum gap is filled with a gas mixture of Ne-Xe at a pressure of $p = 600$ Torr. It should be noted that typical production and prototype cells are more typically at 300–400 Torr. The function of the Ne is to generate the secondary flux, while the Xe supplies the excited states which emit the VUV line of interest (147 nm), as well as serving as the majority positive species.

When the discharge is initiated, the gap region contains only a trace density of charged

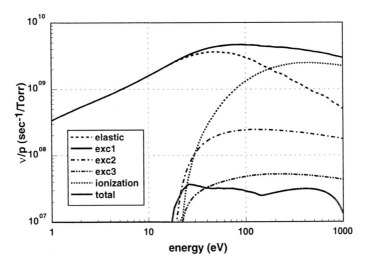

Figure 15.5: Electron collision frequencies for neon.

particles due to cosmic rays, neighboring discharge cells, or secondary emission due to impact of metastable atoms on the MgO surfaces. Electrons in the vacuum region are accelerated in the gap by the applied vacuum electric field, given by $E_0 = V_0/d$. As the electrons exceed the threshold energy for electron-impact ionization in the gas constituents, a new electron-ion pair corresponding to the target gas atom is created. The electrons continue to accelerate in the field, each undergoing additional ionization collisions, until they reach the dielectric surface in front of the anode. The collision frequencies for electron impact in neon and xenon are shown in Fig. 15.5 and Fig. 15.6, respectively.

Meanwhile, the ions are accelerated toward the cathode. They make numerous collisions with the background gas, transferring significant momentum due to the large cross section for ion-neutral charge exchange. The ions which impact the MgO surface in front of the cathode generate a secondary current as given by the secondary coefficient, Eq. 15.3.

If the applied voltage is greater than the breakdown voltage for the system, the electron and ion densities increase. Taking the electron and ion velocities to be mobility limited, they quickly reach the drift velocity,

$$\nu_d = \mu E = \frac{e}{m\nu_m} E \,, \tag{15.13}$$

where μ is the electron mobility, e is the electron charge, m is the electron mass, and ν_m is the total momentum transfer frequency for the gas mixture. The ion velocity is symmetric with this relation, replacing m by the ion mass M, and using the corresponding mobility and total momentum transfer frequency. Applying this approximately constant drift velocity to the flux given by Eq. 15.2, we obtain the density at early times:

$$n = n_0 \, \exp(\alpha\chi) \,, \tag{15.14}$$

where n_0 is the density at the cathode dielectric face, where $x = 0$.

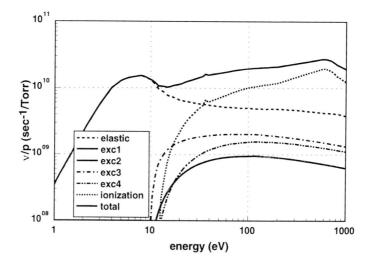

Figure 15.6: Electron collision frequencies for xenon.

Hence, the density is increasing most rapidly at the anode side of the gap at early times. This increase continues until space charge forces cause a flattening of the potential at the anode side of the gap. This low-field region, or positive column, expands gradually towards the cathode in time, resulting in a cathode fall with increasing field but decreasing width.

The current flowing to the dielectric surfaces results in charging of these surfaces, with the resulting surface charge given by

$$q_s(t) = \int_0^t J(t')dt' \, , \tag{15.15}$$

where $J(t)$ is the time-dependent net current density into the dielectric coating. Note that at the cathode dielectric, both the ion current and the generated secondary electron current have the same sign, and result in increased positive charging of that surface.

As the surface charge increases, the net voltage across the gap decreases according to the relation:

$$V_g(t) = V_0 - 2q_s(t)\frac{d_d}{\varepsilon_d} \, , \tag{15.16}$$

where d_d is the thickness of the dielectric, and ε_d is the permittivity of the dielectric. Note that the factor of 2 arises because both electrodes are coated with dielectric.

The gap voltage decreases relatively slowly early in the discharge, and undergoes a rapid decrease as the current increases exponentially. This results in charging of the dielectric layers, leading to a rapid drop in V_g. As the gap voltage drops, the discharge is extinguished and the discharge current decreases rapidly. The later stages are a glow discharge, in which little net current is drawn, and the discharge is decaying by ambipolar diffusion to the walls. The temporal evolution of the current density and the gap voltage is shown in Fig. 15.7.

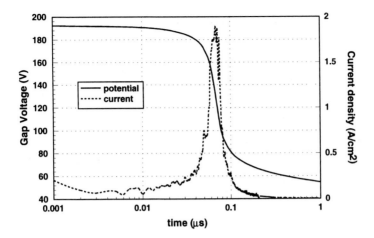

Figure 15.7: Temporal dependence of current density and gap voltage in a 1d AC-PDP cell.

The spatially averaged number densities of the electrons, Ne^+ and Xe^+ are shown in Fig. 15.8. The build-up of the plasma occurs quite rapidly in time (note the logarithmic time scale), while the decay occurs much more slowly. This is a consequence of the temporal decrease of the gap voltage, which results in a slowly decaying glow discharge after the current drops. Corresponding to the plasma build-up, excited states of Xe decay rapidly emitting VUV. The VUV stimulates emission of visible light from the phosphors, which must propagate through layers of dielectric, MgO, and transparent electrodes.

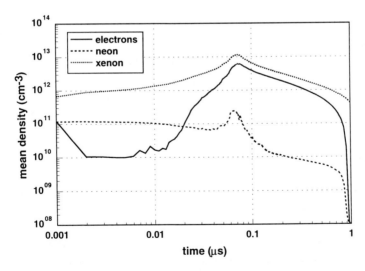

Figure 15.8: Mean number density for a Ne-Xe (90-10%) AC-PDP at 600 Torr with 200 V applied.

15.3.3 Two-dimensional AC-PDP models

15.3.3.1 The matrix and the surface discharge AC-PDP

There are two types of AC-PDPs, the matrix type [Fig. 15.1(a)] and the surface discharge type [Fig. 15.1(b)]. The DC-PDP will be discussed in section 15.4. The PDP research started with a monochromatic matrix-type AC-PDP by Slottow and Bitzer of the University of Illinois [1]. In the matrix type of Fig. 15.1(a), the top side (the front side toward the viewer) has transparent ITO electrodes for sustaining and scanning, which are in contact with non-transparent Cr/Cu/Cr bus electrodes. These electrodes are covered with protective materials such as MgO, which has a high coefficient of secondary electron emission γ_{se}. The bottom side (the rear side to the viewer) has an address electrode perpendicular to the sustaining and scanning electrodes. These two sides are coated with dielectric materials such as PbO and separated by a barrier rib of height 100-150 μm. Along the barrier rib one of the red, green, and blue phosphors is applied. The gas mixture (He, Ne, and Xe) in the cell is ionized by a high AC (or pulsed DC) voltage in the range of 200–300 volts to form a plasma between the top and the bottom electrodes. This matrix type suffers from the discharge stability and the short lifetime of the phosphor, since the phosphor covers the discharge electrodes and is bombarded by the charged particles from the discharge. The color surface discharge PDP cells have been studied since 1979.

15.3.3.2 The discharge characteristics in the AC-PDP cell

A high voltage difference (e.g., of 300 volts) is applied between the sustaining and the scanning electrodes in the surface-type AC-PDP cell. This voltage difference accelerates a small number of seed electrons present to energy large enough to ionize the gas mixture in the cell (Figure 15.9).

As the number of charged particles increases, especially near the anode, the potential profile which has initially a constant slope flattens from the anode leaving the large slope (i.e., the large electric field) in the cathode sheath region. This evolution of the potential is shown at two separate times (0.45 μs and 0.55 μs, the latter at the time of the electron density peak) in Figs. 15.10 (a) and (b). At density peak, the electron density profile and the Xe excited population (Xe*) profile are shown in Figs. 15.10 (c) and (d). The double humped structure of Xe* reveals a VUV emission profile similar to that experimentally observed recently [12]–[14].

The profiles of the potential, and the electron and ion densities at two different times, 22 and 37 ns (prior to the Xe* density peak), are shown in Fig. 15.11 for the kinetic simulation [15] of a matrix-type AC-PDP cell. Here, the anode and the cathode have a voltage difference of 250 volts. The cathode sheath decreases from almost one half of the cell size to 20 μm from 22 to 37 ns. The voltage difference arises between the bulk plasma and the MgO region at the dielectric. This voltage difference decreases to a mere 50 volts, the upper limit of the energy ions obtain during the acceleration from the bulk to MgO. The actual ion energy arriving at the MgO layer is smaller than 50 eV because the ion mean free path, as estimated in section 15.3.1, is much shorter than the cathode sheath length. This property is well reproduced in the kinetic simulation [15], which reveals that 90 % of the ions arriving at MgO have energies

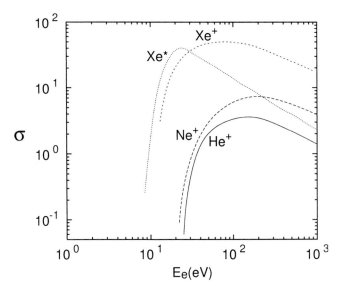

Figure 15.9: Excitation and ionization cross sections (in units of 10^{-21} m^2) of He, Ne, and Xe.

below 25 eV. The secondary electron emission coefficient is usually measured [16, 17] by using an ion beam of energy exceeding 50 eV, which must be lowered to obtain future data.

The general analysis of this problem is very difficult even in a 1d geometry, especially for mixtures. As it is an inherently transient discharge, the electric field rapidly changes in time and space. The dynamics of the discharge depend on the gas composition in the mixture and the population of excited species (metastable atoms, excited atoms, excited molecules).

A simulation study should be able to predict characteristics of a discharge, such as space distributions and time dependence of electric field and current during the discharge, excitation rates and luminosity as functions of time, gas composition, pressure, geometry, applied voltage, and other parameters. In addition, a simulation study should estimate the build-up of charge on the side walls of the barriers and predict the effects of the geometry (electrode size and gap distance) on the efficiency.

A very common simulation approach for evaluation of rates for electron-driven processes is to use a local field approximation (LFA), in which the rates are assumed to be related to the ratio of the local electric field to the neutral particle density, E/n. However, for a more accurate modeling, we should also consider the kinetic effects, including the nonlocal nature of the distribution function of the electrons. Furthermore, we must be careful in dealing with the influence of the boundaries, as either absorbers or emitters of electrons, on the rates.

The self-extinguishing discharges of an AC-PDP cell are particularly sensitive to parameters which can affect its threshold firing conditions. These often lead to dramatic changes in the magnitude and duration of the current pulse, voltage transfer, and population of various species. Many groups performed 2d fluid simulation for Ne-Xe mixtures with five kinds of ions, LFA, and the drift diffusion approximation [8]–[11]. They calculated the gap volt-

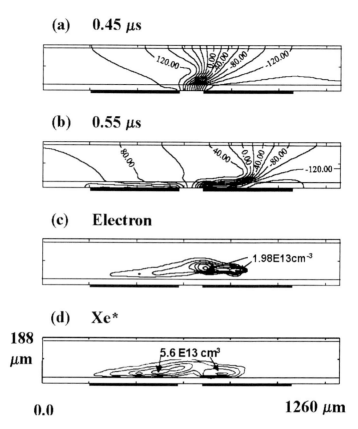

(a)   0.45 μs

(b)   0.55 μs

(c)   Electron

(d)   Xe*

Figure 15.10: (a) Potential profile at 0.45 μs, (b) at 0.55 μs, (c) electron and (d) Xe* profile at peak density.

age, current density, and radiated power as a function of time, and the spatial distributions of electric field, species density, and production rate at a fixed time. Using more sophisticated boundary conditions, they calculated the voltage drop due to the charge build-up at the dielectric layer. They obtained the spatio-temporal distributions of the potential and various species densities. Also they calculated the applied voltage-current relation curves for the discharges.

15.3.4   Driving voltage for the AC-PDP

In this section, the electrical properties of the driving circuits are described, including a discussion of the write and erase pulses, sustain pulses, and the memory effect. In addition, the voltage and frequency requirements of the driving circuitry are outlined. The voltage pulses applied to an individual plasma display panel cell include the write pulse, the erase pulse, and the sustain pulse. The purpose of the write pulse is to initiate a discharge in a cell which previously had none. The sustain pulse is used to continue the AC operation of a cell which was previously turned on. The erase pulse generates a discharge which just cancels the wall

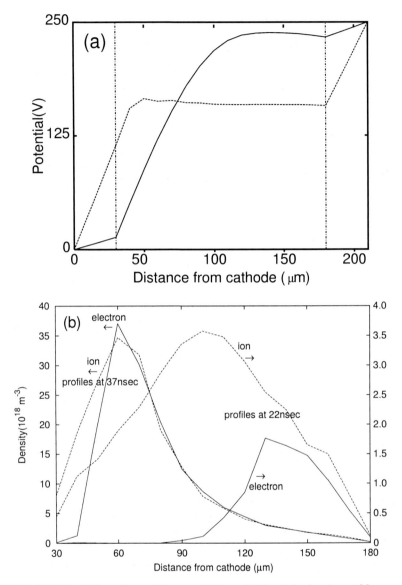

Figure 15.11: (a) Potential profiles at 22 ns and 37 ns. (b) Particle density at 22 ns and 37 ns.

charge, returning the cell to the uncharged (off) state. The write pulse voltage exceeds the breakdown voltage for the cell, $V_{0w} > V_b$, so a discharge occurs even when there is no wall charge on the dielectrics. This results in wall charge accumulating on the dielectrics. The model of section 15.3.2 applies to the write pulse, taking $q_s = V_w = 0$ initially.

Sustain pulses are scanned for all cells, one row at a time. For a sustain pulse, there exists an initial charge on the dielectric surfaces due to the previous discharge in the cell.

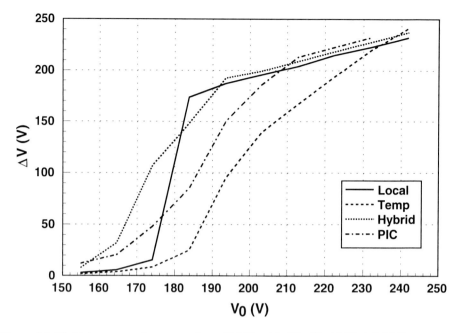

Figure 15.12: Voltage transfer curves for a 600 Torr Ne-Xe (90-10%) AC-PDP (see text).

This is often referred to as the memory effect, since cells which have experienced a previous discharge will breakdown while the others do not – the cells *remember* that they are in the on- or off-state. The applied voltage adds to the voltage due to the wall charge to give $V_g = V_0 + 2q_s d_d/\varepsilon_d > V_b$, where V_b is the breakdown voltage of Eq. 15.10. After the discharge, the gap voltage goes to zero, so the final wall voltage becomes $V_w = 2q_s d_d/\varepsilon_d = -V_0$. This gives a minimum sustain voltage, $V_s > V_b/2$.

The voltage transferred as a function of the initial gap voltage is shown in Fig. 15.12. Included in the comparison are a particle-in-cell Monte Carlo collision model (PIC-MCC) [18], a fluid model with reaction rates obtained from a local field approximation [19], a mean energy (temperature) approximation, and a Monte Carlo hybrid model [20]. The models agree reasonably well, and all are asymptotic to the slope of 1 at high voltages, as they must for a stable discharge.

Cells in the on-state remain in that state until they are turned off by applying an erase pulse of a polarity such that it induces a reverse discharge in the cell which cancels the wall charge in the cell. The cell is then in the off-state, and the sustain pulses do not cause a discharge to occur since the total gap voltage in the absence of wall charge is below the breakdown threshold. The write and erase sequences are shown schematically with the sustain waveform, wall voltage, and light output in Fig. 15.13. The cell is in the on-state between the write and erase pulses, and in the off-state outside that range. Note that the sum of the wall voltage and the sustain voltage exceeds the breakdown voltage only during the on-state, at the time when the light output is indicated.

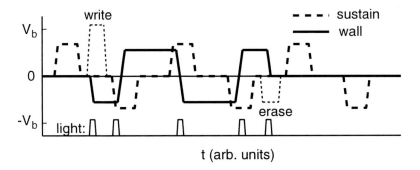

Figure 15.13: Write, sustain, and erase waveform schematics for an AC-PDP with wall voltage and light output indicated.

Color is produced in the AC-PDP by modulating the intensity of the light output of the red, green, and blue pixels. This is done by turning the pixel on for a specified number of sustain pulses during a single frame, called duty cycle modulation. Hence, the light output of a single sustain cycle determines the discrete levels of intensity available, and determines the addressing frequency required.

The sustain waveform is applied to an entire row, sequentially scanning through each row in the display. The sustain frequency is given by $f_s = gNf_r$, where N is the number of rows in the display, g is the color or gray scale bit depth per color ($g = 8$ for 256 intensity levels), and f_r is the full frame refresh rate for the display. Typical refresh rates are between 30–60 Hz, so the refresh circuit typically operates in the 240–480 kHz range for $N = 1000$ and $g = 8$. It is important to note that even with schemes which allow multiplexing to reduce the frequencies of the driving waveforms, the driving circuitry remains the most costly part of the PDP to manufacture.

15.3.5 Research status and remaining issues

The research advances made in recent years in various aspects of PDP are remarkable. From 1996 (when the VGA class of 640 × 480 pixels became a reality) to 1999 (with the XGA class of 1365 × 768), the peak luminance has progressed from 200 to 600 cd/m², the luminous efficiency from 0.6 to 1.6 lm/W, and the power consumption has dropped from 400 to 300 W. The early stage of PDP development was led by the USA (University of Illinois, IBM, and others) but the recent progresses are mainly made by Japan (Fujitsu, NEC, Pioneer, NHK, Matsushita) and Korea (LG, Samsung, Orion, Hyundai Electronics). For the HDTV class (1920 × 1080) for the use in early 2000's, the luminous efficiency and the power would have to reach 5 lm/W and 200 W, respectively.

There are other research goals to be met for the commercialization of PDP; among these are the high contrast by using efficient filters and black stripes. Overall, the reduction of the manufacturing cost is the most constraining issue for immediate commercialization. The protective layer such as MgO with a large secondary electron emission coefficient and with less absorption of 147 nm, needs to be further developed. Improved phosphor materials efficient for conversion of the 147 and 172 nm lines are also in great need. More development is needed to reduce the cost of the circuit that drives the cells with a high voltage exceeding around 200 volts at a frequency above 100 kHz. The drive circuit cost is closely related to the cell properties; the circuit cost decreases dramatically, if the address and sustain voltages can be lowered even slightly.

The remaining issues in the physics point of view are the improvement of efficiency and the reduction of power consumption. These might be accomplished by the optimization of PDP cells, materials, and driving circuits. The luminous efficiency is

$$\eta(\text{lm/W}) = \frac{\pi BS}{P} \quad \text{or} \quad \eta(\%) = \frac{\pi BS}{2.5P},$$

where B is the luminance (in cd/m^2), S is the display area (in m^2), and P is the power (in W). Improvement to 5 lm/W seems feasible by comparison with the success of the closely related fluorescent lamp discharge. The fluorescent lamp can reach 80 lm/W, but its dimension is much larger than PDP cells (by a factor of 100). The increase (from the natural abundance 0.15 % to the optimum 2.6 %) of mercury-196 isotope enhanced the efficiency of the fluorescent lamp significantly by reducing resonant absorption of the dominant 254 nm VUV line as well as the weaker 185 nm line. The use of mercury is not feasible in the PDP due to the high pressure of the PDP, which would limit the evaporation of the mercury in the lamp at reasonable temperatures.

Other methods of improving the PDP efficiency involve the optimization of the cell structure, the materials (protective, secondary emissive, and phosphors), the buffer gas mixture, the electron temperature, the driving pulse, and others.

The radiation transport and trapping in PDP cells are an issue of importance, which recently became the subject of intense research. The VUV 147 nm is the resonant line of Xe excited state; it has a high resonant absorption rate (up to 99 %), with the VUV absorbed by ambient Xe ground-state atoms and re-emitted later with some loss and broadening. To reduce this resonant absorption and emission, higher pressure gas and the 173 nm VUV are used. In a typical PDP cell, the driving circuit loss is around 30 %, the heat and radiation loss in the cell is around 98 %, the VUV loss to walls is around 40 %, and the phosphor heat loss is around 40 %. Surviving all these losses, the remaining power of 0.5 % of the total is available for visible light generation. This efficiency is much lower than that of a fluorescent lamp, where as much as 60 % of the energy goes into the generation of the VUV.

15.4   Other PDP types

Since 1968, the direct current (DC) type of PDP has been studied intensely by Philips, NHK, and Matsushita. The DC-PDP has electrodes in direct contact with plasma particles, and the high energy ion bombardment can cause damage and sputtering of the electrode surfaces. Since there is no dielectric between the electrodes and the plasma, it requires an external resistor to limit the current in each cell; this is a disadvantage in reducing the pixel sizes as well as an added loss term. The current limiting resistor is required because without it the current would continue to grow until the electrodes melt due to ohmic heating. There are two types of DC-PDP, the negative glow and the positive glow type. The present efficiency of the DC-PDP is in the range of 1 lm/W but it has a prospect for 3 lm/W with the positive column matrix DC type. The remaining issues for DC types are the search for good cathode materials (that have a large secondary electron emission coefficient and do not contaminate the plasma) other than the present aluminum, and ways to eliminate the external resistor to limit the current. DC-PDPs are currently not widely under development for commercial applications due to the recent advances in the AC-PDPs.

The plasma addressed liquid crystal (PALC) display is a hybrid combining attributes of a liquid crystal display (LCD) with a three-electrode AC-PDP. One of the problems in the scalability of LCDs to larger size required in HDTV and other applications is the difficulty in manufacturing large panels with a low fault rate for the millions of transistors embedded in the large panel. The fault rate is due to the statistical probability of transistor failure at the large panel sizes as well as to the difficulty to achieve uniform processing of the large substrates.

PALC displays replace the transistor with a plasma discharge switch, using the ten-order of magnitude change in conductivity possible for nonlinear plasma discharges. They take advantage of the cheap, uniformly scalable PDP technology to replace the transistor with a gas filled channel with an electrode pair on the bottom and a transparent data electrode perpendicular at the top (viewable face), similar to Fig. 15.1 with the phosphor removed and the electrode configuration inverted top to bottom. In addition, there is no dielectric coating on the cathode and anode, only on the data electrode. The cathode, anode, and data electrodes play the roles of the transistor gate, source, and drain respectively. Voltage applied to the cathode results in breakdown of the gas, just as in a standard PDP. The plasma electrically connects the data electrode to the anode, until the wall charge on the cathode reduces the voltage below the plasma sustain voltage and the plasma is extinguished. The plasma sustain voltage plays a role analogous to the gate-source threshold voltage in a field-effect transistor (FET) switch.

The liquid crystal material is above the data electrodes. When the plasma is in the on state, the cell can be written or erased; the charge on the top dielectric provides a sample and hold in conjunction with the data signal. The strength of the analog data signal determines the twist of the LCD layer, which in turn determines the color or grayscale level.

Power efficiency is high because the plasma is on only for a short time, with each row activated sequentially. The figure of merit is the switching speed of the PALC cell, which is related to the product of the plasma resistance and the cell capacitance. Lower pressures and higher plasma densities lead to higher conductivity and hence faster switching times. The plasma decay time also has an impact on switching speed.

Recent 16 inch PALC prototypes have achieved 4096 colors in a VGA format, with contrast of 40:1, 40 ms response time, and brightness of 140 cd/m^2 with an input power of 140 W [21].

15.5 Conclusions

A number of types of plasma display panel technologies have been described. The basic physics of operation of the AC-PDP has been addressed in detail, as well as the most important parameters affecting the operation of the device. PDPs have demonstrated commercial viability, as they are now in limited production runs for HDTV applications. Cost remains a significant barrier to commercial success, with progress continuing to be made as production increases.

Many physics and engineering issues are not completely resolved. Kinetic effects of the electrons require additional study, including methods of optimizing that may lead to improvements in power efficiency and brightness. Other areas for improvement include lifetime, MgO secondary emitter coating, gas mixture optimization, and phosphor improvement. Cost and efficiency of driver circuitry and pulse waveforms are also of great importance. Considerable room for improvement exists, as long as PDPs remain an order of magnitude less efficient than the physically similar fluorescent lamp discharge.

Acknowledgements

Helpful contributions from C.K. Birdsall, Y.K. Shin, J.J. Kim, and C.H. Shon are gratefully acknowledged.

References

[1] H.G. Slottow, IEEE Trans. Electron Devices ED-23 (7), 760 (1976).

[2] A. Sobel, IEEE Trans. Plasma Science 19 (6), 1032 (1991).

[3] L.F. Weber, Flat-Panel Displays and CRTs, 332 (ed. L.E. Tannas): Van Nostrand Reinhold (1985).

[4] J.P. Michel, Display Engineering, 185 (ed. D. Bosman): Elsevier Science Publishers (1989).

[5] H. Uchiike and S. Mikoshiba, Plasma Display: K-Book (in Japanese) (1997).

[6] M.A. Lieberman and A.J. Lichtenberg, Principle of Plasma Discharges and Materials Processing, John Wiley and Sons, Inc.:New York (1994).

[7] Y.D. Korolev, G.A. Mesyats, Physics of Pulsed Breakdown in Gases, URO-Press (1998).

[8] Y.K. Shin, C.H. Shon, H.S. Lee, W. Kim, J.K. Lee, Asia Display'98, 609 (1998); IEEE Trans. Plasma Science 27 (1), 14 (1999).

[9] J.P. Boeuf, C. Punset, A. Hirech, J. Phys. IV France 7, C4-3 (1997).

[10] S. Rauf and M.J. Kushner, J. Appl. Phys. 85, 3460 (1999).

[11] R. Veerasingam, R.B. Campbell, R.T. McGrath, IEEE Trans. Plasma Science, 26, 1532 (1998).

[12] H.S. Jeong, et al., J. Appl. Phys. 85 (6), 3092 (1999).

[13] A. Okikawa, T. Yoshioka, K. Tori, SID Digest, 276 (1999).

[14] K. Tachibana, S.J. Feng, T. Sakai, Bull. Amer. Phys. Soc. 44 (4), 18 (1999).

[15] Y.K. Shin, J.K. Lee, C.H. Shon, W. Kim, Jpn. J. Appl. Phys. 38, L174 (1999).

[16] E.H. Choi, et al., Jap. J. Appl. Phys. 37 (12B), 7015 (1998).

[17] K.S. Moon, J.H. Lee, K. Whang, J. Appl. Phys. 86, 4049 (1999).

[18] J. P. Verboncoeur, M. V. Alves, V. Vahedi, and C. K. Birdsall, J. Comp. Phys. 104 (1993).

[19] R. A. Stewart, P. A. Vitello and D. B. Graves, J. Vac. Sci. Technol. B 12, 478 (1994).

[20] G. J. Parker, J. P. Verboncoeur, B. M. Penetrante, P. A. Vitello, and J. Shon, Bull. Amer. Phys. Soc. 40, 1575 (1995).

[21] T. S. Buzak, Digest of Tech. Papers, 1990 SID Int. Sym., Soc. for Inform. Display, 420 (1990).

16   Low-pressure discharge light sources

G.G. Lister

OSRAM SYLVANIA Development Inc., 71 Cherry Hill Drive, Beverly, MA 01915, U.S.A.

16.1   Introduction

The underlying principle of low-pressure discharges for light sources is the excitation of constituent atoms of the discharge into radiation emitting states by electrons which are far from local thermal equilibrium with the other species in the discharge. This paper will discuss three of the principle types of low-pressure discharge lamps: *fluorescent lamps, low-pressure sodium lamps*, and *rare gas discharges*.

Fluorescent lamps are filled with a rare gas, typically argon at around 3 Torr pressure, with a minority of mercury (typically a few mTorr). Mercury has the advantage that at room temperature it has the highest vapor pressure of any of the elements suitable for producing radiation. Between 60–70% of electrical power in these discharges is converted to *ultraviolet (UV)* radiation by mercury atoms. A phosphor is then used to convert the *UV* to visible light, resulting in a total conversion efficiency of electrical power to visible light of about 25%. Today, fluorescent lamps light over 50% of the building floor space in North America [1], and it has been estimated that 80% of the world's artificial light is fluorescent [2].

Low-pressure sodium lamps operate on a similar principle to fluorescent lamps, with mercury replaced by sodium, such that the principal emission is in the visible spectrum, with neon as buffer gas at around 8 Torr pressure. The absence of the phosphor means that these lamps are much more efficient in converting electrical power to visible light. However, the optimum vapor pressure can only be reached at 260 °C, which means the lamp requires good thermal insulation and glass which is resistant to sodium at the operating temperature. Further, the low-pressure sodium lamp produces monochromatic yellow light, and can therefore only be used for applications where color discrimination is not important, such as in street lighting [3].

Rare gases atoms emit radiation in the vacuum ultraviolet (*VUV*) and the visible spectrum, but the efficiency of converting electrical energy to radiation is significantly less than in fluorescent and low-pressure sodium lamps. However, there are special applications for which rare gas discharges have some advantages. Neon lamps are used for some automobile applications, in which the red visible lines may be used directly for stop signals [4], or combined with radiation converted from the *VUV* by a green phosphor to provide an amber light for other traffic signals [5]. Excimer radiation can also be excited efficiently in barrier discharges for flat panel lighting [6].

Some encouraging results have been reported recently for efficient light sources using barium in place of mercury or sodium in a rare gas discharge [7]. The temperature required to obtain an optimum vapor pressure of barium is higher than for sodium (500–600 °C), with the consequent difficulty of finding materials resistant to corrosion. The principal atomic radiation is the green 553 nm resonance line, but excitation of barium ions can lead to a white light source [7].

There are a number of factors which determine the suitability of a light source for a particular application. *Luminous efficacy* is a measure of how efficiently a lamp converts electrical energy to visible light and is measured in *lumens per watt (lpw)*. Efficacy is defined in terms of the response of the average eye to the visible spectra in bright viewing conditions. The eye response is at maximum for green light at 555 nm, and 1 watt of radiative energy at this frequency is defined to be 683 lumens. A typical incandescent lamp operates at 12 lpw, a conventional fluorescent lamp at 80–100 lpw, and a low-pressure sodium lamp at 200 lpw. *Color rendering* is a measure of how well the light source reproduces the colors of objects, and this is defined by a color rendering index (*CRI*) where *CRI*=100 represents *perfect* color rendition. *Correlated Color Temperature (CCT)* is the temperature of the black body whose spectrum most closely represents the spectrum of the light source and is therefore a measure of the apparent color of the light source. Clearly, the relative importance of these factors to a light source will depend on the application for which it is intended. For a more detailed treatment of these and other issues relating to vision and quality of light, the reader is referred to Ref. [8], Part I.

In conventional discharge lamps, the energy required to supply the light source is provided by direct or low frequency alternating electric currents, which require electrodes within the lamp to maintain the discharge. The presence of electrodes places severe restrictions on lamp design and is a major cause of failure, limiting lamp life. Recent developments in fluorescent lamp technology have led a number of companies to introduce electrodeless products, in which the power is introduced in the discharge by inductive coupling of radiofrequency (*rf*) power [9, 10]. This culminates a century of development since Tesla demonstrated in 1891 that light could be generated in the presence of a radiofrequency electromagnetic field [11]. The full potential of these lamps has yet to be demonstrated, but they represent possibly the most significant development in the lighting industry in the last decade, and a discussion of these lamps forms a significant part of this chapter.

In the following sections, the physics of low-pressure light sources is reviewed, with particular emphasis on issues of current interest. For a detailed description of the physics of discharge lamps, the reader is referred to the classic work by Waymouth [12], while recent technological advances in lighting are reviewed in Ref. [8].

16.2   The physics of low-pressure discharge lamps

The conventional (electroded) low-pressure discharge lamp is essentially a glow discharge. Electrons emerging from the cathode are accelerated through the cathode fall into a region of relatively weak electric field, the negative glow. There is an over-production of ions in the negative glow, which is compensated by a dark region of low ionization, the Faraday dark space. Following the Faraday dark space is a region of constant electric field, the positive column, which produces almost all the light in a low-pressure discharge lamp. A bright anode glow is separated from the positive column by an anode dark space. All regions in the neighborhood of electrodes contribute to inefficient use of power; this effect can be minimized by ensuring that the positive column is much longer than the other regions of the discharge.

16.2.1 Collisional processes

Collisions of electrons with atoms and molecules are the dominant mechanisms in low-pressure discharge lamps. Inelastic collisions excite atoms to metastable or radiative states and de-excite them to states of lower energy (super-elastic collisions), strongly influencing the high energy tail of the electron energy distribution function (*EEDF*). Atoms may also be ionized, either directly from the ground state, or from excited states (two-step ionization). Elastic collisions of electrons with atoms and ions couple the electric power to the discharge through the electrical conductivity, and provide gas heating, while electron-electron collisions redistribute the electron energy, and for a sufficiently high degree of ionization lead to a Maxwellization of the *EEDF*. Collisions between atoms in excited states and other atoms and molecules (such as chemi-ionization) can also play an important role in the discharge, quenching radiative states and providing an extra channel for ionization, which influences the current-voltage characteristics.

Atoms with a radiative transition to the ground state (resonance transition) from an energy level which is close to half the ionization energy can provide efficient radiation in low-pressure discharges. Electron impact excitation to the radiative state or neighboring metastable states provides channels for both radiation and two-step ionization to maintain the discharge. Mercury is an efficient radiator because the energy level of the first excited state for resonance radiation is 4.89 eV, and the ionization energy is 10.4 eV. Sodium (2.1 eV and 5.1 eV) and barium (2.2 eV and 5.2 eV) have similar properties, but require higher temperatures to maintain the required vapor pressure. Rare gas atoms have radiative states which are much closer to the ionization level and therefore are less efficient radiators. For example, neon has states for *VUV* resonance radiation at 16.7 and 16.9 eV and an ionization level of 21.6 eV. However, excimer radiation from excited rare gas molecules, such as Xe_2^* can make a valuable contribution to radiation output [6].

16.2.2 Radiation transport

Spectral lines emitted by atoms are broadened and shifted by three different processes; pressure (or collisional) broadening, Doppler broadening, and natural broadening. Radiation emitted at one point in the discharge may be absorbed and re-emitted many times before reaching the walls, broadening the radial density profiles of radiating states and consequently influencing the spectral output and power balance of the lamp. Further, radiation from the wings of each spectral line is less trapped than that from the center of the line, and for strongly absorbed lines the major contribution to the radiation emitted from the discharge is in the line wings. A detailed discussion of resonance radiation transport in low-pressure discharges is given in Ref. [13], and for particular application to fluorescent lamps in Ref. [14].

16.2.3   Ambipolar diffusion

Due to their higher mobility, electrons diffuse more rapidly to the wall than ions, and the ambipolar space charge field is established to maintain an equal radial flow of ions and electrons. Ions are thus accelerated away from the center of the discharge, while the electron motion is retarded. In a discharge containing a minority species with an ionization energy less than that of the atoms of the buffer gas, ambipolar diffusion leads to a depletion of the minority species at the center of the discharge (*cataphoresis*). Cataphoresis can play a role in fluorescent lamps but is a dominant process in low-pressure sodium lamps [15].

16.2.4   Electron energy balance in the positive column

The electron energy balance in the positive column of a gas discharge may be summarized by

$$IE_z = P_{rad} + P_{el} + P_{wall}, \tag{16.1}$$

where I is the discharge current, E_z the maintenance axial electric field, P_{rad} the power per unit length converted to radiation, P_{el} represents power losses due to elastic processes, which result in gas heating, and P_{wall} represents the total power losses to the wall, including the diffusion of metastable atoms, ions and electrons and their recombination at the wall.

16.3   Electroded fluorescent lamps

The low-pressure Hg-rare gas discharge has been studied extensively over more than half a century, and there is an extensive literature on the subject. The underlying physics is covered in Ref. [12], a comprehensive review of low-pressure mercury-rare gas discharge research prior to 1990 is given in Ref. [16], and the current status of fluorescent lighting technology is given in Ref. [2].

The major component of radiation in a fluorescent lamp is *UV* resonance radiation from the 6^3P_1 level of mercury, with a wavelength of 254 nm, and considerable research activity has been devoted to maximizing the contribution of this line to the power balance in the discharge. A second resonance state (6^1P_1) emits *UV* radiation at 185 nm, and there are a number of visible emissions (mainly in the blue and green) from higher excited states radiating down to lower levels. These additional lines are particularly important in discharges at high power loading, and attention must be given to the choice of phosphor to ensure optimum efficacy and color temperature. The situation is complicated by the presence of 5 isotopes in natural mercury, and the hyperfine structure of the resonance lines plays an important role in radiation transport, reducing the trapping of the 254 nm radiation compared to the case of a single isotope [16].

The most thorough experimental analyses of Hg-Ar fluorescent lamps were performed by Philips researchers in the early 1960's [17]–[19]. All experiments were conducted on discharges corresponding to a T12 lamp, so called because of its internal diameter (12/8"=3.6 cm), which represented at that time the size of the conventional fluorescent lamp. Electric fields and electron temperatures and densities were measured using Langmuir probes [17], while detailed spectral analyses were used to determine the energy balance of emitted radiation [18], and the densities of the 6^3P metastable and radiative states [19]. Measurements

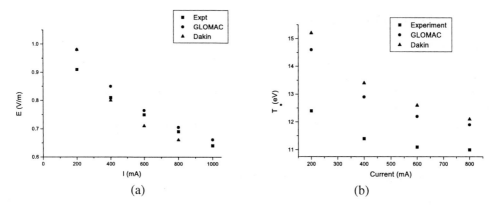

Figure 16.1: Comparisons between experimental measurements and numerical calculations from Dakin [22] and GLOMAC [24] models for T12 lamps (ID = 3.6 cm, argon pressure 3 Torr, mercury pressure 7 mTorr) as a function of discharge current. (a) electric field, (b) electron temperature.

were performed over a wide parameter range, for buffer gas pressures of 1–20 Torr, discharge currents of 0.2–1 A, and mercury pressures of 0.6 to 90 mTorr, corresponding to a minimum temperature on the glass wall (*cold spot*) of 10–80 °C. Langmuir probe measurements were also conducted at lower buffer gas pressures, including pure mercury, at 7 mTorr mercury pressure, but unfortunately there are no accompanying spectral data.

Numerical models [16], [20]–[24] in which the full collisonal-radiative equations have been solved, together with the power balance outlined in section 16.2.4, have reproduced many of the results from these experiments (electric field, emitted radiation, electron density) despite uncertainties regarding fundamental processes, particularly cross sections. Most models use the electron impact cross sections for mercury excitation and ionization obtained from swarm data by Rockwood [25]. These data are restricted to collisions with ground state atoms and must be augmented by realistic estimates of interaction with atoms in excited states. Associative processes are important at low current densities [26], or high mercury pressures. Calculations of the charged particle balance in these discharges is found to be sensitive to associative ionization involving two $^3P_2$ mercury atoms, and there is some discussion regarding the correct choice of cross section [10]. Excitation of rare gas atoms is unimportant at standard operating temperatures, but becomes important for low mercury vapor pressures, corresponding to low *cold spot* temperatures.

Comparisons of experimentally measured electric fields and electron temperatures with computed values [22, 24] at standard operating temperatures (42 °C or 7 mTorr mercury vapor pressure) are shown in Figure 16.1. The electric field calculations agree well, but in common with other reported calculations, the computed electron temperature is consistently larger than the experimentally measured value. This discrepancy is of concern, since an understanding of the *EEDF* is essential in determining the emitted radiation spectra and electrical characteristics of discharges over the wide range of operating parameters in contemporary fluorescent lamps. Operating buffer gas pressures in electrodeless lamps are often a few tenths of Torr (cf. section

16.4.4), fluorescent lamps with diameters as small as 5 mm are used for back lighting of laptop computers, and highly loaded lamps may operate at current densities 10 times or more than for conventional fluorescent lamps.

A number of positive column models [16, 23, 24] have used an approximation for the *EEDF* proposed by Lagushenko [16], in which the main assumptions are

(i) Electron-electron collisions dominate all other collision processes for $\varepsilon \leq \varepsilon_1$, where ε_1 (in eV) is the energy of the first excitation level of mercury, and the *EEDF* may be approximated by a Maxwellian distribution in this range.

(ii) For $\varepsilon > \varepsilon_1$, the only inelastic collisions that play a significant role in the EEDF are excitation and ionization from the ground state and super-elastic collisions to the ground state. The depletion of the high energy tail is compensated by electron-electron collisions (cf. section 16.2.1), such that for sufficiently high electron density the EEDF becomes Maxwellian even for high electron energies.

Further details of this approximation are documented elsewhere [10, 16], but the essential result is an analytic formula describing the depletion of the high energy tail of the *EEDF*. In conventional fluorescent lamps, however, the assumption that the *EEDF* is Maxwellian for electron energies below the first excited state is contradicted by experiments in Ar-Hg and Ne-Hg discharges [27].

An alternative approach to the Lagushenko approximation for the *EEDF* is the two electron group model, developed by Morgan and Vriens [28], and extended by Dakin [22]. This model assumes that the bulk electrons ($\varepsilon \leq \varepsilon_1$) may be described by an electron temperature $\varepsilon_e^{(1)}$, while the tail electrons ($\varepsilon > \varepsilon_1$) may be described by a second electron temperature $\varepsilon_e^{(2)}$.

The approximations described above were made because of the extra computational time required to solve the complete Boltzmann equation in conjunction with a detailed collisional-radiative model. However, this argument no longer applies, since Boltzmann codes now run on desktop PCs in a few seconds or less. Further, the *EEDF* calculations to date have been based on the *local* approach, where the *EEDF* is assumed constant across the discharge. The influence of the local ambipolar electric field on the *EEDF* in low-pressure discharges for lighting is not clear. The *local* theory is only strictly applicable for an electron energy relaxation length $\lambda_\varepsilon \ll R$, where R is the discharge radius. In fluorescent lamp discharges, for elastic collisions, we have $\lambda_\varepsilon \gg R$, and *non-local* theory is applicable, while for *inelastic* collisions $\lambda_\varepsilon \leq R$ holds. Non-local theory is of current interest in modeling for plasma processing [29], and calculations by Feokistov *et al.* [30] indicate that non-local effects are significant if the product of pressure and diameter is $pd < 10$ Torr cm, which is generally the case for fluorescent lamps. Attention to this aspect of modeling may help to resolve the discrepancies between measured and calculated electron temperatures. See also the discussion in chapter 2 of this book.

Discrepancies between experimental measurements and numerical computations of highly loaded discharges have been reported in the case of small diameter (ID = 3.2 mm) lamps [31]. In that paper, there was a discrepancy of factor 2 between the computed and measured electric fields at normal operating temperatures (40–50 °C). More recent results [32] have shown reasonable agreement with experiments at these temperatures, but show strong discrepancies at lower mercury pressures. It is important to resolve these discrepancies, in the light of industry trends towards increased power loading, such as in many electrodeless lamps.

The above discussion has concentrated on Hg-Ar discharges operated with direct currents (*dc*). Lamp voltage and performance can be influenced by choice of a different rare gas, or mixture of gases, since the electrical conductivity depends to a large extent on the electron-momentum collision cross section with the buffer gas. Most fluorescent lamps operate on an alternating current (*ac*) (50/60 Hz) ballast. Power losses near the electrodes can be reduced by operating the lamp at a frequency above the ambipolar diffusion frequency [33], typically 1 kHz. High frequency operation also enhances the high energy tail of the *EEDF* [34], leading to increased radiation efficiency in the positive column. In practice, the gain of efficacy with increasing frequency appears to saturate at about 20 kHz [35].

16.4 Electrodeless fluorescent lamps

The prospect of developing *electrodeless* lamps has challenged lighting researchers for more than a century. The introduction of lamps based on the electrodeless principle for general purpose lighting has been hampered by the lack of readily available, compact, cheap and reliable driving electronics. However, in the last decade, progress in semiconductor electronics and power switching technology have made this approach commercially feasible.

The first electrodeless fluorescent lamps were introduced by Philips (*QL*) and Matsushita (*Everlight*) in 1991, followed by GE Lighting (*Genura*) in 1994 and OSRAM SYLVANIA (*ENDURA* in Europe, *ICETRON* in North America) in 1998. All of these lamps are based on the principle of inductive coupling, described in section 16.4.4 below.

16.4.1 Benefits of electrodeless discharges for lighting

There are a number of potential benefits to lighting to be obtained from electrodeless operation of lamps [9, 10], and many of these have been realized in commercial products, while others are the subject of active research.

Lamp life: As noted in section 16.1, a major limitation on the operating life of a lamp is the presence of electrodes. Fluorescent lamps operate with thermionic cathodes, i.e., an electron emitting material (such as barium oxide) is coated onto the electrode. The evaporation of emitter material during life causes unsightly darkening of the end of the lamp and results eventually in lamp failure. Conventional fluorescent lamps have rated lifes of around 5,000–20,000 hrs, while electrodeless fluorescent lamps such as *QL* and *ENDURA* are rated at 100,000 hrs.

Lamp design: For many general lighting applications the source of light should be unobtrusive while providing optimum illumination, and there is a premium for developing smaller lamps. Very short fluorescent lamps are inefficient because the energy loss due to the electrodes is significant compared to the energy converted to radiation [2]. Compact fluorescent lamps are made by folding the discharge tube, and are equipped with electronics which enable them to fit into a regular incandescent lamp socket, but they have limited aesthetic appeal. Inductively coupled fluorescent lamps have been developed with shapes resembling the incandescent lamp, which they may eventually replace.

Electronic control: An attractive feature of electrodeless fluorescent lamps is virtually instant starting and restarting, similar to incandescent lamps. Unlike either incandescent or conventional fluorescent lamps, however, they perform well under rapid switching.

16.4.2   Electromagnetic interference and safety

Electrodeless lamps are required both to be both safe and to avoid interference with radio communications [9]. International regulations on limits of electromagnetic interference (*EMI*), which are generated by plasma, coils and circuits in electrodeless lamps, mean that for practical purposes the available frequencies are limited to the Industrial, Scientific and Medical (*ISM*) bands allocated for non communications use. Of these, the two bands used for commercially available electrodeless fluorescent lamps are the industrial frequency 13.56 MHz and 2.65 MHz, which is within a band with reduced restrictions on *EMI* allowed for lighting. High frequency currents in the mains supply due to *rf* radiation, the *rf* potential of the lamp relative to ground, and *rf* energy fed back to the *ac* mains from the ballast must be eliminated by a suitable combination of filtering and shielding. There are no safety issues with regard to *rf* applications for lighting [9].

16.4.3   The physics of electrodeless fluorescent lamps

The absence of electrodes enables fluorescent lamps to operate under conditions which would be impractical in conventional fluorescent lamps:

(i)   Electrodeless lamps often operate at much higher current densities and electrical power than would be possible in conventional lamps.

(ii)   In electroded lamps, the buffer gas protects the electrodes. Electrodeless lamps can operate at lower gas pressures, giving better efficacy, particularly at high power loading.

The influence of buffer gas pressure and discharge current on efficacy in a Hg-Ar fluorescent lamp of similar dimensions as *ENDURA* (ID = 4.8 cm) is illustrated in Figure 16.2. Results shown are from computer calculations using the models described in section 16.3, but they are consistent with experimental observations for electroded fluorescent lamps. At low values of discharge current (< 1 A), the maximum efficacy is obtained for argon pressure between 0.5 and 1 Torr, and the optimum pressure is even lower for higher discharge currents. Electrode maintenance in standard fluorescent lamps, however, requires a minimum buffer gas pressure of 2.5–3 Torr, where efficacy is much poorer, especially for high discharge currents.

The computed power balance as a function of pressure in fluorescent lamps at different power loading is illustrated in Figure 16.3. For a conventionally operated fluorescent lamp at 400 mA discharge current, the total fraction of power emitted as radiation increases as the buffer gas pressure is increased from 0.2 to 1 Torr, the increase in elastic losses at higher pressure being more than compensated by a reduction in wall losses. For a discharge current of 10 A, typical of some of the new electrodeless lamps (cf. section 16.4.4 below) the fraction of power lost in elastic processes is higher than for conventional lamps, due to increased Coulomb interactions resulting from significantly higher electron densities. The fraction of power lost to the walls (chiefly ion flux) remains fairly constant as pressure is increased,

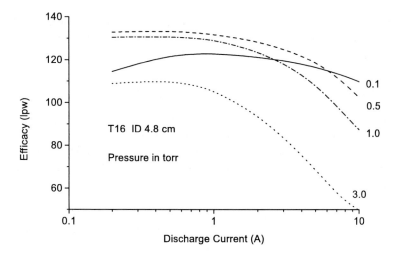

Figure 16.2: Variation of lamp efficacy as a function of discharge current for argon gas pressures 0.1, 0.5, 1.0, and 3.0 Torr (mercury pressure 7 mTorr, ID = 4.8 cm).

so total losses increase and efficacy is reduced. In conventional discharges, the maintenance electric field at constant current increases with gas pressure, while at higher discharge currents, where two step ionization processes dominate, the maintenance field decreases and hence the total discharge power is reduced. Thus, in highly loaded discharges, the decrease in the absolute value of the power lost to the walls is compensated by a reduction in total discharge power, and the ratio of wall losses to discharge power remains constant.

All electrodeless fluorescent lamp discharges are *over dense* – i.e. the applied frequency is smaller than the plasma frequency ω_{pe}, with electron densities $10^{17} - 10^{18}$ m^{-3}, for which $\omega_{pe} \geq 2 \times 10^{10}$ s^{-1} (~ 3 GHz). These discharges also operate with electron momentum transfer collision frequencies ν_m which are higher than the applied frequency (typically $\nu_m \approx 10^8$ to 10^9 s^{-1}.). Therefore, the particle kinetics of these discharges can be adequately treated by steady state *dc* models, while their electrodynamics must take into account the inhomogeneity of the electromagnetic field, mostly due to the *skin effect*. The collisional skin depth δ_c is [36]

$$\delta_c = \frac{c}{\omega_{pe}} \left(\frac{2\nu_m}{\omega} \right)^{1/2} \tag{16.2}$$

In all existing electrodeless lamps, the depth is of the order of the lamp dimensions.

16.4.4 Inductive fluorescent discharge lamps

The plasma in an inductive (or *H*) discharge forms a single turn secondary to an exciter coil which may be placed in or around the discharge. In common with inductively coupled discharges (*ICD*) for other industrial applications, the discharge starts as a capacitively coupled *E* discharge (cf. section 16.4.5), until sufficient poloidal electric field is present to create the high density *H* discharge required to provide adequate light output. The starting phase in

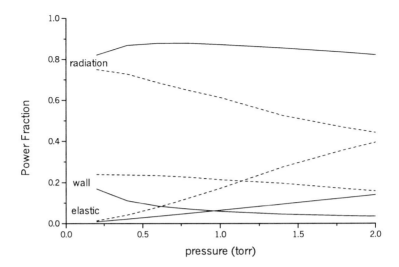

Figure 16.3: Energy balance in the positive column of a fluorescent lamp as a function of buffer gas pressure; discharge current 400 mA (full lines), and 10 A (dashed lines), ID = 3.6 cm, mercury pressure 7 mTorr.

these lamps lasts only a few milliseconds and is therefore effectively instant for the user [9]. Provided sufficient power is applied to maintain the *H* discharge, high coupling efficiencies may be achieved at low frequencies, with the benefits of reduced electromagnetic interference (*EMI*) and cheaper electronics.

ICDs are currently the only form of electrodeless fluorescent lamps to be exploited commercially, and a summarizing description of these lamps and their underlying physics is given in the following sections.

16.4.4.1   Re-entrant cavity lamps

QL and *Genura* lamps operate on the re-entrant cavity concept introduced by Bethenod *et al* [37], which is illustrated schematically in Figure 16.4. An insulated coil is wound several turns around a ferrite core and placed within the re-entrant cavity. The application of an *ac* current activates the discharge. The high permeability of the ferrite core increases the magnetic flux surrounding the discharge and leads to more efficient coupling, but power losses in the ferrite increase at high temperatures and high electric fields. Additional phosphor coating on the re-entrant is used to enhance the production of visible light from *UV* transported to the inside of the lamp. The maintenance electric field can be reduced by using krypton in place of argon, but the choice of buffer gas must also be balanced by efficacy considerations.

The *QL* lamp resembles a regular incandescent bulb [38], although somewhat larger (the 55 W lamp has a 8.4 cm diameter and is 15 cm tall). *QL* operates at 2.65 MHz, with argon at a gas pressure of a few hundred mTorr. Heating of the ferrite is reduced by separating the lamp and electronics via a shielded coaxial cable. The main feature of the *QL* lamp is its exceptional life (100,000 hrs) making it ideal for situations where maintenance is difficult.

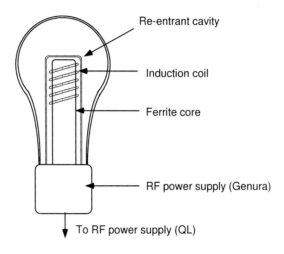

Re-entrant cavity

Induction coil

Ferrite core

RF power supply (Genura)

To RF power supply (QL)

Figure 16.4: Schematic of a re-entrant cavity lamp (*QL* and *Genura*).

Genura is the first compact electrodeless fluorescent lamp with integrated electronics, intended as a high-efficacy direct replacement for incandescent reflector lamps [9]. *Genura* also operates at 2.65 MHz with krypton at a gas pressure of a few hundred mTorr. In view of the application, life is not the major issue, but the lamp offers a fourfold improvement in efficacy compared to the incandescent reflector lamp it replaces. *Genura* has an electrically transparent *ITO* coating to reduce *EMI* as well as the usual phosphor, and a further titania reflector coating on the re-entrant and neck of the bulb, which gives it the appearance of an incandescent reflector bulb.

The fundamental properties of *ICD* in a re-entrant cavity lamp are illustrated in Figure 16.5a, obtained from a *1D* model [39]. The inductive coil is assumed to be infinitely long, and the plasma forms the secondary loop of a transformer. Figure 16.5a was obtained by integrating Maxwell's equations, assuming a simple Schottky model for the ambipolar diffusion, with zero electron density at both the re-entrant and the outer wall. In contrast to conventional fluorescent lamps, the electron density has a maximum near the center of the annular ring forming the discharge, and the electrical power deposited in the discharge is concentrated near the re-entrant wall. Experimental measurements in *QL* lamps show relatively high gas temperatures near the re-entrant [40], leading to mercury depletion in this region.

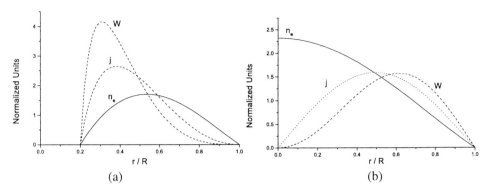

Figure 16.5: Radial profiles of normalized electron density n_e, current density j and power density $W = jE_\theta$ in *ICD* discharges calculated using a *1D* (infinite length coil) model [39] with (a) internal and (b) external coils. All quantities are normalized to their volume averaged value across the radius.

16.4.4.2 Lamps with outer coils

The fundamental properties of *ICD* in a lamp with an external coil such as *Everlight* are illustrated in Figure 16.5b, obtained from a *1D* transformer model [39, 41] similar to that described in the preceding section. In this case, the electron density is peaked at the discharge axis, as in the positive column of conventional fluorescent lamps, but power deposition is now peaked towards the *outer* wall, due to the skin effect.

Everlight [42] was introduced in 1991 in the Japanese market only. The coil is wound on the outside of a 4.5 cm diameter bulb (cf. Figure 16.6) requiring increased electromagnetic screening compared to other inductively coupled lamps, which is provided by a mesh screen construction outside the lamp. The lamp operates at the industrial frequency of 13.56 MHz using neon as a buffer gas to provide visible light during starting.

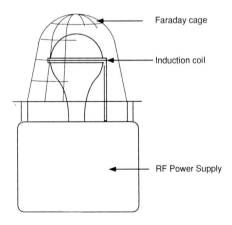

Figure 16.6: Schematic of *Everlight* lamp.

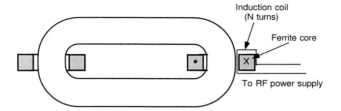

Figure 16.7: Schematic of the *ENDURA* lamp.

16.4.4.3 Toroidal lamps

Anderson [43] was the first to demonstrate that a fluorescent lamp could be operated at low frequency (100–500 kHz), by designing a lamp based on the principle of a ring discharge, similar to that of a tokamak used in fusion research. The tube ring penetrated a ferrite core, wound with a primary winding to which *rf* power was applied, the ferrite ring providing the single turn secondary. The voltage in the lamp was thus induced by a closed magnetic path. Low frequency operation of electrodeless lamps is attractive because of the low cost of electronics and easier restrictions on *EMI*. However, the lamp required a large amount of ferrite and considerable power was dissipated in the ferrite ring, resulting in unacceptably low efficacy.

Recent research has shown that ferrite losses can be minimized provided the power loading is sufficiently high [44]. The reason for this are firstly the negative $V - I$ characteristics of the fluorescent lamp (Figure 16.1a), and secondly the scaling of ferrite losses as a steep function of core magnetic induction, and hence of the discharge current I. For a typical fluorescent lamp $V \propto I^{-0.3}$ holds, and typical power losses in the low frequency ferrite core are given by $P_c \propto V^{2.8}$, such that the ratio of core losses to power in the discharge P_d follows

$$P_c/P_d \propto I^{-1.5} . \tag{16.3}$$

ENDURA (*ICETRON* in Northern America), illustrated schematically in Figure 16.7, is based on a concept similar to that of Anderson [43], and has the lowest applied frequency (250 kHz) of all commercially available electrodeless lamps. *ENDURA* is thus a high power, high brightness lamp developed with emphasis again on life and use in inaccessible areas. The lamp is 35 cm long with a discharge path of 72 cm, and operates with krypton at a pressure of a few hundred mTorr. Operation of the *ENDURA* lamp at 7.1 A, compared to 500 mA in the Anderson lamp, resulted in a reduction in ferrite losses by more than a factor of 50 (cf. Eq. 16.3). Since the high permeability closed magnetic core completely encloses the discharge current, coupling of *rf* power to the discharge is close to 100%.

16.4.5 Capacitively coupled fluorescent lamps

The simplest capacitively coupled (or *E*) discharge (*CCD*) consists of a gas filled vessel placed between the plates of a capacitor. For electrodeless operation, these plates are situated outside the discharge vessel. Coupling of the applicator to the discharge is principally through the sheaths next to the electrodes, resulting in a discharge current proportional to ω^2. In contrast to *H* discharges, higher powers can only be transferred to electrons at higher frequencies, where cost of electronics and *EMI* issues are less advantageous.

One of the first electrodeless fluorescent lamps was based on a capacitive discharge operating at 915 MHz [45], developed at GTE Laboratories. However, commercial exploitation of this device was limited by low efficacy and the complicated microwave power supply required. Beneking [46] has analyzed the impedance and emission properties of *CCD* in Hg-Ar mixtures, for a range of frequencies, *rf* power and gas pressure. His conclusions, supported by results from a numerical model [47], are that much of the non-radiative power loss can be explained by the space-charge limited ion current across the sheaths. Excitation of Ar at the sheath boundary and dielectric losses in the walls of the lamp provide additional loss mechanisms. Results confirm that *CCD* can compete with conventional fluorescent lamps only at relatively low power loading and high frequency (100 MHz and above).

16.4.6   Surface wave fluorescent discharge lamps

Electromagnetic surface waves can be used to sustain plasma columns [48, 49]. A device referred to as a *wave launcher* is used to excite the plasma at one end of the discharge, and the wave travels along the plasma column which it sustains, as well as along the dielectric media surrounding it.

Considerable interest was shown by the lighting community in the late 1980s in the potential of surface wave discharges (*SWD*) for general light sources. Experiments showed *SWD* to have the same efficiency as the positive column of a fluorescent lamp [50, 51]. However, the efficiency gain due to the absence of losses in the electrode region of a conventional fluorescent lamp was more than offset by the losses in conversion of mains to *rf* power.

The ability to dim an *SWD* lamp by lowering discharge power and consequently the discharge length, in direct analogue to turning down the wick of a kerosene lamp, was perceived as an attractive feature of these lamps. In practice, the axial non-uniformity of the discharge proved a disadvantage, leading to non-uniform light output. It was hoped that the non-uniform radial profile of the maintenance electric field would lead to a concentration of radiating states close to the wall and a reduction in radiation trapping. However, in practice the density profiles of the radiative states are controlled by radiation transport and is peaked in the center, as in conventional fluorescent lamps.

16.5   Low-pressure sodium lamps

The low-pressure sodium (LPS) lamp was introduced in the 1930s and is still the most efficient light source available. The physics and operation of LPS lamps is discussed in detail by Waymouth [12] and Denneman [52], and the current status of LPS lamp technology is reviewed in Ref. [3].

Low-pressure sodium lamps emit radiation principally in two resonance lines, 589.0 nm and 589.6 nm (*D* lines), which are near the peak of the eye sensitivity curve, corresponding to about 520 lpw. In fluorescent lamps, the conversion efficiency of 254 nm radiation to visible light using current phosphors is about 160 lpw, so an LPS lamp with the same energy efficiency in converting electrical power to resonance radiation as a fluorescent lamp would have more than three times the efficacy, i.e. about 300 lpw [12]. In the 1930s, LPS lamps achieved 50 lpw, but significant improvements since that time have led to efficacies as high as 200 lpw [3]. However, the radiation produced is principally yellow, which gives very poor color rendition, and the use of these lamps is mainly restricted to outdoor applications, e.g., on highways and parking areas, where brightness is more important than color definition.

As noted in the introduction, LPS lamps contain a minority of sodium vapor in a buffer gas which is typically neon. The optimum vapor pressure of sodium in these lamps is 3 mTorr, which is attained at a temperature of 260 °C. Below the optimum temperature there is insufficient vapor to provide radiation, and above this temperature, increased trapping of resonance radiation causes loss of efficacy. In order to maintain the lamp at the required temperature, additional means for heating the lamp are necessary. In early LPS lamps, this was achieved by including a neon buffer gas at a pressure of 20 Torr, to provide gas heating by elastic scattering from electrons, and enclosing the arc tube in a vacuum sealed outer jacket to reduce losses due to convection [12]. In the current generation of LPS lamps, a coating on the inside of the outer jacket, which reflects infrared radiation back to the arc tube, reduces the heat losses still further, enabling the lamps to operate with neon pressures below 10 Torr.

The first detailed experimental investigation of LPS lamps was by Druyvesteyn in 1933 [53]. He used Langmuir probes to measure the electron temperature, and absorption spectroscopy to measure the relative densities of excited sodium atoms and sodium ions. Druyvesteyn was the first to observe sodium depletion at the center of the discharge, finding that at his conditions 83% of the sodium atoms were ionized in this region. Extensive numerical and experimental investigations of low-pressure Cs- and Na-rare gas discharges were conducted by van Tongeren [54]; they showed good qualitative agreement, despite the relatively simple three-level atomic model adopted. Later calculations by Vriens [55], which included multistep ionization processes, led to better agreement with experiment, particularly for the electric field.

Depletion of sodium at the center of the discharge increases with decreasing sodium vapor pressure and increasing discharge current [52, 54]. Waymouth [12] has estimated that the depression of atomic density at the center is approximately 10 times the ion density, meaning that an ion density of 5% of the atomic density at the walls leads to a 50% reduction of sodium atom density in the center. Because of its lower ionization energy, sodium ion density is considerably higher than the mercury ion density in a fluorescent lamp. In passing, it may be noted that this is also true for cesium [54] and barium [7] rare gas discharges. In the case of barium, the presence of a number of ionic lines can augment the dominant green line of atomic barium to produce *white* light [7].

The main difficulty in attaining the theoretical efficacy limit of the LPS is to achieve both the optimum current density and the wall temperature simultaneously [12]. The discharge power may be increased for constant current by raising the discharge voltage or lengthening of the lamp. The latter solution has the further advantage, in common with the fluorescent lamp, of reducing the fraction of input power which is lost due to the electrodes. The discharge in *SOX* lamps is lengthened by bending the tube into a U shape [3, 12]; in a 90 W *SOX* lamp, 30% of input power is output as visible radiation, 5% as infrared radiation, 22% of losses are due to electrodes, and the remaining 43% are due to other losses, such as gas heating [3].

Conventional LPS lamps operate on a 50/60 Hz *ac* ballast. The effect of sodium depletion can be reduced by use of a square wave ballast, to reduce the peak value of the current, but this has limitations due to the control gear requirements. Increases in efficacy of 10–20% have been reported in discharges at operating frequencies of 100–400 kHz [56]. This is well above the frequency associated with sodium ion diffusion (1 kHz) and of the order of the decay of sodium atoms (100 kHz) in a typical LPS discharge. In contrast to fluorescent lamps, operation much below 100 kHz can lead to a reduction in efficacy compared to 50/60 Hz operation, whilst a monotonic increase in efficacy is found as the operating frequency is increased from 100 to 400 MHz.

A further problem affecting the efficacy of LPS lamps is associated with radiation transport. Radiation trapping of the *D* lines is more effective than trapping of the 254 nm line in mercury discharges, and most of the radiation from the lamp is emitted near the surface. In a U-shaped lamp, some of the radiation emitted in one leg may be trapped in the other [3]. Changes in the shape of the lamp, to replace the circular cross section with one with an increased ratio of surface area to volume (such as a cross or crescent shape [12]) have proved successful in increasing the efficacy of these lamps.

During the starting phase before the lamp has heated sufficiently to vaporize the sodium, the discharge is essentially a rare gas plasma. This phase lasts typically 10–15 minutes. A small amount (0.5–1%) of argon is added to the neon, forming a so-called Penning [57] mixture, to reduce the starting voltage. Penning ionization occurs when ground state argon atoms collide with neon metastable atoms to produce argon ions. During the long starting phase, however, argon ions diffuse to the wall and may become embedded in the glass arc tube, leading to the removal of argon from the gas mixture and a rise in starting voltage. The glass used in LPS lamps must therefore be both non-absorbing for argon and resistive to chemical interaction with sodium.

16.6   Rare gas discharges for lighting

Rare gas discharges have been the subject of extensive research for more than a century. However, low-pressure discharges in rare gases have not been used for general lighting. The most readily excited radiation spectrum is in the *vacuum ultraviolet (VUV)*, and the phosphor conversion efficiency to visible light is inefficient. Rare gas discharges have been used in special light sources such as street signs, photocopier lamps, as well as plasma displays [58]. These discharges have the advantage that their performance is independent of the ambient temperature, unlike those in fluorescent and LPS lamps, and are therefore suited to outdoor applications. The efficacy of these discharges would be greatly improved if phosphors with quantum conversion efficiencies to visible light which are substantially greater than 1 could be developed.

Heavy rare gases such as neon, argon, krypton, and xenon have an identical structure for the major energy levels. Therefore, the elementary processes in the discharge plasma of these gases are identical and may be described in a similar way. The major excited levels are *1s* (4 levels) and *2p* (10 levels). Resonance radiation from two of the *1s* levels is in the VUV, while visible radiation is from the *2p* to the *1s* levels. The principle visible lines are red in neon, near-infrared in argon, yellow in krypton, and blue in xenon.

Neon discharges are the most efficient of all rare gas discharges in producing visible light, and the bright red neon line is finding application as the third stop light on some automobiles

[4]. The narrow bore (4 mm outer diameter) and uniform color make it ideal for this purpose. The lamp runs on *dc* electricity supply and starts virtually instantly, so that, in common with *LEDs*, the human response is 150 msec shorter than for a comparable incandescent lamp with a red filter. Efficacy is 13 lpw, comparable to incandescent lamps. These narrow bore lamps run with a discharge current of 8 mA, the current density being comparable with (or higher than in) standard fluorescent lamps. In the 2–20 Torr pressure range, neon discharges are also the best of the rare gas discharges in providing visible light from *VUV* when converted by currently available phosphors [5]. Use of a green emitting phosphor, together with the visible radiation can provide an amber light suitable for turning signals on automobiles [5].

Dielectric barrier discharges are finding increasing applications as *UV* sources [59]. Recently, *OSRAM* introduced a new lamp (*PLANON*) based on this principle [6] which, through pulsed excitation, has achieved an efficiency of 60% conversion from electrical power to *VUV* xenon excimer radiation. The lamp is filled with 100 Torr xenon, is extremely thin (10 mm) with large surface area ($\sim$ 800 cm^2).

The dielectric barrier discharge (see chapter 13) is a *non-thermal* plasma, in which the *EEDF* is very far from *local thermodynamic equilibrium (LTE)*. A dielectric coating covers at least one electrode to limit the electron current. The discharge is filled with high current density (100–1000 A cm^{-2}) micro discharges of radii $\sim$0.1 mm. In the *PLANON* lamp, the duty cycle of the voltage pulse must be adjusted to minimize electron density (and hence the high energy tail of the *EEDF*) while maximizing the production of metastable Xe* atoms [6]. The metastable atoms then combine with ground state Xe in a three-body interaction to produce the Xe$_2^*$ excimer which has a broad band emission centered at 172 nm.

16.7 Conclusions

Research into low-pressure discharges for lighting application has a long and distinguished history. It has only been possible to give a flavor of the subject in this paper, and there are many challenges remaining for the development of new and more efficient light sources.

The introduction of electrodeless lamps into the market heralds an exciting new phase for the lighting industry. The current generation of electrodeless lamps have been mainly directed towards specialist applications, such as those for which long life is a premium, but with the ever increasing developments in circuit technology and improved understanding of discharge behavior in high-frequency electromagnetic fields, the range of applications should increase and perhaps lead to a replacement for the household lamp.

There are also other avenues showing promise for the future. A better knowledge of the electron distribution function and how it can be influenced in a gas discharge can lead to new ways of generating plasmas with improved radiation output, as in the case of the dielectric barrier discharge (see Ref. [6] and chapter 13). The development of multi-photon phosphors would also revolutionize the use of low-pressure discharges for lighting. The great advances in computer technology, together with parallel advances in *state of the art* diagnostics provide the opportunity for numerical modeling and experimental investigations of these discharges to make significant advances in our understanding and make an impact on product development.

Acknowledgments

I am most grateful to Valery Godyak and Robert Piejak for many stimulating discussions.

References

[1] The Lighting Research Center Electrodeless Lamps: The Next Generation, Lighting Futures, Vol. 1, No. 1, Rensselaer Polytechnic Institute, Troy (1995)

[2] M.G. Abeywickrama, in J.R. Coaton and A.M. Marsden (Eds.), Lamps and Lighting, Arnold: London, pp. 194-215 (1997)

[3] M.W. Kirby, ibid., pp. 227-234 (1997)

[4] F.A.V. Ligthart and J. Geboers, Proc. 7th Int. Symp. on the Science and Technology of Light Sources, Kyoto, The Illuminating Engineering Society of Japan, pp. 267-8 (1995)

[5] C. Roozkranz, F. Ligthart and J. Geboers, Proc. 8th Int. Symp. on the Science and Technology of Light Sources, Greifswald, pp. 138-9 (1998)

[6] F. Vollkommer and L. Hitzschke, ibid., pp. 51-60 (1998)

[7] J.J. Curry, H.M. Anderson, J. MacDonagh-Dumler, J.E. Lawler and G.G. Lister, J. Appl. Phys. 87, 2058 (2000)

[8] J.R. Coaton and A.M. Marsden (Eds.), Lamps and Lighting, Arnold: London (1997)

[9] D.O. Wharmby, IEE Proceedings-A **140**, 485 (1993); also in J.R. Coaton and A.M. Marsden (Eds.), Lamps and Lighting, Arnold: London, pp. 216-226 (1997)

[10] G.G. Lister, in H. Schlüter and A. Shivarova (Eds.), Advanced Technologies Based on Wave and Beam Generated Plasmas, NATO ASI Series, Kluiver Academic Publishers (1999), p. 65

[11] Electrical Engineer 7, 549 (1891)

[12] J.F. Waymouth, Electric Discharge Lamps, MIT Press: Cambridge (1971)

[13] J.E. Lawler and J.J Curry, in U. Kortshagen and L.D. Tsendin (Eds.), Electron Kinetics and Applications of Glow Discharges, NATO ASI Series, Plenum, pp. 471-488 (1998)

[14] F. Vermeersch and W. Wieme, in Optogalvanic Spectroscopy, Inst. of Physics Conf. Ser. 113, 109 (1991)

[15] A.G. Jack and L.E. Vrenken, IEE Proceedings 127A, 149 (1980)

[16] J. Maya and R. Lagushenko, Advances in Atomic, Molecular and Optical Phys. 26, 321 (1990)

[17] W. Verweij, Philips Res. Rep. Sup. 2, 1 (1961)

[18] M. Koedam and A.A. Kruithof, Physica 28, 80 (1962)

[19] M. Koedam, A.A. Kruithof and J. Riemens, Physica 29, 565 (1963)

[20] J.F. Waymouth and F. Bitter, J. Appl. Phys. 27, 122 (1956)

[21] M.A. Cayless, in Proc. Vth Int. Conf. On Ionization Phenomena in Gases (ICPIG), Munich, pp. 263-277 (1962)

[22] J.T. Dakin, J. Appl. Phys. 60, 563 (1986)

[23] G. Zissis, P. Bénétruy and I. Bernat, Phys. Rev. A 45, 1135 (1992)

[24] G.G. Lister and S.E. Coe, Computer Physics Communications 75, 160 (1993)

[25] S.D. Rockwood, Phys .Rev. A 8, 2348 (1973)

[26] T.J. Sommerer, Phys. Rev. Lett. 77, 502 (1996)

[27] V.A. Godyak (1997) private communication

[28] W.L. Morgan and L. Vriens, J. Appl. Phys. 51, 5300 (1980)

[29] U. Kortshagen, C. Busch and L.D. Tsendin, Plasma Sources Sci. Technol. 5, 1 (1996)

[30] V.A. Feokistov, A.M. Popov, A.T. Popovicheva, T. Rhakimov, V. Rhakimova and E.A. Volkova, IEEE Trans. Plasma Sci. 20, 66 (1991)

[31] D.Y. Fang and C.H. Huang, J. Phys. D: Appl. Phys. 21, 1490 (1988)

[32] G.G. Lister (1998), unpublished

[33] J.W.F. Dorleijn and A.G. Jack, J. Illum. Eng. Soc. 15, 75 (1985)

[34] J. Polman, Physica 54, 305 (1971)

[35] R.A. Guest and E.J.P. Mascarenhas, in J.R. Coaton and A.M. Marsden (Eds.), Lamps and Lighting, Arnold: London, pp. 292-335 (1997)

[36] E.S. Weibel, Phys. Fluids 10, 741 (1967)

[37] J. Bethenod et al., US Patent #2,030,957 (1936)

[38] A. Netten and C.M. Verheij, QL lighting product presentation storybook, Philips Lighting, Eindhoven, Product Literature (1994)

[39] G.G. Lister and M. Cox, Plasma Sources Sci. Technol. 1, 67 (1992)

[40] J. Jonkers, M. Bakker and J.A.M. van der Mullen, J. Phys. D: Appl. Phys. 30, 1928 (1997)

[41] J. Denneman, J. Phys. D: Appl. Phys. 23, 293 (1990)

[42] M.Shinomaya, K. Kobayashi, M. Higashikawa, S. Ukegawa, J. Matsuura and K. Tani-
gawa, J. Illum. Eng. Soc. 44 (1991)

[43] J.M. Anderson, US Patent # 3,500,118 (1970)

[44] V. Godyak and J. Schaffer, Proc. 8th Int. Symp. on the science and technology of light
sources, Greifswald, pp. 14-23 (1998)

[45] J.M. Proud and R.K. Smith, US Patent # 4,266,166 (1981)

[46] C. Beneking, J. Appl. Phys. 68, 5435 (1990)

[47] C. Beneking, J. Appl. Phys. 68, 4461 (1990)

[48] M. Moisan, J. Hubert, J. Margot, G. Sauvé and Z. Zakrewski, in C.M. Ferreira and
M. Moisan (Eds.), Microwave discharges, fundamentals and applications, NATO ASI
Series; Plenum Press, pp. 1-24 (1993)

[49] M. Moisan, J. Hubert, J. Margot and Z. Zakrewski, in H. Schlüter and A. Shivarova
(Eds.), Advanced Technologies Based on Wave and Beam Generated Plasmas, NATO
ASI Series; Plenum (1998)

[50] C. Beneking and P. Anderer, J. Phys. D: Appl. Phys. 25, 1470 (1992)

[51] A.T. Rowley and D.O. Wharmby, in L. Bartha and F.J. Kedves (Eds.), Proc. 6th Int.
Symp. on the Science and Technology of Light Sources, Technical University of Budapest (1962)

[52] J.W. Dennemann, IEE Proc. 128A, 397 (1981)

[53] M.J. Druyvesteyn, Physica 1, 14 (1933)

[54] H. van Tongeren, Philips Res. Rep. Suppl. No. 3 (1975)

[55] L. Vriens, J. Appl. Phys. 49, 3814 (1978)

[56] J.J. de Groot, A.G. Jack and H. Coenen, J. Ilum. Eng. Soc. 14, 188 (1984)

[57] F.M. Penning, Z. Phys. 46, 335 (1926)

[58] H. Murakami, in Proc. 7th Int. Symp. on the Science and Technology of Light Sources,
Kyoto, The Illuminating Engineering Society of Japan, pp. 283-292 (1995)

[59] U. Kogelschatz, Pure and Appl. Chem. 62, 1667 (1990)

17 High-pressure plasma light sources

Klaus Günther

Osram GmbH, Nonnendammallee 44–61, 10625 Berlin, Germany

17.1 Introduction and basic equations

In contrast to light sources with low-pressure plasmas far from local thermal equilibrium (LTE) conditions, a typical high-pressure discharge lamp is characterized by a density of power transformation from electric energy to light several orders of magnitude higher accompanied by a corresponding higher particle number density (HID lamp = **High Intensity Discharge lamp**). These high densities are connected with very high collision frequencies of all particles and a high energy transfer between the particles of the same kind and between the different particle types. The result is a thermalization within each type of particles as well as between all types of particles, i.e. the kinetic energy distribution $f(E)$ of ground state neutrals, ions and electrons is Maxwellian with corresponding temperatures T_n, T_+ and T_e and given by

$$f_{n,+,e}(E) = \frac{2}{\sqrt{\pi}} \left(\frac{E}{kT_{n,+,e}} \right)^{3/2} \exp\left(-\frac{E}{kT_{n,+,e}} \right) \frac{dE}{E} \ , \tag{17.1}$$

where E is the kinetic energy, while the population of the excited levels of the heavy particles is given by a Boltzmann distribution of the temperature T_{ex}:

$$n_j = n\frac{g_j}{Z} \exp\left(-\frac{E_j}{kT_{ex}} \right) \ , \tag{17.2}$$

where g_j and Z are the statistical weight of the level j and the partition function, respectively, and n is the number density of particles that occupy the energy level E_j. The chemical equilibrium between the neutral (density n_a) and charged particles (densities n_+ and n_e) is defined by the Saha equation

$$\frac{n_+ n_e}{n_a} = S(T_S) = \frac{2Z_+}{Z_a} \left(\frac{2\pi m_e kT_S}{h^2} \right)^{3/2} \exp\left(-\frac{\chi - \Delta_\chi}{kT_S} \right) \tag{17.3}$$

with χ and Δ_χ as ionization energy and its correction for plasma interaction. The radiation field is described by Kirchhoff's and Planck's laws

$$\frac{\varepsilon_\lambda}{\kappa_\lambda} = B(\lambda, T_r) \quad \text{and} \quad B(\lambda, T_r) = \frac{2c_1}{\lambda^5} \left[\exp\left(\frac{c_2}{\lambda T_r} \right) - 1 \right]^{-1} \ , \tag{17.4}$$

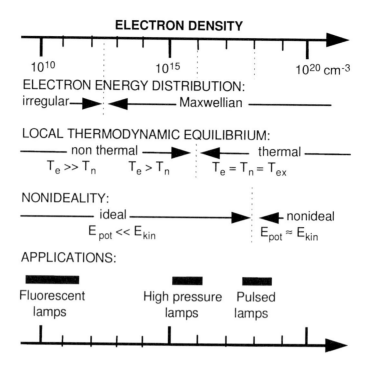

Figure 17.1: Plasma conditions of typical discharge lamps.

where ε_λ is the emissive power, κ_λ the absorption coefficient, λ the wavelength of the emitted radiation, $2c_1 = 1.19 \times 10^{-12}$ W cm^{-2} sr^{-1}, $c_2 = 1.439$ cm K, and T_r the corresponding temperature. The intensity emitted by a space element of this temperature having a diameter d is

$$L_\lambda = [1 - \exp(-\kappa_\lambda d)] \cdot B(\lambda, T_r) \tag{17.5}$$

with the special cases

$$L_\lambda = B(\lambda, T_r) \qquad \text{for } \kappa_\lambda d \gg 1$$
$$L_\lambda = \varepsilon_\lambda d \qquad \text{for } \kappa_\lambda d \ll 1,$$

where $\kappa_\lambda d$ denotes the optical thickness. LTE here means that all these temperatures are equal, i.e,

$$T_n = T_+ = T_e = T_{ex} = T_S = T_r. \tag{17.6}$$

Fig. 17.1 demonstrates these features of discharge lamps on a scale of the charge carrier density which is mainly responsible for the thermalization of the plasma. The indication for

Table 17.1: Non-ideality effects for HID lamps.

	n_e	T	γ	
	cm^{-3}	K		
High-pressure Hg	5×10^{15}	6000	0.06	ideal
High-pressure Na	10^{16}	4000	0.1	weakly non-ideal
Pulsed Na	10^{18}	5000	0.4	weakly non-ideal

non-ideality refers to the effect of the electrostatic particle interaction to the plasma properties as plasma composition, electrical and thermal conductivity or radiation properties (see, e.g., Refs. [1, 2]). The importance of this effect can be estimated from the non-ideality parameter γ

$$\gamma = \frac{E_{pot}}{E_{kin}} = \frac{e^2(2n_e)^{1/3}}{kT}, \qquad (17.7)$$

which describes the ratio of the mean potential energy to the mean kinetic energy of a particle. It is assumed that the plasma is ideal for $\gamma < 0.1$ and non-ideal for $\gamma > 0.5$. Table 17.1 shows the situation for some types of HID lamps.

Thus, all essential properties of the plasma of a HID lamp can be described by only one parameter T in combination with the partial pressures of the lamp gas without referring to elementary processes or rate equations (this is not valid for critical zones near the electrodes or near the walls). On the other hand, there is no independent mechanism like the transformation of ultraviolet (UV) into visible radiation which can be used to optimize the spectral distribution by a proper mixture of phosphors, or a special non-Maxwellian energy distribution of particles suitable for the preferred excitation of the resonance transitions in a fluorescent lamp. In HID lamps the photometric characteristics of the lamps are determined mainly by the elementary radiation processes in the plasma; it can be influenced only by the choice of the radiating particles and optimum temperature distributions. Objective of a lamp development is to select the different radiation processes in such a way that the efficacy of the lamp is maximized and the colour temperature and the colour rendering correspond to the demands of the application field.

17.2 Application demands

17.2.1 Photometric properties

Of all the parameters describing the photometric properties of a lamp, the efficacy, the correlated colour temperature (CCT), and the colour rendering index (CRI) are the most essential.

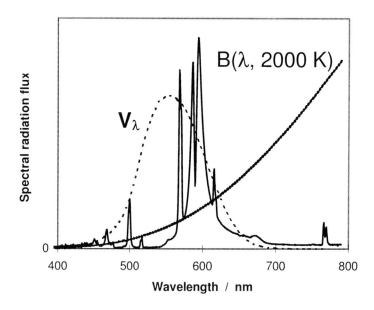

Figure 17.2: Spectrum of a high-pressure sodium lamp together with the luminous efficiency function V_λ and the blackbody function $B(\lambda, T)$ at $T = 2000$ K.

Efficacy. The efficacy η of a lamp should be distinguished from its radiation efficiency η_{rad} considering the definitions:

$$\eta_{rad} = \frac{\int_0^\infty \Phi_\lambda d\lambda}{P_{in}} \quad \text{and} \quad \eta = 683 \,\text{lm/W} \times \frac{\int_0^\infty V_\lambda \cdot \Phi_\lambda d\lambda}{P_{in}}. \tag{17.8}$$

While the radiation efficiency describes the fraction of the input power P_{in} which is transformed into radiation of any wavelength, the efficacy is calculated by multiplication of the spectral radiant flux Φ_{rad} by the luminous efficiency function V_λ before integration over λ. V_λ describes the wavelength dependent effect of the radiation on the human eye. For this function, there are different definitions depending on the illumination level and the observation angle (see, e.g., Ref. [3]). Fig. 17.2 shows, for example, a comparison of V_λ and the spectral distribution of the radiant flux of a high-pressure sodium lamp. It can be seen that the high efficacy of such lamps is due to the vicinity of the sodium resonance lines to the maximum of V_λ.

Correlated colour temperature CCT. The second important information about a light source is the correlated colour temperature. It is determined by a comparison of the spectral radiation distribution of the lamp with that of the most similar relative blackbody distribution within the visible region (as a reference, see Ref. [4]). In the example of Fig. 17.2 the CCT is 2000 K. The optimum CCT depends on the conditions of the special application field, for general illumination it varies between 2500 K and 6000 K.

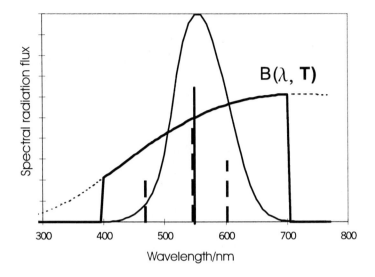

Figure 17.3: Three different levels of colour rendering.

Colour rendering index CRI. The CRI describes the ability of a light source to present different spectral reflection coefficients of an illuminated surface as different colours. To realize that, all wavelengths reflected from the surface must be contained in the radiation of the source. Per definition, the blackbody distribution is considered perfect with a CRI = 100, and deviations from this distribution decrease it. The generally accepted procedure to determine the CRI of a given spectral distribution of a source is described in Ref. [5].

Fig. 17.3 presents two extreme cases which explain the conflict in the development of light sources: A perfect colour rendition requires a blackbody radiation distribution in the visible wavelength region that produces an efficacy of at most 250 lm/W. On the contrary, a maximum efficacy 683 lm/W is realized with a single line at 550 nm. As a compromise, three lines at 470, 550 and 610 nm produce CRI = 82 and an efficacy $\eta = 375$ lm/W. These numbers can be considered as the theoretical maximum values: they assume a 100 % transformation of the input energy into radiation, while heat conduction and electrode losses are neglected. Thus, the essential problem of lamp development is the proper positioning of the emission processes within the range of the V_λ curve. In a fluorescent lamp, this is done by the choice of the phosphor blend. In HID lamps, it must be achieved by an optimization of the discharge conditions and the gas fill. Fig. 17.4 illustrates the state of the art and the scope to the theoretical limit. It can be seen that

- one has the choice between high efficacy and good colour rendition, and that

- there is yet a considerable gap between the actual data and the optimum.

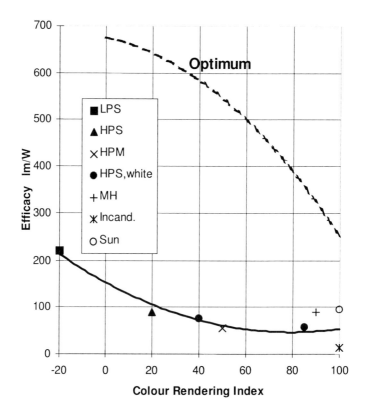

Figure 17.4: Typical HID lamps at a power level of 100 W in the efficacy – CRI diagram. The different lamp types are introduced in section 17.3.

17.2.2   Operation requirements

In addition to the photometric properties, certain operational requirements must be considered. For instance because of the **warm-up phases**, the application of HID lamps is not possible when instantaneous light is required (the extremely short warm-up of the modern car head light HID lamps is due to a very high xenon pressure which requires special ignition circuits). In the same connection, the **high reignition voltage** or a time delay of some minutes after a switch off is problematic in many applications. Finally the **lamp life** and the **lumen depreciation** during life, as well as the **stability of the colour temperature** and of the **colour rendition**, determine the choice of a lamp type for a given illumination application. In this sense, the technical progress in the field of illumination does not mean automatically the development of lamps having higher efficacy. Sometimes the removal of one of these barriers enables us to introduce a high efficacy HID lamp into an application field where until now incandescent lamps are in use.

17.2.3   Costs and environmental aspects

The development and application of light sources are primarily determined by economical considerations. Thus, besides the photometric properties the costs of the lamp, the costs of the consumed energy over life, the replacement costs and, to an increasing degree, the costs for disposal of waste lamps influence the decision for or against a lamp system. In certain situations, this decision is influenced by the actual energy costs and wages, changing according to political and regional conditions.

Environmental pollution has been considered in the R & D divisions for many years, and some studies have addressed the use of critical substances: mercury, compounds of heavy metals, radioactive isotopes including thorium or krypton 85. However in many cases, the substitution of such substances results in higher production costs and/or worse maintenance of the lamps. A real pressure has raised only when legislative and economical measures were applied.

17.3   High intensity discharge (HID) lamps and their operational principle

17.3.1   The plasma of HID lamps

The condition of the plasma of a HID lamp is determined by the **balance of heating and loss processes**. For a unit volume this balance is written as

$$\sigma E^2 - \operatorname{div} F_c - \operatorname{div} F_r - \frac{\partial u}{\partial t} = 0 . \tag{17.9}$$

Here σE^2 is the heating term depending on the electrical conductivity σ and the electric field strength E applied to the plasma,

$$F_c = -\lambda_T \operatorname{grad} T \tag{17.10}$$

describes the energy loss due to heat conduction with the conductivity λ_T, and

$$\operatorname{div} F_r = \int_0^{4\pi} \int_0^\infty \kappa_\lambda (B_\lambda(T) - L_\lambda) d\lambda d\omega \tag{17.11}$$

is the energy loss from the volume unit as the result of radiation. The expressions in Eq. 17.11 are explained in Eqs. 17.4–17.6. With

$$u = \frac{3}{2} kT(n_a + 2n_e) + \chi n_e \tag{17.12}$$

as the internal energy for a singly ionized plasma the changes in time are taken into account. In the stationary case we get the Elenbaas-Heller equation in cylindrical coordinates for the specific power balance of a volume element

$$\sigma E^2 + \lambda_T \frac{\partial^2 T}{\partial r^2} + \frac{\partial \lambda_T}{\partial r} \frac{\partial T}{\partial r} + \frac{\lambda_T}{r} \frac{\partial T}{\partial r} - \operatorname{div} F_r = 0 . \tag{17.13}$$

This equation can be solved numerically without problems as long as this balance contains coefficients which depend on the condition of the plasma at this point only. Eq. 17.11 introduces the spatial distribution of the emission and absorption processes in the whole arc in the balance at this point via the quantity L_λ which depends on the distribution of the emission and absorption coefficient in the whole discharge volume. This aspect complicates the quantitative treatment of our discharges, and some approximations were considered to overcome the problem (see, e.g., Ref. [6]). One of the first concepts was the introduction of the so-called net emission coefficient by Zollweg et al. [7], which corrected the emission coefficient semi-empirically by the absorbed power at this point. A second way is to distinguish roughly between two or more isothermal concentric zones in the discharge column whose radiation interaction is considered.

A further complication is the effect of the inhomogeneous zones of the discharge near the electrodes showing strong axial gradients in addition to the radial distribution. In the last years some models were presented which describe the behaviour of the lamps as a whole. Some of them include simplified models of the electrodes, and the agreement with the measurements from real lamps is rather good (see, e.g., Refs. [8, 9]).

Concerning the coefficients used in the above mentioned equations, for the calculation of the **electrical conductivity** σ the Chapman-Enskog formalism is recommended [10], non-ideal effects should be taken into account – where necessary – according to Ref. [2]. Sometimes the choice of reliable transport coefficients for electron-atom collisions can become a problem [11].

The **absorption coefficient** κ_λ in Eq. 17.11 can be derived from the emission coefficient according to Kirchhoff's law (Eq. 17.4) resulting as the sum of all emission processes occurring in the plasma. In the following the essential emission processes describing the spectrum of a HID lamp and their dependence on the thermodynamic properties of the discharge plasma are presented in a simplified form and discussed with respect to their relations with each other:

$$\text{Line radiation:} \qquad \varepsilon_L \propto n_a \exp(-\frac{E_m}{kT}) \qquad\qquad (17.14)$$

$$\text{Bremsstrahlung:} \qquad \varepsilon_{ff} \propto n_a \frac{T}{\lambda^2} \frac{\exp(-\frac{\chi}{kT})}{\exp(\frac{hc}{\lambda kT}) - 1} \qquad\qquad (17.15)$$

$$\text{Recombination radiation:} \qquad \varepsilon_{fg} \propto n_a \frac{T}{\lambda^2} \exp(-\frac{\chi}{kT}) \xi(\lambda, T) \qquad\qquad (17.16)$$

$$\text{Molecular radiation:} \qquad \varepsilon \propto n_{AB^*} \exp(-\frac{E}{kT}). \qquad\qquad (17.17)$$

Here E_m is the excitation energy, and the factor $\xi(\lambda, T)$ accounts for the special electronic structure of the radiating atom in comparison to the hydrogen atom (see, e.g., Ref. [12], and for numerical data of some elements Ref. [13]).

The dependence of all these radiation processes on the plasma condition can be described as a product of the number density of the emitting particles and an exponential expression containing the energy of the upper level of the transition (the additional factors in Eqs. 17.15 and 17.16 do not influence this general relation essentially). While the exponential expression is an increasing function of temperature, the density factor decreases due to the decay of the emitting particles according to their chemical equilibrium, Eq. 17.3. Thus, the emission coefficient shows a sensitive temperature dependence with low values at both low and high temperatures and a so-called **norm maximum**. Its position depends on the actual ionization or dissociation energy of the radiating species and the excitation energy of the upper level of the transition. If the plasma contains more than one element with different ionization energies the chemical equilibrium equations are coupled by means of the electron particle number density, and the result can be more than one maximum of the emission coefficient.

An important feature is the relation of the different emission processes one to each other. Thus, the relation of the recombination continuum to the bremsstrahlung continuum in the Kramers-Unsöld approximation (see, e.g., Ref. [14]) is

$$\frac{\varepsilon_{fg}}{\varepsilon_{ff}} \approx \exp\left(\frac{c_2}{\lambda T}\right) - 1 . \tag{17.18}$$

For $T = 6000$ K at $\lambda = 500$ nm this relation is about 100, i.e. bremsstrahlung usually can be neglected in HID lamp physics. For the corresponding relation between the recombination radiation ε_{fg} and the optically thin emission ε_L of a line

$$\frac{\varepsilon_{fg}}{\varepsilon_L} \propto T \exp\left(\frac{E_n - \chi}{kT}\right) \tag{17.19}$$

holds, where E_n is the upper level energy of the line transition and χ the ionization energy. Obviously, at low temperatures the line radiation is dominating while at higher temperatures the continuum radiation is stronger.

The relations developed above should be used to optimize the spectral distribution of the lamp plasma with respect to the demands of section 17.2.1 by means of a corresponding selection of the radiating elements or molecules and their concentrations, and by choosing an optimal temperature in relation to the norm temperatures of the selected transitions. Of course, there is no direct possibility to influence the temperature and its spatial distribution in a discharge. This temperature is the result of the equalization between heating and dissipating processes according to Eq. 17.9: all terms of this equation are strongly temperature dependent, so it can be seen as the mathematical condition for the equilibrium temperature. To shift the plasma temperature towards the optimum of a wanted emission process or to suppress a parasitic process means to modify Eq. 17.9 by changing the heating and the loss processes.

Let us neglect for a moment the expressions describing the heat conduction and the time behaviour in Eq. 17.9. The equation reduces to a comparison of Ohmic heating and cooling through radiation: In the optical thin case it can be written as

$$E^2 \geq \frac{4\pi\varepsilon_N}{\sigma} , \tag{17.20}$$

where ε_N is the net emission according to Ref. [7]. The equalization in Eq. 17.20 happens in the axis of the arc and describes the temperature maximum. The shape of the radial temperature distribution is essentially determined by the relation ε_N/σ which can be written as a function of the temperature and the particle number densities. Lowke discusses some realistic situations and finds for the arc a tendency to constrict, if

$$E_m < \frac{\chi}{2} \tag{17.21}$$

and vice versa [17]. E_m and χ have the same meaning as in Eqs. 17.3 and 17.14. To prevent this mostly unwanted constriction an element must be added to the plasma which meets the condition $E_m > \chi/2$.

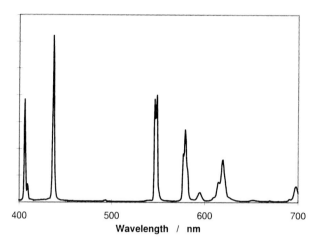

Figure 17.5: Spectrum of a HPM lamp.

17.3.2   High-pressure mercury (HPM) lamps

The first HID lamp of technical importance was the high-pressure mercury lamp (HPM). It consists of a more or less tubular discharge vessel made of quartz glass filled with a starting gas and an amount of mercury which evaporates and produces an operation pressure of some bars. The electrodes usually consist of tungsten which is doped by barium compounds to lower the work function. The electrical feedthrough consists of a molybdenum foil which is compatible with the expansion coefficient of the quartz glass. The discharge plasma of a HPM lamp is close to LTE and produces a temperature profile with an axis temperature of 6000 … 6500 K. The spectrum consists of a number of separated lines in the UV and in the visible range (see Fig. 17.5). Most of the UV lines are rather strong and more or less optically thick, the resonance lines are self-reversed. The visible lines are distant with large gaps which are responsible for the low CRI; the severe deficit in the red range is filled in part by the radiation of a phosphor which is excited by the UV lines. However, the gain of radiation power is not enough to raise the efficacy much over 50 lm/W. For the physical description of the HPM lamp the classical publication of Elenbaas is recommended [15], but recent papers report considerable improvements in the understanding of the electrical properties and the radiation balance [16].

Despite the moderate photometric data (see Fig. 17.4), the HPM lamp is an important design for two reasons: simplicity and low costs. It is the origin of modern metal halide lamps.

17.3.3   Metal halide (MH) lamps

There are three important features of a HPM lamp which have been improved during the last thirty years:

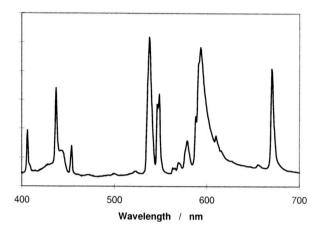

400   500   600   700

Wavelength / nm

Figure 17.6: Spectrum of a MH lamp with a CCT = 3300 K and CRI = 74.

- the intense UV radiation of the mercury plasma has been suppressed to raise the efficacy,

- the wide gaps of the visible spectrum were filled to increase the colour rendering, and

- the spectral distribution was modified to meet the different demands of the applications for modified colour temperatures.

As mentioned above, in a HID lamp this can be done only by optimizing the partial pressures of the heavy particles and the temperature distribution of the discharge. The simple addition of new radiating elements to the mercury plasma fails because of the chemical incompatibility and/or of the by many orders of magnitude too low saturation vapour pressures of the candidates. In the early sixties the introduction of the metal halides (MH) as additives in mercury high-pressure discharges solved the problem [18]: they have a sufficiently high vapour pressure at the permissible temperature of the wall materials and dissociate at the higher temperatures near the core of the discharge. The partial pressures of the metal halides exceed the equilibrium pressures of the pure elements related to the actual temperature of the coldest spot in the discharge vessel by many orders of magnitude. The condensation of the metal vapours at the walls is prevented by the recombination with the halogen ions in the cooler zones of the arc before the impact at the wall. The selection of the additives depends primarily on

- the chemical compatibility with the wall and the electrode materials,

- the vapour pressures of all components, and

- the excitation energies and the spectral distribution of the oscillator strengths of all radiation processes in consideration of the desired colour temperature and colour rendition index.

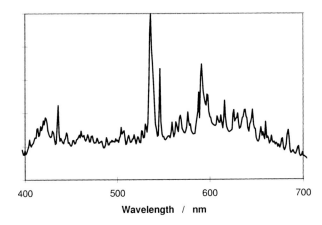

Figure 17.7: Spectrum of a ceramic MH lamp with CCT = 4200 K and CRI = 95.

The first halide mixtures used in MH lamps consisted of the iodides of Na, Th, In, and Li (see Fig. 17.6). It can be seen that the Hg line spectrum with HgI 405, 436, 546, 577 and 579 nm is improved mainly by strong lines of a mixture of metal halides. The relation between them and to the Hg lines determines the colour temperature CCT = 3300 K and a CRI = 74. The lower ionization energies and excitation energies of the metals compared to the data of mercury causes a decrease of the arc temperature as a result of the enhanced radiation cooling. This temperature decrease depopulates the mercury levels and lowers the mercury line spectrum. The benefit of this lowering is a decrease of the strong ultraviolet Hg lines with a corresponding rise of the efficacy from about 50 lm/W to 75 ... 90 lm/W. By the introduction of the halides of the rare earths which emit systems of many lines the photometric data were improved still more.

In the last years the use of ceramic instead of quartz glass arc tubes increased considerably the allowed wall temperature, and the concentration of radiating metal atoms could be raised in comparison to that of the quartz lamps. The result was a more dominating radiation of the additives compared to the mercury radiation and a broadening of their lines forming a quasi-continuum with CRI > 90 and an efficacy > 90 lm/W (see Fig. 17.7). A further positive feature is the weaker sensitivity of the spectrum as against changes of the cold spot temperature and of the arc temperature: the colour temperature is more stable over the lamp life and as to variations of the environment of the lamp (see, e.g., Ref. [19]).

17.3.4   High-pressure sodium (HPS) lamps

The high-pressure sodium (HPS) lamp has secured its strong position especially in the outdoor illumination due to some unique properties of the sodium atom. As a member of the alkali metals it contains only one electron outside a closed shell, and such atoms exhibit an extremely strong resonance transition compared to the other lines. In addition, this resonance transition

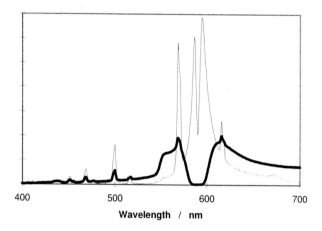

400   500   600   700

Wavelength / nm

Figure 17.8: Spectra of HPS lamps at different sodium pressures. Thin line: low sodium pressure with CCT = 1900 K, CRI = 10, h = 89 lm/W; solid line: high sodium pressure with CCT = 2500 K, CRI = 84, and h = 48 lm/W.

at 589 nm is situated near the maximum of the eye efficiency curve (see Fig. 17.2), and it can be easily broadened by the pressure of sodium and mercury to limit the saturation in the range of the resonance lines, so it produces a very high efficacy up to 150 lm/W at a colour rendition CRI = 20.

All high-pressure sodium spectra show a pronounced self-reversal of the resonance lines because of the high optical thickness near the line centre: The optical path length decreases to a few hundredth of a mm, and from outside of the lamp one can see only the outermost layers of the discharge having blackbody intensities corresponding to the wall temperature (see Eq. 17.6).

In order to introduce this promising system into general lighting applications the problem of chemical incompatibility of alkalines to the usual discharge tube materials as glass or quartz had to be solved. The use of discharge tubes made of polycrystalline alumina in combination with a heat resistent glass solder to electrode feedthroughs from niobium started the development of the well-known HPS lamps for outdoor illumination [20]. By the way, this invention related primarily not to sodium but exclusively to cesium as the radiating system because of the very attractive recombination continuum of CsI in the visible range. Unfortunately, this continuum is accompanied by strong resonance lines in the infrared (IR) which limit the efficacy of such systems to about 10 lm/W.

In order to improve the colour rendition, i.e. to raise the colour temperature and the CRI, the sodium and/or mercury partial pressure may be increased more. This results in a higher optical thickness due to the higher number density of the radiating particles especially in the range of the resonance lines. That means the self-absorption gap becomes broader, and the reverse maxima near the gap increase less than the optically thin spectral regions in the blue which are dominated by the transitions from high levels of the sodium atom. The result is an increased colour temperature (see Fig. 17.8). Unfortunately the far wings of the broadened resonance lines shift a considerable fraction of the resonance line intensity to the infrared, and the efficacy deteriorates (see Ref. [21]).

17.3.5   Technical applications

The optimization conditions vary enormously for HID lamps which are destined for the illumination of a technical apparatus as a laser or a photolithographic machine. Here the spectral emission characteristics should correspond to the absorption characteristics of the materials to be excited or to be treated. In application fields where optical parts have to be illuminated the radiation efficiency is less important than an extremely high radiance over a limited volume. As an example, a new ultra-high-pressure mercury lamp was developed for the illumination of modern video projection devices [22].

17.3.6   New developments

Besides the step by step improvement of the established HID lamps new principles of operation or new radiating mechanisms were proposed which may open new areas in some fields of application. An attractive new principle of energizing HID lamps is the **electrodeless** introduction of the electric energy into the discharge vessel. The fundamental benefit of this principle is the elimination of all complications originating from metallic electrodes as electrode corrosion due to aggressive gas fills, or wall blackening. The power supplies are operated at frequencies above 1 MHz to some GHz in the allowed industrial, scientific, and medical bands. The choice of the frequency depends on the availability of economic power supplies with high conversion efficiency and of the desired dimensions of the radiating volume reaching from some millimetres to centimetres. The coupling into the discharge vessel is accomplished by purely magnetic or electromagnetic fields (see, e.g., Ref. [24]). A serious problem is the matching between the output circuit of the power supply and the lamp at widely changing plasma impedance during starting, run up, and the stationary mode [23].

A recently proposed system excites a ball-shaped quartz glass vessel filled with **sulphur vapour** emitting a very intense spectrum of S_2 molecules in the visible range with a colour temperature of 6000 K which decreases rapidly towards the UV and IR [25]. Thus, the conversion efficiency of the absorbed microwave power into visible radiation is extremely high ($\approx$ 130 lm/W) considering the good colour rendering index of 80, but unfortunately there are no efficient and reliable magnetrons in the range of moderate power outputs. The efficacy of a system operated at 2.4 GHz is about 95 lm/W for a luminous flux of 130 klm.

Besides the radiation of molecules with its numerous vibrational and rotational transitions there is another radiation system with a broad spectral distribution resulting in an excellent colour rendition: the thermal radiating surface of compact bodies. The poor luminous efficacy because of the limited operational temperature of solid surfaces was improved by using the surfaces of microscopic droplets which were created by oversaturating metallic vapours in the discharge. This is achieved by dissociating metal halide vapours of sufficient high-pressure in the warmer zones of the discharge. This droplets or **clusters** produce a blackbody-like radiation with a considerably higher efficacy than solid surface radiators [26]. Such systems can be operated in electrodeless lamps only, because the very reactive halogen ions which prevent the deposition of the metal droplets on the inner surface of the discharge vessel through recombination would destroy metallic electrodes immediately.

Presently much effort is spent to **reduce the use of mercury** in HID lamps to avoid environmental pollution. The problem is that Hg combines some very attractive features which can be found hardly in other substances:

- it is compatible with all materials used in lamp construction as quartz, alumina, tungsten, molybdenum, etc.,

- it has a very low vapour pressure at room temperature and does not need high ignition voltages,

- the heavy mercury atoms prevent heat loss from the hot arc to the cold wall by means of elastic collisions,

- the metal lines are broadened effectively by pressure broadening; on this way a decrease of efficacy due to saturation effects is prevented,

- its excitation energy levels are high compared to those of the most metals used in MH and HPS lamps; so it can be used to adjust the electrical conductivity for getting a lamp voltage of about 100 V (half of the mains voltage) without affecting appreciably the radiation flux.

The attempts to replace mercury by xenon are not satisfactory for two reasons: xenon has a high pressure at room temperature and needs high ignition voltages, and it has a very small transport cross section for collisions with electrons of a kinetic energy of about 0.5 eV because of the Ramsauer effect. As a consequence, lamp voltages tend to be rather low at the expense of high lamp currents. The use of metal halides with high vapour pressures instead of mercury suffers from an extraordinary heat conduction at the boundary zones of the arc which extracts non-radiating energy from the lamp and increases the thermal wall load. Another concept is the substitution of mercury by zinc which has similar electric and energetic properties [27]. However, the electrical and photometrical features of such mercury-free HID lamps are not yet comparable to the traditional systems until now.

17.4   Lamp operation

17.4.1   Starting of HID lamps

The transition of a HID lamp from the cold insulating discharge vessel to the burning lamp with hot electrodes and the design data of the electrical and photometric properties is characterized by some phases which may be well distinguished from each other. The first step is an **electrical breakdown** between the electrodes which proceeds corresponding to the well-known Townsend and streamer mechanisms (see, e.g., Ref. [28]). It starts from random electrons present in the discharge volume which are accelerated by the electric field of the high-voltage ignition pulse. This start is delayed in relation to the rise of the ignition voltage by a lag time the length of which depends on the number of initial electrons. This time lag may be shortened by irradiating the discharge tube with photons which produce photoelectrons

[29]. It is not quite clear whether these electrons arise from the electrodes by photoeffect or by means of a cascade effect starting with the excitation of the gas atoms. These electrons collide with neutrals and excite or ionize them. The result are new charge carriers and photons which produce secondary electrons on the surface of the electrode. On the other hand, electrons are lost due to collisions with electro-negative heavy particles forming negative ions. The charge carrier balance of all these elementary processes decides whether the breakdown process ends in a glow discharge between the electrodes which are completely cold. The lamp voltage in this condition is rather high because of the high voltage drop at the cold electrodes while the voltage in the discharge space is rather small. However, in wall stabilized lamps having an electrode distance which is large compared to the tube diameter the walls can play a very important role for the breakdown process: we find a breakdown between electrode and the wall forming a charge at the inner wall surface which moves along the wall towards the other electrode [30]. The mechanism is exactly that of the well-known dielectric barrier discharge.

The next phase describes the heating of the electrode by means of the **glow discharge** until the essential part of the current can be produced by the thermal emission of the electrode surface. During this transition so-called vapour arc modes can be detected which are highly erosive and burn in the vapour of electrode material [31]. Experimental and theoretical work was done to find conditions for shortening this phase [32, 33]. The thermal capacity of the electrode should be small, the open circuit voltage should be high, and the impedance of the lamp circuit should be not too high.

The last phase is the **heating of the discharge vessel** to the stationary temperature distribution and the corresponding build-up of the partial pressures of the filling components. This process can be observed by the increase of the lamp voltage after the minimum which is characterized by the breakdown of the cathode fall as result of the transition of the cathode from the glow mode to the thermionic arc.

17.4.2   Conventional operation

Because of the falling characteristics the HID lamps must be operated in series with an impedance which turns the sum impedance to positive values. The typical construction contains an inductance in series with an ignition circuit which generates ignition pulses either by interrupting the short circuit current of the inductance or by transfer of a high-voltage pulse into the lamp circuit using a transformer. The matching between the ballast and the lamp can be well considered using a plot of the lamp power versus the lamp voltage (Fig. 17.9). The **ballast curve** describes the response of the ballast circuit (output power and voltage) at different lamp impedances. It depends on the construction of the ballast and on the voltage-current characteristics of the lamp. The **lamp characteristics** connects all points on the power-voltage plot where the lamp can be operated stationary and stable, i.e. at a given power the temperature distribution along the arc tube, the vapour pressures, and the resulting lamp voltage are in equilibrium. It can be easily seen that the stability of the system depends sensitively on the angle between both curves at the intersection point. This system is cheap, simple and robust, but there are also severe disadvantages:

- the light emission is strongly modulated with the doubled mains frequency,

- the magnetic and Ohmic losses of the ballast decrease the system efficacy considerably,

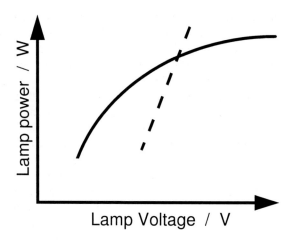

Figure 17.9: Lamp characteristics (dashed curve) and ballast curve (solid line) of a HID lamp system.

- due to the deep modulation of the lamp power at comparably low frequencies the temperatures of the lamp components oscillate, and the lamp life can be affected in a negative way.

17.4.3 Electronic operation

17.4.3.1 General considerations

The electronic operation of HID lamps is not yet as familiar as that of fluorescent lamps mainly for two reasons: the increase of the lamp efficacy at high frequency operation is not so relevant, and the standardization of HID lamps is not as advanced as that of fluorescent lamps. The interface between lamp and ballast cannot be defined precisely to ensure that lamps and power supplies from different suppliers can be operated without difficulty. Nevertheless, the possibility to modify or to stabilize the photometric properties of a lamp and the demand to control the system power to match it to the actual necessary illumination level, e.g., by means of a bus system, will promote the electronic operation of HID lamps. To determine the operation frequencies and the control time constants we should have a general idea about the time constants ruling the relaxation processes in HID lamps in Table 17.2. It can be seen that the LTE is realized in times which are very short compared with any changes of parameters of the power supply, i.e. the plasma of a HID lamp is homogeneous, isotropic and neutral. A more critical aspect is the response time of the discharge to changes of the input power. The time constant describing the response of the plasma temperature after a change of the lamp power is determined by the ratio of the plasma enthalpy and the radiation power. For typical HID plasmas it is in the order of magnitude of $10^{-3} \dots 10^{-4}$ s. The consequences are demonstrated in Fig. 17.10 for the simple case of a ballast with a linear Ohmic characteristics.

Table 17.2: Time constants for the relaxation processes in a HID lamp.

Establishment of Local Thermodynamic Equilibrium:	
– Maxwell distribution of electron energy	$10^{-11} \dots 10^{-9}$ s
– Gas temperature = Electron temperature	$10^{-7} \dots 10^{-6}$ s
Power balance of discharge and circuit	10^{-4} s
Power balance of the arc tube	$10^{1} \dots 10^{2}$ s

Similar as in Fig. 17.9 the intersection point between ballast curve 1, and the lamp curve 2 is the stable operation point of the system (the higher intersection at lower current is unstable as can be seen from the diagram). If the open circuit voltage U_0 changes with frequencies ≫ 1 kHz, the HID lamp plasma cannot follow the modulation of the input power: the plasma temperature and, as the result, the electrical conductivity does not change. The operation point of the lamp moves on curve 3, the lamp behaves like an Ohmic resistor. The long-time stability of the operation point depends on the question whether the average power is equal to that power which is needed to maintain the plasma condition defined by this point. For slow changes of the voltage U_0 with frequencies ≪ 1 kHz the plasma condition follows the

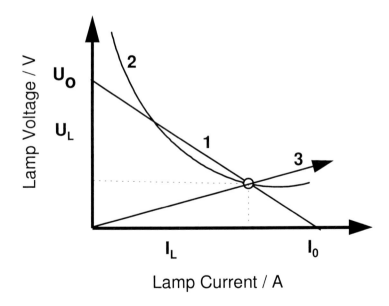

Figure 17.10: Voltage-current characteristics of a HID lamp. 1 ballast curve, 2 lamp characteristics for supply frequencies ≪ 10 kHz, 3 lamp characteristics for supply frequencies ≫ 10 kHz.

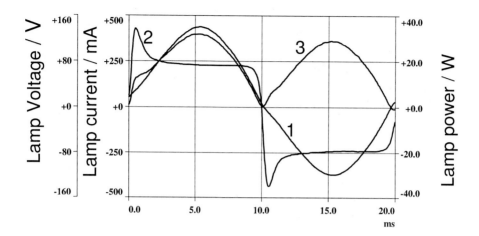

Figure 17.11: Time behaviour of lamp current 1, lamp voltage 2, and lamp power 3 for a 50 Hz operation.

changing power, and the operation point moves along curve 2. As an illustration, in Fig. 17.11 the well known waveform of a 50 Hz-operated lamp is shown. It can be seen that at low currents after the zero point the voltage increases to the so-called reignition peak; the absense of this peak at the current decay indicates that the cooling of the plasma cannot follow the decrease of the lamp power: the electrical conductivity remains high enough to prevent the build-up of a second peak. In contrast, at operation frequencies of, say, 50 kHz, voltage and current show exactly the same waveform.

The third time constant listed in Table 17.2 refers to a lamp containing a discharge medium whose vapour pressure depends on the temperature of the unevaporated liquid material. A change of the lamp power will influence the temperature distribution of the discharge tube and, in this way, the temperature of the **cold spot**. As a result, the vapour pressure responds with a time constant which is determined by the ratio of the heat capacity of the discharge tube and the additional heating power. The equilibrium of Ohmic heating and the dissipating processes adjusts to a new level according to Eq. 17.13 resulting in a different plasma temperature and, as a response, in a changed electrical conductivity. This change lasts typically 10...100 seconds and modifies the lamp characteristics 2 in Fig. 17.10.

17.4.3.2 Electronic control and new discharge conditions

In the last section the electrical interactions of the ballast and a lamp including the stabilization conditions were discussed. Of course, the user of such a system is interested mainly in the photometric properties. Fig. 17.12 shows a schematic interaction pattern considering external and internal influences on a HID system including a power supply and a high-pressure discharge lamp. Here we discuss the typical case of a fill system containing partially evaporating

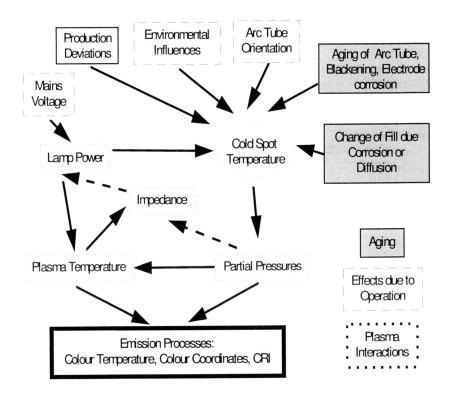

Figure 17.12: Interactions in the lamp system influencing the photometric properties.

additives, whose partial pressures depend on the composition of the fill and on the temperature
T_{cs} of the coldest area in the arc tube. This extremely important temperature is affected by a
series of influences:

- deviations in manufacturing,

- external influences: environment (e.g., tightly closed luminaire), arc tube orientation (can
  shift the temperature distribution of the arc tube), mains voltage, ...

- lamp aging: absorbing layers near the electrodes, electrode corrosion, change of fill composition due to diffusion or corrosion, ...

- lamp power.

By means of a suited lamp design these effects can be diminished but not completely avoided;
so the partial pressures are influenced, and the power balance of the discharge changes the
plasma temperature and the emission processes, i.e. the photometric properties. The numerous feedback loops in the diagram can enforce or suppress these shifts in dependence on the

actual plasma conditions. The corresponding change of the electrical conductivity σ may even influence the power fed from the power supply into the lamp, building an additional feedback loop.

Electronic operation in principle opens the following options for the development of modern HID systems:

Stabilization of the lamp power: the feedback loop including the lamp impedance is interrupted, so changes of the lamp cannot influence the input power. Possible changes of the photometric properties due to internal aging processes cannot be suppressed, for this reason this most common kind of control is not satisfactory.

Stabilization of the photometric properties of the lamp: To stabilize the spectral emission distribution of a lamp it is not enough to operate it at a constant power. As can be seen in Fig. 17.12 there are many parameters influencing the temperature distribution of the discharge tube, e.g., deviations during the lamp production, the cooling conditions around the burning lamp, and effects of aging as wall blackening or electrode corrosion. The result is a change of the densities of the additives which influence the intensity and spectral distribution of the radiation and in this way also the colour temperature and the colour rendering of the lamp. To stabilize these features it is necessary to meet two conditions: (i) the photometric data can be adjusted by means of a modification of the lamp power or power waveform, (ii) the power supply must get information describing the deviations from the set point. This information should be preferably an electrical signal, e.g., the voltage or the impedance of the lamp. In [34] the chances are discussed to derive such a signal from the lamp. For this purpose it is necessary that the essential radiation processes and the electrical conductivity are functions of the same particle densities. For the radiation processes let us choose a generalized line strength as proposed by Schirmer [35]

$$P_{rad} = n_M \, \bar{K}_M \, \exp\left(-\frac{\bar{E}_M}{kT}\right) , \tag{17.22}$$

where n_M is the number density of the radiating particles, and $\bar{K}_M$, $\bar{E}_M$ are an averaged line constant and excitation energy, respectively. As a representative for the electrical properties we consider the electrical conductivity. To be more transparent we use the simple Lorentz formula for partially ionized plasmas

$$\sigma = C \, \frac{n_e}{\sqrt{T}} \, \frac{1}{n_{Hg}Q_{Hg} + n_M Q_M + n_e Q_i} \tag{17.23}$$

with $C = const$, T the plasma temperature, and n_{Hg}, n_M, n_e the particle number densities of mercury, of the component M, and of electrons, respectively. The Q_j are the transport cross sections for electron collisions of the indicated species. Let us apply these expressions on different lamp types. Table 17.3 shows the application of equations 17.22 and 17.23 to the special conditions and density relations of the listed lamps. It can be seen that there is a strong correlation between the electric and the photometric properties in the case of **high-pressure xenon** lamps and **mercury-free HPS** lamps, i.e. the radiation properties of the lamps can be detected by the power supply by means of a test of voltage and/or current. For **standard HPS** lamps this correlation also holds when a constant relation between the vapour pressures of mercury and sodium in the case of a given amalgam can be assumed. Obviously it must fail when the amalgam ratio changes due to sodium corrosion or diffusion.

Table 17.3: Dependence of the electrical and optical properties of the different lamp types on the plasma temperature and the particle number densities.

Lamp	Conditions	Conductivity σ	Radiation P_{rad}
High-pressure Xe	$n_{Hg} = 0$ $n_{Xe}Q_{Xe} \ll n_eQ_i$	$\frac{c}{\sqrt{T}Q_i}$	$n_{Xe}\,\bar{K}_{Xe}\,\exp\left(-\frac{\bar{E}_{Xe}}{kT}\right)$
Standard HPS	$n_{Hg} \gg n_{Na}$ $n_{Hg}Q_{Hg} \gg n_eQ_i$	$\frac{c}{\sqrt{T}}\,\frac{n_e}{n_{Hg}Q_{Hg}}$	$n_{Na}\,\bar{K}_{Na}\,\exp\left(-\frac{\bar{E}_{Na}}{kT}\right)$
Hg free HPS	$n_{Hg} = 0$ $n_{Na}Q_{Na} \gg n_eQ_i$	$\frac{c}{\sqrt{T}}\,\frac{n_e}{n_{Na}Q_{Na}}$	$n_{Na}\,\bar{K}_{Na}\,\exp\left(-\frac{\bar{E}_{Na}}{kT}\right)$
Metal halide	$n_{Hg} \gg n_M$ $n_{Hg}Q_{Hg} \gg n_eQ_i$	$\frac{c}{\sqrt{T}}\,\frac{n_e}{n_{Hg}Q_{Hg}}$	$n_M\,\bar{K}_M\,\exp\left(-\frac{E_M}{kT}\right)$

The correlation is completely lost in the case of **metal halide** lamps where the electrical properties are determined nearly exclusively by the mercury, and the radiation distribution by the halide additives. Because they coexist without any interaction it is extremely difficult to get an information about the radiation from static electrical tests.

A chance to carry out the task even for metal halide lamps may be a dynamic test by applying a fast power step to the lamp: the discharge can be considered as an energy reservoir whose energy is mainly determined by the temperature and density of mercury vapour. The relaxation of the system is given mainly by the radiation which depends on the particle number densities of the metal vapour. In this way the decay or rise time of the electrical conductivity contains information about the radiation distribution. Dynamic tests of this kind may provide also other useful information about the lamps as, e.g., excess of halogen in case of increased reignition peaks.

Modification of the photometric properties of the lamp: it was found that the colour temperature of a HPS lamp can be increased not only by means of an increased optical thickness (see section 17.3.4) but also by a pulsed operation of the system [36, 37]. During the pulse phase the population of higher lying energy levels of the sodium system is increased, and spectral transitions including the recombination continuum of the 2^2P level in the blue region are excited resulting in a raised colour temperature. Thus, we use the fact that the time constant of the plasma is much shorter than that of the arc tube which feels the average power only. The lamp is fed from an electronic control gear containing a microcontroller that monitors the electrical behaviour including ignition, warm-up and switch-off. All data of the power waveform as pulse length, pulse height, repetition frequency, and keep alive power are computed and controlled to ensure the designed correlated colour temperature and colour rendering index kept within very narrow limits regardless of individual deviations of the lamp, environmental influences, and aging effects. The lamp voltage is used as reference signal because of the strict correlation between the electrical and spectroscopical data. The design data of the system can be modified by program, of course; so the colour temperature can be switched between two fixed levels at 2700 and 3100 K without changing the lumen output or the colour rendering index. For outdoor illumination the lumen efficacy of the lamp can be increased at the expense of the colour rendering index without changing the hardware.

Establishment of new discharge conditions: here we refer to the electrodeless sulphur and cluster lamps mentioned above.

17.5   Conclusions

It is shown that the electronic operation enables the stabilization and modification of the photometric properties of HID lamps. By electrodeless transfer of the lamp power it is possible to introduce new radiating elements and compounds generating very attractive and economic spectral distributions. Furthermore, the use of intelligent electronic circuits will allow to control all essential operation modes of a HID lamp as:

- control of the ignition,

- control of the warm-up phase from glow to arc,

- stop of the power supply after a given time, if there is no ignition,

- test of the lamp condition using the signals discussed in the last section and compare them with reference parameters to readjust the waveform of the electric power supplied to the lamp until the reference data are realized,

- control of the lamp power in dependence on the actual needs, and

- stop of the power supply, if the allowed control range of the reference parameters is exceeded, and indicates end of life for the lamp.

References

[1] W. Ebeling, A. Förster, V.E.Fortov, V.K. Gryaznov, A. Ya. Polishchuk, Thermophysical Properties of Hot Dense Plasmas, Teubner-Texte zur Physik, Stuttgart, Leipzig (1991).

[2] K. Günther, R. Radtke, Electric Properties of Weakly Nonideal Plasmas, Birkhäuser Verlag Basel, Boston, Stuttgart (1984).

[3] M.S.Rea, Proc. of the 8th Int. Symp. on the Science & Techn. of Light Sources, Greifs-wald (1998), p. 120.

[4] CIE Technical Report: Colorimetry (1986).

[5] CIE Technical Report: Colour Rendering Index for Light Sources, Draft 6/1996.08.27.

[6] J.F. Waymouth, IEEE Transact. on Plasma Sciences Vol.19, 1003 (1991).

[7] R.J. Zollweg, J.J. Lowke, R.W. Liebermann, J. Appl. Phys. 46, 3828 (1975).

[8] E. Fischer, E. Schnedler, Proc. of the 5th Int. Symp. on the Science & Techn. of Light Sources, York (1989), p. 99.

[9] H. Wiesmann, PhD Dissertation, University of Karlsruhe (1997).

[10] S.Chapman, T.G. Cowling, The mathematical theory of non-uniform gases, Cambridge University Press, New York (1952).

[11] H. Hess, G. Hartel. H. Schoepp, L. Hitzschke, Proc. of the 8th Int. Symp. on the Science & Techn. of Light Sources, Greifswald (1998), p. 200.

[12] L.M. Biberman, G.E. Norman, Usp. fiz. nauk 91, 193 (1967).

[13] D. Hofsaess, Z. Phys. A 281, 1 (1977).

[14] K. Günther, Der Energiehaushalt strahlungsbestimmter Plasmen und seine Bedeu-tung für die Optimierung selektiv strahlender Impulslichtquellen. Habilitation thesis, Akademie der Wissenschaften, Berlin (1982).

[15] W. Elenbaas, High Pressure Mercury Vapour Lamps and their Applications, Philips Technical Library, Eindhoven (1965).

[16] G. Hartel, H. Schoepp, H. Hess, L. Hitzschke, Proc. of the 8th Int. Symp. on the Science & Techn. of Light Sources, Greifswald (1998), p. 196.

[17] J.J. Lowke, J. Appl. Phys. 41, 2588 (1970).

[18] B. Kühl, H. Krense, German Patent DBP 184008 (1960).

[19] S. Carleton, P.A. Seinen, J Stoffels, J. Illum. Eng. Soc. 26, 139 (1997).

[20] K. Schmidt, US Patent No. 2971 110 (1960).

[21] J. de Groot, J. van Vliet, The high pressure sodium lamp, Philips Technical Library, Deventer (1986).

[22] E. Fischer, Proc. of the 8th Int. Symp. on the Science & Techn. of Light Sources, Greifs-wald (1998), p. 36.

[23] A. Inouye, Proc. of the 8th Int. Symp. on the Science & Techn. of Light Sources, Greifs-wald (1998), p. 99.

[24] W. Lapatovich, Proc. of the 7th Int. Symp. on the Science & Techn. of Light Sources, Kyoto (1995), p. 139.

[25] B.P. Turner, M.G. Ury, Y. Leng, W.G. Love, J. Illum. Eng. Soc. 26, 10 (1997).

[26] B. Weber, R. Scholl, J. Appl. Phys. 74, 607 (1993).

[27] M. Born, J. Phys. D: Appl. Phys 32, 2492 (1999)

[28] H. Hess, Der elektrische Durchschlag in Gasen, Akademie-Verlag Berlin (1976).

[29] W.W. Byszewski, A.B. Budinger, J. Illum. Eng. Soc. 19, 70 (1990).

[30] A. Lembcke, K. Günther, Proc. of the 8th Int. Symp. on the Science & Techn. of Light Sources, Greifswald (1998), p. 198.

[31] A. Anders, B. Jüttner, Lighting Res. & Technol. 22, 111 (1990).

[32] G.M.J.F. Luijks, J.A.J.M. van Vliet, Lighting Res. & Technol. 20, 87 (1990).

[33] L.C. Pitchford, I.Peres, K.B. Liland, J.P. Boeuf, H. Gielen, J. Appl. Phys. 82, 112 (1997).

[34] K. Günther, Proc. of the 7th Int. Symp. on the Science & Techn. of Light Sources, Kyoto (1995), p. 93.

[35] H. Schirmer, Habilitation thesis, Berlin (1960).

[36] P.D. Johnson, T.H. Rautenberg, J. Appl. Phys. 50, 3207 (1979).

[37] K. Günther, H.G.Kloss, T. Lehmann, R. Radtke, Contrib. Plasma Phys. 30, 715 (1990).

18   Plasma etching in microelectronics

Harald H. Richter and André Wolff

IHP–Innovations for High Performance Microelectronics, Im Technologiepark 25,
D–15236 Frankfurt (Oder), Germany

18.1   Characterization of plasma etching

In the early days of silicon integrated circuit (IC) technology, circuit dimensions were large
(10 μm and above) and wafer sizes were small (2 inches and below). Mask patterning was usually transferred by direct contact exposure of photoresist. After development, the photoresist
pattern, i.e. the material not covered by resist, was removed by different aggressive liquids.
These so-called wet chemical etching processes were highly selective. However, wet etching
is in principle isotropic, so that the pattern etched was wider than the photoresist pattern [1].

Due to the decreasing feature size and increasing complexity of integrated circuit devices,
the performance of wet etching processes became inadequate. In the late 1970s, wafer sizes
increased from 2 to 4 inches, and feature sizes shrank down to 3 μm. At these device dimensions, a precise pattern control, e.g., a 1:1 transfer process from the photoresist pattern to the
underlying material, became a dominant demand [2]. The isotropic etching process had to be
replaced by an anisotropic one. That meant, etching had to take place in one direction only,
perpendicular to the sample surface.

The discovery that the most important semiconductor materials such as polysilicon or silicon nitride can be etched by exposure to an radio-frequency (rf) glow discharge with halogen-
containing gases was a major achievement for integrated circuit processing [3]. The first
plasma etchers (*barrel* reactors) introduced in technology lines led to real improvements in
process control, but the etching profiles were still isotropic. The time had come for a next
decisive milestone in dry processing, distinguished by the development of alternative etching
systems. The most prominent of these new etching tools was the reactive ion etching (RIE)
system. RIE is a plasma-based dry etching technique characterized by a combination of physical sputtering with the chemical activity of reactive species. This enables material-selective
etch anisotropy [4].

Since etch rates could be increased from 1 nm/s to 10–20 nm/s in these RIE systems,
the use of dry etching tools also became advantageous from the economical point of view.
Moreover, a great deal of effort was focused on the introduction of new gas mixtures. In the
first barrel reactors simple gases like CF_4 were used. Presently the situation is more complex:
nowadays each etching step is developed with its own etching chemistry.

From the mid to the late 1980s new high density (i.e. $> 10^{11}/cm^3$) plasma sources operating at low pressure ($< 10^{-1}$ Torr) were developed to meet the production demands of
ever smaller feature devices. The high uniformity of etching across the increasing wafer sizes,

minimal particle contamination, and sufficiently high etch rates for the desired throughput of these sources has ensured the introduction of these etching tools in very large scale integrated (VLSI) circuit production lines.

The last two decades have seen tremendous speed in the introduction of dry etching in microelectronics technology. The first commercial parallel plate reactor was launched in 1976, and the first cassette-to-cassette system followed three years later. By the middle of the 1980s, dry etching was dominating all wafer patterning steps in VLSI production lines [1]. Modern wafer fabs now use wet chemical processing only in clean and strip steps.

In the process of introducing dry etching into technology it became increasingly clear that etching systems are extremely complicated. Plasma etching has always been more of an art than a science, more black magic than engineering [5]. All species produced in the discharge can be classed as two types which have special importance for etching:

(i) *reactive neutrals* for chemical reactions to form volatile compounds or to establish surface coatings by polymerization;

(ii) *positive ions* which are responsible

- for material removal by sputtering,
- for suppression of polymerization processes,
- for maintaining the surface reaction with reactive neutrals.

The most important parameters in technology-oriented dry etching are etch rate, selectivity, anisotropy, and non-uniformity. These fundamental terms can be defined with reference to figure 18.1, which shows typical etching profiles.

The quantity etch rate (R) does not require an explanation. Selectivity (S) is defined as the etch rate ratio of two different materials under identical plasma conditions. The term anisotropy (f) characterizes the ratio of lateral to vertical etch rate. Consequently, a perfectly isotropic etching process (Fig. 18.1a) has an anisotropy of 0, while $f = 1$ indicates a perfectly anisotropic process (Fig. 18.1b). However, the most frequent profiles (Fig. 18.1c) are a

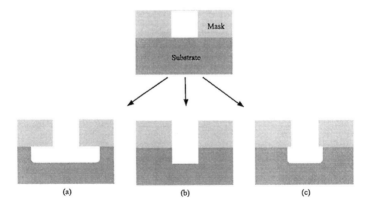

Figure 18.1: Typical etching profiles.

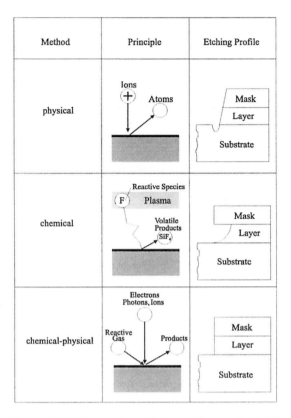

Method	Principle	Etching Profile
physical	Ions (+) Atoms	Mask / Layer / Substrate
chemical	Reactive Species (F) Plasma / Volatile Products (SiF₄)	Mask / Layer / Substrate
chemical-physical	Electrons Photons, Ions / Reactive Gas Products	Mask / Layer / Substrate

Figure 18.2: Comparison of dry etching methods [6].

combination of these two extreme cases. The last important feature, the non-uniformity (i), is a degree of etching evenness across the wafer. An ideal process is characterized by $i = 0$. Process development requires that these key parameters be balanced while at the same time a variety of other important processes are to be considered.

18.2 Etching techniques

Dry etching is defined as the removal of material by interaction with glow discharges. This removal of material by a plasma-based process can have many names, for instance reactive ion etching (RIE), magnetically enhanced reactive ion etching (MERIE), reactive ion beam etching (RIBE), and so on.

In summary, the diverse etch processes are different methods by which a substrate surface is etched either physically by ion bombardment, or chemically by a chemical reaction of reactive species generated in a plasma at the surface, or both physically and chemically by an ion-, electron-, or photon-induced chemical reaction at the surface [6]. The various dry etching methods are illustrated in Fig. 18.2.

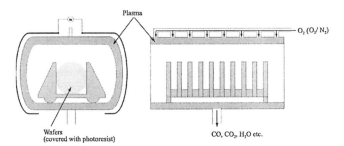

Figure 18.3: Classical principle of plasma ashing in a *barrel* reactor (as an example for isotropic etching) [10].

18.2.1 Physical etching

Physical dry etching processes are characterized by a surface bombardment with ions, electrons or photons. The interaction of electrons and photons with surface atoms leads to a vaporization of etching material. Laser-induced physical etching is a common method for marking silicon wafers.

An ion bombardment of the surface (ion sputtering) results in an ejection of atoms from the surface. This procedure makes etching of all semiconductor materials possible. On the other hand, there are many disadvantages which considerably limit the use of ion etching in microelectronics:

- small etching rates,

- small selectivities between etched materials and masks,

- high damage caused by high ion energies (*trench effect*) [7].

For these reasons, pure physical etching is of no importance in microelectronics technology. Sputtering can only be used for cleaning procedures [8].

18.2.2 Chemical etching

Chemical etching in a plasma results from gasification reactions between etchant species and the material which is to be etched. The role of the plasma is the production of etchants. This type of etching can be equated to isotropic etching [9]. Chemical etching was first applied to semiconductor processing during the 1960s as plasma ashing, also known as plasma stripping (Fig. 18.3). This is a technique for the removal of organic photoresist material consisting essentially of carbon, hydrogen, nitrogen and oxygen. In an oxygen or oxygen/nitrogen plasma (characteristic equipment: *barrel* reactor), solid carbon is converted to gaseous CO and CO_2. These gases and other volatile reaction products are then pumped away. In the case of plasma ashing, the purpose of the discharge is the generation of atomic oxygen which (in contrast to molecular oxygen) reacts with photoresist at room temperature.

Purely chemical (isotropic) etching processes are unsuitable for the patterning of small structures. Therefore, these processes nowadays are only used for the removal of blanket layers and for cleaning procedures. A current example of a purely chemical process is downstream microwave plasma etching, which completely separates the wafer from the plasma and achieves a totally isotropic etch by the action of the reactive chemical species alone [5].

18.2.3   Chemical-physical etching

Most etching processes in use today rely to some degree on both chemical and physical etch mechanisms. These chemical-physical etching processes make directional etching possible, thus meeting many dry etch requirements in VLSI technologies. There are two types of directional etching mechanisms which are stimulated by a vertical ion flux bombardment; they lead to (Fig. 18.4)

- energetic ion-induced anisotropy, and

- inhibitor ion-induced anisotropy [11].

In the first case, impinging ions increase the reactivity of the surface. The second type, inhibitor ion-induced etching, requires two conceptually different species: etchants and inhibitors. The substrate and the etchants react spontaneously, and etching proceeds isotropically without inhibitor species. The energetic ion bombardment removes the inhibitors and allows the direct substrate contact of the reactive species for a following reaction. However, vertical surfaces are only exposed to a weak ion bombardment; this leads to a sidewall passivation by a thin inhibitor film. This has the advantage of helping to block etching of the sidewalls, and leads to increased anisotropy.

In general, chemical-physical etching is characterized by anisotropic profiles, high etch rates, and high selectivities to underlying layers and masks. Because of these characteristics, chemical-physical etch procedures are used extensively in integrated circuit fabrication for a variety of materials and etching applications. There are two strategies of process development for meeting the high demands on dry etching in advanced submicron technologies:

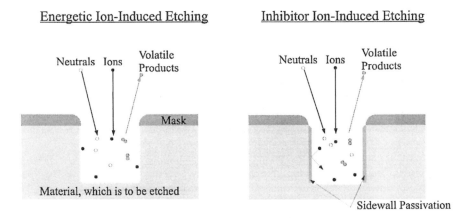

Figure 18.4: Two kinds of chemical-physical etching.

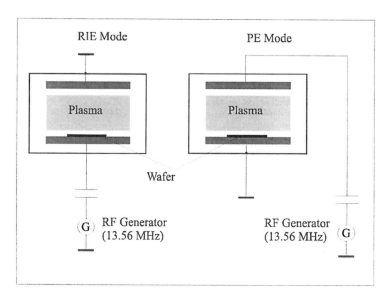

RIE Mode

PE Mode

Plasma

Plasma

Wafer

G   RF Generator
    (13.56 MHz)

RF Generator
(13.56 MHz)   G

Figure 18.5: Parallel plate reactor in two variants: *Reactive Ion Etch* (RIE) Mode and *Plasma Etch* (PE) Mode.

(i) development of new plasma equipment, especially of new low-pressure, high-density plasma sources (section 18.3),

(ii) use of new dry etch chemistries (section 18.4).

18.3   Equipment-related topics

There are basically two types of dry etching systems: barrel (Fig. 18.3) and planar systems. In the 1970s and early 1980s the standard tools of semiconductor processing were barrel etchers. The reactors were designed to process batches of 5 to 25 wafers simultaneously. The disadvantage of these configurations is that the plasma and chemical conditions are always somewhat different from one wafer position to another. As wafer sizes increased, uniformity became a serious concern. It quickly became evident that the wafer to wafer variability could be reduced by processing only one wafer at each time in a constant environment. Consequently, the equipment market shifted to single-wafer reactors, especially for processing larger wafers [12].

The classical parallel plate reactor (see Fig. 18.5) is the most popular single-wafer planar system. Here the wafer is placed with the backside down on the lower electrode. At this point it is sensible to introduce the terms *reactive ion etching* (RIE) and *plasma etching* (PE).[1] Historically, these terms described the difference in the electrode on which wafers were placed. RIE denotes a class of dry etching processes in which ion bombardment of the substrates

---

[1]In the title of this chapter, the term plasma etching is used more generally, as a synonym for dry etching.

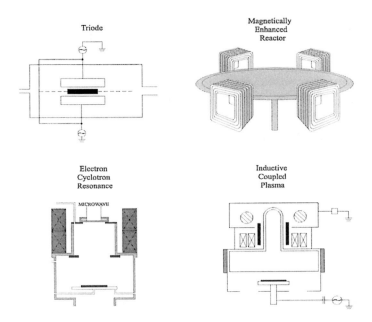

Figure 18.6: Different types of plasma processing equipment.

contributes considerably to the etch rate and etch profile characteristics, in contrast to the PE processes where etching is mainly chemical. A difference in the etch process performance results from the different areas of the electrodes. In the RIE mode the wafer is placed on the small RF-powered electrode. The ground electrode typically is the chamber itself. RIE systems are distinguished by low pressures ($\leq 10^{-1}$ Torr), higher ion and electron energies, and higher electrode voltages compared with PE etching.

In spite of the successful application of traditional parallel plate sources for semiconductor manufacturing new plasma sources were developed. In the mid-eighties, several companies improved the classic parallel plate design. The crucial point was the partial separation of control of plasma generation and ion bombardment energy. Important examples of these *enhanced* planar etchers [12] are the triode design (introduction of a grid electrode) and magnetically enhanced (MERIE) etch reactors (see schematic drawings in Fig. 18.6). Here, magnets have been added to the RIE system to control the path of electrons and ions with the final aim of increasing the ion density. It was argued that this equipment could achieve higher etch rates with less damage than conventional parallel plate etchers.

A new direction in etch design is based on the replacement of the rf power source by a magnetically enhanced source driven by microwave excitation that resonates at the orbital frequency at which electrons circle in the magnetic field. In practice, the electrical frequency is 2.45 GHz. Thus the magnetic field strength required to generate resonance of the electrons is 875 Gauss. This new generation of so-called electron cyclotron resonance (ECR) plasma etchers (Fig. 18.6) is characterized by low-pressure (10^{-5} to 10^{-2} Torr), high-density plasmas [12]. However, the ECR plasma etching technology has not yet been widely accepted

Table 18.1: Operating pressure of different reactor configurations [15].

Reactor Configuration	Etching Method	Pressure Range [Torr]
Barrel Reactor	chemical	$10^{-1} \ldots 1$
RIE	chemical-physical	$10^{-3} \ldots 10^{-2}$
Triode	chemical-physical	$10^{-3} \ldots 10^{-2}$
MERIE	chemical-physical	$10^{-4} \ldots 10^{-3}$
ECR	chemical-physical	$10^{-4} \ldots 10^{-3}$
ICP	chemical-physical	$10^{-4} \ldots 10^{-3}$

in production lines because of poor uniformity, low etching rate, and poor anisotropy [13]. Market data indicate that inductively coupled plasma (ICP) technology is today's most widely accepted high-density source for etch application (Fig. 18.6) [14]. The relatively simple design of ICP allows independent control of ion density and ion energy. This minimizes damage, and enables optimum selectivities and high etch rates. The ICP source controlled by physical means creates a very uniform ion density across the wafer, even in the case of scaling to large wafer diameters. In Table 18.1, the pressure ranges of the most common plasma sources are compared.

In summary, a variety of new sources are capable to deliver high-density plasmas at low pressures. However, it is not clear which of them is best suited for production applications. Many experts believe the requirements of quarter micron technology etching mark the point of no return for conventional plasma etch equipment [14], but it seems unlikely that one source technology will prove to be universally applicable. Moreover, there is the discrepancy that much of the work on new sources is focused on plasma physics, with little or no attention to chemistry. Modern low-pressure sources often allow the use of simpler etch chemistries [16].

18.4   Etch chemistries

In semiconductor technology many different chemical compounds are used in dry etching processes. The gases are selected for their ability to produce reactive species upon decomposition that will react chemically with a surface to be etched, forming volatile compounds. In addition, the equipment must be able to use the selected gases without corroding the inside of the chamber and thus destroying the reactor. Another phenomenon that is shaping etch gas selection relates to environmental and health aspects. Table 18.2 provides a list of common semiconductor materials and the most common corresponding primary etchant gases today used. The dominance of elements of the 7th main group is obvious. Often, the benefits of different families of gases are combined to balance etch rate, anisotropy and selectivity. It is nearly impossible to get one single gas or gas mixture that gives a high etch rate, high selectivity and high directionality. For this reason, multi-step processing has become commonplace [16].

In addition to equipment changes there has also been a trend towards using different etch

Table 18.2: Common materials used in IC fabrication and corresponding etching gases [17]. The symbol /+ indicates that the listed gas is often used with additives for etching.

Material	Etch Gases	Reaction Products
Si, Poly-Si	BCl_3/+, Cl_2/+, HBr(/+), HCl(/+), SF_6/+ , CF_4(/+)	SiF_4, $SiCl_4$
SiO_2	C_2F_6, CHF_3(/+), CF_4(/+), C_4F_8/+	SiF_4, $SiCl_4$, CO
Si_3N_4	C_2F_6, CHF_3(/+), CF_4(/+), CH_3F/+, SF_6(/+)	SiF_4, NF_3
Metal (AlCu)	BCl_3(/+), HBr(/+), Cl_2(/+), $SiCl_4$(/+), HI	$AlCl_3$, Al_2Cl_6
Photoresist	O_2(/+), CF_4/+	CO, CO_2, H_2O, HF

chemistries, either for performance reasons (profile control, selectivity, or etch rate) or environmental reasons (converting to CFC alternatives). In 1991, it was noticed that etching chemistry, as the single most critical factor for current processes, has been neglected in recent development work. It could become the focal point for gaining major advantages in future [12].

18.5 Dry etching in advanced technologies (selected examples)

Etching processes in advanced submicron technologies have to meet the following requirements [7]:

(i) anisotropic etch profiles with controlled taper capability,

(ii) controlled endpoint detection,

(iii) etch uniformity better than 1%,

(iv) high selectivity to the photoresist,

(v) high selectivity to the underlying layer,

(vi) good process reproducibility,

(vii) low microloading effect (i.e. only small etch rate differences between tight spaces and wide spaces),

(viii) limited radiation or ion damage.

In addition to these prerequisites for improved etch performance there is also a new and powerful push to minimize cost of ownership.

The materials that need to be etched in a typical Si-based integrated circuit generally fall into one of the following three areas: polysilicon, dielectrics, and metals. Each of these areas require a dedicated etch process and equipment set-up. These applications are primarily driven by the push to smaller geometries, but also by new transistor design, dynamic random access memory (DRAM) cell structure, isolation technologies, planarization, and interconnect schemes [18]. According to the market data from industry analysts, polysilicon etching represents 25% of the total available market, oxide etching 44%, and the etching of aluminum and its alloys 31% [5].

18.5.1 Silicon dry etching

18.5.1.1 Trench etching

Deep trenches etched in Si substrates are used for device isolation or for the creation of capacitor structures for DRAM chips. They typically are more than 5 μm deep and less than 1 μm wide (Fig. 18.7). Sloped trench sidewalls are needed to prevent void formation during following trench refill. Moreover, rounded corners (U-grooves) and smooth surfaces are desirable to avoid electrical field breakdown. Several gas combinations based on chlorine or bromine chemistry have been found to produce high quality trench etches. During the etch process, the top surface should remain clean and free of damage. For this reason, the etch mask is usually made by dry etching a thermally grown top silicon oxide and stripping the photoresist, prior to starting the actual trench etch process. Therefore, an etch process with a high selectivity silicon/oxide is required. An additional advantage of a hard oxide mask is that it enables the use of high power densities, which are necessary to keep the etch time within reasonable limits [19]. For typical trench etch processes, the selectivity of the oxide mask is > 50:1, with an etch uniformity on 200 mm wafers of $\leq$ 3%. In order to achieve a successful trench etch, free of residues, *in-situ* wafer pre-clean process steps are essential.

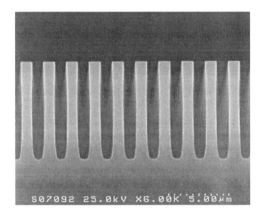

Figure 18.7: Trench structure after dry etching in a HBr/NF$_3$/He-O$_2$-plasma.

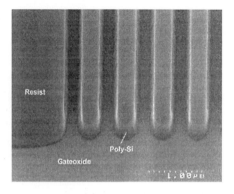

Figure 18.8: Top view of a polysilicon gate structure before (left) and after (right) dry etching in a hydrogen iodide (HI) plasma.

18.5.1.2 Polysilicon gate etching

The field effect transistor, currently called MOS transistor, is the most widely used device for high density integrated circuit (IC) fabrication. These devices are made of an n or p-type Si substrate. A thin thermal gate oxide serves as an insulator, and a metal contact is termed the gate. One of the most critical process steps in the preparation of MOS devices is the definition of the poly-Si gate. For polysilicon gate etching in advanced device processes, the etch must be anisotropic, highly selective to the underlying oxide, uniform in etch rate, low in ion impact energy at the wafer surface, and uniform in ion flux to the wafer to prevent damage. The need to etch stop over very thin gate oxide layers (3–6 nm thick) has led to an intensive search for poly-Si etch gas mixtures that can attain very high selectivities to oxide and photoresist.

The conventional etch chemistry for poly-Si gate patterning is chlorine or bromine. A significant alternative to these etchants seems to be hydrogen iodide (HI, see Fig. 18.8). Using an iodine based chemistry, selectivities silicon/oxide of 180:1 are realizable [20]. Such high selectivity values minimize the oxide loss, even with a 50–100% over-etch. Long over-etch periods guarantee the complete removal of polysilicon. A transmission electron micrograph demonstrates the influences of plasma etching on the region near the interface silicon/oxide (Fig. 18.9). No microtrenching of the gate oxide exists at the periphery of the gate region. The natural, inherent energy of the ions in HI discharges seems to be low enough for the gate oxide not to be sputtered away. The micrograph above shows the minimal gate oxide consumption during patterning in HI plasma. Even at 100% over-etch, the maximal oxide loss is less than 1 nm.

18.5.2 Oxide etch processes

In the area of isolator etching, the major application is the anisotropic patterning of contact holes to silicon, and of via holes between various interconnect levels in multilevel metalization systems. To realize a high aspect ratio contact etch (Fig. 18.10), high selectivities to underlying silicon or nitride films are required. Profile control is an additional crucial parameter

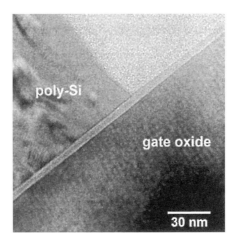

Figure 18.9: TEM micrograph showing the minimal gate oxide consumption during patterning in HI plasma.

in contact etch, especially in quarter micron technologies. If the profile in a 1 μm contact is changed by 1°, the area change at the contact bottom is small. But in a 0.25 μm contact, that change is considerable. To maintain a consistent contact resistance, very good profile control is necessary. Near vertical profiles with a slope > 89° are preferred [18].

Nearly all dry etching processes for SiO_2 are based on fluorocarbon discharges. The chemical endproducts of the reactions are SiF_4 and CO or CO_2 [21]. Because of the high strength of the silicon-oxide bond [22], the etching of SiO_2 is the most difficult of any etch processes. The mechanism for etching SiO_2 is dictated by the ability of CF_x radicals to reduce the oxide and form volatile products. On the other hand, Si is mainly etched by fluorine atoms.

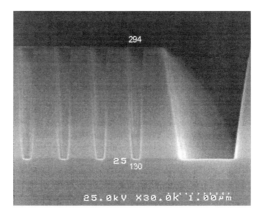

Figure 18.10: Contact holes after dry etching in a CHF_3/C_2F_6 plasma.

Therefore, a high oxide-to-silicon etch rate ratio is realizable through a high CF_x–to–F ratio. A successful etch gas combination which can remove SiO_2 at high rates is C_2F_6/CHF_3 [23]. Increasing amounts of hydrogen-containing CHF_3 scavenge the fluorine atoms; this yields the desired high selectivity. After RIE, the surfaces are characterized by polymeric deposits. For that reason, cleaning procedures are absolutely essential [24].

Recently, the fabrication process of ULSI circuits has been focused on self-aligned contact (SAC). In the fabrication process of SAC, high selectivities of SiO_2 over Si_3N_4 and Si are required. It is quite difficult to realize these isolator etch demands with sufficiently high anisotropy and high etch rate at the same time [25].

18.5.3   Metal etch

Currently, metal etching still means aluminum etching; tungsten is also often used for contact and via fill. In the near future, Al and its alloys (containing Cu and Si) may be replaced by other materials such as Cu or Au. However, up to now, Al has been commonly used as the main metalization material because it meets most of the metalization requirements for microelectronic devices. The nature of Al metalization has changed dramatically as device technology has evolved. Barrier metal layers (Figs. 18.11) have been used to prevent Si migration into Al. Moreover, a thin anti-reflective coating (ARC) may be deposited over the bulk Al layer. This assists the forming of well resolved, fine-line resist images [26]. Satisfactory dry etching of this whole composite layer system requires process conditions that are different from those for etching a single film of aluminum.

Fluorine-containing gases used for silicon and oxide etch are unsuitable for Al etching since the AlF_3 by-product is not volatile. The relatively high vapor pressure of $AlCl_3$ has dictated the use of Cl-based etch chemistries (cf. Table 18.2). To define submicron metal lines without undercutting photoresist, polymerizing agents such as CF_4 have been added to Cl-based discharges. Barrier layers like Ti-W can be etched with either Cl or F mixtures, although fluorine-rich plasmas may lead to undercutting [26].

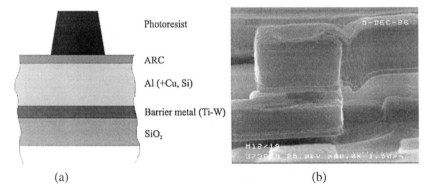

(a)   (b)

Figure 18.11: (a) Composite metalization for ULSI circuits [26]. (b) Multi-level metalization system after dry etching and post-etch treatment.

After etching, a post-etch treatment consisting of surface passivation to prevent corrosion due to residual chlorine on the surface of the wafer and photoresist stripping is absolutely essential.

18.6 Process control

With increasing wafer and decreasing feature sizes, advanced process and equipment control is becoming more and more mandatory for IC manufacturing. The keys to success for future IC fabrication are in maximizing the product quality and minimizing the production costs. Some specific strategies include increasing wafer throughput and wafer yield, establishing more robust processes, reducing monitor wafers, optimizing equipment cleaning procedures, and minimizing wafer scraps. One method for achieving several of these goals is to use *in-situ* and real time optical emission spectroscopy (OES), which is capable to measure the dynamic evolution of a broad range of wavelengths.

The contact hole etch process for 64M DRAM fabrication will be presented as a case study (Fig. 18.12). Optical emission data were recorded every two seconds in the spectral range between 300 and 740 nm. Since the process time is approximately 200 seconds, data filters are necessary to reduce the large number of variables. In this study, the principal component analysis (PCA) technique was used to analyze and reduce the data space [27]. In real samples, there are usually many different variations that make up a spectrum, including at a minimum constituents of the sample gas mixture, instrument variations such as detector noise, and changing environmental conditions that affect the baseline and the absorbency. The eigenvectors of a PCA decomposition represent the spectral variations that are in common with all of the spectroscopic calibration data. The variation spectra, F (eigenvectors), can be used to reconstruct the spectrum of a sample by multiplying each one by a different constant scaling factor and adding the results together. The scaling factors, S, used to reconstruct the spectra are generally known as PC scores.

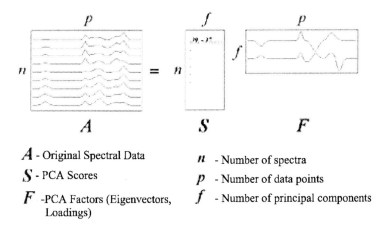

A - Original Spectral Data n - Number of spectra

S - PCA Scores p - Number of data points

F -PCA Factors (Eigenvectors, f - Number of principal components
    Loadings)

Figure 18.12: Principle of the Principal Component Analysis (PCA).

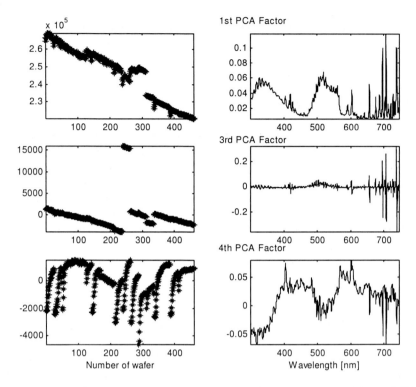

Figure 18.13: Results from data reduction with principal component analysis (PCA). The different PC scores are depicted in the left column, whereas the right column shows the corresponding PCA factors.

Fig. 18.13 shows the results of the measured spectra for the contact hole etch process for a production cycle. The left column shows the PC scores and the right column the PCA factors of the 1st, 3rd and 4th principal components. The PC scores of the 1st component show a significant decrease that can be correlated to an increase in the dirtiness of the chamber. As indicated in Fig. 18.13, the 3rd PC scores show a significant deviation for one misprocessed lot. By analyzing the spectroscopic raw data, it was found that one of the mass flow controllers failed during processing. The 4th PC scores show an initial increase before the values remain constant during processing of one lot. This behavior is correlated to the so-called first wafer effect. A process chamber needs some time to warm up when the idle time of the machine exceeds a certain period.

The results show that on-line, real time, *in-situ* full size spectral analysis of plasma etch processes, combined with intelligent data analysis (e.g., principal component analysis, PCA) allows the characterization of plasma processes in production lines. As discussed above, some potential benefits are determination of chamber cleanness, seasoning effects and detection of faults [28, 29].

18.7   Plasma process-induced damage

Dry etching techniques can result in damage and contamination of the materials used in the device structures. This damage becomes more of a problem as line widths continue to shrink. In reality, many different types of damage have been attributed to plasma etching, ranging from physical damage to the silicon lattice to degradation of minority carrier lifetime. Contamination from etching, caused, e.g., by metal impurities from the etch chamber, by polymer-based residues, and even by hydrogen, is also considered to be damage [30]. In this section, we focus on two types of damage:

(i) damage as a result of changes in substrate characteristics, caused by interaction of plasma-induced particles or radiation with surface atoms,

(ii) charge build-up damage.

18.7.1   Contamination effects

Damage is often inherent in dry etch processes. The ion bombardment can cause bonding damage in semiconductors and insulators. The presence of ultraviolet (UV) radiation is an additional cause of such damage in insulating materials. Contamination resulting from the etch process is also regarded as damage. It can be caused by metallic impurities from the etch chamber, by the presence of residue layers formed by reactant species, and by reaction products.

A general model of silicon surfaces after a conventional RIE process has been proposed [31]. It describes the damage and contamination effects on silicon substrates during contact hole etching in SiO_2. The damage produced can be considered to lie in three layers: a residual layer on the top of the substrate, followed by a heavily lattice-damage film, and a region of light substrate damage. The near surface region can be permeated by impurities, e.g., by carbon and hydrogen.

18.7.2   Charging damage

Plasma etching also offers an increased damage potential because of surface charging of floating gates in MOS devices. With the continuing decrease in gate oxide thickness to improve device performance, this type of damage is becoming more of a concern. The damage can degrade all the electrical properties of a gate oxide. Consequently, all the MOS parameters which depend on the oxide properties can be degraded by charging [32].

As shown in Fig. 18.14, the factors affecting this type of damage can be classified as plasma issues, device structure, and oxide quality. The major problem in characterizing and controlling charging damage has been that the damage changes when any of these three factors is changed. The primary cause of oxide damage during the etch process is charge build-up on conductors. Experiments have shown that plasma non-uniformities across the wafer surface play a major role in this damage. Plasma inconsistencies are caused by hardware or by a poor choice of process conditions. Additional causes include transient surge currents produced by gas chemistry changes in the over-etch step, and changes in plasma exciting power [33]. Plasma non-uniformities produce electron and ion currents that do not balance locally and can

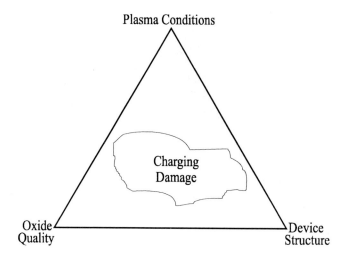

Figure 18.14: The components determining charging damage [32].

generate oxide damage. Correlations were detected between charge build-up and device yield: higher charge build-up results in lower yield [34].

In general, there are three trends in the attempts to get the problem of *contamination/damage* under control: First, there has been a great deal of discussion about alternative plasma sources, such as ECR plasma etchers or ICP sources. These sources seem to reduce at least some types of damage. On the other hand, the potential for charging damage may increase with these new sources due to the higher ion fluxes, UV photon, and x-ray fluxes [35]. A second opportunity for damage reduction may be to choose a cleaner, simpler etch chemistry. This means a renunciation of passivating gases for profile control. The change from conventional gas mixtures to new bromine or iodine etch chemistries has the disadvantage that these gases are often extremely corrosive. At present, it does not appear that the SiO_2 etch process would benefit from the use of alternative gas chemistries. However, the search for suitable etch gases with a smaller ozone depleting potential continues. The third approach to damage control is based on the fact that dry etch damage may not necessarily cause device damage, i.e., a degradation in performance of the final electronic device. It depends not only on the device structure but also on the process technology. Hence, the removal of dry etch damage in subsequent annealing processing steps is a promising procedure in technological practice. However, the large number of dry processing steps in integrated circuit manufacturing preclude the possibility of following each critical dry etching process with a damage healing step. This is not compatible with the process complexity and drastically increases the undesirable thermal stress of future low thermal budget integrated circuits. Moreover, such annealing steps are only effective if the contamination layer has to be removed. Thus, surface cleaning plays an important role in semiconductor fabrication and processing. However, there is no universal post-etch cleaning treatment for damage elimination; only optimized solutions with regard to the special technological demands and objectives are realizable [23].

18.8   Future outlook

In Ref. [5] today's top etch challenges are summarized in ten points since the success of an etch process in deep-submicron technology is measured by ten parameters:

1. critical dimension uniformity,

2. selectivity,

3. high etch rate,

4. etch profile control,

5. low damage,

6. sidewall passivation,

7. residue,

8. stringers (unwanted *decorations* after etching),

9. corrosion (in metal etch),

10. particle control.

As dimensions shrink and aspect ratios grow, dry etch technologies will need to realize complicated trade-offs in order to optimize the mentioned key process parameters. The ability to do so successfully will depend on the width of the process window achieved in current new high density etch systems [5]. It is difficult to forecast what future plasma etch systems will look like. However, it seems clear that they will be multiple chamber systems operating at low pressure with sophisticated diagnostic capabilities.

Acknowledgment

The authors wish to thank Sebastian Dietrich, Steffen Marschmeyer, and Judy Marquardt for their help in the preparation of the manuscript. They are also grateful to Dirk Knobloch from Infineon Technologies München, and Massud-A. Aminpur from AMD Dresden for stimulating discussions of process control and new dry etch chemistries.

References

[1] A.J. van Roosmalen, J.A.G. Baggerman, S.J.H. Brader, Dry Etching for VLSI, Plenum Press: New York and London (1991)

[2] F.H. Bell, Thesis, Université de Nantes (1996)

[3] M.P. Seah, W.A. Dench, Surface Interface Anal. **1**, 1 (1979)

[4] S.M. Rossnagel, J.J. Cuomo, W.D. Westwood, Editors, Handbook of Plasma Processing Technology, Noyes Publications, Park Ridge, New Jersey (1989)

[5] P. Singer, Semiconductor International **19**, 152 (1996)

[6] K.-H. Hellwege, O. Madelung, Editors, Landoldt-Börnstein: Vol. 17c Semiconductors, Technology of Si, Ge and SiC, Springer-Verlag, Berlin (1984)

[7] D. Widmann, H. Mader, H. Friedrich, Technologie hochintegrierter Schaltungen, Springer-Verlag Berlin (1996)

[8] G. Franz, Oberflächentechnologie mit Niederdruckplasmen, Springer-Verlag Berlin (1994).

[9] N.G. Einspruch, D.M. Brown, Editors, VLSI Electronics Microstructure Science: Vol. 8 Plasma Processing for VLSI, Academic Press Orlando, Florida (1984)

[10] B. Chapman, Glow Discharge Processes: Sputtering and Plasma Etching, John Wiley & Sons, New York (1980)

[11] Y.X. Li, Thesis, Technische Universiteit Delft (1995)

[12] D.L. Flamm, Solid State Technol. **34**, 47 (1991)

[13] S. Samukawa, Materials Science Forum **140/142**, 521 (1993)

[14] P. Burggraaf, Semiconductor International **17**, 56 (1994)

[15] M.-A. Aminpur, Thesis, Universität Hannover (1998)

[16] L. Peters, Semiconductor International **15**, 66 (1992)

[17] P.M. Kopalides, Thesis, University of Rochester, New York (1993)

[18] P. Singer, Semiconductor International **16**, 50 (1993)

[19] R.N. Carlile, V.C. Liang, M.M. Smadi, Solid State Technol. **32**, 119 (1989)

[20] H.H. Richter, M.-A. Aminpur, H.B. Erzgräber, A. Wolff, D. Krüger, A. Dehoff, M. Reetz, Jpn. J. Appl. Phys. **36**, 4849 (1997)

[21] A.J. van Roosmalen, Vacuum **34**, 429 (1984)

[22] P. Singer, Semiconductor International **20**, 109 (1997)

[23] H.H. Richter, A. Wolff, K. Blum, K. Höppner, D. Krüger, R. Sorge, Vacuum **47**, 437 (1996)

[24] G.S. Oehrlein, Y. Zhang, D. Vender, O. Joubert, J. Vac. Sci. Technol. **A12**, 333 (1994)

[25] K. Miyata, M. Hori, T. Goto, J. Vac. Sci. Technol. **A15**, 568 (1997)

[26] P.E. Riley, S.S. Peng, L. Fang, Solid State Technol. **36**, 47 (1993)

[27] I.T. Jolliffe, Principal Component Analysis, Springer-Verlag Berlin (1986)

[28] F.H. Bell, D. Knobloch, A. Steinbach, J. Zimpel, 8. Bundesfachtagung Plasmatechnologie, Dresden (1997)

[29] D. Knobloch, F.H. Bell, A. Steinbach, J. Zimpel, 45th International AVS Symposium, Baltimore (1998)

[30] P.H. Singer, Semiconductor International **15**, 78 (1992)

[31] S.J. Fonash, J. Electrochem. Soc. **137**, 3885 (1990)

[32] J.P. McVittie, 1st International Symposium on Plasma Process-Induced Damage, Santa Clara (1996), p. 7

[33] C.T. Gabriel, J.P. McVittie, Solid State Technol. **35**, 81 (1992)

[34] C.T. Gabriel, SPIE: Microelectronic Processes, Sensors and Controls, Bellingham (1993), p. 239

[35] P.K. Gadgil, T.D. Mantei, X.C. Mu, J. Vac. Sci. Technol. **B12**, 102 (1994)

19   Low-temperature plasmas for polymer surface modification

Jürgen Meichsner

Institut für Physik, Ernst-Moritz-Arndt-Universität Greifswald, Domstraße 10a
D-17487 Greifswald, Germany

19.1   Introduction

Low-temperature plasmas are widely used for surface processing of polymers like etching and patterning, or chemical modification of a thin surface layer in the fields of adhesion, biomaterials, barrier materials, optical or protective coatings and selective interaction with surrounding media [1]–[3]. Applying the low-temperature plasma of an electrical gas discharge in reactive, molecular gases, the discharge has to be described by a large number of elementary processes in the bulk plasma, in the transition region to the material surface, and on the material surface. In the bulk plasma, the electron-neutral collisions in a molecular process gas generate many fragment ions and reactive neutrals which can produce new chemical compounds in the gas and at the surrounding surface [4]. Therefore the diagnostics of the plasma and the plasma sheath in front of the material surface together with the characterisation of the thin surface layer of the immersed polymer material are of fundamental interest for a better understanding of complex plasma-surface interactions [5]. In this paper the interaction of different plasma species like ions, electrons, neutrals, and photons is taken into consideration, and their physical and chemical effects on the polymer material will be investigated. That means different plasma diagnostic methods as well as surface and thin film analysis must be used to investigate the plasma-polymer interaction.

19.2   Low-temperature plasma and plasma-polymer interaction

19.2.1   Characterisation of low-pressure electric gas discharges

The investigation of the plasma-polymer interaction were performed using a 50 kHz middle-frequency discharge or an unconfined capacitively coupled radiofrequency (rf) discharge at a frequency of 13.56 MHz. The 13.56 MHz rf discharge arrangement consists of two circular stainless steel electrodes (diameter 9 cm) with variable separation between 1 and 10 cm, as shown in figure 19.1. By changing the connections to the two electrodes externally, the upper and the bottom electrode act as powered and grounded electrode, respectively. The applied peak to peak voltage (V_{PP}) amounted to 1000 V at a total pressure between 5 and 100 Pa. In

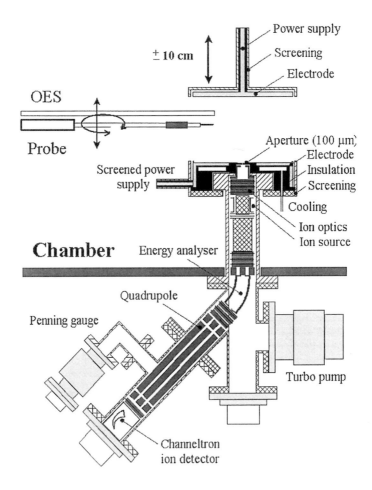

Figure 19.1: Scheme of the rf discharge arrangement with integrated energy selective mass spectrometer, plasma diagnostics using electrical probe, and optical emission spectroscopy.

all cases the powered discharge electrode was smaller than the effective area of the grounded discharge electrode. This results in an asymmetric discharge with a strong negative direct current (dc) self-bias voltage (V_{SB}) at the powered electrode. The self-bias voltage V_{SB} is coupled to the applied rf voltage and can have its maximum value at half of the peak-to-peak rf voltage V_{PP}. The flow and kinetic energy of charged particles impinging on the discharge electrodes is controlled by the plasma sheath in front of the surface. Especially the self-bias voltage and the strong sheath potential modulation, see figure 19.2, as well as collision processes (charge transfer, elastic collisions) influence the ion transit to the powered rf electrode and the resulting ion energy distribution function at this electrode.

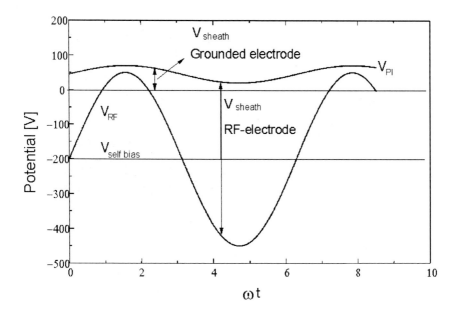

Figure 19.2: Characteristic plasma potential and sheath voltage at the powered and grounded electrode of the capacitively coupled asymmetric rf discharge.

$$V_{SB} \leq \frac{1}{2}V_{PP}, \qquad (19.1)$$

$$V_{RF} = V_{SB} + \frac{1}{2}V_{PP}\sin(\omega t), \qquad (19.2)$$

$$V_{Pl} = V_{fl} + \frac{1}{2}(V_{SB} + \frac{1}{2}V_{PP}) \times (1 + \sin(\omega t)), \qquad (19.3)$$

where V_{Pl} and V_{fl} are plasma and floating potential, respectively. In dependence on the total pressure and the peak-to-peak voltage the powered electrode is bombarded with energetic positive ions of some hundred eV. At the grounded electrode the sheath voltage remains almost constant at low values. The maximum kinetic energy of the ions is in the order of a few 10 eV. The positive ion spectrum and ion energy distribution function were measured by extracting the ions through a 100 μm aperture plate, integrated in the bottom electrode, into an energy selective mass spectrometer (EQP 300 HIDEN Analytical [6], see figure 19.1). The ion energy distribution functions of O_2^+ measured at the powered discharge electrode are shown in figure 19.3. At low pressure the well-known saddle shaped and multiple peak structures can be observed in the ion energy distribution function on the powered electrode; they have their origin in sheath potential modulation and charge transfer collisions.

Additionally, electrical double probe measurements and optical emission spectroscopy were used to characterise the rf plasma. The axial plasma ion density profile of the oxygen rf plasma shows a dominant density maximum of a magnitude between 10^9 and 10^{10} cm^{-3}

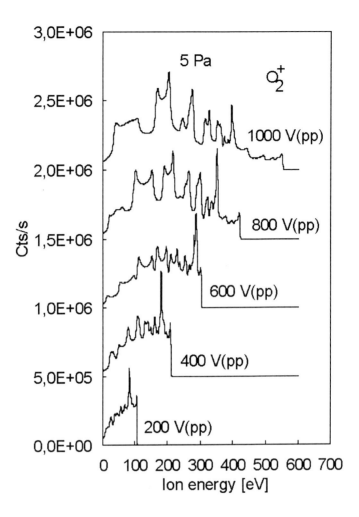

Figure 19.3: Ion energy distribution functions of the molecular oxygen ion (O_2^+) in a pure oxygen rf discharge at 5 Pa. For clarity, the curves are shifted upwards. Parameter is the peak-to-peak voltage $V(pp)$.

near the powered electrode, see figure 19.4. The estimated electron temperature (Maxwell approximation) leads to values between 1 and 2 eV that did not change significantly within the measured axial positions. Using the SPEX monochromator 750M together with a charged coupled device (CCD) detector the optical emission spectrum of the gas discharge was studied. In the oxygen rf plasma the emission intensity of the atmospheric band at 760 nm from which the rotational temperature is calculated was measured (figure 19.5). In dependence on the peak-to-peak voltage the temperatures range between 400 and 550 K in steady state [5].

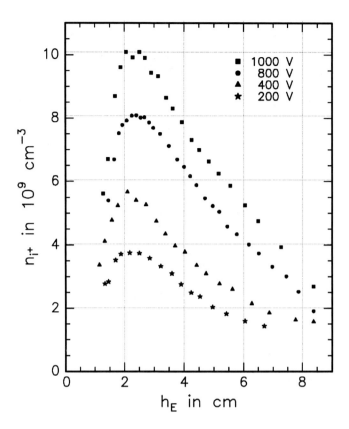

Figure 19.4: Axial profile of the O_2^+ ion plasma density of an oxygen rf discharge, where h_E is the distance from the powered electrode ($p = 5$ Pa, parameter is the peak-to-peak voltage).

Generally, the rotational temperature can be used as a good approximation for the gas temperature. The vacuum ultraviolet (VUV) emission ($\lambda < 200$ nm) of gas discharges are of special interest for polymer surface modification by photochemical reactions. In figure 19.6 the corresponding spectra of H_2, N_2, O_2, and Ar rf discharges are plotted taken at the same discharge parameters. The cut-off in the spectrum is determined by the MgF_2 window of the spectrometer (Acton Research VM–504). It is clearly seen, that in hydrogen discharges the VUV radiation (e.g., Lyman–α at 121.6 nm from H atoms, and the H_2 continuum) is significantly higher in comparison with the discharges in the other process gases. In the rf discharge described above, the polymer surface modification was studied for samples placed on both, the powered and the grounded discharge electrode which involves significantly different ion bombardment in the surface treatment. The investigation of the polymer surface modification in the plasma bulk were carried out in a gas discharge at 50 kHz excitation frequency. The discharge was maintained between two parallel circular electrodes of 10 cm in diameter and a separation of 3.5 cm. Here, the discharge voltage at the electrodes (up to 1000 V peak-to-

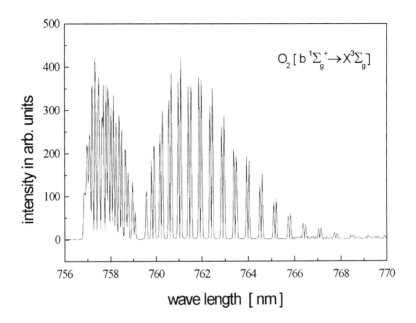

Figure 19.5: Rotationally resolved emission spectrum of the atmospheric band in the rf oxygen discharge.

peak) was applied symmetrically against ground potential. The effective discharge current amounted to 20 mA at a total pressure of 20 to 30 Pa. In this case a well confined plasma region in the space between the electrodes was realised. The 50 kHz discharge can be described by a quasi-stationary direct current (dc) discharge. The bulk plasma consists of the negative glow. In this region the plasma density should weakly depend on time because of the loss of charged particles by ambipolar diffusion to the walls, and the low recombination per time unit in respect to the period time of the used excitation frequency. On the other hand, the space charge region in the cathode layer alternates between the two discharge electrodes with excitation frequency.

Electrically insulated samples immersed in the bulk plasma between the discharge electrodes are charged on floating potential (V_{fl}). The current density j depends on the plasma parameters (plasma density, electron temperature, Bohm criterion). The ion kinetic energy as estimated from the difference of the plasma potential and the floating potential (see Fig. 19.2),

$$V_{Pl} - V_{fl} = V_{sh} + \Delta V_{Bohm} , \tag{19.4}$$

where

$$\Delta V_{Bohm} = \frac{1}{2} \frac{k_B T_e}{e} , \tag{19.5}$$

and where the sheath potential

$$V_{sh} = \frac{1}{2} \frac{k_B T_e}{e} \ln \left(\frac{m_+}{2\pi m_e} \right) \approx 3 \ldots 4 \times \frac{k_B T_e}{e} \tag{19.6}$$

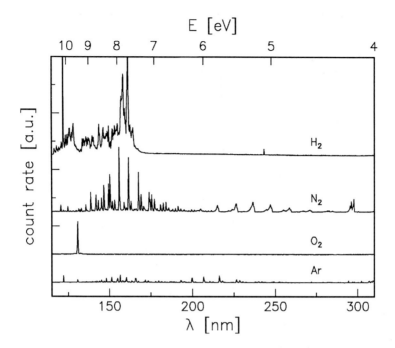

Figure 19.6: A comparison of the VUV spectra of rf discharges in different gases at the same process parameters ($p = 15$ Pa, $V_{PP} = 425$ V).

remains lower than 10 eV. Here m_e and m_+ are electron and ion mass, respectively, e is the electron charge, T_e the electron temperature, and k_B the Boltzmann constant. Besides the plasma physical description of the discharge, the production of chemically reactive atomic or molecular species and the plasma chemical gas conversion must be taken into consideration. The neutral reactive species (for example atomic oxygen or fluorine) that diffuse to the surface may effectively contribute to the polymer surface modification as well as the charged particles and photons. Moreover, the chemically active radicals with sufficiently long live times can diffuse to the reactor walls and react with the polymer material remote from the active plasma region.

The polymer surface modification was studied in detail by use of electric gas discharges in different gases like Ar, H_2, O_2, N_2 and CF_4.

19.2.2 Plasma species and expected effects in polymer surface treatment

The different plasma species like charged particles (ions and electrons), reactive neutrals and photons contribute in different ways and efficiencies to polymer surface modification. The ions or fast neutrals with kinetic energies in the range of some hundred eV can modify the polymer material within a thin surface layer. The penetration depth of such ions estimated by TRIM simulations is limited to a few nanometers [7]. As a result, the sputtering of surface material is either connected with the breaking-up of chemical bonds and cross-linkings, or the chemical reaction of ions with the polymer material, by which new functional groups and low molecular compounds are produced. On the other hand, ions with low kinetic energy will interact with the top monolayer only. These ions that form the low energy part of the ion energy distribution function either stem from collision dominated ion transport in the sheath in front of the powered electrode or result from ions crossing the plasma sheath in front of samples that are placed on the grounded electrode or on a floating potential in the bulk plasma. In dependence on the plasma gas and the polymer material those ions are responsible for the desorption of low molecular products, or react chemically with polymer material.

Electrons and negative ions should have no significant influence for polymer surface modification because of the (with respect to the plasma potential) negatively charged polymer surface. The negatively charged particles are retarded to low kinetic energies (electrons), or they are repelled from the surface and trapped in the plasma bulk (negative ions).

The reactive neutrals diffuse to the sample and can be adsorbed on the polymer surface. In dependence on the reactivity of the neutrals to the polymer material, new chemical compounds are formed (oxidation, fluorination) at the surface.

The plasma radiation, especially in the VUV region, may have influence on the polymer modification due to photochemical reactions, e.g., hydrogen abstraction, formation of free macroradicals and cross-linking. In particular, the used discharge type and the nature of the gas must be taken into consideration. The typical penetration depth of the VUV photons, depending on the absorbance of the polymer materials is estimated to be in the order of some 10 nm [8, 9]. Generally, the plasma surface modification by energetic photons in the investigated gas discharges requires longer treatment times (some ten minutes or more) in respect to typical time constants in the order of seconds for surface modification by particle interaction. In table 19.1 the main groups of plasma species in low-temperature plasmas are listed, and their expected characteristic penetration depth and efficiency on polymer surfaces are specified.

19.2.3   Methods for characterisation of plasma treated polymers

In situ surface and thin film diagnostics were applied for the study of plasma treatment of samples placed on the discharge electrodes and immersed in the bulk plasma on floating potential. According to the penetration depth of the different plasma species and the resulting changes in atomic composition and molecular structure in a surface layer of a few nm thickness, sufficiently sensitive techniques must be applied to study the plasma-polymer interaction. The techniques are partially integrated within the process chamber. Generally, no single diagnostic tool will satisfy all requirements. That means a combination of different methods will provide results that contribute to a better understanding of the plasma-surface interaction. In most cases an appropriate sample preparation is required in order to apply the chosen diagnostic tool.

To characterise the chemically modified surface layer of polymer materials two *in situ* Fourier transform infrared (FTIR) sampling techniques were applied, the infrared reflection absorption (IRRAS), and the attenuated total reflection (ATR) spectroscopy [10]. These techniques are used for analysing the molecular structure of specially prepared thin polymer films.

Table 19.1: Plasma components and their efficiency in polymer modification.

Plasma component	Kinetic energy	Processes and effects in the polymer material	Depth of interaction
Ions, fast neutrals	100–500 eV	Elastic collisions, sputtering, incorporation, chemical reactions	2–5 nm
	∼ 10 eV	Adsorbate sputtering, chemical reactions	Monolayer
Electrons	5–10 eV	Inelastic collisions, surface dissociation, surface ionisation	∼ 1 nm
Reactive neutrals	Thermal, 0.05 eV	Adsorption, chemical surface reactions, formation of functional groups, low molecular (volatile) products	Monolayer
		Diffusion and chemical reaction	bulk
Photons	> 5 eV (VUV)	Photo chemical processes	10 nm – 50 nm
	< 5 eV (UV)	Secondary processes	μm range

In IRRAS experiments the infrared (IR) beam of the spectrometer is reflected by a metal surface coated with the polymer material. It has been shown that a maximum absorption of a
given film thickness $d \ll \lambda$ requires parallel polarisation and grazing incidence [11]. The
IRRAS set-up, used for the experiments, consists of a vacuum chamber which contains two
parallel plate electrodes. This vacuum chamber is integrated in the standard compartment of a
commercial FTIR spectrometer (BRUKER). The IR beam of the spectrometer is deflected by
$15°$ using prismatic windows made of chalcogenide glass. The resulting incident angle at the
IRRAS sample placed at the electrode is $75°$.

For ATR measurements the IR beam was coupled to an internal reflection element (IRE)
with an incident angle of $45°$ and 34 reflections. The technique of ATR spectroscopy (evanes-
cent wave spectroscopy) is described in detail in Ref. [12]. The IRE plate is made of chalcogenide glass and is integrated into the rf electrode of the process chamber [10, 13]. One side
of the IRE was coated with the sample by a dip coating procedure. The effective number
of reflections in the sample is restricted to 13. The used ATR technique for film thicknesses

<100 nm permits to apply the thin film case of ATR spectroscopy (d $\ll$ λ). In this case the polymer film can be plasma-treated from one side and probed by FTIR spectroscopy from the other side.

The described FTIR techniques are *insitu*, that means the signal sampling is made in the process chamber. But these techniques are not real time techniques because of the relatively high sampling time needed for high enough signal intensity, in comparison with the characteristic time constants for the change of the polymer surface properties. Therefore, the pulsed mode plasma treatment had to be applied, and the FTIR spectrum was taken in the pulse pause.

The change in the polarisation state of light reflected non-normally from a sample surface during the ellipsometric measurement gives information about thickness and refractive index of the plasma modified polymer surface (see chapter 9). The reflection properties of a plane substrate and deposited and plasma modified layers with plane interfaces are characterised by the complex reflection coefficients R_p and R_s (p and s describe the reflection for the parallel and perpendicular electric field vector to the plane of incidence). They include the Fresnel reflection coefficients at all interfaces, and multiple interference by the layers on the substrate [14]. The shape of the polarisation ellipse depends on the ratio R_p/R_s which is related to the ellipsometrically measured angles Δ and Ψ. Generally, the ratio R_p/R_s is a transcendental function of the film thickness, complex refractive index of the films, substrate and ambient conditions, as well as of the angle of incidence and the wavelength. Principally, the interpretation of the measured angles Δ and Ψ in relation to the refractive index and the film thickness depends strongly on the chosen optical model (substrate, one–layer, multi–layer) and has to fit the physical situation. It has been shown by model calculations that ellipsometry is very sensitive, if the substrate material, the angle of incidence, and the wavelength region are chosen properly [15, 16].

The *in situ* ellipsometric measurements are carried out at a wavelength of 632 nm and an angle of incidence of 70° by means of a commercial ellipsometer (PLASMOS) which is adapted to the plasma process chamber. In this arrangement the polarised light was reflected from a sample which was placed on the powered electrode of a 13.56 MHz discharge. The ellipsometric angles can be measured in real time during the plasma processing with a temporal resolution of about 1 s.

The film thickness and the dispersion relation of the refractive index are determined *ex situ* by use of a spectroscopic ellipsometer (RUDOLPH RESEARCH, wavelength range 300–800 nm).

The mass change of polymer samples was measured in real time with a time resolution of about 1 s by means of an electronic vacuum microbalance (SARTORIUS). The microbalance was placed into the plasma chamber [17]. The freely suspended sample was arranged symmetrically between the parallel plate discharge electrodes of the 50 kHz gas discharge. At a maximum sample mass of < 2 g a total mass change $\geq$ 1 μg can be measured during the plasma treatment.

Although the obtained information addresses the mass change only, the real-time measurements give information about the first steps in plasma modification on the top surface layer, if the used sample surface is large enough. That means that it is possible to study the concurrence between the incorporation of atoms and molecules from the process gas and the etching processes. This microgravimetric technique needs no special sample preparation and can be used for investigations of technical polymer foils, fibers or fabrics.

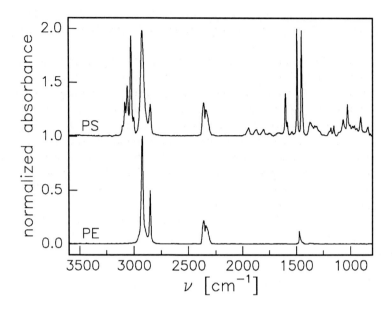

Figure 19.7: FTIR absorption spectrum of thin polyethylene (PE) and polystyrene (PS) films, prepared by dip coating.

19.2.4 Polymer samples and thin film preparation

Synthetic polymers, like polystyrene and low-density (LD) polyethylene foils without any additives, were used as received from Goodfellow Co. Special thin polymer films have been prepared from polymer solutions (polyethylene–toluene and polystyrene–chloroform) by dip or spin coating procedures. The measured FTIR spectra of the prepared thin polymer films show the typical asymmetric and symmetric stretching vibrations of the CH_2 group at 2900 cm^{-1} and the deformation vibration at 1450 cm^{-1} (Fig. 19.7). For polystyrene (PS), the characteristic C–H stretching vibrations of the aromatic ring structure at 3050 cm^{-1} appear in addition to the stretching vibrations of the aliphatic structure, as well as the two strong ring deformation vibrations at 1500 cm^{-1}.

The total polymer film thickness of less than 50 nm reached by dip coating was suitable for the characterisation of plasma treatment by the FTIR spectroscopy and ellipsometry, due to the appropriate thickness ratio between expected modified surface layer and untreated polymer material.

The plasma treatment of polyethylene and polystyrene foils as well as of fabrics from natural polymers like wool and cellulose fibers (linen) were studied by *in situ* microgravimetry. A scanning electron (SEM) micrograph of a sample of wool fibers as typically used in the experiments is shown in figure 19.8.

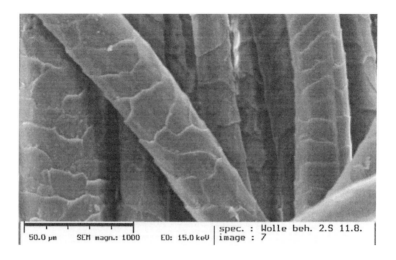

| | | | spec. : Wolle beh. 2.S 11.8. |
| 50.0 µm | SEM magn.: 1000 | EO: 15.0 keV | image : 7 |

Figure 19.8: SEM micrograph of a fabric with wool fibers as used in the experiments.

19.3 Plasma modification of polyethylene and polystyrene

The chemical surface modification of polyethylene films by plasma treatment in different process gases is presented in figure 19.9. The difference spectra (untreated against plasma treated polymer film) show the changes in the chemical structure of the polymer film, and the degradation of the polymer. The positive peaks indicate newly formed compounds, and the peaks negative with respect to the base line the corresponding removed or destroyed molecular structures of the polymer. The bands that characterise the CH_2 structures of polyethylene are reduced by hydrogen loss and polymer degradation due to plasma etching. In particular, after the plasma treatment in oxygen or tetrafluoromethane the spectra represent new functionally groups assigned to CO, OH or CF_x absorption bands. The plasma treatment of polystyrene shows quite similar results in the FTIR analysis. Additionally, the degradation of the aromatic ring structure is observed. Especially in hydrogen discharges, new absorption bands were observed in the range of C–H stretching vibration (2900 cm^{-1}). This new bands should have their origin in cross-linking and/or destroyed aromatic rings due to the plasma treatment.

Generally, with increasing plasma treatment time a saturation effect in the formation of functional groups is observed. For the used discharge parameters the oxidation of the polymer surface during plasma treatment typically is finished in less than 5 s treatment time, whereas in tetrafluoromethane plasmas the time for fluorination is about one order of magnitude higher. Simplified, the polymer surface modification can be described by two processes:

(i) the incorporation of plasma species and formation of new functional groups, and

(ii) the degradation of the polymer material by hydrogen abstraction, formation of volatile products, sputtering, and ion assisted desorption of low molecular products.

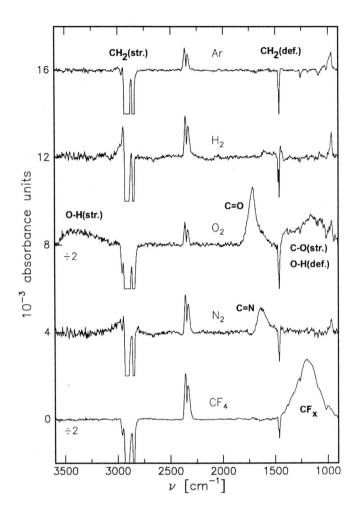

Figure 19.9: Difference spectra of plasma treated thin polyethylene films in rf plasmas with different gases (FTIR–ATR-analysis, samples on the powered discharge electrode, $p = 5$ Pa, $V_{PP} = 635$ V, $t = 3$ s).

In steady state both processes occur simultaneously and lead to a nearly constant modified surface layer, thickness a few nm, and a continuous material loss that reduces the total polymer film thickness. This dynamics of the polymer surface modification at the beginning of CF$_4$ plasma treatment can be shown by *in situ* measurements of the mass change of the polyethylene sample, placed in the plasma bulk on floating potential (Fig. 19.10).

At the beginning of the plasma treatment, the sample mass of the polyethylene foil increases by incorporation of fluorine. After formation of a thin fluorinated surface layer the

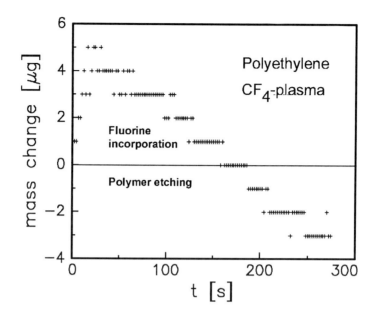

Figure 19.10: Mass change of a polyethylene foil in the plasma bulk of the 50 kHz discharge ($p = 30$ Pa, discharge current $I = 20$ mA).

total sample mass is reduced continuously with time by plasma etching. In the steady state the incorporation of plasma species and etching processes are observed simultaneously.

Taking into account this model for the interpretation of ellipsometric measurements, the two-layer optical model (plasma modified surface layer and underlying unmodified polymer material) can be applied successfully. As a result, the plasma modified surface layer could be separated with the expected thickness of a few nanometer and a changed refractive index. For example, the oxygen plasma treatment of thin polyethylene films results in an increased re-fractive index for the modified surface layer of about 3 nm in thickness (Fig. 19.11), whereas in tetrafluoromethane plasma a diminished refractive index is observed. The thickness of the modified surface layer in both cases was in the same order. The polymer degradation and etching is influenced by the ion flux to the sample. This can be demonstrated by comparison of oxygen plasma treatment of polystyrene samples placed on the powered or on the grounded electrode (Fig. 19.12). At the same discharge parameters the FTIR difference spectra indi-cate the stronger polymer degradation at the powered electrode, but the formation of oxygen containing groups (C=O, OH) is observed in similar way.

The neutral gas mass spectra offer conformable results by the various intensities of typi-cal fragment ions of the etch products during the polymer treatment on the powered and the grounded electrode of the rf discharge.

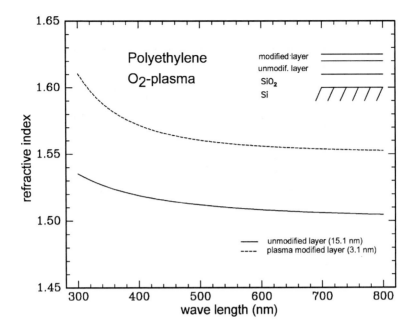

Figure 19.11: Dispersion of the refractive index of the plasma treated polyethylene film in rf oxygen plasma ($p = 30$ Pa, $P = 30$ W, $t = 10$ s). The separated plasma modified surface layer of 3.1 nm thickness is characterised by a higher refractive index with respect to the untreated bulk material.

19.4 Plasma modification of wool and cellulose fabrics

The plasma modification of fabrics from natural or synthetic polymer fibers were investigated by many groups with more or less intensity since about 20 years [18]. The activities were stimulated by the possibilities to create new interesting surface properties for applications to technical or clothing textiles (wettability, coloring, surface conductivity, adhesion, ...) and by environmental aspects, e.g., replacement of the traditionally used wet chemical methods by dry plasma techniques. Currently, the plasma treatment of fabrics preferably is done with gas discharges at atmospheric pressure (corona discharge, silent or barrier discharge, see chapter 13). Nevertheless, low-pressure discharges in chemically reactive gases and vapours imply great technological potentialities in the plasma surface modification. Especially, the more controlled plasma treatment and low material insert are advantages in low-pressure technologies, though the vacuum technical equipment is necessary.

For investigation of the plasma modification of fabrics the symmetric 50 kHz low-pressure gas discharge was applied. The plasma modification is studied by *in situ* microgravimetric measurement, as described above. In the experiments the particle incorporation and etching process on the polymer surface were observed in a way comparable to the FTIR analysis of synthetic polymer films. The mass changes of the wool fibers in argon, oxygen or tetrafluoromethane plasma indicates the different physical and chemical efficiency of the plasmas with

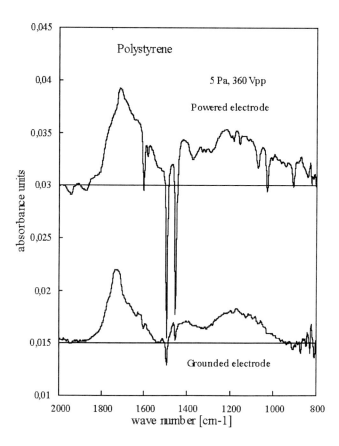

Figure 19.12: Comparison of FTIR difference spectra of rf oxygen plasma treated polystyrene samples on the powered and grounded discharge electrode ($p = 5$ Pa, $V_{PP} = 360$ V). The significant stronger polymer degradation at the powered electrode is indicated by the negative bands of the deformation vibration of the aromatic ring structure at 1500 cm^{-1}.

respect to the polymer material. Whereas in argon discharges the degradation of the wool fibers is characterised by a low etch rate, the production of volatile compounds by plasma chemical etching in oxygen plasmas results in a significantly higher loss of sample mass. In the tetrafluoromethane plasma, the incorporation of fluorine (e.g., by the exchange of hydrogen by fluorine) dominates at the beginning of the plasma treatment, shown by the increasing sample mass as displayed in figure 19.13. After the formation of a (quasi-stationary) modified surface layer the total mass is reduced continuously. The particle incorporation at the beginning of the plasma treatment could be well separated by microgravimetric measurements because of the longer fluorination time of the polymer surface and the relatively low etch rate of the polymer material. In comparison, samples placed in the active plasma region of an

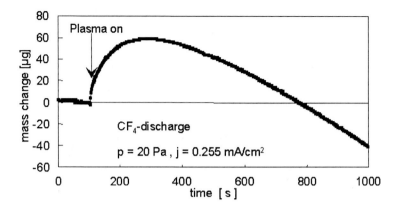

Figure 19.13: Fluorine incorporation and plasma etching of a wool fabric during the treatment in a 50 kHz tetrafluoromethane discharge ($p = 20$ Pa, discharge current $I = 20$ mA, sample area 2.25 cm^2).

oxygen discharge are oxidised within a short time (a few seconds) and the plasma etching, assisted by ions, dominates at the beginning, see figure 19.14. On the other hand, the oxygen remote plasma treatment of a linen fabric, that means the treatment outside the active plasma region without strong charged particle flux, results in an increasing sample mass by oxygen incorporation (e.g., atomic oxygen) at the very beginning, see figure 19.15. The mass change in dependence on the treatment time is qualitatively similar to that of the sample in active tetrafluoromethane plasma. Due to diffusion of reactive radicals (atomic oxygen, fluorine) the

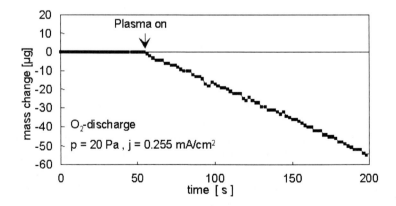

Figure 19.14: Plasma etching of a wool fabric during the treatment in a 50 kHz oxygen discharge ($p = 20$ Pa, discharge current $I = 20$ mA, sample area 2.25 cm^2).

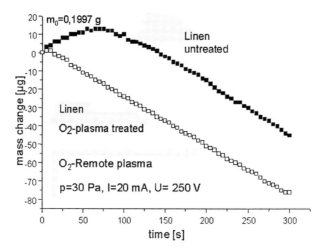

Figure 19.15: Oxygen incorporation and plasma etching of cellulose fibers (linen fabric) during the treatment remote from the 50 kHz oxygen discharge ($p = 30$ Pa, $U = 250$ V, discharge current $I = 20$ mA).

fibers on the back side and inside of the fabric can be modified also. A repeated treatment of this sample in an oxygen plasma shows a material loss from the beginning due to the already existing oxydised fibre surface.

19.5   Summary

The polymer surface treatment using the non-thermal plasma of low-pressure gas discharges provides multiple changes in a thin surface layer of the polymer material. Generally, the thermal effects due to the treatment in such cold plasmas can be neglected. The low-temperature plasma as well as the plasma sheath in front of the material surface and the physical and chemical modification of the polymer material were studied with the help of a set of different diagnostic tools (mass spectrometry, optical emission spectroscopy, electric probe measurements, Fourier transform infrared spectroscopy, ellipsometry, microgravimetry) that allow to characterise the efficiencies of different plasma species in polymer surface interaction and the physical and chemical changes of a thin polymer surface layer. For example, synthetic polymers like polyethylene and polystyrene are taken into consideration as model systems. In this case thin polymer films could be prepared in order to apply special analysing techniques for *in situ* measurements. The incorporation of plasma components in the polymer surface and the formation of new functional groups, cross-linking and degradation of the polymer, and material loss by sputtering and plasma etching were observed simultaneously. The characteristic thickness of the chemically modified surface layer is in the order of a few nanometers. That means the physical and chemical properties of the polymer bulk material remain unchanged. The modified surface layer was formed within a short plasma treatment time between 1 and

50 s depending on the nature of the plasma gas and the discharge parameters. In general, low-temperature plasmas are suitable for a sparing treatment of the polymer material. The change in atomic composition and molecular structure is connected with the change of the macroscopic properties of the surface like wetting behaviour (hydrophilic, hydrophobic, oleophobic surface) or the refractive index. The formation of free macroradicals or specific functional groups permit to initiate defined chemical reaction with molecules from the surrounding gas phase or liquid after the plasma treatment (grafting). On the other hand the natural polymer fibres like wool and cellulose in the form of fabrics can be chemically modified in low-temperature plasma, too. In this case the plasma treatment of textiles can replace partially the prevalent wet chemical steps in textile treatment. That means the low-temperature plasmas may contribute to lower expenditure of energy and of processing materials which involves a reduced strain on the environment.

References

[1] C.-H. Chan, Polymer Surface Modification and Characterisation, Hanser (1994)

[2] R. d'Agostino, Plasma Deposition, Treatment, and Etching of Polymers, Academic Press (1990)

[3] E.M. Liston, L. Martinu, J. Wertheimer, J. Adhes. Sci. Tech. 7 (1993) 1091

[4] H.-U. Poll, J. Meichsner, M. Arzt, M. Friedrich, R. Rochotzki, E. Kreyßig, Surf. Coat. & Techn. 59 (1993) 365

[5] J. Meichsner, M. Zeuner, B. Krames, M. Nitschke, R. Rochotzki, K. Barucki, Surf. Coat. & Techn. 98 (1998) 1565

[6] M. Zeuner, H. Neumann, J. Meichsner, J. Appl. Phys. 81 (1997) 2985

[7] J. Ziegler, J. Biersack, U. Littmark, The Stopping and Range of Ions in Solids, Pergamon Press: New York (1985)

[8] L.R. Painter, E.T. Arakawa, M.W. Williams, J.C. Ashley, Radiation Res. 83 (1980) 1

[9] J.G. Carter, T.N. Jelinek, R.N. Hamm, R.D. Birkhoff, J. Chem. Phys. 44 (1966) 2266

[10] M. Nitschke, PhD. Thesis, Infrarotspektroskopische Charakterisierung plasmamodifizierter Polymeroberflächen, TU Chemnitz (1996)

[11] R. Greenler, J. Chem. Phys. 44 (1966) 310

[12] N.J. Harrick, Internal Reflection Spectroscopy, Interscience: New York (1975)

[13] J. Meichsner, M. Nitschke, R. Rochotzki, M. Zeuner, Surf. Coat. & Techn. 74-75 (1995) 227

[14] W. Fukarek, Ellipsometrische Dickenbestimmung, in D. Herrmann, Schichtdickenmessung, München: Oldenbourg (1993)

[15] R. Rochotzki, M. Nitschke, M. Arzt, J. Meichsner, phys. stat. sol. 145 (1994) 289

[16] R. Rochotzki, PhD. Thesis, Ellipsometrische Untersuchungen an Plasmapolymeren und plasmamodifizierten Polymeroberflächen, TU Chemnitz (1996)

[17] J. Meichsner, N. Hille, H.-U. Poll, Wiss. Z. TH Karl-Marx-Stadt 26 (1984) 672

[18] J. Meichsner, H.-U. Poll, Acta Polymerica 32 (1981) 203

20   Plasma-enhanced deposition of superhard thin films

Achim Lunk

Institut für Plasmaforschung, Universität Stuttgart, Pfaffenwaldring 31, D–70569 Stuttgart, Germany

20.1   Characterization of superhard materials

In the following the term *superhard* will be used in the mechanical sense for a material with high mechanical hardness (Vickers hardness $HV > 40$ GPa). Currently there are two main routes for deposition of superhard thin films. The first route is characterized by plasma-enhanced deposition of materials with intrinsic high hardness. In the second route nanoscaled thin films of materials with low intrinsic hardness are deposited. The resulting hardness of the film is based on its microstructure in the form of superlattices or nanocrystalline composites (extrinsic hardness). According to Cohen and coworkers the intrinsic hardness of a material is proportional to its bulk modulus [1, 2]. They derived a semi-empirical formula for the bulk modulus B depending on the ionicity λ of the bond and the bond length d in the form

$$B = \frac{<N_c>}{4} \frac{1971 - 220\lambda}{d^{3.5}} , \qquad (20.1)$$

where B is given in GPa, $<N_c>$ is the average coordination number, and d is the bond length in Å. In diamond $\lambda = 0$ holds, while in III/V- and II/VI-compounds $\lambda = 1$ and 2, respectively. Another access to the calculation of the intrinsic hardness is given by Clerc. It is based on the screened electrostatic and elastic shear (SEES) model and allows the calculation of the hardness in terms of crystal structure geometry and intrinsic atomic-level properties [3]. From equation 20.1 and also according to the results of Clerc [3] there are only a few materials with an intrinsic hardness larger than 40 GPa. Ionicity should be low and the bond length also. These conditions are only realized in materials in the B-C-N triangle shown in Fig. 20.1. Carbon exists in two superhard modifications – in crystalline configuration as diamond and in an amorphous form as tetrahedral amorphous carbon (ta-C). The material with hardness next to diamond is cubic boron nitride (c-BN). The hardness of B_4C corresponds to 30 to 40 GPa and therefore it is often rejected to denote it as superhard.

In 1989 Liu and Cohen predicted that the crystalline form of C_3N_4 might be superhard [4]. Since then a large number of papers reported on plasma-enhanced deposition of carbon nitride (CN_x) films. Today, after more than 10 years of intensive investigations, certain doubts that plasma-enhanced synthesis of C_3N_4 thin films should be possible seem to be justified (see, e.g., Refs. [5, 6]).

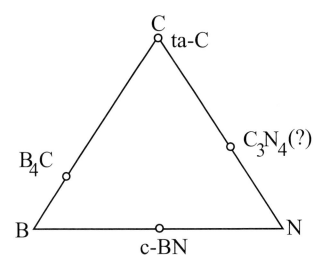

Figure 20.1: Intrinsic superhard materials in the B-C-N system.

Intensive systematic investigation on the composition and the phase inside of the C-B-N triangle started in the last five years. Characterization of boron carbonitride thin films produced by plasma-enhanced physical vapor deposition (PEPVD) are presented in Ref. [7]. From Fourier transform infrared spectroscopy (FTIR) and x-ray photoelectron spectroscopy (XPS) the authors found B–N and B–N–B bonds as well as C–C bonds but no B–C and B–B bonds. Pictures from high resolution transmission electron microscopy (HRTEM) indicate also that carbon is phase separated in a BN matrix. The chemical route of boron carbonitride thin film deposition by plasma-enhanced chemical vapor deposition (PECVD) starts with single-source or double-source precursors. As an example the reader is referred to Ref. [8] where the influence of the bias voltage as well as of the single source precursors pyridineborane and triazaborabicyclodecane on the composition of the boron carbonitride thin films is discussed. In conclusion it can be stated that thin film deposition of superhard materials is established for diamond, ta-C and c-BN. It is expected that methods of carbon nitride and boron carbonitride film deposition will be extended in the future. We will restrict ourselves in the following to the plasma-enhanced deposition of diamond, ta-C and c-BN. The hardness of thin films deposited from materials with intrinsic superhardness corresponds to the hardness of bulk material. Over the past ten years superhard multilayer films have been found deposited from materials whose intrinsic hardness is lower than 40 GPa. The multilayer consists of very thin (2–10 nm) layers mostly of nitride materials. The extreme hardness of these nanoscaled superhard materials is based on the concept that dislocation motion can be reduced in a multilayer composed material. The hardness received by this method is denoted as extrinsic hardness. A review on the state of the art in the deposition of superhard multilayers is given in Ref. [9]. The concept of multilayer structured films (superlattices) was extended to superhard nanocrystalline composite materials also in Ref. [6]. In the framework of this contribution nanoscaled (extrinsic) superhard materials will not be considered. For more details

Table 20.1: Properties of the superhard materials diamond, ta-C and c-BN in comparison with graphite and h-BN.

	diamond	ta-C	graphite	c-BN	h-BN
Lattice structure	zinc blende			zinc blende	
Lattice constant (Å)	3.567		a: 2.46 c: 6.71	3.615	a: 2.504 c: 6.660
Density (g cm^{-3})	3.515	2.47–3.1	2.27	3.47	2.28
Bond length (Å)	a: 1.54	1.52	a: 1.41 c: 3.35	1.57	a: 1.45 c: 3.33
Dielectric constant	5.5			7.1	5.5
Specific resistance (Ω cm)	$> 10^{13}$	10^{10}	a: 10^{-3} c: 1	10^{10}	a: 10^{10} c: 10^{12}
Band gap (eV)	5.45	0.65–0.9	≤ 0.04	≥ 6.2	5.2
Refractive index	2.42	2–3		2.12	a: 2.10 c: 1.75
Heat conductivity (W cm^{-1} K^{-1})	20	1–7	a: 0.1	13	a: 0.7 c: 0.008
Graphitization temperature (°C)	1073–1400 ?	200		> 1550	
Oxidation resistance (°C)	600			1200	
Vickers Hardness (GPa)	100	> 40		45–55 20 (at 1000 °C)	10
Bulk modulus (GPa)	380–440			330–400	
Young's modulus (GPa)	920–1100		a: 18.8 c: 5.2	650–900	a: 87 c: 34
Poisson's ratio	0.069–0.107			0.121–0.176	0.3
Thermal expansion (10^{-6}/K)	1.1	< 5		4 (at 700 K)	a: 2.7 c: 3.7

the reader is referred to the review articles [6, 9, 10].

In table 20.1 some properties of the intrinsic superhard materials diamond, ta-C and c-BN are summarized. For comparison the properties of the hexagonal forms of carbon (graphite) and boron nitride (h-BN) are shown too. It should be noted that only the data of ta-C refer to thin films, while the data of the other materials refer to those of bulk material. Often the data presented in the literature differ. In this cases the range of these data is presented. It is noted from table 20.1 that diamond as well as c-BN are very interesting materials for optoelectronics as well as mechanical applications.

20.2   Plasma-enhanced deposition of diamond and diamond–like carbon

In the following we summarize the methods of plasma-enhanced deposition of intrinsic superhard materials based on carbon. The methods applied can be divided somewhat arbitrarily in chemical and in physical methods. As decision parameter between chemical and physical methods the local position of the chemical reactions can be used. In the case that the chemical reactions take place mainly on the substrate surface we will denote this deposition process as a physical vapor deposition (PVD) method. The combination of chemical reactions in the volume and on the surface should be denoted as chemical vapor deposition (CVD).

20.2.1   Deposition of diamond

Before discussing different methods of deposition the basic mechanisms of diamond growth during plasma activation should be mentioned. According to currently used models, on the carbon surface competing reactions between etching by atomic hydrogen and growth by $CH_x^{\bullet}$ radicals occur. In the volume the atomic hydrogen generation as well as the formation of radicals are the most important processes. In a first step three different mechanisms in plasma-enhanced deposition of diamond can be studied separately [11]:

  (a) formation of atomic hydrogen and radicals in the volume,

  (b) nucleation processes,

  (c) diamond growth.

The volume processes can be enhanced by high temperature as well as by processes in low-temperature plasma. The gas temperature is therefore a decisive parameter in diamond deposition. Fig. 20.2 shows schematically the dependence of the diamond growth rate as a function of the gas temperature [12]. In hot radiofrequency (rf) and direct current (dc) discharges the growth rates can become very high. The deposition area in these devices is usually small. In low-pressure dc or rf discharges the growth rates become four orders of magnitude smaller. Deposition by flames, microwave discharges and hot filament is in the intermediate range. A more detailed survey on the applied methods is presented in table 20.2.

A detailed discussion of the applied methods is beyond the scope of this contribution. Here we refer the reader to several review articles and books [13, 14, 15, 16, 17]. In diamond deposition with high gas and substrate temperatures applications are restricted to temperature-insensitive substrate materials. Low-temperature growth is an essential technique required for applications in electronics and for deposition on substrates with low melting points. Reduction of gas and substrate temperature can be obtained by reduction of the pressure and the power input, i.e., typical parameters of low-pressure low-temperature plasma processing. By application of low-temperature plasma processing (LTPP) the substrate temperature could be reduced below 100 °C. By the low pressure and energy input the growth rates in LTPP are naturally low. Fig. 20.3 shows typical data of growth rates in dependence on substrate temperature and the type of LTPP used [18]. Curves labeled (a), (b), (c), and (d) represent data from hot filament CVD, microwave plasma CVD in CO/H_2, microwave plasma CVD

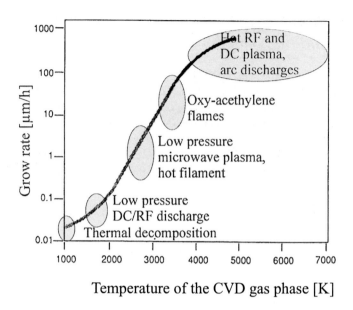

Figure 20.2: Dependence of diamond growth rate on gas temperature [12].

in $CO/O_2/H_2$, and ECR-plasma CVD in CH_3OH/H_2, respectively. Activation energies obtained from Fig. 20.3 are: (a) and (b) $E_A = 21$ kJ/mol at temperatures higher than 400 °C, while $E_A = 4$ kJ/mol below 400 °C, (c) $E_A = 29$ kJ/mol, and (d) $E_A = 21$ kJ/mol. These values are considerable lower than those of conventional high-temperature deposition ($E_A = 92$ - 100 kJ/mol) [19]. Fig. 20.3 indicates that deposition of diamond can be performed in plasma containing carbon/hydrogen and oxygen. Fig. 20.4 shows the data known for dia-

Table 20.2: Methods of diamond deposition.

Thermal CVD	dc plasma	rf plasma	microwave plasma
Thermal decomposition	low-pressure dc plasma	low-pressure rf-discharge	low-pressure plasma (0.915, 2.45, 8.2 GHz)
Chemical transport reactions	medium pressure dc plasma	thermal rf plasma CVD	electron cyclotron resonance (ECR) plasma
Hot filament technique	hollow cathode discharge		atmospheric pressure plasma torch
Oxyacetylene torch	dc arc plasma		
Halogen assisted CVD	dc plasma jet		

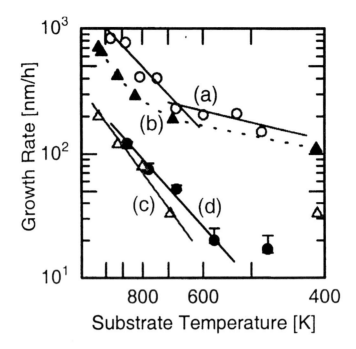

Figure 20.3: Dependence of growth rate on substrate temperature and the type of LTPP (see text) [18].

mond deposition in a summarized form (Bachmann diagram) [20]. In the figure the regions in the neighborhood of the CO-line where diamond can be grown are shown. In regions 1 and 2 oxygen-free gases were used, while in regions 3 and 4 methane/hydrogen/oxygen or methane/oxygen mixtures were applied. Higher oxygen concentrations were used in regions 5 (methane/O_2) and 6 (acetone/O_2). Region 7 represents film growth by combustion flames in C_2H_2/O_2 mixtures. Recently this diagram was refined using steady-state computations of gas-phase and gas-surface chemistry [21]. In the high-temperature regime gas phase reactions can be described approximately by equilibrium thermodynamics using for example CHEMKIN calculations [22, 23]. Other authors include heterogeneous reactions, too [24, 25, 26].

In Ref. [25] the growth rate and defect density was modeled in terms of the concentration of atomic hydrogen and $CH_3^\bullet$ at the substrate. Based on the model of Harris [27], the authors derived the following equation for the diamond growth rate $G_{<100>}$ along the <100> crystallographic direction in dependence on the atomic hydrogen concentration [H] and the CH_3 concentration [CH_3]

$$G_{<100>} = \frac{K_1[H][CH_3]}{K_2 + [H]},\qquad(20.2)$$

where K_1 and K_2 are constants which depend on the substrate temperature. The concentrations of atomic hydrogen and $CH_3^\bullet$ are calculated at equilibrium and quasi-equilibrium ther-

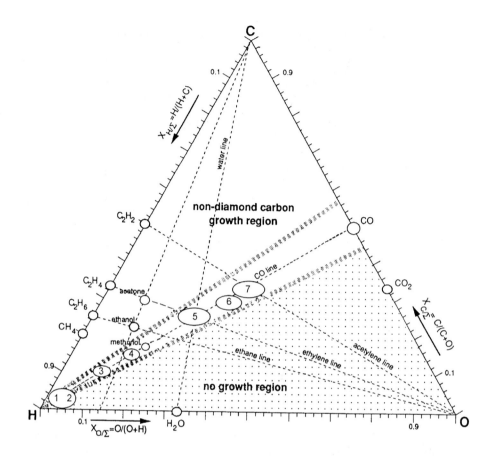

Figure 20.4: Growth domain of diamond in the C-O-H diagram [12].

modynamic conditions. In LTPP the conditions are far from equilibrium, and therefore the kinetics of the electrons and heavy particles should be taken into consideration.

In the literature only a few articles can be found which describe diamond deposition in the framework of electron kinetics in combination with particle transport equations. In Refs. [28] and [29] the electron energy distribution function (EEDF) as well as the rate and transport coefficients have been calculated by solving the stationary Boltzmann equation in $H_2/H/CH_4$ and H_2, respectively. Results show a deviation of the EEDF from a single Maxwell distribution, but can be approximated by two Maxwellian distributions. Fig. 20.5 shows the influence of the atomic hydrogen mol fraction on EEDF in a hydrogen plasma [29]. The data used are: reduced field strength $E/N = 30$ Td, vibrational temperature $T_{vib} = 2000$ K, and mol fraction of electronically excited states $c^* = 0$. It shows that the increase in the mol fraction [H] causes an increase of EEDF in the energy range higher than 3 eV. The rates of chemical reactions and of energy transfer between electrons and heavy particles are estimated using the

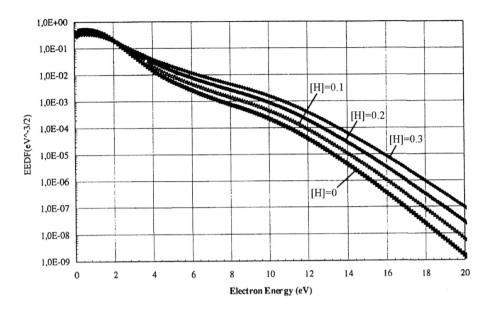

Figure 20.5: Electron energy distribution function in H in dependence on mol fraction [H] (see text) [29].

approximation by a bi-Maxwellian distribution of the electrons [29, 30]. The calculated rates together with the use of a one-dimensional transport model allow the estimation of plasma species concentrations and temperatures along the axis. Applying the model of Harris [27], the growth rate depends on the atomic hydrogen concentration [H] and the $CH_3^{\bullet}$ concentration (Eq. 20.2). Fig. 20.6 shows a satisfactory agreement between calculated and experimentally obtained growth rates [30]. At low methane percentage ([CH_4] < 2%) the model predicts a linear dependence of the growth rate on CH_4 concentration, which is confirmed by experimental results. At higher percentage of [CH_4] the experiments show a maximum at 3% and than a sharp decrease. It is assumed, that this behavior is mainly due to the enhancement of secondary nucleation and of sp^2 phase formation, which is not included in the Harris model.

Beside the modeling of low-pressure low-temperature plasma processes in diamond deposition also progress has been made in simulation and development of microwave plasma reactors for diamond deposition. In Ref. [31] a numerical model was developed which includes an electromagnetic field model as well as a fluid plasma model. The microwave electric field interaction with the plasma is described using a finite difference solution of the momentum transport equation. A simulation program was also applied to the design and optimization of microwave plasma reactors used for diamond deposition [32]. Based on these program, a novel microwave reactor was developed employing an ellipsoid cavity.

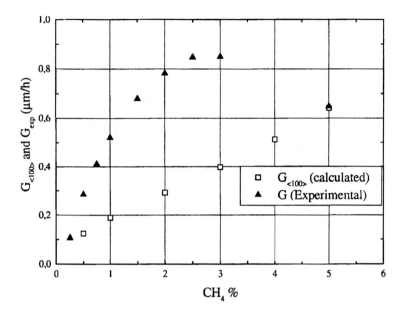

Figure 20.6: Calculated and measured growth rates in dependence on CH_4 gas feed for a microwave discharge (microwave power $P_{MW} = 1$ kW, total gas pressure $p_{tot} = 5.2$ kPa, substrate temperature $T_{sub} = 1200$ K) [30].

20.2.2   Plasma-enhanced deposition of diamond-like carbon

Carbon can exist in sp^3, sp^2, and sp^1 hybridizations. The allotropes have different properties. Diamond is a sp^3-bonded, hard and wide band gap semiconductor. Graphite is sp^2 bonded and shows metallic properties. Amorphous carbon (a-C) and hydrogenated amorphous carbon (a-C:H) can also form films of high hardness. The hardness of the film depends on the content of sp^3-bonding and of the coordination. Amorphous carbon films with high sp^3-content are denoted as tetrahedral amorphous carbon films (ta-C). The ternary phase diagram of sp^2- and sp^3-carbon and hydrogen is shown in Fig. 20.7 [33]. The hardness of hydrogenated ta-C films (ta-C:H) is relatively low, and therefore will not be denoted as superhard. According to the definition of *superhardness* only the ta-C films without hydrogen content are superhard. Deposition of ta-C films is achieved by plasma-enhanced deposition in low-temperature low-pressure plasma. The hybridization of sp^3-bonding during deposition of ta-C is obtained by ion bombardment in the ion energy range of about 100 eV. The deposition methods applied are based on physical deposition and can be divided in two groups: methods, where ions and neutral particles consist of carbon (direct deposition), and methods, where ions and neutral particles consist of a mixture of carbon and a noble gas (indirect deposition) [34]. Direct deposition of ta-C by plasma-enhanced techniques can be achieved by cathodic vacuum arc evaporation. To prevent the formation of macroparticles during deposition a magnetic filtering field is applied between particle source and substrate [35, 36]. The measurement of the

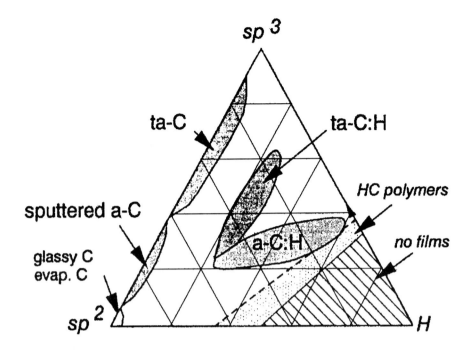

Figure 20.7: Ternary phase diagram of carbon [33].

plasma parameters during deposition was performed by Langmuir probe and Faraday cup for ion energy measurements [36]. Indirect plasma-enhanced deposition is based on sputtering systems as well as laser ablation or carbon evaporation in a plasma [37]. Experiments show a strong dependence of the Young's modulus and the internal compressive stress on ion energy and substrate temperature. Concerning the ion energy, where a maximum of sp^3-bonding in ta-C films can be obtained, the results of different authors show a broad spectrum. In filtered arc technique the corresponding ion energies are in the range of 20 eV to 600 eV [37]. The substrate temperature in ta-C thin film deposition is also a key parameter because at temperatures higher than about 140 °C graphitization of the ta-C film occurs. Probably in many experiments the influence of the ion energy and substrate temperature on ta-C formation cannot be distinguished clearly. Detailed measurements of the plasma parameter and modeling of the energy and particle flux to the substrate in relation to the properties of the ta-C films are presented in [38]. Films of ta-C with high thickness are candidates for tribological applications. One of the actual problems is the poor adhesion of the films. One way to overcome this problem is the deposition of gradient layers. In hardmetal (WC-Co) samples the adhesion of ta-C films was improved by chemical etching of the surface before deposition [39].

20.3   Plasma-enhanced deposition of cubic boron nitride films

Cubic boron nitride (c-BN) is one of the materials that does not exist in nature. Usually c-BN is produced by a high-temperature high-pressure process converting hexagonal boron nitride (h-BN) into its cubic form. The first $p - T$ phase diagram of boron nitride at thermal equilibrium was proposed in 1963 [40]. More recently an adapted pseudo-Debye model was proposed which allows more realistic extrapolations to the low-temperature range [41]. The $p - T$ phase diagram shown in Fig. 20.8 suggests that c-BN should be the thermodynamic stable phase at room temperature [41].   According to Fig. 20.8 the transformation c-BN to h-BN at low pressure occurs at 1650 K. The formation of c-BN starting from the gas phase at low temperatures and pressure is forbidden by the Oswald-Volmer rule. According to this rule the less-dense phase (h-BN) is formed first during nucleation of BN at thermodynamic equilibrium. Plasma-enhanced deposition of boron nitride started in the beginning of the eighties with the aim to overcome the high-temperature high-pressure conditions for c-BN production. First successful results are published in Refs. [42, 43, 44]. At non-equilibrium conditions the growth of c-BN films consists of two steps: nucleation followed by growth. In the first step the conditions for c-BN nucleation should be realized. These conditions are not fully known yet. Preferentially nucleation can be achieved on a h-BN layer on the substrate, when the c-axis of the h-BN layer is orientated parallel to the substrate surface. This orientation of the h-BN crystals is obtained by ion bombardment of the growing h-BN film. The growth of the

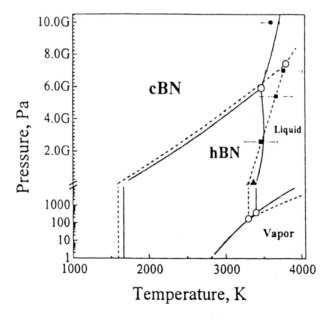

Figure 20.8: Phase diagram of boron nitride in thermal equilibrium [41].

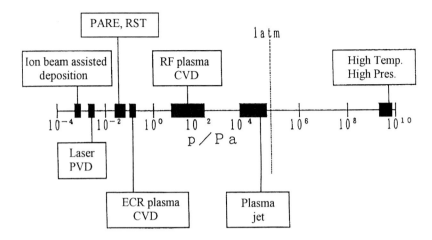

Figure 20.9: Plasma-enhanced deposition techniques of c-BN formation, according to Ref. [45].

c-BN film after nucleation as the second step only occurs in a specific range of the ion current density and of the ion energy transfer to the growing film. Therefore in plasma-enhanced deposition of c-BN the substrates operate at sufficiently high negative bias voltage. It attracts and accelerates ions from the plasma to the surface at an amount which is necessary for the formation of c-BN. An overview of different plasma-enhanced deposition techniques for c-BN film synthesis scaled on the pressure axis is shown in Fig. 20.9 [45].

The methods currently used are: ion beam assisted deposition (IBAD), laser-PVD, reactive sputtering technique (RST), and plasma activated reactive evaporation (PARE). These methods are denoted as physical vapor deposition (PVD), while plasma jets, radiofrequency (rf) and microwave excited plasma deposition (ECR) are denoted as chemical vapor deposition (CVD).

20.3.1   Physical vapor deposition

In the following the different PVD methods will be categorized with respect to the different methods to provide boron and nitrogen and, as well, to generate high energetic ions. Fig. 20.10 shows the different plasma-PVD methods currently used for c-BN deposition taken from the early publications on these techniques. In PVD boron is provided from solid state. In IBAD and PARE pure boron is used, while for RST and laser PVD h-BN or boron carbide (B_4C) act as boron source. Plasma-enhanced PVD will be discussed in the following sections in more detail.

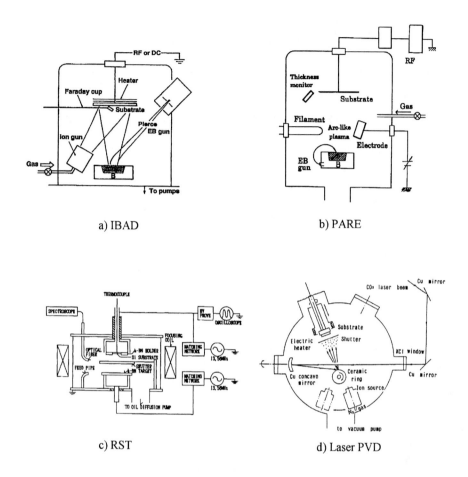

a) IBAD   b) PARE

c) RST   d) Laser PVD

Figure 20.10: Methods of plasma-enhanced physical vapor deposition of c-BN, a) ion beam assisted deposition (IBAD) [46], b) plasma activated reactive evaporation (PARE) [47], c) reactive sputtering technique (RST) [48], d) plasma-enhanced laser deposition (PLD) [49].

20.3.1.1 Ion beam assisted deposition (IBAD)

Fig. 20.10a shows the principle of the IBAD method. It consists of an electron beam evaporation source and an ion source producing ions with appropriate energy, up to several keV. It should be emphasized that IBAD is not a plasma-enhanced deposition technique as only ions are used. Nevertheless, the application of IBAD is very helpful to show the influence and also the importance of ion bombardment for the formation of cubic boron nitride. From literature are known two crucial experiments, which quantified the role of the ion bombardment during c-BN growth. In Ref. [50] a second additional ion source was used. One source was used for nitrogen ion production while the other could be used to produce ions of inert gases as argon, krypton, xenon. The authors showed that there is one major controlling factor in the c-BN formation, the momentum per deposited atom. With p, the momentum transferred into the film, a, the number of atoms deposited per unit area per unit time, J_i, the number of ions of species i bombarding the film unit area per unit time, and m_i and E_i the masses and energies of bombarding ions, respectively, the controlling factor is given by

$$\frac{p}{a} = \sum_i \frac{J_i}{a}(2m_i\gamma_i E_i)^{\frac{1}{2}} .$$ (20.3)

The maximum energy transfer factor γ_i is given by (see chapter 4)

$$\gamma_i = \frac{4m_i M}{(m_i + M)^2} ,$$ (20.4)

with M the mass of the atoms being bombarded in the film. Furthermore it was found that for a specific substrate temperature a sharp threshold value for c-BN formation exists. By variation of the nitrogen flux it could be confirmed that c-BN is formed only at a boron to nitrogen ratio of 1:1 in the film. At high values of momentum per atom resputtering of the film is achieved. The data obtained are summarized in Fig. 20.11 [50]. The figure shows the boron nitride phase as a function of ion bombardment in terms of momentum per arriving boron atom, given in $(\text{eV} \times \text{amu})^{1/2}$, substrate temperature, and boron/nitrogen ratio (chemistry). Fig. 20.11 shows that at 300 °C for c-BN formation a momentum per arriving atom of $p/a = 350$ $(\text{eV} \times \text{amu})^{1/2}$ is necessary. In the hatched region a resputtering of the film is observed. The results of these experiments are confirmed in many papers of plasma-enhanced deposition provided the momentum of the ion species could be measured. The phase diagram of Fig. 20.11 has been refined in an experiment using mass selected ion beam deposition [51]. The authors used beams of B^+ and N^+ at different ion energies and substrate temperatures. The phase diagram of c-BN formation obtained shows a shift to lower temperatures in relation to Fig. 20.11. Furthermore indications have been found that after nucleation at temperatures higher than the threshold value the growth of a c-BN film can continue at lower temperatures [52]. The layer sequence during growth of BN was first observed for films deposited by the IBAD technique [53, 54]. This technique was also used to investigate the influence of ternary particles on c-BN growth. The formation of boron-aluminum nitride alloys were investigated in Ref. [55] while in Ref. [56] the effects caused by titanium and aluminum incorporations on the structure of BN films were studied.

20.3.1.2 Plasma activated reactive evaporation (PARE)

Fig. 20.10b shows the principle of PARE corresponding to one of the first realizations for c-BN deposition by this method [47]. It consists of a crucible containing metallic boron and a source producing the plasma. The plasma source used is a dc low-voltage arc consisting of a heated filament as cathode, and an anode. Because of the isolating properties of BN the anode should be protected against deposition . The substrate is powered by negative bias voltage and attracts ions from the plasma. In Fig. 20.10b a biased rf voltage is applied. At high substrate temperatures also a dc voltage can be applied. By using a parallel magnetic field the plasma density between the hot cathode and the anode can be enhanced [57, 58, 59]. Instead of activation by hot filament dc discharge also rf discharges at 13.56 MHz are used.

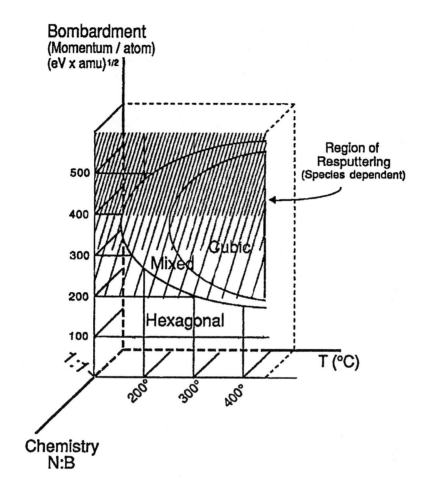

Figure 20.11: Phase diagram of BN in ion beam assisted deposition (see text).

In Ref. [60] a specially shaped antenna within an axial magnetic field is applied to excite helicon waves in a chamber between the electron gun and the substrate. It is well-known that in plasma sources using helicon waves high ionization degrees can be achieved. Evaporation of boron in an Ar-N_2-plasma can be also performed using a hollow cathode arc. This method was first introduced in c-BN deposition [61] and developed further in Ref. [62]. In hollow cathode arc evaporation beam-like electrons generated in the cathode are deflected to a water cooled crucible containing the metallic boron. The energy of the beam-like electrons is about 25 eV to 30 eV, currents are between 50 A $\leq I \leq$ 200 A, and current densities 50 A/cm^2 $\leq j \leq$ 100 A/cm^2 [63]. Therefore the energy density in the crucible becomes high enough for evaporation of the material. The isotropically distributed electrons in the plasma as well as the beam-like electrons ionize the gas/vapor components very effectively.

Parallel to the evaporation technique by electron beams also cathodic vacuum arc evaporation has been developed for c-BN deposition [64, 65]. The heated boron target acts as cathode,
where the cathodic spot moves randomly. In the spot boron evaporates and becomes ionized
to relatively high ionization degrees (up to 86 % B^+). Measurement of the ions species by
energy resolved mass spectrometry showed that boron was ionized up to B^{3+} while nitrogen
exists as N^+ and N^{2+}. The kinetic ion energy at low pressure was measured as 45 eV for
B^+ and 90 eV for B^{2+} [61]. To prevent the emission of macroparticles, which is usually an
accompanying symptom in vacuums arcs, a magnetic filter was mounted between the boron
source and the substrate. In principle the deposition conditions for c-BN formation in IBAD
and PARE are similar. In both cases only high energy ions bombard the growing thin film. In
IBAD electrons are absent, and in PARE the negative bias voltages applied are so high that the
electrons are reflected. One important difference between both methods is the possibility of
handling the ion flux. In IBAD the energy and ion current density can be influenced directly
by the parameters of the ion gun. In PARE the ion flux to the substrate depends on the plasma
parameters indirectly. In a mixture of different ion species with mass m_i and density n_i the
partial ion current density j_i is given by

$$j_i = 0.6 n_i \sqrt{\frac{kT_e}{m_i}}, \tag{20.5}$$

where k is the Boltzmann constant and T_e the electron temperature. In a mixture the different ion densities can be measured by appropriate mass spectroscopy (see chapter 8). The
electron temperature or electron energy distribution function (EEDF) can be evaluated from
Langmuir probe measurements (see chapter 6). Alternatively, the ion density can be calculated
by solving simultaneously the Boltzmann equation for the electrons and the particle transport
equations. A combination of experimental and theoretical methods can be applied as well.
In Ref. [66] the electron temperature was measured by Langmuir probe and the ion densities
were calculated applying the equation of continuity.

In deposition devices with a complicated geometry the transport equation can be solved
numerically with great expense. To obtain experimentally an overview of the plasma conditions in the deposition device, it is helpful to use a two-step procedure. First the plasma
parameter kT_e (respectively the EEDF) and the ion density n_i of the device will be measured spatially resolved for only one ion species. This measurement can easily performed
by spatially resolved Langmuir probe measurements. It provides important information on
the spatially dependent ion density and the electron temperature in the deposition device in
dependence on pressure and electrical power. Secondly the reactive components are added
and the influence on electron temperature and electron density can be obtained. By analyzing
this influence the partial ion densities can be estimated. In the following the application of
this method to c-BN deposition is demonstrated by results of measurements of the author's
group [67]. Fig. 20.12 shows the dependence of the ion current density on the position in the
deposition chamber. On the bottom of the deposition chamber a hollow cathode arc evaporation system is mounted consisting of the hollow cathode, a water cooled crucible (anode,) and
the transverse magnetic field for the beam deflection. At heights more than 300 mm from the
bottom (200 mm far from the crucible) in argon electron temperature as well as ion density
were measured. The ion current density is calculated from Eq. 20.5. The outer deposition parameters applied correspond to those which are used in reactive deposition with Ar as buffer

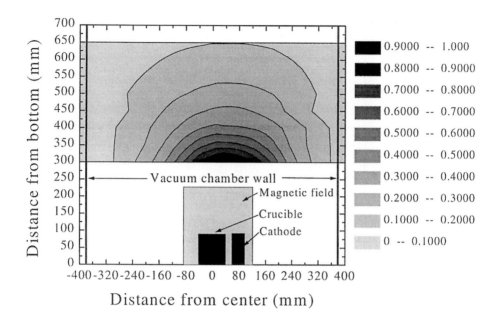

Figure 20.12: Ion current density in dependence on the position in the deposition chamber (see text). The crucible is water-cooled. Plasma parameters: $j_0(r = 0 \text{ mm}, z = 300 \text{ mm}) = 3.2$ mA/cm^2, argon gas pressure $p_{Ar} = 0.6$ Pa, argon gas flow $Q_{Ar} = 80$ sccm, discharge power $P = 5.2$ kW.

gas. Fig. 20.12 shows that the ion current density varies about one magnitude between the axial distances 300 mm to 650 mm. On the other hand it can be seen that at an axial distance of $z = 550$ mm the current density remains nearly constant over a broad radial distance $r = \pm 240$ mm. At $z = 475$ mm the current density remains constant inside of $r = \pm 160$ mm. In this region one condition for an uniform phase of BN is met corresponding to Fig. 20.11. In a second step the locally resolved boron deposition rate was measured. This can be done by a movable quartz monitor and also by film thickness measurements as a function of deposition time at different positions. The boron deposition rates measured under the same conditions as in Fig. 20.12 at $z = 325$ mm and $z = 475$ mm are shown in Fig. 20.13. The data can be fitted by a cos$^{3.5}$ ϕ–dependence, with ϕ the angle with respect to the surface normal, as shown. Evaluating the data from Figs. 20.12 and 20.13 a surface in the deposition chamber can be calculated where the ratio of the argon ion flux to the boron flux J_i/a becomes constant. By applying the appropriate bias voltage to the substrate in relation to the plasma potential also the momentum transfer per deposited atom is kept constant. In the second part of step two the influence of boron and nitrogen addition on the plasma parameters was measured. Fig. 20.14 shows the dependence of the electron density (Fig. 20.14a) and mean electron energy (Fig. 20.14b) on argon or nitrogen addition at different Ar-base flows. The position in the deposition chamber was fixed at a distance $z = 600$ mm on the axis ($r = 0$ mm). In the experiments

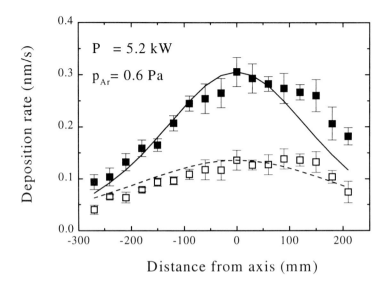

Figure 20.13: Dependence of boron deposition rate on the position in the deposition chamber, (■) at $z = 325$ mm and (□) $z = 475$ mm. Solid and dashed lines are corresponding fits assuming a $r_0^{-2} \cos^{3.5} \phi$–dependence (see text).

the pumping speed was held constant during addition of gas, and the increasing total pressure results from the addition of argon or nitrogen. Experiments started at a pressure $p_{Ar} = 0.6$ Pa in pure argon. Fig. 20.14 shows that at high Ar-base flow the influence of a small additional nitrogen flow on electron density as well as on mean electron energy is relatively low. In contrast to this at low Ar-base flow the mean electron energy decreases dramatically with increasing nitrogen flow. For c-BN deposition the appropriate nitrogen flow depends on the boron deposition rate corresponding to Fig. 20.11. Applying a high Ar base flow (150 sccm $< Q_{Ar} < 200$ sccm) and evaluating the data of Fig. 20.13 the authors showed that in the first-order approximation the data from Fig. 20.12 can be used to find in the chamber an extended region where c-BN deposition occurs. This approximation was confirmed experimentally.

20.3.1.3 Reactive sputtering for c-BN deposition (RST)

Sputtering techniques are well established in industrial applications. There are numerous activities reporting on c-BN deposition by this method. Fig. 20.10c shows the principle scheme of the reactive sputter technique (RST). The target is sputtered by high energetic ions bombarding the surface. The ions are produced by radiofrequency (rf) or direct current (dc) discharges. A dc voltage can be only applied if electrically conducting targets are used. In c-BN deposition the application of electrically conducting boron carbide targets (B_4C) is reported [68, 69]. The target acts as the cathode of the discharge, and due to the cathode drop the ions are accelerated. High plasma densities at the cathode can be achieved applying an additional magnetic field (magnetron sputtering). The targets mostly used consist of hexagonal

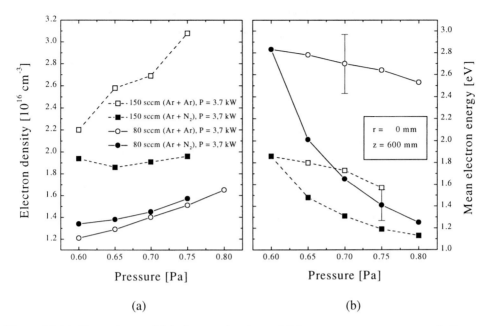

Figure 20.14: Dependence of (a) electron density and (b) mean energy on argon and nitrogen pressure (see text). Indicated flow rates give the Ar-base flow and hold for both figures (a) and (b).

boron nitride (h-BN), but also pure boron can be used [70, 71]. Because of the high specific resistance of boron and h-BN the plasma is excited by rf. The acceleration of ions from the plasma to the substrate is realized in the same manner as described in the previous section 20.3.1.2. The sputtering technique can be used for c-BN deposition if the ion density near the substrate becomes high enough to fulfill the conditions shown in Fig. 20.11. There exist different methods to increase the ion density. In Ref. [48] an external transverse magnetic field is applied as shown in Fig. 20.10c. The magnetic induction near the substrate is high enough to magnetize the electrons, which move on helix paths. At low electric field strength the electrons drift parallel to the magnetic field lines. The electrons attract the ions, and therefore the plasma is concentrated at the center of the deposition chamber. At high electric field strength, for example near the target, a drift component $\vec{v}_\perp$ perpendicular to the electric field strength $\vec{E}$ and the magnetic induction $\vec{B}$, the so-called $\vec{E} \times \vec{B}$ drift, must be taken into account [72]. The drift velocity $\vec{v}_\perp$ is given by

$$\vec{v}_\perp = \frac{\vec{E} \times \vec{B}}{B^2} .$$
(20.6)

Several authors report on magnetically enhanced rf sputtering in the so called magnetron device. At the backside of the target permanent magnets are mounted. The magnetic flux between north and south pole penetrates the target and is concentrated near the target surface in the so-called *balanced* mode. In the case that the magnetic field lines penetrate far into the transport space between target and substrate the magnetron is called *unbalanced*. The unbalanced mode is often used in c-BN thin film deposition to enhance ion density near the substrate. Applying an unbalanced magnetron the magnetic induction at the substrate was measured to amount to 1 mT with an electron density $n_e = 10^{10}$ cm^{-3} and an electron temperature of $kT_e = 5$ eV [73]. The momentum per arriving boron atom was calculated to $p/a = 500$ (eV $\times$ amu)$^{1/2}$. Using h-BN targets in a sputtering device the question of the transport of the nitrogen from the target to the substrate is discussed controversely. Most of the authors found an excess of boron in the film using argon as sputtering gas without any additional nitrogen. In Ref. [74] it was shown that nitrogen can be completely transferred from the target to the substrate using pyrolytic h-BN targets. The excess of boron in films sputtered from hot pressed h-BN targets may be caused by two reasons. Hot pressed h-BN targets are porous and may contain impurities. Also the thermal conductivity is low, and therefore the target can be heated to higher temperatures than the pyrolytic h-BN. Non-porous, high density targets of pure boron are also applied for unbalanced magnetron sputtering [70]. The authors describe a special procedure to obtain boron targets with a density of 2.34 g/cm^3 and a porosity < 0.15 %. The sputtering of targets made from SiB$_6$ and AlB$_{12}$ in a nitrogen atmosphere is reported in Ref. [75]. The plasma is produced by a hot filament discharge, and a dc voltage is applied to the target. For the deposition of BN-multilayer films several other targets are used. Multilayer films of CN$_x$/BN:C were deposited by an unbalanced dual magnetron with two targets made from pyrolytic graphite and B$_4$C [76]. In Ref. [7] the application of BC$_x$N$_y$ targets with different compositions is reported. The targets were produced by hot pressing of powder mixtures. The authors found that the mobility induced by ion bombardment is sufficient to cause phase separation into c-BN and carbon.

The combination of magnetron sputtering of h-BN with electron cyclotron resonance (ECR) plasma generation is proposed in Ref. [77]. The substrate was immersed in the ECR plasma and well removed from the plasma confinement region of the sputtering source. This technique was also used in Refs. [78, 79] in combination with reflection high energy electron diffraction (RHEED) as *in situ* diagnostics.

In the literature many references focus on the effects of substrate bias voltage, substrate temperature and rf power on the growth of c-BN films in reactive sputtering. The range of the bias voltage, wherein the c-BN phase grows, is emphasized in Refs. [7, 48, 70, 73, 74, 77, 80, 81]. Besides in Refs. [70, 80, 81] special attention is paid to the substrate temperature which can be applied for c-BN deposition. Low-temperature deposition of c-BN is referred in Ref. [81], where the authors found a very small bias voltage range (−70V to −130V) at a substrate temperature of about 400 K. The substrate tuning technique was applied to a rf magnetron sputtering system to obtain a variable substrate bias voltage without an additional substrate power supply [82]. The authors studied the dependence of the bias voltage on pressure, gas composition and rf power. A Langmuir probe was used to measure plasma potential, electron temperature and electron density. The maximum ion energy available by this technique was 150 eV. A quite different deposition technique is proposed in [83]. The authors reported deposition of c-BN in rf magnetron sputtering with the assistance of a simultaneous electron bombardment of the growing surface. An additional low voltage discharge between a hot filament and the substrate generates an electron flux to the substrate. At an electron current density of 140 mA/cm^2 films are deposited whose FTIR spectra correspond to those of c-BN. According to the published literature this experiment could not be repeated up till now.

In the literature there are only a few references concerning plasma diagnostics of sputtering sources for c-BN film deposition. In Ref. [84] measurements of the energy distribution of ions and neutrals generated during sputtering of boron nitride are discussed. In dependence on the primary ion energy the energy distribution of boron neutrals was measured and correlated to the surface binding energy. Sputtered nitrogen or nitrogen compounds from the BN-target could not be detected. The simulation of the boron nitride (h-BN and c-BN) sputtering process by the transport of ions in matter (TRIM) code and its comparison with experimental data is reported in Ref. [85]. In Ref. [86] measurements are reported on the mass and the energy of the ions and neutrals impinging on the substrate surface during BN deposition. The ion energy distribution showed asymmetric bimodal shapes with Ar^+ as the most dominant ion. The flux ratio of Ar^+/B^+ was estimated as 100:1; therefore it could be concluded that Ar^+ primarily contributes to the energy and momentum transfer to the growing film. A mean threshold value for the momentum transfer $(p/a)_{th} = 138$ (eV × amu)$^{1/2}$ (Eq. 20.3) was found for c-BN growth. Mostly argon is used as sputter gas in a magnetron sputtering device. In Ref. [87] neon was applied instead of argon. The metastable energy of neon is higher than the ionization energy of N_2, and therefore in a Ne/N_2 mixture the nitrogen molecules are ionized effectively to N_2^+ (Penning ionization). The relative density of N_2^+ ions was measured by optical emission spectroscopy. The enhancement of N_2^+ ion density by the Penning effect was demonstrated clearly. The results showed that N_2^+ enhancement additionally has to be accompanied by an appropriate bias voltage at the substrate to form c-BN films. Optical emission spectroscopy in an Ar/N_2-mixture is reported in Ref. [88]. The authors measured the relative intensity of N_2^+ (390.5 nm) and B (499.5 nm) lines. They found a maximum of c-BN concentration in the film in dependence on the line intensity ratio.

Comparative studies of composition, structure and elastic properties of boron nitride films deposited by magnetron sputtering and IBAD technique are reported in Refs. [89, 90], while in Ref. [91] a study comparing sputtering and pulsed laser deposition (PLD) is carried out. The authors found significant differences in the film properties, depending on the deposition processes. The hardness as well as the density of the films were higher in films processed by IBAD [89].

20.3.1.4 Plasma-enhanced laser deposition (PLD)

First successful experiments to deposit c-BN by laser ablation were published in 1990 [49, 92]. In Ref. [49] the authors used a high-power continuous wave CO_2 laser of 200–1000 W in combination with a Kaufman-type ion source. In Ref. [92] a pulsed excimer laser (248 nm, KrF) was applied in a nitrogen atmosphere at a pressure of $p = 6.7$ Pa without an additional ion source. Fig. 20.10d shows the principle of a laser ablation device [49]. The laser input is focused on the peripheral surface of the rotating, sintered h-BN ring target. The ion source produces nitrogen ions with an energy up to 2 keV. A model of processes in plasma-enhanced laser deposition is shown in Fig. 20.15 [93]. By interaction of the laser beam with the target surface the material is (I) evaporated and dissociated into atoms. In the second step (II) clusters are formed successively from atoms in plasma chemical reactions. (III) The neutral clusters and the atoms/molecules are ionized by electron impact ionization and (IV) are transported to the substrate. In the growing film small crystalline particles can be observed which have the original structure of the target material. Processes I to III are located in the so-called plasma

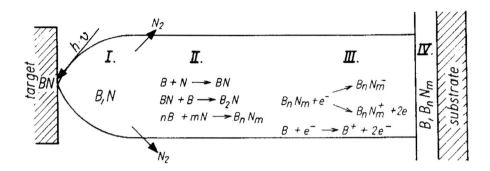

Figure 20.15: Model of processes in plasma-enhanced laser deposition of BN [93].

plume in front of the target, while outside of the plume the particle density corresponds to those of the background pressure (10^{-5} Pa to 1 Pa). There exist only a few publications referring to c-BN deposition in a nitrogen atmosphere without additional ion bombardment [94]. The formation of c-BN films is more successful, if the growing films are bombarded with ions from an additional ion source. Systematic investigations of the dependence of c-BN formation on the parameters of PLD are reported in Ref. [95]. Results from Ref. [95] show that the parameter range appropriate for c-BN formation is comparable with that obtained in the other plasma-enhanced deposition techniques. Substrate effects in ion-assisted PLD of c-BN are reported in Ref. [96]. The substrates used were metallic Ag, Cu, and Ni, metallic-like TiN on MgO, and silicon. The authors found that the adhesion of the BN films on metals was in general not as good as on Si substrates. The c-BN content in the films on metallic substrates was less compared to those on Si substrates. A hybrid technique consisting of a capacitively coupled rf discharge and PLD is described in Ref. [97]. The electron temperature and electron density were measured by Langmuir probes and were found to lie in the range of several eV for kT_e and between $10^7/\mathrm{cm}^3$ to $10^9/\mathrm{cm}^3$ for n_e. Using a rotating h-BN target and a silicon substrate parallel to the target surface, the film properties varied laterally because of the small size of the plume. The ion-to-boron arrival ratio was varied by changing the laser pulse repetition rate, keeping the ion energy constant. The evaluated data for ion-to-boron arrival ratio for c-BN deposition agree with data resulting from IBAD and PARE techniques.

More recently the processes inside as well as outside of the plume were investigated by mass spectroscopic studies [98], by high-speed photography [99], and emission spectroscopy [100]. In Ref. [98] the authors measured the ionized as well as the non-ionized species produced by laser ablation of a pyrolytic boron nitride target. The results of the ion mass spectroscopy showed that at low laser fluences (< 0.52 J/cm^2) only B$^+$ and a small amount of B^{2+} were generated. N$^+$ ions appeared in the ion mass spectra at laser fluences above 1 J/cm^2, but the intensity of the N$^+$ ion peak was much smaller than that of B$^+$. Neutral N$_2$ molecules were detected up to a laser fluence of 1.7 J/cm^2. BN molecules were not detected, but a small amount of B$_2$ dimers was found. The lack of nitrogen supports the experimental results that PLD requires plasma activation of nitrogen for c-BN deposition. A more detailed picture of the processes in PLD can be derived from the measurements of the plume expansion by high-speed photography [99]. The authors propose that in the PLD process the growing film is first bombarded by the fastest species belonging to the plume front. Then this bombardment is followed by the arrival of the plume core with relatively low velocity and energy. Experiments showed that the velocities of the front of the plume and of the mass center depend on laser fluence and on gas pressure. The velocities of the mass center are lower by a factor of two in comparison with the velocities of the front. Therefore the plasma plume can be considered as a luminous cloud whose time evolution is slow in comparison with the displacement of its front. In Ref. [98] the laser fluences applied were in the range from 0.25 J/cm^2 to 3 J/cm^2, while in Ref. [99] in the range from 7 J/cm^2 to 200 J/cm^2. Relatively high laser fluences (1.5 kJ/cm^2) were applied in Ref. [100]. Under these conditions the authors found in the plasma plume highly ionized boron (B^{3+} and B^{4+}) and nitrogen (N^{4+} and N^{5+}). The electron temperatures measured were in the order of 25 eV to 40 eV and the average electron density was about 10^{19}/cm^3 at a distance of 1 mm in front of the target.

20.3.2   Plasma-enhanced chemical vapor deposition

The formation of c-BN in physical vapor deposition can only occur, if the momentum transferred by the ions to the growing film is high enough. The bombardment by ions causes dislocations and interstitial defects resulting in the formation of intrinsic stress in the film. The intrinsic stress amounts to about 5 GPa or higher, and therefore the films peel off from the substrate. One way to overcome the problems of PEPVD is the application of plasma-enhanced chemical vapor deposition (PECVD) at very low bias voltage in order to prevent the defects. In the literature different plasma-enhanced chemical methods are reported for c-BN deposition. The methods differ by the gases used as well as by the type of plasma activation. In the following we will discuss the different methods related to the type of plasma generation applied. Fig. 20.16 shows a schematic sketch of different plasma-CVD methods currently used in c-BN deposition. The devices shown in Fig. 20.16 are taken from one of the first publications on these techniques. Fig. 20.16a shows a deposition device driven by rf at 13.56 MHz [101], while in Fig. 20.16b the plasma is generated by ECR at 2.45 GHz [102]. One of the applications of a dc plasma jet, as shown in Fig. 20.10c, is reported in Ref. [103]. Details of plasma-enhanced CVD will be discussed in the following sections in more detail. An overview of different compounds used in PECVD is given in table 20.3. The materials applied contain boron as well as nitrogen, often together with hydrogen. But the role of hydrogen in c-BN deposition is still unclear. Chlorine and fluorine are used to support chemical etching of h-BN.

20.3.2.1   Plasma-enhanced chemical vapor deposition in radiofrequency (rf) discharges

The principle of PECVD using rf for plasma generation is shown in Fig. 20.10a [101]. The lower electrode is driven by the capacitively coupled 13.56 MHz power supply through a matching network. The samples may be clamped to the grounded electrode (shown in Fig. 20.10a) or to the powered ones. If the powered electrode is blocked by a capacitor both electrodes are self-biasing. The value of the bias voltage depends on the ratio of the area of the powered electrodes to the area of the grounded ones [117]. The ions will be accelerated

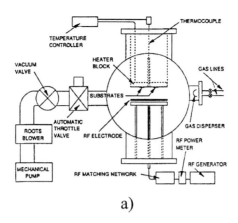

a)

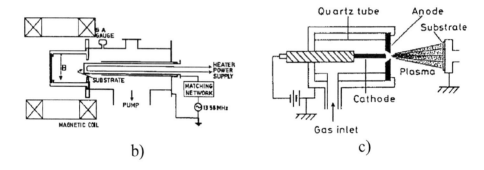

b)   c)

Figure 20.16: Methods of plasma-enhanced chemical vapor deposition of c-BN, a) radiofrequency (rf) driven plasma [101], b) microwave driven plasma (electron cyclotron resonance, ECR) [102], c) direct current (dc) plasma jet reactor [103].

by the potential difference between plasma and bias potential acquiring the corresponding energy at low pressure without collisions. At high pressures the ion energy is reduced mainly by charge exchange collisions in the plasma sheath in front of the substrate. The authors of Ref. [101] used BF_3, N_2 and H_2. A comparison of the properties of h-BN films deposited on the powered electrode and on the grounded electrode is given in Ref. [118]. The optical and structural properties as well as the film stability showed different behavior in dependence on the gas mixtures used ($B_2H_6/H_2/NH_3$ and B_2H_6/N_2) and the electrode applied.

Instead of capacitive coupling of the rf also inductively coupled plasma generation (ICP) was applied. In most of the cases the authors used the combination of inductively coupled plasma with capacitive coupling to the substrate. The inductive coupling generates the plasma at an input power in the kW-range, and the capacitive coupling to the substrate generates the bias voltage at an input power in the range of several dozen Watt [105]. By this technique a breakthrough to lower bias voltages was achieved in rf-PECVD using a low-pressure inductively coupled rf discharge at 13.56 MHz [105]. The gas mixture used contained B_2H_6, N_2,

Table 20.3: Reactants and deposition techniques applied in PECVD of c-BN (rf, radiofrequency plasma; ICP, inductively coupled plasma; ECR, electron cyclotron resonance plasma; dc, direct current plasma).

Reactants	Technique	c-BN	Bias voltage (V)	Pressure (Pa)	Temperature (K)	Ref.
B(s)/H_2/N_2	ICP	No	0	11...43	378, 743	[104]
B_2H_6/N_2/Ar/He	ICP	Yes	30...60	0.1	1173	[105, 106]
B_2H_6/N_2/H_2/He/Ar/Xe	rf	Yes	500...800	2	≤ 723	[107]
HBN$(CH_3)_3$ (Trimethylborazine)	ICP	Yes	−200...0	2	550...773	[108]
$(BH_3 \cdot NH_3)$/N_2 (Boraneammonia)	rf + hot filament	Yes	—	26-90	623	[109]
$(CH_3)_2$-NH-BH_3/H_2/Ar (Dimethylamineborane)	rf	Yes	—	50	1073	[110]
$B_3N_3H_6$/Ar (Borazine)	helicon wave	Yes	350	0.04	<1073	[111]
B_2H_6/Ar/N_2	ECR	Yes	0...400	0.04...0.27	873	[102]
BH_3-NH_3	ECR	Yes	10...100	0.07	523...848	[112]
$B_3N_3H_6$/NH_3/Kr	ECR	No	10...165	0.2...0.7	1073	[113]
$(C_2H_5)_3$N·BH_3/N_2 (Boranetriethylamine)	microwave	No	—	0.3	870...1170	[114]
BCl_3/N_2/Ar	dc arc jet	Yes	60...90	27...133	770	[115]
BF_3/Ar/N_2/H_2	dc arc jet	Yes	70...150	6650	1253	[116]

He, Ar. The authors measured the plasma potential in front of the substrate by an emissive Langmuir probe to derive the sheath potential and thus the energy of the ions striking the growing film. It was found that the formation of c-BN occurs only in the small sheath potential range between 68 V and 98 V. Investigation of the growth of boron nitride in a hot filament assisted rf discharge were also performed in a BCl_3–NH_3–H_2-mixture [119, 120]. The substrate holder used in Ref. [120] was floating and the maximum ion energy was expected to lie in the range of 10 eV.

To achieve higher electron densities helicon wave plasma sources were applied [111]. Helicon waves are bounded whistler waves in the frequency range below the electron cyclotron frequency and above the lower hybrid frequency [121]. In Ref. [111] the plasma density along the radial direction was measured by a double Langmuir probe. The electron densities were $n_e = 10^{13}$/cm^3 in the plasma source tube and $n_e = 10^{11}$/cm^3 in the reaction chamber. The working pressure of 4×10^{-2} Pa for the deposition process was maintained in an Ar/Borazine ($B_3N_3H_6$) atmosphere. Experiments showed a correlation between electron density and film growth rate. The substrate was biased by additional rf power.

Detailed plasma diagnostics in rf-driven PECVD processes used for c-BN deposition are reported in Refs. [122] and [123]. In Ref. [122] optical emission spectroscopy at different rf powers in nitrogen with different admixtures of borane-ammonia (BH_3–NH_3) searching for nitrogen lines was performed. In the investigation of neutral nitrogen the second positive system and for N_2^+ the first negative system was considered. Both, N_2 as well as N_2^+, lines showed a strong dependence on rf power with a threshold power of 50 W for the appearance of the N_2^+ lines. At about 160 W the intensities of the ion lines increased considerably, changing the color of the plasma from pink to blue. At this power the measured self-bias voltage ranges between –100 V and –200 V (depending on the pressure). These values are responsible for the c-BN growth. Hydrogen as well as boron lines could not be detected. In Ref. [123] the authors combined optical emission spectroscopy with mass spectrometry in a plasma of Ar/H_2, $Ar/H_2/N_2$, $Ar/H_2/BCl_3$, and $Ar/H_2/N_2/BCl_3$ mixtures. The authors also compared the properties of films deposited on the grounded electrode with those of films deposited on the powered electrode. On the grounded electrode no c-BN films could be obtained, while on the powered electrode c-BN growth was obtained when the rf power was higher than 400 W. By measuring the self-bias voltage as a function of the rf power the authors found a bias voltage of –500 V at 400 W. By comparing the bias voltage and the ion/atom ratio, necessary for c-BN deposition, the authors found values comparable to those obtained in PEPVD and described by Eq. 20.3 and Fig. 20.11.

20.3.2.2 Plasma-enhanced chemical vapor deposition in ECR discharges

One of the first papers reporting the formation of c-BN in PECVD was presented already in 1987 [102], Fig. 20.16b. Gases used were Ar, N_2, and diborane (B_2H_6), and an electron cyclotron resonance (ECR) discharge at 2.45 GHz was applied. The plasma chamber is set between the magnetic coils, as shown in Fig. 20.16b. In this configuration the magnetic field strength is inhomogeneous so that the condition for electron resonance ($\omega_{rf} = \omega_{res} = eB/m_e$, $B = 87.5$ mT for $f = 2.45$ GHz) is fulfilled only in some part of the chamber. The substrate was biased by an additional 13.56 MHz generator. In this experiment c-BN could only be formed at bias voltages of –200 V to –300 V, similar to values applied in PEPVD. In a following paper [124] the authors showed that also at lower bias voltages (–60 V to –100 V) c-BN could be deposited using diamond as substrate material.

In Ref. [113] mass spectrometric diagnostics were performed in an ECR-generated plasma in a borazine $B_3N_3H_6/Kr$ as well as in a $B_3N_3H_6/NH_3$ mixture. In both mixtures the authors found high concentrations of H^+, H_2^+, and H_3^+, very small amounts of B^+ and N_2^+ or (because of the same mass) $B_2H_6^+$, but no N^+ ions. No evidence of c-BN deposition was observed, and the authors concluded that due to the high concentrations of H^+, H_2^+, and H_3^+ ions the momentum transfer to the growing film according to Eq. 20.3 was not high enough. In Ref. [112] a combination of mass spectrometry and optical emission spectroscopy for ECR-plasma diagnostics is used in borane-ammonia vapor (BH_3–NH_3). The mass spectrum showed BH_x, NH_x as well as BNH_x decomposition products, while by emission spectroscopy only BH, N_2, N, NH, and H lines were detected. Only a small amount of the wurtzitic and/or cubic phase of BN was found.

More recently the downstream technique was successfully applied for c-BN deposition [125]. Ar and N_2 were fed in the ECR-plasma generation zone and B_2H_6 was added in the down stream zone. Langmuir probe measurements showed that the plasma was uniform over a range of ± 100 mm with an electron density of about 5×10^{11} cm^{-3}. Deposition of c-BN was obtained in a bias voltage range between –100 V and –150 V with a maximum c-BN content of 55%.

20.3.2.3 Plasma-enhanced chemical vapor deposition by direct current (dc) plasma jet

The successful application of the arc jet to diamond synthesis led to the attempt to explore the possibility of c-BN synthesis, too. In Ref. [115] the conditions for the c-BN synthesis are described in greater detail than in Ref. [103]. A dc arc discharge operating with an Ar/N$_2$ mixture in the 1–2 kW power range was expanded to low densities (10 to 100 Pa) by a converging-diverging nozzle to obtain supersonic velocities. BCl$_3$ was injected into the dissociated jet with a flow ratio of BCl$_3$ to N$_2$ in the range of 0.1% to 1%. More recently BF$_3$ together with H$_2$ was introduced instead of BCl$_3$ [126]. In both reports and in the following paper the bias voltage applied ranges between –75 V to –100 V. In Ref. [126] the deposition rate achieved was 5 nm/s, which is more than one magnitude larger than obtained in PEPVD and in the other PECVD techniques mentioned. In Ref. [115] the momentum per atom according to Eq. 20.3 was assumed to be on the order of $p/a \approx 2000$ (eV $\times$ amu)$^{1/2}$ – one order higher than discussed in the case of Fig. 20.11. Comparing the deposition rates in Ref. [126] with the values obtained in Ref. [115] in the latter the rate is about 50 times higher, and therefore the p/a ratio should be reduced by this factor. The authors of Ref. [126] assume that the c-BN deposition in the plasma jet with BF$_3$/H$_2$ may be a combined process via both *physical* and *chemical* routes.

Today, plasma-enhanced deposition seems to be a promising tool for the production of superhard materials such as c-BN. At present it is still an open question what kind of deposition method (e.g., PEPVD and /or PECVD) can be transferred to industrial applications in the nearest future.

References

[1] M.L. Cohen, Phys. Rev. B 32 (1985) 7988.

[2] M.L. Cohen, Mater. Sci. Eng. A 209 (1996) 1.

[3] D.G. Clerc, Journal of Materials Science Letters 17 (1998) 1461.

[4] A.Y. Liu, M.L. Cohen, Science 245 (1989) 841.

[5] S. Muhl, J. M. Mendez, Diamond & Related Materials. 8 (1999) 1809.

[6] S. Veprek, Journal of Vacuum Science & Technology A 17 (1999) 2401.

[7] S. Ulrich, A. Kratzsch, H. Leiste, M. Stüber, P. Schlossmacher, H. Holleck, J. Binder, D. Schild, S. Westermeyer, Surface & Coat. Technol. 116–119 (1999) 742.

[8] D. Hegemann, R. Riedel, C. Oehr, Thin Solid Films 339 (1999) 154.

[9] P.C. Yashar, W.D. Sproul, Vacuum 55 (1999) 179.

[10] P.M. Anderson, C. Li, Nanostructured Materials 5 (1995) 349.

[11] L.S.G. Plano, in: L.S Pan, D.R. Kania (Eds.), Diamond: Electronic Properties and Applications, Kluwer Academic Publishers (1995), p. 61

[12] P.K. Bachmann, D. Leers, H. Lydtin, Diamond & Related Materials 1 (1992) 1.

[13] R.E. Clausing, L.L. Horton, J.C. Angus, P. Koidl (Eds.), NATO ASI Series B 266, Diamond and Diamond-like Films and Coatings, Plenum Press (1991).

[14] P.K. Bachmann, W. v. Enckevort, Diamond & Related Materials 1 (1991) 1021.

[15] L.S. Pan, D.R. Kania (Eds.), Diamond: Electronic Properties and Applications, Kluwer Academic Publishers (1995).

[16] B. Dischler, C. Wild (Eds.), Low-Pressure Synthetic Diamond: Manufacturing and Application, Springer-Verlag: Berlin Heidelberg New York (1998).

[17] W. Kulisch, Deposition of Diamond-like Superhard Materials, Springer Tracts in Modern Physics 157, Springer Verlag: Berlin Heidelberg (1999).

[18] A. Hatta, A. Hiraki, in: B. Dischler, C. Wild (Eds.), Low-Pressure Synthetic Diamond: Manufacturing and Application, Springer-Verlag: Berlin Heidelberg New York (1998).

[19] K.A. Snail, C.M. Marks, Appl. Phys. Lett. 60 (1992) 3135.

[20] P.K. Bachmann, H.J. Hagemann, H. Lade, D. Leers, D.U. Wiechert, H. Wilson, D. Fournier, K. Plamann, Diamond & Related Materials 4 (1995) 820.

[21] S.C. Eaton, M.K. Sunkara, Diamond & Related Materials, 9 (2000) 1320.

[22] I. Schmidt, C. Benndorf, P. Joeris, Diamond & Related Materials 4 (1995) 725.

[23] C. G. Schwärzler, O. Schnabl, J. Laimer, H. Störi, Plasma Chemistry and Plasma Processing, 16 (1996) 173.

[24] M. Sommer, K. Mui, F.W. Smith, Solid State Communication 69 (1989) 775.

[25] M. Sommer, F.W. Smith, High Temperature Science 27 (1990) 173.

[26] D.G. Goodwin, J. Appl. Phys. 74 (1993) 6888 and 6895.

[27] S.J. Harris, Appl. Phys. Lett. 56 (1990) 2298.

[28] M. Capitelli, G. Colonna, K. Hassouni, A. Gicquel, Plasma Chemistry and Plasma Processing 16 (1996) 153.

[29] K. Hassouni, C.D. Scott, S. Faraht, A. Gicquel, M. Capitelli, Surface and Coatings Technology, 97 (1997) 391.

[30] K. Hassouni, O. Leroy, S. Farhat, A. Gicquel, Plasma Chemistry and Plasma Processing 18 (1998) 325.

[31] W. Tan, T.A. Grotjohn, Diamond & Related Materials 4 (1995) 1145.

[32] M. Füner, C. Wild, P. Koidl, Surface and Coatings Technology 116–119 (1999) 853.

[33] J. Robertson, Phil. Mag. B76 (1997) 335.

[34] J. Schwan, S. Ulrich, H. Roth, H. Ehrhardt, S.R.P. Silva, J. Robertson, R. Samlenski, R. Brenn, J. Appl. Phys. 79 (1996) 1416.

[35] H. Scheibe, B. Schultrich, Thin Solid Films 246 (1994) 92.

[36] B.K. Tay, X. Shi, L.K. Cheah, D.I. Flynn, Thin Solid Films 308–309 (1997) 199.

[37] Y. Lifshitz, Diamond & Related Materials 5 (1996) 388.

[38] U. Bothe, Deposition of hard, hydrogen free carbon films by magnetron sputtering, Ph.D. thesis, Stuttgart (2000).

[39] M. Hakovirta, Diamond & Related Materials. 5 (1996) 186.

[40] F. P. Bundy, R. H. Wentorf, J. Chem. Phys. 38 (1963) 1144.

[41] V. L. Solozhenko, V. Z. Turkevich, W. B. Holzapfel, J. Phys. Chem. B 103 (1999) 2903.

[42] M. Sokolowski, J. Cryst. Growth, 46 (1979) 136.

[43] C. Weissmantel, K. Bewilogua, K. Breuer, D. Dietrich, U. Ebersbach, H.-J. Erler, B. Rau, G. Reisse, Thin Solid Films 96 (1982) 31.

[44] K. Inagawa, K. Watanabe, H. Ohsone, K. Saitoh, A. Itoh, Proc. 10th Symp. ISIAT, Tokyo (1986), p. 381.

[45] Y. Ichinose, H. Saitoh, Y. Hirotsu, Surf. Coat. & Techn. 43/44 (1990) 116.

[46] T. Ikeda, Appl. Phys. Lett. 61 (1992) 786.

[47] T. Ikeda, Y. Kawate, Y. Hirai, J. Vac. Sci. & Technol. A 8 (1990) 3168.

[48] M. Mieno, T. Yoshida, Japan. J. Appl. Phys. 29 (1990) L1175.

[49] S. Mineta, M. Kohata, N. Yasunga, Y. Kikuta, Thin Solid Films 189 (1990) 125.

[50] D.J. Kester, R. Messier, J. Appl. Phys. 72 (1992) 504.

[51] H. Hofsaess, H. Feldermann, M. Sebastian, C. Ronning, Phys. Rev. B 55 (1997) 13230.

[52] H. Feldermann, R. Merk, H. Hofsaess, C. Ronning, T. Zheleva, Appl. Phys. Lett. 74 (1999) 1552.

[53] D.J. Kester, K.S. Ailey, R.F. Davis, K.L. More, J. Mat. Res. 8 (1993) 1213.

[54] D.J. Kester, K.S. Ailey, R.F. Davis, Diamond & Related Materials 3 (1994) 332.

[55] J.H. Edgar, D.T. Smith, C.R. Eddy, C.A. Carosella, B.D. Sartwell, Thin Solid Films 298 (1997) 33.

[56] A. Kolitsch, X. Wang, D. Manova, W. Fukarek, W. Moeller, S. Oswald, Diamond & Related Materials 8 (1999) 386.

[57] S. Watanabe, S. Miyake, M. Murakawa, Surf. & Coat. Technol. 49 (1991) 406.

[58] S. Watanabe, S. Miyake, M. Murakawa, Vacuum 45 (1994) 1009.

[59] J. Tian, L. Xia, X. Ma, Y. Sun, E. Byon, S.H. Lee, S.R. Lee, Thin Solid Films 356 (1999) 229.

[60] W.D. McFall, D.R. McKenzie, R.P. Netterfield, Surf. & Coat. Technol. 81 (1996) 72.

[61] K. Inagawa, K. Watanabe, I. Tanaka, S. Saitoh, A. Itoh, Proc. 9th Symp. ISIAT, Tokyo (1985), p. 299.

[62] K.-L. Barth, A. Lunk, J. Ulmer, Surf. & Coat. Technol. 92 (1997) 96.

[63] G. Rohrbach, A. Lunk, Surf. & Coat. Technol. 123 (2000) 231.

[64] G. Krannich, F. Richter, J. Hahn, R. Pintaske, V.B. Filippov, Y. Paderno, Diamond & Related Materials 6 (1997) 1005.

[65] F. Richter, S. Peter, V.B. Filippov, G. Flemming, M. Kuhn, IEEE Transact. Plasma Sci. 27 (1999) 1079.

[66] A. Lunk, R. Basner, Contrib. Plasma Phys. 30 (1990) 637.

[67] M. Hock, Optimisation of the Conditions in the c-BN Deposition, Diploma thesis, Stuttgart (1999).

[68] H. Luethje, K. Bewilogua, S. Daaud, M. Johansson, L. Hultman, Thin Solid Films 257 (1995) 40.

[69] M.P. Johansson, H. Sjostrom, L. Hultman, Vacuum 53 (1999) 451.

[70] A.F. Jankowski, J.P. Hayes, D.M. Makowiecki, M.A. McKernan, Thin Solid Films 308/309 (1997) 94.

[71] M. Wakatsuchi, Y. Takaba, V. Ueda, M. Nishikawa, Japan. J. Appl. Phys. 37 (1998) L1082.

[72] F. Chen, Plasma Physics and Controlled Fusion, Plenum Press, New York (1984).

[73] W. Otano-Rivera, L. J. Pilione, J.A. Zapien, R. Messier, J. Vac. Sc. Technol. A 16 (1998) 1331.

[74] V.Yu. Kulikovsky, L.R. Shanginyan, V.M. Vereschaka, N.G. Hatynenko, Diamond & Related Materials. 4 (1995) 113.

[75] N.V. Novikov, M.A. Voronkin, N.I. Zaika, Diamond & Related Materials. 7 (1998) 1693.

[76] M.P. Johansson, N. Hellgren, T. Berlind, E. Broitman, L. Hultman, J.E. Sundgren, Thin Solid Films 360 (2000) 17.

[77] S. Kidner, C.A. Taylor, R. Clarke, Appl. Phys. Lett. 64 (1994) 1859.

[78] D. Litvinov, R. Clarke, Appl. Phys. Lett. 71 (1997) 1969.

[79] D. Litvinov, R. Clarke, Appl. Phys. Lett. 74 (1999) 955.

[80] G. Chen, X.W. Zhang, B. Wang, H. Yan, Surf. & Coat. Technol. 113 (1999) 25.

[81] A. Bonizzi, R. Checchetto, A. Miotello, P.M. Ossi, Europhysics Letters 44 (1998) 627.

[82] A. Lousa, S. Gimeno, J. Vac. Sci. Technol. A 15 (1997) 62.

[83] Y. Tzeng, H. Zhu, Diamond & Related Materials 8(1999) 1402.

[84] E. Franke, H. Neumann, M. Zeuner, W. Frank, F. Bigl, Surf. & Coat. Technol. 97 (1997) 90.

[85] M. Chen, G. Rohrbach, A. Neuffer, K.L. Barth, A. Lunk, IEEE Transact. Plasma Sci. 26 (1998) 1713.

[86] O. Tsuda, Y. Tatebayashi Y. Yamadatakamura, T. Yoshida, J. Vac. Sci. Technol. A 15 (1997) 2859.

[87] R.B. Heil, C.R. Aita, J. Vac. Sci. Technol. A 15 (1997) 93.

[88] J.M. Mendez, E. Ramirez, A. Gaonacouto, Diamond & Related Materials 7 (1998) 1184.

[89] M.G. Beghi, C.E. Bottani, A. Miotello, P.M. Ossi, Thin Solid Films 308/309 (1997) 107.

[90] S. Logothetidis, C. Charitidis, P. Patsalas, T. Kehagias, Diamond & Related Materials 8 (1999) 410.

[91] J.L. Andujar, E. Pascual, R. Aguiar, E. Bertran, A. Bosch, J.L. Fernandez, S. Gimeno, A. Lousa, M. Varela, Diamond & Related Materials 4 (1995) 657.

[92] G.L. Doll, J.A. Sell, L. Salamanga-Riba, A. Ballal, Mat. Res. Soc. Symp. Proc. Vol 191 (1990) 55.

[93] S. Becker, H.J. Dietze, G. Keßler, H.D. Bauer, W. Pompe, Z. Phys. B 81 (1990) 47.

[94] S. Acquaviva, G. Leggieri, A. Luches, A. Perrone, A. Zocco, N. Laidani, G. Speranza, M. Anderle, Appl. Phys. A 70 (2000) 197.

[95] T.A. Friedmann, P.B. Mirkarimi, D.L. Medlin, K.F. McCarty, E.J. Klaus, D.R. Boehme, H.A. Johnsen, M.J. Mills, D.K. Ottesen, J.C. Barbour, J. Appl. Phys. 76 (1994) 3088.

[96] P.B. Mirkarimi, K.F. McCarty, G.F. Cardinale, D.L. Medlin, D.K. Ottesen, H.A. Johnsen, J. Vac. Sci. Technol. A 14 (1996) 251.

[97] T. Klotzbücher, M. Mergens, D.A. Wesner, E.W. Kreutz, Diamond & Related Materials 6 (1997) 599.

[98] H. Chae, S.M. Park, Appl. Surf. Sci. 127–129 (1998) 304.

[99] B. Angleraud, J. Aubreton, A. Catherinot, Eur. Phys. J. Appl. Phys. 5 (1999) 303.

[100] T. Atwee, L. Aschke, H.J. Kunze, J. Phys. D: Appl. Phys. 33 (2000) 2263.

[101] J.M. Mendez, S. Muhl, M. Farias, G. Soto, L. Cota-Araiza, Surf. & Coat. Technol. 41 (1991) 422.

[102] A. Chayahara, H. Yokoyama, T. Imura, Y. Osaka, Japan. J. Appl. Phys. 26 (1987) L1435.

[103] Y. Ichinose, H. Saitoh, Y. Hirotsu, Surf. & Coat. Technol. 43/44 (1990) 116.

[104] K. Akashi, T. Yoshida, S. Komatsu, Y. Mirsuda, J. Korean Institute Metals 24 (1986) 712.

[105] T. Ichiki, T. Momose, T. Yoshida, J. Appl. Phys. 75 (1994) 1330.

[106] S. Amagi, D. Takahashi, T. Yoshida, Appl. Phys. Lett. 70 (1997) 946.

[107] W. Dworschak, K. Jung, H. Ehrhardt, Thin Solid Films 254 (1995) 65.

[108] M. Kuhr, S. Reinke, W. Kulisch, Diamond & Related Materials 4 (1995) 375.

[109] M.Z. Karim, D.C. Cameron, M.S.J. Hashmi, Surf. & Coat. Technol. 60 (1993) 502.

[110] J. Loeffler, I. Konyashin, J. Bill, H. Uhlig, F. Aldinger, Diamond & Related Materials 6 (1997) 608.

[111] I.-H. Kim, K.-S. Kim, S.-H. Kim, S.-R. Lee, Thin Solid Films 290/291 (1996) 120.

[112] C.R. Eddy, B.D. Sartwell, J. Vac. Sci. Technol. A 13 (1995) 2018.

[113] S.M. Gorbatkin, R.F. Burgie, W.C. Oliver, J.C. Barbour, T.M. Mayer, M.L. Thomas, J. Vac. Sci. Technol. A 11 (1993) 1863.

[114] C. Rohr, J.H. Boo, W. Ho, Thin Solid Films 322 (1998) 9.

[115] D.H. Berns, M.A. Cappelli, Appl. Phys. Lett. 68 (1996) 2711.

[116] W.J. Zhang, S. Matsumoto, Appl. Phys. A 71 (2000) 469.

[117] K. Köhler, J.W. Coburn, D.E. Horne, E. Kay, J.H. Keller, J. Appl. Phys. 57 (1985) 59.

[118] J.L. Andujar, E. Bertran, M.C. Polo, J. Vac. Sci. Technol. A 16 (1998) 578.

[119] H. Saitoh, K. Yoshida, W.A. Yarbrough, J. Materials Research 8 (1993) 8.

[120] Y. Guo, Z. Song, Y. Zhang, F. Zhang, G. Chen, phys. stat. sol. (a) 143 (1994) K13.

[121] F.F. Chen, C.D. Decker, Plasma Physics Control. Fusion 34 (1992) 635.

[122] M.Z. Karim, D.C. Cameron, M.S.J. Hashmi, Diamond & Related Materials 3 (1994) 551.

[123] F. Rossi, C. Schaffnit, L. Thomas, H. del Puppo, R. Hugon, Vacuum. 52 (1999) 169.

[124] M. Okamoto, Y. Ursumi, Y. Osaka, Japan. J. Appl. Phys. 29 (1990) L1004.

[125] M. Ye, M.P. Delplancke-Ogletree, Diamond & Related Materials 9 (2000) 1336.

[126] S. Matsumoto, W. Zhang, Japan. J. Appl. Phys. 39 (2000) L442.

21 Markets for plasma technology

Hans Conrads

Wolfshovener Str. 195, D–52428 Jülich, Germany

21.1 Introduction

Gas discharges and their applications have been the issue of numerous conferences over the last two decades.[1] Asking for actual applications, the previous chapters of this book define to a large extent the line of impact between plasma physics on the one hand and products and processes on the other hand.

Perspectives of a global market for plasma technology is a young issue introduced for the first time 5 years ago by an investigation concerning the status and perspectives of plasma technology [1]. About 200 scientists from 12 countries contributed to the related field work and drew their conclusions. The market potential was estimated to be in the order of 250 billion (1 billion $\equiv 10^9$) Euro per year worldwide.

Some investigations on the actual market situation considered the worldwide per annum trade volume of plasma sources only, which is about a factor of 20 smaller than the figure mentioned above. The global impact of such a market is very small, similar to that of specialized equipment like particular tooling machines or highly sophisticated production processes for luxurious goods. A fair number of business units can serve such a market successfully but are economically important locally only.

21.2 Existing markets

Looking into the rich world of plasma technology and plasma techniques a large number of various sources and applications can be identified in almost all fields of science, engineering, and in particular in production technology. There is no power network without plasma switches, no cost-effective general lighting without plasmas, no brilliant large flat panel displays without low-pressure plasma discharges, no coated glass for heat control in homes without plasma-magnetrons, no cutting, welding or thick coating of metals without a plasma torch, no cost-effective production of microelectronics without low-pressure plasma reactors, etc. Each of these areas represents a market of 50 to 100 billion Euro per year worldwide. Here one has to point out, however, that in some of these areas, as in the last mentioned, a real market does not exist because of the lack of free trade. The economic relevance of these technologies to the market is more the one of a hidden treasure by which profitable goods

---

[1]Most of this article has been presented as invited paper during the XIII International Conference on Gas Discharges and their Applications, Glasgow, September 2–8, 2000.

are produced, rather than that of a standardized process accessible to everybody who would like to use it. Present-day details of the plasma processes are unknown outside the factory. A micro-monopoly structure of enterprises is created by the collaboration of large companies with small businesses, a process similar to the cooperation of defense agencies with small and medium size companies on research and development (R & D) programs for the development of weapons in the past.

Plasma assisted production and operation of flat panel displays could become another example for this kind of business relation. The micro-monopoly business structure is quite profitable for the participants but does not create a major economic impact globally. This kind of plasma technology is highly personalized, a designer tool but not a monetary tool. The latter property is a prerequisite for the quick spread of a technology in a global market like the stock market.

21.3   Impacts to push the spread of plasma based products and processes

Incentives for developing plasma based products and processes were and are the unique properties of plasmas as there is an open scale in terms of temperature, heat, and current density, low electrical conductivity, and fluid-like properties, if required, far away from thermal equilibrium, free electrons interacting with bound ones in molecules and atoms, and ions being accelerated straight in the boundary between plasma and solids.

The plasmas considered in this chapter are created and sustained either by electric current, and/or by electromagnetic radiation or by particle beams. Where necessary, magnetic fields are applied to confine or to condition the plasma. Powering plasma sources by electric current, electromagnetic radiation or particles the investment costs for power conditioning range from 0.1 Euro to 5 Euro per Watt installed. From these figures it is obvious, that the kind of source has to match the required plasma properties perfectly. In case the plasma based product or process should substitute existing technology an adaptation to the existing infrastructure is mandatory. Plasma technology can easily comply with many constraints, but an existing infrastructure dictates, what to develop. This shall be demonstrated by the example of light sources.

Since 10 % of the world's electricity production is used for lighting, the huge economic and ecological gain is obvious, if the efficiency of plasma light sources could be pushed by a few tenths of a percent only. The efficiency of the present products is by a factor 2 to 3 down from the theoretical limit. The above mentioned increase in efficiency is not sufficient to promote the development of novel light sources for general lighting, which do not fit into existing installations or which would require additional power conditioning. The microwave driven sulfur light source is an example of a plasma source with excellent efficacy but little chance to be introduced in general lighting.

In the following I would like to consider plasmas as means of operation: mercury plasmas for lighting, nitrogen/oxygen plasmas for switching off electric current, methane plasmas for producing diamond coatings, hexamethyldisiloxane plasmas for silicon coatings, diesel exhaust plasmas for eliminating micro-particles and their polycyclic aromates, NH_3 plasmas for amino-functionalization of the contact zone between polymers and cells in biotechnology.

It is distinguished between sealed systems and open ones with no recycling, and closed and open systems with recycling flows. Light sources, purification of exhausts, coating of surfaces and sterilization of recycled packing materials in the food industry are each examples related to these four categories, respectively.

All the plasma systems considered here require the input of energy to be created and sustained, because the plasma state of matter does not exist under terrestrial conditions. The action of electrical current, electromagnetic radiation, or particle beams on the plasma and on the gas, respectively, is instantaneous. This is due to the high velocity of free electrons even at energies below 1 eV. It may take some time to create a particular density of free electrons, but those exposed to an electric field pick up velocity at once until they lose some velocity due to collisions. Electrical conductivity or temperature rise rapidly and are limited in their rise time by the impedance of the driving circuit or the time it takes to tune a ringing circuit to resonance conditions depending on the instant parameters of the plasma acting as a load for the whole circuit.

The quickest release of energy from the plasma is by electromagnetic radiation. Heat conduction is a slow process. As long as the plasma remains optically thin up to the transparent walls of the containing vessel, energy fed to the plasma is radiated away to a large extent before it reaches the wall by heat conduction. Modern low-pressure plasma light sources produce about 100 lumen per Watt. The best incandescent light sources are by a factor of 8 worse compared to the plasma light sources.

In plasma arcs used, e.g., for cutting, welding, and spraying temperature can rise beyond the sublimation point of carbon, the solid, which can stand by far the highest temperature. Therefore it is not surprising, that the related plasma technologies have spread in all countries around the world.

Coatings created by physical or chemical vapor deposition have been gradually replaced by plasma processes, if the coating has to have particular optical, decorative, adhesive, etc. properties, or if deposition is combined with etching during the production. Etching down to sub-micrometer structures is a domain of low-pressure plasmas, because the self-created electric field between the plasma boundary and the substrate is strictly perpendicular to the substrate and allows etching of ditches by the accelerated ions. Aspect ratios of depth to width of the ditches between 25 to 50 can be achieved even when the width is in the sub-micrometer area. Progress in further miniaturization for the production of microelectronics and therefore also in communication technology depend on further achievements in plasma technology. For these and numerous other applications, highly specialized plasma sources have been developed over the last decades. Special journals are devoted to this issue [2].

Since the art of chemical engineering for coatings in microelectronics is developed to a very high level and many papers of conferences are treating this issue [3], I will not address this very fine technology here any further. The next generations of microelectronics call for even more sophisticated plasmas, being cold and free of electromagnetic radiation of wavelengths shorter than 400 nm. But in the not too distant future (in 10 to 15 years) a new kind of computers – quantum computers – may enter the market. The role of plasma technology for the production of this kind of computer is not yet clear to the author. However, he believes that for the production of PC – like computers plasma technology will be essential for a long time.

21.4   New markets

In chemistry the throughput of masses and the reaction path are two essential features. In connection with the production of microelectronics, the masses added from outside and released from the process are very small, and so is their impact on costs. There is no problem related to supplies or environment, which would call for the development of new plasma-chemical processes. In the following discussion processes are considered, where this is different.

After high flying expectations concerning a positive impact of plasma chemistry on the national economy of the former GDR that were followed by disappointment, plasma chemistry had lost its reputation as pusher for major economic impacts.

Thermal processes as mostly employed in chemical technology are the optimum way to feed energy to as many particles (molecules and atoms) as possible for a given energy input. This wide distribution of energy can turn out to be a problem, because unwanted reaction paths are promoted as well. Catalysts are used to start reactions, to enhance their rates, to retard others, or to quench them, if possible at reduced energy flows and process temperatures.

Converting crude oil to middle distillates or changing hydrocarbons to polymers catalyst are essential in petrochemical technology, serving one of the largest markets worldwide. Plasma has the capability to serve as catalyst as well. Not to use this possibility but merely using it in thermal chemical processes for mass production is not feasible because plasma is an expensive agent. Each cubicmeter of plasma at 1000 K and standard conditions has an energy content of 25 MJ. At a consumption of 100 liters per second the installed power has to be at least 2.5 MW, in reality it has to be the triple of this number with respect to process efficiency and losses.

Plasma can crack or convert every chemical compound and this at low gas temperatures. This has the obvious draw-back that the number of species of reaction products is increased significantly, when plasma is present during a chemical process. In a pure methane plasma, more than 200 species are formed. It is not easy to force a particular reaction path. On the other hand, plasma can easily dissociate all the molecules, which are very stable in wet chemistry. Dissociation of some hydrocarbons, fluorinated ones in particular, or odors are examples, where plasma chemical processes are very effective today already.

Plasma technology will play an important role in avoiding, reconditioning, and destruction of unwanted substances released into our environment. A doctoral thesis concerning the incorporation of waste management into production theory gives a base for considering different strategies of production methods and of the treatment of associated waste problems under different legal constraints [4]. For a particular product, plasma based processes and plasma based waste management can be evaluated with the goal to optimize the costs of production. The thesis includes the treatment of open recycling too, a problem common to some packing materials, e.g., in the food industry. This kind of recycling has a time-dependent nature on a scale much larger than the time of a production cycle.

The legal constraints for waste management are different around the world. General fees and penalties are common tools for control. For the first time, the thesis includes into the production theory *certificates for the release of waste*, that allow the control of emission of waste by means of a market mechanism. This *certification theory* is close to Coase's theorem, who was Nobel laureate in 1991. If these certificates are prized by an international organization (concerned, e.g., with climate changes), particular plasma technologies for production and waste management will have another huge global market. The reduction of the release of methane and carbon dioxide are two examples for the global relevance of waste management. The solution of one of the two problems could create a market of 100 billion Euro p.a. worldwide. I will restrict myself to the methane problem.

Methane, if not burned, is released from the deposit to the open air together with crude oil. It is a by-product of oil production and has an energy equivalent of 25% which is wasted when it is burned to protect the atmosphere against the accumulation of methane. In addition there are huge fields of natural gas – methane is the dominating hydrocarbon – inaccessible to networks serving highly populated areas. The incentive to develop a technology for converting this methane off-shore in middle distillates is huge, because such a technology suitable to be mounted on a drilling rig off shore could be also used for single bore holes in the desert and in other remote areas with unsecured connection to customers. Small conversion-units serving the capacity of a single bore hole have to be developed. In Canada, France, Germany, Norway, Russia, Switzerland, and the USA, researchers are working on this problem intensively, because of the huge size of the market. No break-through has been achieved so far. Plasma-based processes are presently tried out and under consideration for the future.

Advantages in general production technology could have a significant impact on the market of the future. Plasma technology is easily integrated into existing production lines because of its simple interfaces: conducts to power supplies, leads to computers, fittings to gas lines. Plasma is versatile in its action, i.e. it allows the reduction of steps in processing compared to conventional production, it is scalable to any dimension, it is acting highly parallel, it is active down to atomic dimensions, it is highly specific in acting on free molecules and surfaces. Chemical reactivity and concentrations can be controlled by the plasma conditions, and no solvents are required. This helps to minimize the production of waste.

Due to its low inertia plasma can be accelerated and stopped much easier than fluids or gases having the same reaction rates. Atoms being in a plasma always in the *status nascendi* in a large number and radicals created by impacts of free electrons with molecules have a high reactivity, and electrons, carriers of the energy that is coupled into the plasma, have a low mass. Plasma can be created repetitively in large volumes and/or over large surfaces by modulating the electric or electromagnetic energy fed to the plasma. Progress during the last decades and in the time to come in developing digitized power electronics has and will have a big impact on plasma technology in general and plasma chemistry in particular. Here tailoring the reaction path is an important feature.

If the energy is fed to the plasma continuously all particles will be thermalized. A part of the energy will be lost, however, to the environment via the walls of the containment. A substantial part of the reaction path will be similar to that of conventional chemistry. But since the number of species will grow due to the plasma, unwanted reaction channels are opened up as well. Therefore it is better to feed short pulses (10 ns–100 ns) of energy to the gas repetitively in order to create and sustain a plasma for a short time rather than to feed the energy in a time comparable to the reaction time (0.1 μs–100 ms) of molecules. Even if one has to wait until a particular reaction, which is possible in a plasma only, is fully developed, it might be advisable to turn off the power to the plasma, in order to cut the number of species that reduce the density of the wanted one. This has been demonstrated experimentally with Xenon excimers some years ago [5]. Similar considerations might be valid in connection with plasma borne radicals as CH_x and NH_y for automotive and biotechnology related products, respectively.

One of the technological goals is to develop plasmas lean in species. Besides pulsing the power, gaseous and solid quenchers (dust) in the plasma can contribute to this goal as well.

Last but not least it has to be pointed out, that there are special bond electronic states inaccessible to conventional chemistry to which free electrons couple very effectively. Examples are electronic states of mercury and excimers of noble gases, to which the electric energy can be coupled very effectively resulting in suitable light sources for general lighting and other applications.

21.5   Features to enlarge the potential of new markets

The second and final points have similarity with catalysis in common chemistry. Plasma catalysis could be one of the main drivers of the plasma technology to come. There are 4 main routes to follow:

homogeneous catalysis by plasma borne radicals,

plasma borne generation of novel catalysts for carriers in heterogeneous catalysis,

plasma borne novel carriers (dusty plasmas) for heterogeneous catalysis, and

in situ regeneration of solid or gaseous catalysts by plasma.

Two large fields of applications are the petrochemical industry and the waste management in production, which will become increasingly important because of the growing world population and increasing demands for energy and industrial products per capita.

In developing a plasma technology for these markets the following constraints should be considered:

1. Pulsed plasma reactors for various applications have to be developed. This is more an issue of inexpensive, reliable power supplies, and insulators withstanding chemicals and gradients of temperature rather than of plasma physics.

2. Plasma has to be created and sustained under atmospheric pressure and beyond, which requires high-voltage power supplies and narrow reaction channels.

3. Since the reduced field strength E/n governs the reaction path – E is the electric field strength and n the gas density – the mechanical design of the reaction channels has to be tailored to the reaction path. High mechanical precision in terms of absolute numbers (< 0.1 mm) is required. This is not trivial for huge structures relying on ceramics and being subject to temperature gradients.

4. The properties of viscous flows will determine the layout of the reactor too, where the physics of boundaries between the plasma and solids (catalysts) have to be addressed in depth.

5. Cross flow reactors have to be developed, which have porous walls coated by catalyst. These reactors must be able to stand heat waves at elevated temperatures (1200 K). The latter is important not only to optimize reaction rates for particular reaction paths but also to allow pumping of hydrogen in the reactor.

6. Plasma dust feeders have to be developed to load reactors with nano-particles.

7. Large reactors require distributed power supplies and controls because of impedance problems. Power electronics and controllers, which could stand elevated temperatures would offer a substantial advantage in terms of costs and leverage in design.

8. Standardization of the plasma processes would help to penetrate the market globally.

9. These reactors should have standardized interfaces to other reactors, in order to prolong, shorten, or change reaction paths by either adding or eliminating components.

10. The data base of cross sections for plasma-chemical reactions has to be broadened significantly and modeling the plasma has to be extended to further dimensions, in order to ease the scaling of plasma processes.

Points 5–9 relate to stationary reactors. Automotive systems have more requirements. Here the repetitive storage or sudden release of substances are essential. Plasma technology for reconditioning filters when driving or the instant release of fuel require new approaches. A filter for catching soot and polycyclic aromates emitted by diesel engines and a plasma based integrated reconditioning system have been presented recently during the HAKONE VII in Greifswald [6].

The growing demand for communication and navigation satellites during the next decade and beyond will call for plasma thrusters for orbit corrections in the first place. The useful lifetime of the satellites today could be about 5 times longer, if the spacecraft would not become short in fuel for correcting its orbit. In case the mass equivalent of this fuel could be fed to plasma thrusters, about 5 times more thrust could be produced by, e.g., using $\vec{E} \times \vec{B}$-plasma thruster having about 5 times the temperature of chemical thrusters. Since additional equipment for the plasma thruster may burden the pay load a bit, an economic gain of the equivalent of two satellite missions per rocket launched might be realistic. Depending on the annual demand for new satellites the economic value of billions of Euro per year by plasma technology could be gained soon. $\vec{E} \times \vec{B}$-thrusters have been under investigation since the sixties of the last century.

References

[1] H. Conrads, ed., *Stand und Perspektiven der Plasmatechnologie*, Vol. I+II, BMFT, Bonn (1995)

[2] N. Hershkowitz, chief editor, Plasma Sources Science and Technology

[3] H. Conrads, ed., Proc. 18th Int. Display Res. Conf., Seoul, Korea (1998)

[4] H.E. Klingelhöfer, *Betriebliche Entsorgung und Produktion* Wiesbaden, Gabler Verlag (2000)

[5] E. Kindel, C. Schimke, *Density of Excited Atoms and VUV-Radiation in the Pulsed Xenon Medium Pressure Discharge*, Proc. VIIth Int. Conf. Science and Technology of Light Sources, Kyoto (1995), p. 181

[6] S. Müller, H. Conrads, W. Best, *Reactor for Decomposing Soot and other Harmful Substances Contained in Flue Gas*, VIIth Int. Symp. High Pressure Low Temperature Plasma Chemistry, Greifswald (2000), p.340

Index

isotropic distribution, 36–38, 40–42, 44–46, 48–50
isotropic etching, 436

Kiessig fringes, 254, 256
kinetic energy, 29, 31, 34, 35, 44, 47, 80, 82, 86, 92, 94, 95, 98, 100, 101, 106–108, 114, 115, 117, 126, 199, 200, 203, 206, 213, 223, 407, 409, 421
Kirchhoff's law, 407
Knudsen number, 144, 149–151
Kramers-Unsöld approximation, 415

lamp, *see* also discharge lamp
    electrodeless, 391, 393–396, 399, 400, 403, 420, 429
    high-pressure mercury (HPM), 409, 416, 420, 427
    high-pressure sodium (HPS), 409, 410, 418, 419, 427
    high-pressure xenon, 427, 428
    low-pressure sodium (LPS), 419
    mercury-free, 421, 427, 428
    metal halide (MH), 416, 418, 428
    sulphur, 429
lamp characteristics, 409, 422–425
Langmuir
    oscillation, 17
    paradox, 21
    probe, 131, 132, 134, 136, 140, 149, 151, 153–158, 161, 163, 190, 191, 390, 391, 401, 482, 488, 492, 494, 497, 498
laser diagnostics, 301
laser-induced fluorescence (LIF), 67, 71, 178, 188, 359–361, 363
lighting, 287, 387, 388, 390, 392–394, 400, 402, 403, 419, 507, 508, 512
local thermal equilibrium (LTE), 407, 408, 416, 423
Lorentz force, 16, 18, 203, 285
low-pressure plasma, 16, 183, 285, 477, 480
low-temperature plasma, 15–17, 55, 56, 68, 74, 79, 113, 125, 131, 153, 174, 180, 283, 453, 460, 476, 480
luminance, 368, 369, 381, 382

macroscopic kinetics, 305, 306, 311, 322, 323, 325–329

magnetohydrodynamics (MHD) model, 17
magnetron, 153, 235, 271, 296, 492
magnetron sputtering, 490, 492
Maxwell distribution, 16, 18, 20–22, 40, 127, 142, 185, 479
mean free path, 134, 135, 150, 151, 154
Michelson interferometer, 174
microdischarge, 331, 332, 334, 335, 337, 339, 345, 348
microelectronics, 284, 285, 295, 434, 507, 509, 510
microgravimetry, 462, 463, 467, 468, 470
micromechanics, 284
microwave discharge, 177, 296, 298, 300, 305, 320, 476
microwave plasma, 175, 177, 180–182, 187, 190, 286, 296–298, 300
mobility, 19, 27
molecular dynamics model, 79, 88
Monte Carlo simulation, 16, 88, 153

nano-particle, 513
Nernst-Townsend-Einstein relation, 19
non-equilibrium, 483
non-equilibrium plasma, 182, 183, 188, 287, 335
non-ideal plasma, 283
non-radiating energy, 421
non-thermal plasma, 15–17, 29, 30, 52, 283, 285, 286, 290, 365
nucleation, 276, 480, 483

optical multi-channel (OMA) analyser, 232
orbital motion limited (OML) model, 149

particle density, 16, 35, 136, 138, 309, 315, 336, 407, 415, 419, 427
Penning ionisation, 493
photoionisation, 203
physical etching, 436
physical vapour deposition (PVD), 476, 484, 495
planar probe, 131, 138, 139, 147
Planck's law, 407
plasma
    arc, 16
    beam, 299
    catalysis, 512
    chemistry, 63, 173, 175, 181, 284, 305, 306, 309, 365, 366, 510, 511